# SYNAPTIC TRANSMISSION

# SYNAPTIC TRANSMISSION

### Stephen D. Meriney
*Department of Neuroscience, University of Pittsburgh, Pittsburgh, PA, United States*

### Erika E. Fanselow
*Department of Neuroscience, University of Pittsburgh, Pittsburgh, PA, United States*

ACADEMIC PRESS
An imprint of Elsevier

Academic Press is an imprint of Elsevier
125 London Wall, London EC2Y 5AS, United Kingdom
525 B Street, Suite 1650, San Diego, CA 92101, United States
50 Hampshire Street, 5th Floor, Cambridge, MA 02139, United States
The Boulevard, Langford Lane, Kidlington, Oxford OX5 1GB, United Kingdom

Copyright © 2019 Elsevier Inc. All rights reserved.

No part of this publication may be reproduced or transmitted in any form or by any means, electronic or mechanical, including photocopying, recording, or any information storage and retrieval system, without permission in writing from the publisher. Details on how to seek permission, further information about the Publisher's permissions policies and our arrangements with organizations such as the Copyright Clearance Center and the Copyright Licensing Agency, can be found at our website: www.elsevier.com/permissions.

This book and the individual contributions contained in it are protected under copyright by the Publisher (other than as may be noted herein).

**Notices**
Knowledge and best practice in this field are constantly changing. As new research and experience broaden our understanding, changes in research methods, professional practices, or medical treatment may become necessary.

Practitioners and researchers must always rely on their own experience and knowledge in evaluating and using any information, methods, compounds, or experiments described herein. In using such information or methods they should be mindful of their own safety and the safety of others, including parties for whom they have a professional responsibility.

To the fullest extent of the law, neither the Publisher nor the authors, contributors, or editors, assume any liability for any injury and/or damage to persons or property as a matter of products liability, negligence or otherwise, or from any use or operation of any methods, products, instructions, or ideas contained in the material herein.

**British Library Cataloguing-in-Publication Data**
A catalogue record for this book is available from the British Library

**Library of Congress Cataloging-in-Publication Data**
A catalog record for this book is available from the Library of Congress

ISBN: 978-0-12-815320-8

For Information on all Academic Press publications
visit our website at https://www.elsevier.com/books-and-journals

*Publisher:* Nikki Levy
*Acquisition Editor:* Natalie Farra
*Editorial Project Manager:* Barbara Makinster
*Production Project Manager:* Bharatwaj Varatharajan
*Cover Designer:* Mark Rogers

Typeset by MPS Limited, Chennai, India

# Contents

Preface ix
Acknowledgments xi

## 1. Introduction

Hypothesis Development 1
The Use of Animal Model Systems to Study Synapses 2
References 4

## Part I
## SYNAPTIC BIOPHYSICS AND NERVE TERMINAL STRUCTURE

### 2. The Formation and Structure of Synapses

How Do Neurons Send Signals to One Another? 7
Synapse Structure and Organization 13
How Does the Neuron Assemble the Cellular Components Required to Create Synapses? 15
Construction of Active Zones During Synapse Development 16
References 18

### 3. Basics of Cellular Neurophysiology

Neurons are Excitable Cells 19
Movement of Ions Across the Cell Membrane 23
References 34

### 4. Ion Channels and Action Potential Generation

Ion Channels 35
Voltage-Gated Ion Channels 37
Ionic Currents Through Voltage-Gated Ion Channels 47
Action Potentials 53
References 62

### 5. Electrical Synapses

History of Electrical Synapses 65
Structure and Physiological Characteristics of Electrical Syapses 69
Roles of Electrical Synapses 76
Electrical Synapse Plasticity 85
References 89

## Part II
## REGULATION OF CHEMICAL TRANSMITTER RELEASE

### 6. Function of Chemical Synapses and the Quantal Theory of Transmitter Release

Costs and Advantages of Chemical Communication 95
Electrical Footprints of Chemical Transmitter Release 97
Spontaneous Release of Single Neurotransmitter Vesicles 102
The Quantal Theory of Chemical Transmitter Release 105
Quantal Analysis of Chemical Transmitter Release at the Neuromuscular Junction 109
Quantal Analysis of Chemical Transmitter Release at Central Synapses 113
Optical Quantal Analysis 116
Summary 117
References 118

## 7. Calcium Homeostasis, Calcium Channels, and Transmitter Release

Calcium as a Trigger for Neurotransmitter Release 121
Control of Neurotransmitter Release by Calcium Ions 130
Voltage-Gated Calcium Channels in Nerve Terminals 141
References 151

## 8. Cellular and Molecular Mechanisms of Exocytosis

Discovery of the Mechanisms of Neurotransmitter Release 155
Biochemical Mechanims of Calcium-Triggered Synaptic Vesicle Fusion 163
References 185

## 9. Cellular and Molecular Mechanisms of Endocytosis and Synaptic Vesicle Trafficking

Retrieval and Reuse of Synaptic Vesicle Membrane 189
Endocytosis Occurs Outside the Active Zone 193
Mechanisms of Endocytosis 195
Clathrin-Mediated Endocytosis 195
Bulk Endocytosis 197
Kiss-and-Run 197
Synaptic Vesicle Pools 199
Synaptic Vesicle Trafficking in the Nerve Terminal 201
References 203

# Part III
# RECEPTORS AND SIGNALING

## 10. Introduction to Receptors

Neurotransmitter Receptors Can Be Divided Into Two General Classes: Ionotropic and Metabotropic 209
Comparison Between Ionotropic and Metabotropic Receptors 213
References 214

## 11. Ionotropic Receptors

The Pentameric Ligand-Gated Ion Channel Family (Cys-Loop Receptors) 217
The Glutamate Ionotropic Receptor Family 227
The Trimeric Receptor Family 236
The Transient Receptor Potential Channel Family 239
References 239

## 12. Metabotropic G-Protein-Coupled Receptors and Their Cytoplasmic Signaling Pathways

Common Themes in Receptor Coupling to Heterotrimeric G-Proteins 251
The Four Most Common G-Protein-coupled Signaling Pathways in the Nervous System 255
Other G-Protein-Coupled Signaling Pathways in the Nervous System 265
Specificity of Coupling Between Receptors and G-Protein-Coupled Signaling Cascades 269
References 271

## 13. Synaptic Integration Within Postsynaptic Neurons

Passive Membrane Properties 277
Spines Are Specialized Postsynaptic Compartments on Dendrites 280
Active Membrane Properties 283
References 285

## 14. Synaptic Plasticity

Short-Term Synaptic Plasticity 288
Metabotropic Receptor-Mediated Plasticity of Ionotropic Signaling 299
Habituation and Sensitization 303
Long-Term Synaptic Plasticity 308
Clinical Cases That Focused the Investigation of Long-Term Synaptic Plasticity 308
Long-Term Potentiation 308
Physiological Stimulus Patterns That Can Induce Long-Term Potentiation 311
Associative Long-Term Potentiation 312
Spike Timing-Dependent Plasticity 313
Long-Term Depression 314
Heterosynaptic Plasticity 315

Synaptic Signaling Mechanisms of Long-Term
    Potentiation and Long-Term Depression   316
Metaplasticity   319
Plasticity Modulation   320
Homeostatic Synaptic Plasticity   320
References   323

# Part IV
# CHEMICAL TRANSMITTERS

## 15. Introduction to Chemical Transmitter Systems

Neurotransmitter Versus Neuromodulator   333
Criteria Used to Classify a Signaling Molecule
    as a Neurotransmitter   335
Neurotransmitter Characteristics   339
Types of Neurotransmitters   341
References   343

## 16. Acetylcholine

History of the Discovery of Acetylcholine and Its
    Identity as a Neurotransmitter   345
Synthesis, Release, and Termination of Action of
    Acetylcholine   350
Roles of Acetylcholine in the Nervous
    System   359
Drugs and Other Compounds that Affect
    Cholinergic Signaling   363
References   366

## 17. Monoamine Transmitters

Catecholamine Neurotransmitters   370
Serotonin   376
Histamine   382
Projections of Monoaminergic Neurons and
    Functions of Monoamines in the Nervous
    System   383
Therapeutic Drugs Related to Monoamine
    Neurotransmitters   390
Monoaminergic Drugs of Abuse   395
References   397

## 18. Amino Acid Neurotransmitters

Glutamate   399
GABA   406

GABA and the Neurological Disease
    Schizophrenia   412
Glycine   413
References   417

## 19. Neuropeptide Transmitters

How Do Neuropeptides Differ From Classical
    (Type 1) Neurotransmitters?   422
Neuropeptide Synthesis, Release, and
    Regulation   423
Neuropeptide Y as a Model for Neuropeptide
    Action   430
References   432

## 20. Gaseous Neurotransmitters

Nitric Oxide   435
Carbon Monoxide   443
Hydrogen Sulfide   445
References   446

## 21. The Use of Multiple Neurotransmitters at Synapses

STEPHANIE B. ALDRICH

Overview and Historical Perspective   449
Cotransmission and Corelease of
    Neurotransmitters   452
Neurotransmitter Specification and
    Switching   463
Summary   475
References   476

## 22. Complex Signaling Within Tripartite Synapses

The Role of Astrocytes in Synaptic
    Function   482
Interactions Between Astrocytes: Gap Junctions and
    Calcium Waves   488
Release of Neurotransmitters From Astrocytes   489
Do Astrocytes Play a Role in Information Processing
    Within the Brain?   490
References   492

Glossary   493

Index   499

# Preface

This text is written with the undergraduate neuroscience major in mind. Here, we outline our approach to compiling this textbook, and explain our rationale for the material included. This text is organized into 22 chapters divided into four parts. This text evolved from the lecture notes and materials that have been used to teach an undergraduate course in synaptic transmission at the University of Pittsburgh over the past 20 years.

The manner by which neurons in the brain communicate with one another and the rest of the body forms one of the fundamental building blocks for understanding nervous system function. Every function of the nervous system is controlled, in one way or another, by the rich diversity of synaptic communication mechanisms and the plasticity of these.

Experimental study has driven our understanding of the field over many decades. In the brain, thousands of synaptic communication events can converge onto a single neuron. The cellular and molecular processes that control synaptic transmission occur within a very small compartment in the neuron (usually within one ten-thousandth of an inch, or 2–3 μm), and often occur within one thousandth of a second (1 ms). This makes such processes difficult to study in isolation in an experimental setting. In many cases, these constraints have led scientists to seek model organisms in nature that are easier to study, and we will discuss these model systems as they arise. In addition, technology development has historically driven advancements in the field, as major discoveries usually follow breakthroughs in the development of new tools and techniques.

To emphasize these experimental and conceptual constraints and breakthroughs, this book will cover major topics in the field of synaptic transmission by using specific examples that drive critical concepts and support the hypotheses we discuss. That said, due in part to the difficulty in studying synapses, there are few completely understood molecular and cellular processes and mechanisms in this field. Instead, the field is driven primarily by multiple hypotheses that are each supported by a range of experimental results. Many subtopics in synaptic transmission are enriched each year by new discoveries that add immense complexity and detail to our understanding of how synapses work. We have chosen to keep this textbook focused on the main principles and hypotheses that form the framework for our understanding. As such, we have eliminated discussion of some complexities and details often present in graduate or professional reference textbooks, which might serve to confuse students who are learning the foundational concepts of synaptic transmission.

# Acknowledgments

First and foremost, we would like to thank our families for their never-ending support of this endeavor, as the work behind the generation of this textbook took us away from them for many nights and weekends over a very long period of time.

We would also like to thank several undergraduate student assistants. In particular, Max Myers and Onnaleah Trentini were instrumental in generating and editing many of the drawings used within the text, assembling the references, securing permission for the use of previously published work, and providing invaluable assistance for many other aspects of this work. In addition, Blake Vuocolo stained tissue and collected images for use on the cover and within the text. Lauren Ferrer Pistone, Katy Fisher, Kerryann Koper, Julia Lewis, Lydia Lewis, Giancarlo Speranza, and Shelby Szott assisted with a number of editing tasks.

We are also indebted to Rachel Harding, an undergraduate student studying Scientific Illustration from Arcadia University, who created many of the artistic figures throughout the textbook.

We also thank three graduate students from the Center for Neuroscience at the University of Pittsburgh (CNUP) PhD program (Stephanie Aldrich, Annie Homan, and Kristine Ojala) who had served as teaching assistants for the course that inspired this text, and who edited the contents of this book. In addition, Kristine Ojala authored some of the boxes focused on toxins from nature and treatments from nature, and stained tissue and collected images for use on the cover and within the text. One of these students, Stephanie Aldrich, also authored the chapter on the use of multiple neurotransmitters.

We thank Rozita Laghaei of the Pittsburgh Supercomputing Center, Carnegie Mellon University, for creating some of the figures that were derived from published crystal structures.

In addition, we are thankful for the feedback we received from faculty consultants (Deborah Artim, Barry Connors, Yan Dong, Zachary Freyberg, Guillermo Gonzalez-Burgos, Jon Johnson, Maria E. Rubio, Oliver Schluter, Rebecca Seal, Alan Sved, and Amantha Thathiah) who edited specific chapters.

We also thank the associate dean for Faculty Affairs (Kay Brummond) from the Kenneth P. Dietrich School of Arts & Sciences, University of Pittsburgh for offsetting fees for reproduction of copyrighted material.

Finally, we are thankful for the contributions by numerous laboratories within the CNUP for providing unpublished figures from their research for use as illustrations within the text.

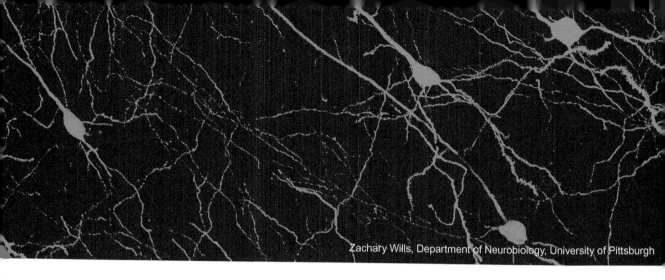

Zachary Wills, Department of Neurobiology, University of Pittsburgh

# CHAPTER 1

# Introduction

The nervous system is essentially an information processing system that allows living organisms to control bodily functions, react to the environment, move, think, and display of emotions. At a basic level, all of these functions are governed by electrical activity within neurons and chemical communication between cells, which is referred to as synaptic transmission. Therefore, the phenomenon of synaptic transmission between cells represents a basic building block for understanding everything the nervous system does.

Take, for example, your reaction to someone throwing a ball toward you. You need to have cells that sense the ball flying toward you, cells to relay that information to other parts of the nervous system, and still other cells that help you decide what to do and then move muscles in your body accordingly. Furthermore, all of these cells respond (hopefully accurately!) within a fraction of a second to let you catch the ball. The mechanisms that underlie these processes include a combination of events that occur within neurons (basic cellular neurophysiology) and between neurons (synaptic transmission). In the chapters that follow, we will discuss many processes that govern how most cells in the nervous system communicate.

## HYPOTHESIS DEVELOPMENT

Synaptic transmission is largely a hypothesis-driven field. This means that much of what we understand about how neurons communicate with one another is simply a proposed explanation based on limited experimental information for a given observation.

This is referred to as a **hypothesis**. Initial hypotheses, in turn, form the basis for planning additional experiments that are designed to support, refine, or refute those and subsequent hypotheses and to provide further details about them. Because the study of synaptic transmission involves this process of repeatedly developing and refining hypotheses, it is still a constantly evolving field.

Therefore, to most convincingly present the material in this text, it is critical not only to outline the major hypotheses, but also to discuss the experimental support for those ideas. Because of this, the study of synaptic transmission is not simply pure memorization of facts, but rather of thinking critically about ideas. Toward this end, we have chosen particular hypotheses we believe drive home the fundamental principles of synaptic transmission, and we will take the student through the various experimental findings that came together to support them.

## THE USE OF ANIMAL MODEL SYSTEMS TO STUDY SYNAPSES

The study of synaptic transmission is focused on understanding the molecular and physiological bases of how neurons communicate with one another in the nervous system or on other cells affected by the nervous system. One of the main applications of this information is to understand and treat human neurological disorders. However, since detailed study of the human nervous system is not usually possible due to ethical, moral, and/or technical limitations, experiments that form the basis for our hypotheses are often performed using model systems. Model systems typically utilize simpler biological tissue and functions that allow us to learn the basic principles that underlie the function of the synapse. Such model systems can employ a specific type of cell and/or animal that is easier to study than mammalian brain synapses themselves. This use of animal model systems is based on the ethical use of these tissues as governed by national and local oversight agencies (https://www.ncbi.nlm.nih.gov/books/NBK24650/). With the accumulation of significant data from animal model systems, occasionally scientists have been able to use computational models to further explore hypotheses about synaptic function. These computational models are most productively employed when they can be used in conjunction with animal experimental data, but to date they do not replace basic animal research.

The use of nonmammalian animal model systems is particularly essential for the study of synapses because neuronal structures in mammals are typically very small and they are especially difficult to study in the mammalian central nervous system (see Fig. 1.1A). For example, a pyramidal cell in the hippocampus receives thousands of synapses, each contained within a small subcompartment of the neuron. When an experimenter is studying the details of the function of one of these mammalian hippocampal synapses, the task is difficult, in part due to the fact that so many other synapses onto that same pyramidal neuron are active at the same time. This might be akin to trying to understand a conversation with a person at a crowded party where everyone is talking at the same time. Some animal models provide the opportunity to clarify the details of synaptic function within one particular synapse in isolation.

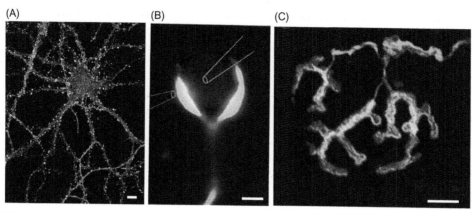

FIGURE 1.1  **Small central nervous system synapses as compared to larger model synapses.** (A) Thousands of synapses onto a cultured hippocampal neuron stained for a postsynaptic glutamate receptor subunit (*green*), a presynaptic vesicular glutamate transporter (*red*), and a marker for the postsynaptic hippocampal neuron cytoskeleton (*blue*). Each of the small red and yellow dots around the blue cell represents single synapses. (B) The large calyx of Held synapse from the auditory brainstem of the rat is so large that it is amenable to direct electrical recording from the nerve terminal. The presynaptic nerve terminal and axon are in yellow and the postsynaptic cell in blue. (C) A single neuromuscular junction of the mouse is very large and there is only one presynaptic nerve terminal onto each postsynaptic cell, making it easier to visualize transmitter release sites and simplifying the interpretation of synaptic transmission data. Scale bar in all images = 5 μm. *Source: (A) Adapted from Journal of Neuroscience Cover art related to Ferreira, J.S., Schmidt, J., Rio, P., Águas, R., Rooyakkers, A., Li, K.W., et al. (2015). GluN2B-containing NMDA receptors regulate AMPA receptor traffic through anchoring of the synaptic proteasome. J. Neurosci., 35 (22), 8462–8479 (Ferreira et al., 2015); (B) Adapted from Borst, J.G., Helmchen, F., Sakmann, B., 1995. Pre- and postsynaptic whole-cell recordings in the medial nucleus of the trapezoid body of the rat. J. Physiol. 489 (Pt 3), 825–840 (Borst et al., 1995); (C) Ojala and Meriney, unpublished image.*

Specific animal model systems are chosen for the experimental advantages they offer. These include larger or simpler synaptic structures and circuits, and/or synapses that are easier to remove and maintain outside the body during experimental study (see Fig. 1.1). That said, in some cases, the model systems used may be from higher-order animals when a synapse being studied has specializations that are not present in lower-order animals. In all cases, the rationale for the study of animal model systems is that basic principles learned in a model system can be applied to our understanding of function in more complex systems or higher-order animal species. The goal in designing these studies is to perform a critical experiment or test a hypothesis in a model system using a detailed approach that would not be possible in higher-order animals.

It is important to keep in mind that experiments in model systems often have limitations that should be taken into account when interpreting data. For example, studying the function of an ion channel when expressed in isolation in a frog egg or cultured kidney cell eliminates other modulatory proteins that might exist in the native environment of the neuron of interest. Such limitations are generally dictated by known differences between the model system and synapses in higher-order animals. However, when experiments using animal model systems are properly designed, the results will apply broadly to our understanding of synapses in higher organisms, including humans. Detailed experiments in simple model systems can sometimes be followed up with limited evaluations of the hypothesis in higher-order animals to confirm the validity of experimental findings in these species. In the following chapters, we will discuss many experiments for which this is the case.

# References

Borst, J.G., Helmchen, F., Sakmann, B., 1995. Pre- and postsynaptic whole-cell recordings in the medial nucleus of the trapezoid body of the rat. J. Physiol. 489 (Pt 3), 825–840.

Ferreira, J.S., Schmidt, J., Rio, P., Águas, R., Rooyakkers, A., Li, K.W., et al., 2015. GluN2B-containing NMDA receptors regulate AMPA receptor traffic through anchoring of the synaptic proteasome. J. Neurosci. 35 (22), 8462–8479.

# PART I

# SYNAPTIC BIOPHYSICS AND NERVE TERMINAL STRUCTURE

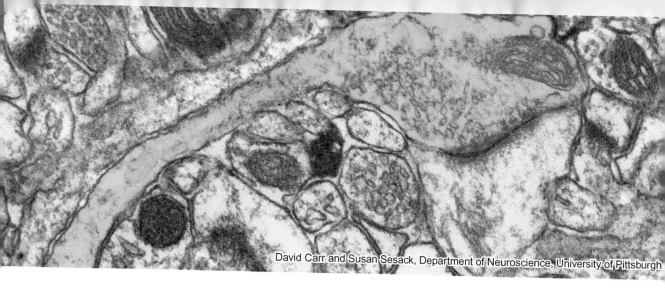

David Carr and Susan Sesack, Department of Neuroscience, University of Pittsburgh

# CHAPTER 2

# The Formation and Structure of Synapses

## HOW DO NEURONS SEND SIGNALS TO ONE ANOTHER?

We will start with an historical perspective on the study of synapses. In the 19th century, scientists developed methods for labeling cells of the nervous system with dyes. Initially, researchers used dyes that labeled only the neuron cell body and sometimes a few cellular compartments near the edge of the cell body. This microscopic cell labeling showed that neurons were not like other cells in the body, which often appeared as isolated round spheres. Instead, it became clear that neurons had long extensions leaving the cell body, which we now know to be axons, as well as overlapping highly branched structures, which we now know to be dendrites. These early observations gave the impression that neurons were directly connected to, and indeed, contiguous with, one another. However, these early images of neurons did not allow scientists to view the very ends of axons or dendrites, or the actual connections between neurons. As such, connections between neurons were only inferred from these early observations of cellular processes extending beyond the cell body.

Based on these early cell labeling techniques, in 1871 Joseph von Gerlach proposed the **Reticular Theory** of connectivity in the brain. He envisioned that all neurons in the brain were connected to one another in a type of reticulum, not unlike the capillary beds of the vascular system (see Fig. 2.1). Other scientists using similar staining techniques proposed

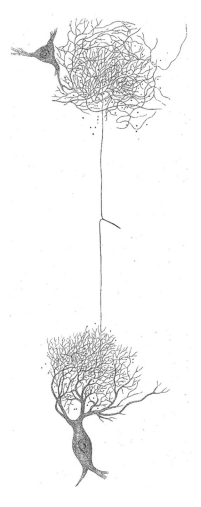

FIGURE 2.1 **Early drawing of neuron connectivity by Joseph von Gerlach in support of the Reticular Theory.** This drawing shows two neurons connected to one another by what appears to be a continuous processes. *Source: CC BY 4.0 via Wikimedia Commons; Fibre net of Joseph von Gerlach.*

an opposing theory—that neurons were individual cells that did not connect in a reticular network, but were instead completely separate from one another. This later theory was later termed the **Neuron Doctrine**.

In 1873, Camillo Golgi (see Fig. 2.2) invented a new staining technique which could provide a detailed microscopic view of the entire extent of a single neuron in the brain. Golgi published this technique in the *Gazzetta Medica Italiana* in a paper entitled "On the structure of the brain grey matter." This new staining technique used silver nitrate, which, when put onto a slice of neurological tissue, impregnates and labels a small subset (1%–5%) of the neurons in the tissue sample with a black reaction product. These neurons are labeled seemingly at random, and each labeled neuron is stained in its entirety, including all parts of its axons and dendrites. If all of the neurons were labeled, it would be impossible to distinguish any one neuron's processes from the tangle of densely packed cells, but because this stain only labels a few neurons in a given area of tissue, the cellular

FIGURE 2.2  Portrait of Camillo Golgi. *Source: Wellcome Images, a website operated by Wellcome Trust, a global charitable foundation based in the United Kingdom. Photo: Anton Mansch. CC BY 4.0 via Wikimedia Commons.*

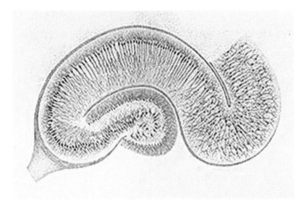

FIGURE 2.3  One of Golgi's drawings of a brain section from the hippocampus after staining with silver nitrate. *Source: WikiCommons, public domain; An old drawing by Camillo Golgi of the Hippocampus from Opera Omnia, 1903.*

processes from a single neuron can be identified and traced from end to end. This labeling technique continues to be used today, although it is still not clear why only a subset of neurons become labeled with this method. This process was eventually named the Golgi stain, and the drawings scientists made based on Golgi-stained neurons (see Fig. 2.3) had a major influence on early hypotheses that were developed to explain how neurons in the brain function.

Camillo Golgi examined brain tissue using his newly developed stain, and based on his interpretation of what he observed, he expressed his support for the Reticular Theory—the theory that the neurons in the brain form a continuous network of interconnected cells. The Reticular Theory was much more popular at the time than the Neuron Doctrine, in part because of underlying theories about how the nervous system might function. At this point in time, it was commonly believed that the nervous system was one continuous structure, whose function was the result of the collective action of many neurons, rather than specific subsets of neurons working independently (Cimino, 1999). Thus, the debate surrounding the Reticular Theory and the Neuron Doctrine was less about the data, and more about general theories of how the nervous system might work: a holistic view (meaning, a view emphasizing the system as a whole rather than its individual parts) versus a reductionist cellular view (meaning, a view that focuses on understanding the

FIGURE 2.4  **Portrait of Santiago Ramón y Cajal.** *Source: The original photo is anonymous although published by Clark University in 1899. Restoration by Garrondo (Cajal.PNG) [Public domain], via Wikimedia Commons.*

system's fundamental components in order to explain how it functions as a whole). At the time, the holistic view prevailed, which meant there was strong support among contemporary scientists for the Reticular Theory.

During the 1800s, news did not travel very quickly, and this included news of the incredible usefulness of the Golgi stain. It was not until 1887, 14 years after the first report of the Golgi stain, that a Spanish scientist named Santiago Ramón y Cajal (see Fig. 2.4) was shown brain tissue prepared using the Golgi stain. These images of Golgi-stained tissue inspired Ramón y Cajal, and he set out to make observations of brain tissue using this new approach. After refining the Golgi staining technique and making his own observations, Ramón y Cajal documented his findings by meticulously drawing what he saw in the labeled tissue he viewed under the microscope. Most importantly, he developed his own interpretations and hypotheses about what he observed.

Ramón y Cajal argued that cells in the brain were entirely separate from one another and that there was space between their endings, a view that challenged the Reticular Theory and supported the Neuron Doctrine. Perhaps the first recognition of an anatomical specialization at the ends of axons came while Ramón y Cajal was using the Golgi stain to observe the specialized connections between two labeled neurons (see Fig. 2.5). In the drawings he made, it is clear that there is a specialized structure at the ends of axons and that this structure is not continuous with neighboring cells. Ramón y Cajal's interpretation of the Golgi-stained brain tissue was a significant piece of evidence in support of the Neuron Doctrine (Ramón y Cajal, 1954; Guillery, 2005).

In 1906, the Nobel Prize in Physiology or Medicine was awarded jointly to Camillo Golgi and Santiago Ramón y Cajal "in recognition of their work on the structure of the nervous system."

While assembling information from many of these early publications, researcher Wilhelm Waldeyer stated that "The nervous system is made up of innumerable nerve cells (neurons) which are anatomically and genetically independent of each other. Each nerve unit consists of three parts: the nerve cell, the nerve fiber and the fiber aborizations (terminal aborizations)" (von Waldeyer-Hartz, 1891). von Waldeyer-Hartz is credited with proposing the term "neuron" to describe the individual cells of the nervous system.

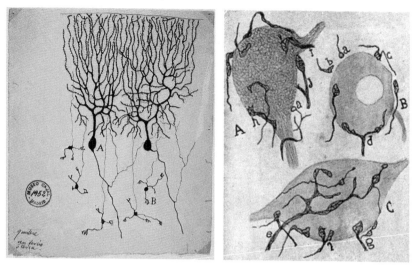

FIGURE 2.5 **Drawings by Ramón y Cajal of Golgi-stained nervous system tissue.** (Left) Individual neurons in the cerebellum that have specialized endings on axons. (Right) Nerve endings (*dark*) onto neurons (*gray*) in the ventral horn of the spinal cord. In some cases, it is possible to observe a gap between the nerve ending and the ventral horn neuron. *Source: CC BY 4.0 via Wikimedia Commons.*

Support for the Neuron Doctrine, and for the idea that there are specialized sites of communication between neurons, grew with observations of the functions of neurons. A neurophysiologist working at the same time, Sir Charles Sherrington, studied reflexes in the spinal cord and showed that electrical signals between dorsal and ventral roots of the spinal cord could travel in only one direction (Fig. 2.6). He hypothesized that within the spinal cord there was a specialized site of information transfer between the dorsal and ventral roots that allowed only unidirectional communication. In an 1897 publication, Sherrington named this site of communication the "synapse" (from the Greek phrase "connection junction"). This evidence for a unidirectional transfer of information was difficult to reconcile with the Reticular Theory, but fit well with the Neuron Doctrine. Because of this, Sherrington is credited with ending the debate over these theories, leading the way for the Neuron Doctrine to be generally agreed upon (Burke, 2007).

With the acceptance of the Neuron Doctrine, scientists began studying these sites of information exchange from the axon of one neuron to dendrites on separate neurons, that is, the synapse. We now know that the synapse is a very specialized structure, often present at the ends of axons, and that it is the site of communication with neighboring cells (Fig. 2.7).

What makes the synapse special? We now know that a reticulum of continuously connected elements would permit electrical activity to flow quickly and continuously through all the cells in the brain, but it would not allow the brain to perform the amazing functions we have come to appreciate (e.g., processing sensory input, learning and memory, cognitive functions, homeostatic regulation of the body). These functions are derived from, and depend upon, tightly controlled sites of unidirectional information transfer that can either

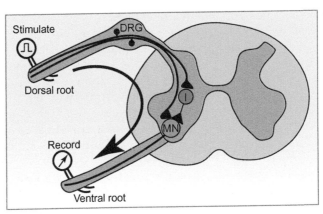

FIGURE 2.6 **Diagram of a reflex arc through the spinal cord.** This drawing of a cross-section through the spinal cord shows sensory axons entering the spinal cord through the dorsal root ganglion (DRG) and forming synapses onto interneurons (I) and motoneurons (MN) in the ventral horn of the spinal cord. These MN then project axons through the ventral root of the spinal cord. If one stimulates the dorsal root and records from the ventral root, a signal can be measured. However, stimulating the ventral root does not generate a signal in the dorsal root. Therefore, communication through this circuit is unidirectional (large arrow).

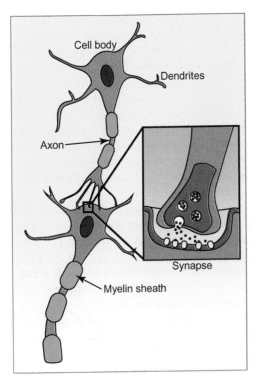

FIGURE 2.7 **Artist's rendering of the specialized compartments of a neuron, including the cell body, dendrite, axon, myelin sheath, and the synapse.** Synapses are specialized regions of the neuron that include the machinery required for communication between neurons.

excite or inhibit connected neurons, and whose strength of communication with neighboring cells can be changed and regulated. In fact, synaptic plasticity, the ability of a synapse to change the strength of communication between cells, is fundamental to most complex functions in the brain (see Chapter 14). Furthermore, malfunctioning synapses are associated with a wide range of neurological diseases (epilepsy, Parkinson's disease, Alzheimer's disease, bipolar disorder, schizophrenia, myasthenia gravis, and many others). Additionally, synapses are often targets of drugs used to treat such diseases (e.g., drugs for epilepsy, mood disorders, hypertension, Parkinson's disease, and Alzheimer's disease), as well as of recreational drugs/drugs of abuse (e.g., cocaine, amphetamine, ecstasy, marijuana, nicotine, alcohol). Given these roles, it could be argued that synapses are the most important fundamental building block of the nervous system. Understanding how synapses function contributes significantly to our understanding of complex nervous system functions, neurological disease, and the mechanisms of drug action.

## SYNAPSE STRUCTURE AND ORGANIZATION

What specific organelles, proteins, and other elements must exist at a synapse to create this special site of communication between cells? There are several experimental approaches to answering this question. This chapter will focus on anatomical studies. Subsequent chapters will focus on biochemical, molecular, and proteomic studies that further our understanding of synaptic specialization.

Because synapses are so small (many brain synapses are 1–2 μm in diameter), scientists had to wait for the invention of the electron microscope to view high-resolution images of these specialized structures. The first electron micrographs of synapses were published by two laboratories nearly simultaneously, in papers by De Robertis and Bennett (1955) and Palay and Palade (1955). These early micrographs confirmed Ramón y Cajal and Sherrington's conclusion that there was a gap between pre- and postsynaptic neurons at these specialized sites of contact. This gap was named the **synaptic cleft**. In addition, these early micrographs showed that there is a large number of **synaptic vesicles** within the presynaptic cytoplasm. These are small, roughly spherical structures whose contents are segregated from the cytoplasm and enclosed within their own lipid bilayer membrane. More recent electron micrographs have provided extremely detailed structural information about the elements that are located at synapses (Fig. 2.8).

The fine structure of a synapse is characterized by a number of other common features. First, almost all synapses contain two types of transmitter-containing synaptic vesicles. The first type is the **small clear vesicles** that are present in great numbers and contain small-molecule chemical transmitters (e.g., acetylcholine, glutamate). The second type of synaptic vesicle is the **dense-core vesicle**. These vesicles are larger than the small clear vesicles, have an electron-dense center (thus their name), and often contain large neuropeptides (e.g., calcitonin gene-related peptide, dynorphin, see Chapter 19).

Also apparent in electron micrographs are electron densities on the pre- and postsynaptic membranes at the site of transmitter release. The **presynaptic density** is often called the **active zone**, since it is the site of transmitter release and is made up of proteins that regulate the fusion of synaptic vesicles with the plasma membrane. The **postsynaptic**

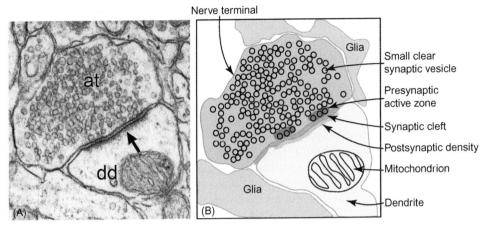

FIGURE 2.8 **Electron micrograph of a synapse in the central nervous system.** (Left) Electron micrograph from the rat striatum illustrating an axon terminal (at, also called the nerve terminal) forming an excitatory-type synapse (*black arrow*) onto a distal dendrite (dd; one that is not close to the cell body). The synapse is characterized by a widened synaptic cleft, filaments within the synaptic cleft, small clear synaptic vesicles within the nerve terminal, and electron-dense material accumulated on the postsynaptic side of the synapse (postsynaptic density). A mitochondrion is also visible in the postsynaptic dendrite. (Right) Drawing of the electron micrograph on the left with labeled synaptic specializations, including the identification of the likely sites for the presynaptic active zone (the site of synaptic vesicle fusion that mediates chemical transmitter release), and glial cells that wrap the synapse. *Source: Micrograph provided by Susan Sesack, Department of Neuroscience, University of Pittsburgh.*

**density**, an electron-dense region located across the synapse from and directly opposite to the active zone, is thought to result from the collection of transmitter receptors and associated anchoring proteins essential for detecting transmitter release and mediating postsynaptic signaling. Importantly, to allow for fast communication between pre- and postsynaptic cells, active zones within the presynaptic terminal are precisely aligned with postsynaptic densities within the postsynaptic membrane. This is an adaptation that increases the speed of neuronal communication by decreasing the distance chemical transmitters must diffuse to mediate signaling between cells. Nerve terminals also typically contain mitochondria, which are important calcium buffers, as well as a source of energy for mechanisms associated with transmitter release.

Interestingly, the synaptic cleft is not empty, but contains a large number of important regulatory proteins anchored in an **extracellular matrix**. This matrix usually appears in electron micrographs as a region of high electron density within the synaptic cleft. In addition, many synapses have a wrapping of glial cells that serves to isolate the synaptic cleft from surrounding cells and fluid. This glial wrapping is formed by Schwann cells at **neuromuscular junctions** (synapses between neurons and muscle cells), and by oligodendrocytes at synapses in the central nervous system (note that the role of glial cells at synapses is different from their role along axons where they form myelin; see Chapter 4). These synaptic glial cells are active participants in the chemical communication between neurons. In fact, it is an oversimplification to consider only the pre- and postsynaptic cells when studying chemical communication at synapses. A collection of glial cells and neurons that are arranged in this way form a **tripartite synapse** (Araque et al., 1999; see Fig. 2.8), which is discussed in detail in Chapter 22.

Synapses at various locations in the nervous system can take a variety of gross morphological forms, but at the electron microscopic level, they look essentially the same. It is a general principle of synaptic transmission that the basic synaptic structure is highly conserved, both across synapses in the human body and across species.

Most of the specialization at synapses cannot be appreciated by observing synaptic structure using an electron microscope. For example, the pre- and postsynaptic densities at synapses are composed of hundreds of proteins that cannot be distinguished without the use of labeling or other identification techniques. Therefore, scientists have used antibodies to label specific proteins at synapses, and have also used mass spectrometry-based proteomics to identify thousands of specific synaptic proteins (Bayés et al., 2017; Lassek et al., 2015). These types of studies are used to catalog the many proteins that are part of the synaptic specialization and to identify the location within the synaptic structure where they are expressed. We will discuss such studies in later chapters as they become relevant within the context of synaptic function.

## HOW DOES THE NEURON ASSEMBLE THE CELLULAR COMPONENTS REQUIRED TO CREATE SYNAPSES?

Given the complex specialization at synapses, one might predict that it takes a long time for developing neurons to acquire the ability to release chemical transmitters, but this is not the case. Before synapses can form, a neuron must extend an axon with a growing tip that explores the extracellular environment, looking for an appropriate target with which to initiate chemical communication. Surprisingly, growing axons already have the capability to release chemical transmitter from the **growth cone** (the growing tip of the axon). It appears that the basic presynaptic specializations that are required for chemical communication between cells are present from the very beginning of neuronal development (Young and Poo, 1983; Hume et al., 1983). Furthermore, postsynaptic cells express transmitter receptors even before communication is initiated with synaptic partners (see Fig. 2.9). Therefore, although the mechanisms for chemical transmitter release are very specialized, neurons express these specialized proteins and begin releasing chemical transmitters early in their development.

The neurotransmitter-filled vesicles described above release neurotransmitter from the presynaptic neuron by fusing the membranes of synaptic vesicles with the neuronal plasma membrane, such that the contents of the vesicle (neurotransmitters and/or neuropeptides) are released into the synaptic cleft. This process will be discussed at length in Chapter 8. Vesicle release from growth cones occurs using the same set of proteins that will later regulate vesicle fusion at the mature synapses.

In conclusion, it appears that neurons can assemble the specialized transmitter release site machinery required for chemical communication even while they are growing and exploring the environment for synaptic partners. In fact, the release of chemical transmitters appears to be an important regulator of axon pathfinding (growth and navigation) and/or synapse formation (see Andreae and Burrone, 2014, for review).

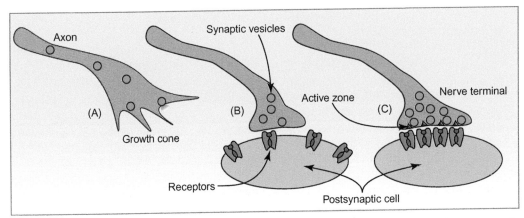

FIGURE 2.9  Diagrams that depict the early formation of a chemical synapse. (A) The growing tip of an axon (growth cone) already has the capability to release chemical transmitter from synaptic vesicles. Further, target cells are already expressing receptors for these chemicals. (B) Upon contact with a target cell, the presynaptic neuron is able to release chemical transmitter using components already present in the growing axon. (C) As the presynaptic neuron assembles a more mature functional transmitter release site (active zone), the postsynaptic cell collects existing receptors under the new synaptic contact.

# CONSTRUCTION OF ACTIVE ZONES DURING SYNAPSE DEVELOPMENT

How do growing neurons assemble the specialized proteins required for chemical transmitter release? The active zone appears to be assembled around many intracellular matrix proteins (Sudhof, 2012; Guldelfinger and Fejtova, 2012). This intracellular matrix includes a variety of molecules that can form scaffold-like structures, including cell adhesion molecules that guide synaptic proteins to future synaptic sites. These intracellular protein complexes can signal across three compartments at the site of future synapse formation: the presynaptic cell, the extracellular matrix, and postsynaptic cells. The formation of a new transmitter release site requires a variety of cell signaling mechanisms to initiate the construction of such a complex site of communication. These mechanisms include adhesion (holding pre- and postsynaptic cells together), enzymatic activity (cleaving some proteins and phosphorylating others), and transmembrane chemical messaging (triggering biochemical cascades within either the pre- or postsynaptic cell). The synapse assembly machine starts working as soon as neurons extend axons, and quickly builds small but functional chemical transmitter release sites (Patel et al., 2006; Wanner et al., 2011).

Neurons do not just transport the raw materials required to build a synapse to the end of the axon as isolated proteins. As synapses develop and mature, the addition of new active zone proteins at synaptic sites begins with the preassembly of many of these protein pieces within the Golgi apparatus of the cell body, where some proteins are

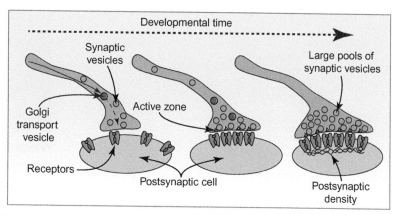

FIGURE 2.10 **Diagrams that depict the maturation of synapses after initial assembly.** After initial contact, proteins and organelles required for transmitter release are rapidly assembled at the site of contact. Growing axons already contain both synaptic vesicles and transport vesicles derived from the Golgi apparatus that contain proteins required for synapse formation and maturation. These Golgi transport vesicles contain "preformed" release sites with many of the proteins required for function. Therefore, upon fusion with the plasma membrane, they can participate in regulating transmitter release very quickly. With time after initial synapse contact, active zones form around sites where these Golgi transport vesicles fuse and insert their contents into the nerve terminal membrane; synaptic vesicles accumulate and form large pools; and postsynaptic receptors increase in density and are supported by postsynaptic scaffolding proteins (postsynaptic density). Over time, many of these and other components come together to form a mature synapse.

copackaged into Golgi transport vesicles. The Golgi apparatus then pinches off a given transport vesicle and targets it for transport along the neuronal cytoskeleton and delivery to the site of synapse formation (see Fig. 2.10). These vesicles contain preassembled active zone units, which are composed of the mixture of proteins required to start a new transmitter release site or add to an existing one. This preassembly facilitates the rapid construction of transmitter release sites (Zhai et al., 2001; Ziv and Garner, 2004).

The preassembled active zone precursor vesicles contain a variety of proteins that regulate synaptic vesicle fusion with the plasma membrane. After their insertion at the nerve terminal plasma membrane, these protein patches recruit other active zone proteins that are important for fast transmitter release and will be discussed in subsequent chapters (Ahmari et al., 2000; Zhai et al., 2001; Patel et al., 2006; Regus-Leidig et al., 2009; Owald and Sigrist, 2009). Once these elements of the active zone are collected at transmitter release sites, extracellular adhesion molecules align presynaptic sites of vesicle fusion with postsynaptic receptor clusters (Nishimune et al., 2004; Nishimune, 2012). Therefore, starting with preformed synaptic modules contained in Golgi transport vesicles, a newly forming synapse can quickly assemble a functional transmitter release site and then refine that site with additional proteins to build a larger and more complete synapse.

# References

Ahmari, S.E., Buchanan, J., Smith, S.J., 2000. Assembly of presynaptic active zones from cytoplasmic transport packets. Nat. Neurosci. 3 (5), 445–451.

Andreae, L.C., Burrone, J., 2014. The role of neuronal activity and transmitter release on synapse formation. Curr. Opin. Neurobiol. 27, 47–52.

Araque, A., Parpura, V., Sanzgiri, R.P., Haydon, P.G., 1999. Tripartite synapses: glia, the unacknowledged partner. Trends Neurosci. 22 (5), 208–215.

Bayés, À., Collins, M.O., Reig-Viader, R., Gou, G., Goulding, D., Izquierdo, A., et al., 2017. Evolution of complexity in the zebrafish synapse proteome. Nat. Commun. 8, 14613.

Burke, R.E., 2007. Sir Charles Sherrington's the integrative action of the nervous system: a centenary appreciation. Brain 130 (Pt 4), 887–894. Available from: https://doi.org/10.1093/brain/awm022.

Cimino, G., 1999. Reticular theory versus neuron theory in the work of Camillo Golgi. Physis. Riv. Int. Stor. Sci. 36 (2), 431–472.

De Robertis, E.D., Bennett, H.S., 1955. Some features of the submicroscopic morphology of synapses in frog and earthworm. J. Biophys. Biochem. Cytol. 1 (1), 47–58.

Guillery, R.W., 2005. Observations of synaptic structures: origins of the neuron doctrine and its current status. Philos. Trans. R. Soc. Lond. B. Biol. Sci. 360 (1458), 1281–1307. Available from: https://doi.org/10.1098/rstb.2003.1459.

Guldelfinger, E., Fejtova, A., 2012. Molecular organization and plasticity of the cytomatrix at the active zone. Curr. Opin. Neurobiol. 22 (3), 423–430.

Hume, R.I., Role, L.W., Fischbach, G.D., 1983. Acetylcholine release from growth cones detected with patches of acetylcholine receptor-rich membranes. Nature 305 (5935), 632–634.

Lassek, M., Weingarten, J., Volknandt, W., 2015. The synaptic proteome. Cell Tissue Res. 359 (1), 255–265. Available from: https://doi.org/10.1007/s00441-014-1943-4.

Nishimune, H., 2012. Active zones of mammalian neuromuscular junctions: formation, density, and aging. Ann. N.Y. Acad. Sci. 1274, 24–32.

Nishimune, H., Sanes, J.R., Carlson, S.S., 2004. A synaptic laminin-calcium channel interaction organizes active zones in motor nerve terminals. Nature 432 (7017), 580–587.

Owald, D., Sigrist, S.J., 2009. Assembling the presynaptic active zone. Curr. Opin. Neurobiol. 19 (3), 311–318.

Palay, S.L., Palade, G.E., 1955. The fine structure of neurons. J. Biophys. Biochem. Cytol. 1 (1), 69–88.

Patel, M., Lehrman, E., Poon, V., Crump, J., Zhen, M., Bargmann, C., et al., 2006. Hierarchical assembly of presynaptic components in defined C. elegans synapses. Nat. Neurosci. 9 (12), nn1806.

Ramón y Cajal, S., 1954. Neuron Theory or Reticular Theory? Objective Evidence of the Anatomical Unity of Nerve Cells. Consejo Superior de Investigaciones Científicas, Instituto Ramón y Ramón y Cajal, S, Madrid.

Regus-Leidig, H., Tom Dieck, S., Specht, D., Meyer, L., Brandstätter, J.H., 2009. Early steps in the assembly of photoreceptor ribbon synapses in the mouse retina: the involvement of precursor spheres. J. Comp. Neurol. 512 (6), 814–824.

Sudhof, T.C., 2012. The presynaptic active zone. Neuron 75 (1), 11–25.

von Waldeyer-Hartz, W., 1891. Ueber einige neuere Forschungen im Gebiete der Anatomie des Centralnervensystems. Georg Thieme.

Wanner, N., Noutsou, F., Baumeister, R., Walz, G., Huber, T.B., Neumann-Haefelin, E., 2011. Functional and spatial analysis of C. elegans SYG-1 and SYG-2, orthologs of the Neph/nephrin cell adhesion module directing selective synaptogenesis. PLoS. One 6 (8), e23598.

Young, S.H., Poo, M.M., 1983. Spontaneous release of transmitter from growth cones of embryonic neurones. Nature 305 (5935), 634–637.

Zhai, R.G., Vardinon-Friedman, H., Cases-Langhoff, C., Becker, B., Gundelfinger, E.D., Ziv, N.E., Garner, C.C., 2001. Assembling the presynaptic active zone: a characterization of an active one precursor vesicle. Neuron 29 (1), 131–143.

Ziv, N.E., Garner, C.C., 2004. Cellular and molecular mechanisms of presynaptic assembly. Nat. Rev. Neurosci. 5 (5), 385–399. Available from: https://doi.org/10.1038/nrn1370.

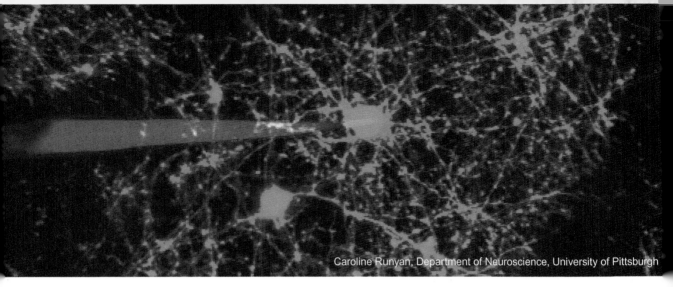

Caroline Runyan, Department of Neuroscience, University of Pittsburgh

# CHAPTER 3

# Basics of Cellular Neurophysiology

## NEURONS ARE EXCITABLE CELLS

Fundamentally, cellular neurophysiology is the study of signaling within individual neurons. One of the most important properties of neurons is that they are "excitable." This property of **excitability** refers to a neuron's ability to use the **flux** of electric current across its cell membrane to trigger a transient, all-or-none voltage spike, referred to as an **action potential**. An action potential can pass a signal very quickly along the length of the neuron, from near the cell body to the axon terminal.

How are neurons built for the fast transfer of electrical information along their length? Excluding body fat, 50–65% of human body weight is water. From a molecular perspective, more than 90% of the molecules in the body are water, and 1% of these are inorganic ions (e.g., sodium, potassium, chloride, calcium). As a result, the body is filled with what we might call "salt water." Salts are charged particles that cells can use to carry electric current. In essence, your nervous system uses electricity to move information along the length of axons.

**Electric current** is defined as the movement of charged particles. By convention, the direction of current flow is defined by the movement of net positive charge. This convention is true in physics and in neuroscience. In neuroscience, we calculate current in terms of the movement of net positive charge between the inside and outside of a cell. This convention is true regardless of the charge on an ion (i.e., whether it is positively or negatively charged). Therefore, if a positively charged ion moves from the extracellular space to the inside of a cell, this is defined as inward current, and if a negatively charged ion moves from the intracellular space to the outside of the cell, this is also defined as inward current.

Because neurons can regulate the flow of charged ions across their cell membrane, they can create tiny electric currents, which is what makes neurons excitable. Currents are measured in **amperes** (A), which is a measure of the rate of charged particle flow in an electrical conductor (cell cytoplasm, for example). One ampere of current represents one **coulomb** of electrical charge ($6.24 \times 10^{18}$ charged ions) moving past a specific point in one second. Most neurons are so small that they only need tiny currents in order to create electrical signals, so they move ions to create currents that measure in the pico-ampere range ($10^{-12}$ A). Because these currents are so small, we can't feel the electricity moving in our bodies.

## Ions in and Around Neurons

There are two major mechanisms by which charged ions move across neuronal cell membranes. First, when there is an unequal distribution of specific charged ions on each side of the membrane, the cell can actively open a protein pore (ion channel) in its membrane that is selective for that charged ion, and the ions will flow down their **concentration gradient**. That is, if the concentration of an ion is higher outside of the cell than inside, those ions will flow into the cell through the pore, and vice versa. Second, the cell can use energy (e.g., from ATP) to transport charged ions across the membrane, independent of the concentration gradient for that ion. These mechanisms will be covered in more detail later on.

In the nervous system and muscles, ions relevant to neuronal signaling are found in both the extracellular and intracellular spaces. Table 3.1 indicates approximate concentrations of sodium ($Na^+$), potassium ($K^+$), chloride ($Cl^-$), calcium ($Ca^{2+}$), magnesium ($Mg^{2+}$), and negatively charged ions in these spaces. These are the major ions we will discuss with regard to signaling in neurons (i.e., for now, we are ignoring ions such as hydroxide and hydrogen ions). By definition, a **cation** is a positively charged particle (so called because it is attracted to the cathode of a battery), and an **anion** is a negatively charged particle (attracted to the anode of a battery). In addition to inorganic ions, we will also consider the large negatively charged proteins that exist in cells, as these are important to balance charges across the cell membrane. These include anions such as amino acids, sulfates, phosphates, and bicarbonate.

## Membrane Potential and Capacitance

If you add up the numbers of anions and cations on each side of the plasma membrane, you will find that the charges roughly balance one another. Therefore, if no ions can flow across the plasma membrane (i.e., move from the intracellular space to the extracellular space or vice versa), there is no electrical difference between the inside and outside of a cell because each side has the same net charge. A difference in the net charge between two

TABLE 3.1  The Approximate Concentrations of Ions on the Inside and Outside of Neurons

| Ion | Inside (mM) | Outside (mM) |
| --- | --- | --- |
| $Na^+$ | 15 | 145 |
| $K^+$ | 140 | 4 |
| $Cl^-$ | 10 | 150 |
| $Ca^{2+}$ | 0.0001 | 2 |
| $Mg^{2+}$ | 0.5 | 2 |
| $Anions^-$ | 150 | 0 |

This table lists the major ions and their concentrations on either side of the neuronal membrane. Some ions are found in a very high concentration in the cell cytoplasm (inside), including potassium ($K^+$), and anions (which include large negatively charged proteins, amino acids, sulfates, phosphates, and bicarbonate). Others are higher in concentration in the extracellular space, including sodium ($Na^+$), chloride ($Cl^-$), and calcium ($Ca^{2+}$). Note that the actual concentrations of these ions differ by type of neuron and by species, but these approximate values and ratios hold for most neurons.

locations is called an **electrical potential**, or **voltage**, which is a force that can be exerted on a charged particle and cause it to move. For example, if the inside of a cell is more negative than the outside, positively charged particles (e.g., sodium ions) are attracted to the inside of the cell. We refer to the voltage across a cell membrane (i.e., the net charge inside the neuron compared to the net charge outside the neuron) as the **membrane potential** ($V_m$). When neurons signal, a typical action potential is approximately 100 mV (or 0.1 V) in amplitude, which is exceptionally small compared to the amount of volts from the outlets in your house (120 V in the US, 220–240 V in many other countries). We will discuss the details of ionic currents in neurons in the sections below.

The plasma membrane of cells is a very thin ($\sim$3–4 nm) lipid bilayer that separates the ions located inside the cell from those outside the cell. In physics and electronics, a structure that can separate (or store) electrical charges is called a **capacitor**, which is defined as two conducting materials separated by an insulating material. In neuroscience, we use the same terminology because cells have essentially the same configuration: the "salt water" inside and outside cells is a good conductor of electricity, and the lipid bilayer of the plasma membrane is a good separator, or insulator, of charge (see Fig. 3.1). The plasma membrane insulates charges effectively because of the hydrophobic lipid tails on the molecules that make up the plasma membrane. Because ions are charged, they are unable to cross through these lipid tails on their own. The unit of capacitance is the **farad** (F), and the larger the insulator (more surface area), the larger the capacitance. A typical cell membrane has a specific capacitance of $\sim 1\ \mu F/cm^2$. Knowing this, experimenters can measure the capacitance of a cell to estimate the total area of the cell membrane. The cell membrane is a good capacitor in part because it is such a thin insulator—the smaller the distance between the charges in the cytoplasm and those

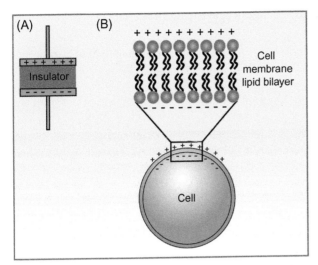

FIGURE 3.1 **The cell membrane is a capacitor (a separator of charges or ions).** (A) An insulator that prevents charge movement between two conductors is a capacitor. (B) The cell membrane lipid bilayer is a capacitor because it is able to separate charges (keep ions on the inside or outside of the cell from moving across the membrane).

in the extracellular saline, the greater the capacity to store charge. We will discuss an experimental approach that uses measurements of capacitance at synapses in Chapter 8.

Critically, neuronal plasma membranes are not completely effective at insulating ionic charges. They do allow some ions to flow across (through) the membrane, but only in a very controlled way. This characteristic makes the plasma membrane a "leaky" capacitor, and it is often referred to as a selective, semipermeable membrane. A porous membrane's **permeability** is the degree to which it permits substances to pass through. The plasma membrane is considered selectively semipermeable because it only allows certain substances to go across, and it carefully regulates which ones do. This regulation of ion flow is carried out by protein pores located in the plasma membrane that are selective for particular ions. The protein pores that allow ions to flow across the membrane are called **ion channels** (see Chapter 4).

When a neuron is at "rest" (defined as the absence of action potential activity or a response to a neurotransmitter), it is primarily permeable to potassium ions. This permeability is due to the fact that all neurons express a type of potassium-selective ion channel called a two-pore-domain potassium channel (K2P; see Chapter 4; Goldstein et al., 2005; Honore, 2007). K2P channels are open at **resting membrane potentials** (also referred to as $V_{rest}$) and create what is referred to as a **potassium leak current** (Talley et al., 2000, 2001; Aller et al., 2005). These channels are open at $V_{rest}$ because the probability that they are open is independent of the voltage across the membrane, but is instead controlled by pH (Mathie et al., 2010). Since these channels tend to be open at physiological pH levels, ion flux through them is driven by the concentration gradient for potassium across the plasma membrane. In addition to these K2P channels, some voltage-sensitive potassium channels might also occasionally open at the resting membrane potential and thus contribute to the resting potassium flux.

# MOVEMENT OF IONS ACROSS THE CELL MEMBRANE

Neurons can be permeable to ions such as potassium and sodium, but the number of ions moving through ion channels is determined by a number of factors. One of these is the resting membrane potential, and another is the concentrations of ions inside and outside the cell. We will first consider potassium ions as an example of how ions are affected by intracellular and extracellular ion concentrations.

If a neuron at its resting membrane potential were only permeable to potassium, what would happen when potassium channels opened? Potassium would move out of the cell, flowing down its concentration gradient from the higher concentration inside the cell to the lower concentration outside the cell. However, negatively charged anions in the cell would not be able to pass through the potassium-selective channels. As the positively charged potassium ions left the cell and the negatively charged anions remained inside, the inside of the membrane would become more negative than the outside (see Fig. 3.2). By convention, we always define the electrical potential across the cell membrane as the summed (net) charge inside the cell membrane, relative to the summed charge on the outside of the membrane. Therefore, when the net charge inside is more negative than the net charge outside, we say that the cell has a negative membrane potential.

However, this negative membrane potential would attract some of the positively charged potassium ions back into the cell. In fact, potassium ions would continue to leave the cell until the electrical potential across the membrane was negative enough to attract

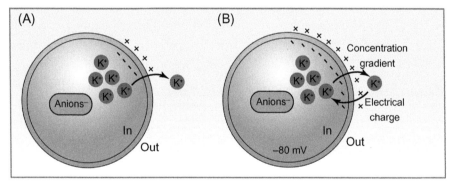

FIGURE 3.2  Establishment of the equilibrium distribution of potassium ions across the cell membrane. Given the resting distribution of potassium ions across the membrane (higher concentration inside the cell), when potassium-selective channels open, potassium will flow down its concentration gradient out of the cell (A). Since large negatively charged anions cannot follow potassium through these channels, this creates an uneven charge distribution across the membrane (negative inside relative to outside). Potassium continues to flow out of the cell down its concentration gradient until the inside becomes so negative that the positively charged potassium ions are attracted back into the cell by the electrical charge distribution (B). When the potassium movement out of the cell driven by the concentration gradient is equal to the potassium movement into the cell driven by the attraction of positive potassium ions to the electrical charge distribution across the membrane, potassium ion movement is at equilibrium. Given the normal distribution of potassium ions across the cell membrane, this equilibrium occurs at a membrane potential of $-80$ mV.

the same number of potassium ions back into the cell by electrical charge attraction. When the number of potassium ions leaving the cell due to the concentration gradient is equal to the number of potassium ions entering the cell due to the electrical charge on the membrane, then the movement of potassium ions is at equilibrium, meaning there is no net flux of these ions (see Fig. 3.2B). The electrical potential across the membrane that creates this equilibrium is called the potassium ion **equilibrium potential**, or $E_{qK}$.

In neurons, the potassium ion equilibrium potential is usually close to $-80$ mV. When a cell's membrane potential is more positive than the $E_{qK}$, there is a net flow of potassium out of the cell. This occurs because the outward force on the potassium ions due to the concentration gradient is greater than the inward force on the ions due to the electrical charge difference across the plasma membrane. Conversely, when the membrane potential is more negative than the $E_{qK}$, there is a net flow of potassium into the cell. This is because the inward force on the ions due to the electrical charge is greater than the outward force on the ions due to the concentration gradient. Because the net flow of ions changes direction, or reverses, when the membrane potential of the cell crosses $-80$ mV ($E_{qK}$), the equilibrium potential is also sometimes referred to as the **reversal potential**.

Very few potassium ions need to leave the cell to create a local negative electrical potential, relative to the absolute number of potassium ions found inside cells. This is because the change in the electrical potential generated by the movement of one potassium ion out of the cell is much greater than the change in the concentration gradient. Therefore, potassium flux that brings the membrane potential to $-80$ mV does not significantly change the potassium ion concentration gradient across the cell membrane.

The same concepts apply to other types of ions. For example, if a neuron were only permeable to sodium ions, sodium would move down its concentration gradient into the cell (since the concentration of sodium is higher in the extracellular fluid than inside cells). This influx of positive ions would make the inside of the cell more positive than the outside. However, this positive membrane potential would repel sodium ions, moving them back out of the cell, and these two forces would reach an equilibrium when the number of sodium ions going into the cell (driven by the concentration gradient) was equal to the number going out (due to repulsion between similar electrical charges). Based on the concentration gradient for sodium, this equilibrium would occur when the membrane potential reached approximately $+60$ mV. Therefore, the equilibrium potential for sodium ($E_{qNa}$) is $+60$ mV.

## Calculating the Equilibrium Potential for an Ion

Each type of ion has its own equilibrium potential. The exact value of a given ion's equilibrium potential is dependent on the concentrations of that ion inside and outside the neuron, which vary by location in the nervous system, as well as by species.

The Nernst equation can be used to calculate the equilibrium potential (which is also called the Nernst potential) for any ion, and it is based on the distribution of that ion across the membrane. This equation provides a quantitative understanding of the forces that influence ion movement and membrane potential. The Nernst equation includes the ideal gas constant $R$ ($R = 8.314$ J/mol-K), the valence of the ion, $z$ (+ or −), Faraday's constant $F$ ($F = 96,485$ C/mol, the charge on a mole of electrons), the absolute temperature ($T$; at 25°C, $T = 298$K), and the concentration of the ion on each side of the neuronal membrane. The Nernst equation is:

$$V = (RT/zF) * \ln (\text{concentration of ions outside}/\text{concentration of ions inside})$$

In order to gain an appreciation for this equation, it is useful to search for an online "Nernst potential calculator." These calculators display an equilibrium voltage based on the valence and ion concentration values you enter, allowing you to see for yourself how changes in ion concentrations impact the equilibrium potential for a given ion.

## Multiple Ions Contribute to the Resting Membrane Potential

If a cell at rest was only permeable to one ion, it would be possible to calculate the resting membrane potential of a neuron simply by using the Nernst equation. For example, if neurons at rest were only permeable to potassium, their typical resting membrane potential would be equal to the potassium ion equilibrium potential, −80 mV. However, when experimenters measured the resting membrane potential of a typical neuron, they noticed that it was roughly −65 mV, not −80 mV (Fig. 3.3), which informed them immediately that neurons could not be solely permeable to potassium.

Indeed, neurons are permeable to more than one type of ion. However, at rest, neurons are ~25–40 times more permeable to potassium ions than to sodium ions (Hodgkin and Katz, 1949). This small but significant amount of resting sodium permeability is likely due to several factors, which include sodium-dependent cotransporters, sodium exchangers (discussed later), persistent sodium current through voltage-gated sodium channels (Crill, 1996), and a class of ion channels termed the "sodium ($Na^+$) Leak Channel, Nonselective" (NALCN; Lu et al., 2007).

NALCN channels are voltage-independent, meaning they can open regardless of the membrane voltage. They are predominately selective for sodium and potassium ions, though they can also pass some calcium ions. The evidence that these NALCN channels are a significant source of the sodium permeability in resting neurons derives from studies of hippocampal neurons from mice in which the NALCN gene was removed, or "knocked out." These NALCN-knockout neurons lack significant leak sodium current at rest, and the **leak current** can be restored by introducing NALCN DNA into these neurons to replace this missing channel (Lu et al., 2007). Thus, although transporters, exchangers, and persistent sodium current through voltage-dependent channels may also contribute to the sodium leak current, it is hypothesized that the voltage-independent NALCN channel is the main source of this current in neurons.

## Calculating the Membrane Potential

An important concept in cellular neurophysiology is that the membrane potential of a neuron (under any condition) is always determined by a combination of two

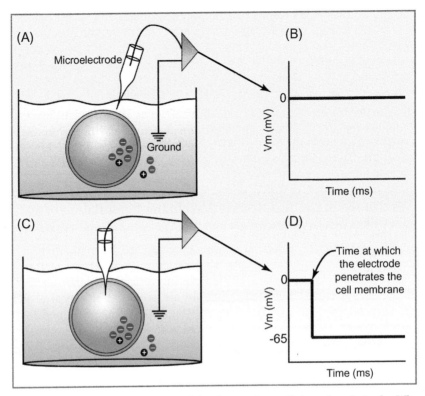

FIGURE 3.3  Recording the membrane potential using an intracellular microelectrode. When an experimenter measures a neuron's membrane potential, the cell under study is bathed in a solution containing ions at concentrations that mimic those in the extracellular space (A). To determine the voltage difference across the cell membrane, one reference (ground) electrode is placed in the extracellular solution, and a second microelectrode is pierced through the cell membrane into the cytoplasm (C). When the microelectrode is still in the extracellular saline, there is no voltage difference between the two electrodes (B; $V_m = 0$ mV). However, when the microelectrode pierces the cell membrane and the tip moves inside the cell, the voltage measured between the two electrodes suddenly jumps to about −65 mV (D).

factors: (1) the equilibrium potentials of all the ions a membrane is permeable to, and (2) the relative permeability of each ion (i.e., how many channels that pass a given ion are open). For example, if the neuronal membrane were permeable to only potassium and sodium, and was equally permeable to each one, the membrane potential would be halfway between the equilibrium potentials for those two ions, or about −10 mV (i.e., halfway between −80 mV and +60 mV). However, in a neuron the situation is rarely this simple, in part because each ion does not have equal permeability. The relationship between the membrane potential, the concentration of each ion, and the relative permeabilities of each ion, can be calculated using the Goldman–Hodgkin–Katz voltage equation (often simply referred to as the Goldman equation; Hodgkin and Katz, 1949).

$$V_m = \frac{RT}{F} \ln\left(\frac{p_K[K^+]_o + p_{Na}[Na^+]_o + p_{Cl}[Cl^-]_i}{p_K[K^+]_i + p_{Na}[Na^+]_i + p_{Cl}[Cl^-]_o}\right)$$

The permeability ($p$) values used in this equation are usually expressed relative to the potassium permeability value since it is usually the largest. Thus, $p_K$ is defined as 1, and the permeability of each of the other ions is expressed as a fraction of the potassium permeability. The concentration ratio for each ion is written with the ion concentration outside the neuron in the numerator and the ion concentration inside the neuron in the denominator for positively charged ions, and the reverse for negatively charged ions. As with the Nernst equation, an appreciation for this relationship can easily be obtained using an online Goldman–Hodgkin–Katz equation calculator.

For example, using the Goldman–Hodgkin–Katz equation (or a web calculator for this equation), one can determine that if extracellular potassium levels increase, the resting membrane potential **depolarizes** due to the increase in the equilibrium potential for potassium. In fact, there is an interesting historical case concerning the use of potassium chloride injections into the blood to dramatically increase extracellular potassium levels and depolarize cells in the body (see Box 3.1).

---

### BOX 3.1

### DR. JACK KEVORKIAN AND ASSISTED SUICIDE

Dr. Jack Kevorkian was in the news in the 1990s for assisting terminally ill patients in ending their lives by administering an injection of highly concentrated potassium chloride (among other agents) into the patient's bloodstream (Fig. 3.4). His critics called him "Doctor Death," although he claimed that his aim was to end suffering, not to cause death. The public exposure of his activity prompted a heated debate surrounding the issue of physician-assisted suicide. He was tried in court for assisted suicide and acquitted several times, before he was eventually arrested for direct participation in the "voluntary euthanasia" of a patient with amyotrophic lateral sclerosis. He was convicted in this case and served eight years in prison.

The neurophysiology behind his method depends on the fact that potassium ions are usually at a very low concentration outside of cells ($\sim$5 mM). What happens to the resting membrane potential when the potassium concentration is raised from 5 mM to 20 mM? Using the Nernst equation to calculate the resulting change in the potassium equilibrium potential, we find that EqK changes from $-80$ mV when the extracellular potassium is 5 mM, to $-45$ mV when the extracellular potassium in 20 mM. How does this change in EqK to a less negative value affect the resting membrane potential? At rest, a cell membrane is most permeable to potassium, which means EqK has the greatest impact on the resting membrane potential. Therefore, an increase in extracellular potassium from 5 mM to 20 mM significantly depolarizes the resting membrane potential and activates voltage-gated ion channels, including voltage-gated sodium channels. However, after opening, voltage-gated sodium channels remain inactive as long as the membrane potential is depolarized, which means the depolarized resting potential persistently

## BOX 3.1 (cont'd)

inactivates these channels. With most of the voltage-gated sodium channels inactivated, neurons and muscle cells cannot fire action potentials. Dr. Kevorkian's potassium chloride injections were lethal because they prevented cardiac muscle cells from firing action potentials, which in turn caused the patient's heartbeat to cease.

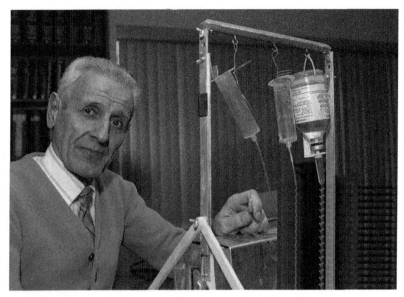

FIGURE 3.4 Dr. Jack Kevorkian pictured with his device used for assisted suicide. *Source: Adapted from <http://www.nytimes.com/2011/06/04/us/04kevorkian.html>, Associated Press.*

In some cells (especially muscle cells), the resting membrane is also permeable to chloride, hence the inclusion of chloride in the Goldman equation. However, the equilibrium potential for chloride is complicated by developmental and tissue-specific changes in the concentration gradient for chloride across the plasma membrane, due to the regulated expression of chloride transporters. Therefore, the discussion of neuronal resting membrane potential in this text will not focus on the role of chloride. For those interested in reading further about chloride channels and their role in resting membrane potential, see the review by Poroca et al. (2017).

## Ion Fluxes at Resting Membrane Potentials

There are multiple forces that influence the amount of current that flows across a cell membrane at a given point in time. A simple way to think about this is that if there are ion channels open in the membrane, ions can move through them, but they will only do so if there is a force acting on them (the **driving force** (DF), discussed below) that will move them through the open channels. In other words, if no channels are open, no current can flow, and if there is no force on an ion, ions will not flow, whether channels are open or not.

## *Driving Force on an Ion*

In the previous section, we discussed the equilibrium potential for each ion and the general concept that the membrane potential determines the direction each ion flows across the cell membrane. This ionic flow is based on the difference between the equilibrium potential for an ion and the membrane potential. It is useful to consolidate this relationship into a single value called the driving force, or **DF**. The DF is simply the membrane potential minus the equilibrium potential:

$$DF = V_m - E_{ion}$$

The sign of DF indicates the direction of ion flow. When the DF on an ion is a negative number, there is inward movement of that ion, and when the DF is a positive number, there is outward movement of that ion. Whether the value of the DF is positive or negative, the larger its absolute value, the stronger the force on the ion.

Using potassium ions as an example, when the membrane potential ($V_m$) is $-80$ mV, the DF is 0 mV (DF = $-80$ mV $-$ ($-80$ mV) = 0 mV; see Fig. 3.5]). This makes sense because $-80$ is the potassium ion equilibrium potential, so there is no net movement of potassium ions at that voltage (i.e., the ion flux is equal in both directions). As such, at $-80$ mV, there is no net DF for the flux of potassium ions. However, when $V_m = +30$ mV, there is a strong outward force on the potassium ions (DF = $+30$ mV $-$ ($-80$ mV) = $+110$ mV).

At any given membrane potential, the DF for potassium ions is different from the DF for sodium ions (see Fig. 3.5). At a $V_m$ of $-80$ mV, there is a very strong inward force on the sodium ion (DF = $-80$ mV $-$ ($+60$ mV) = $-140$ mV), whereas at $+30$ mV, there is a

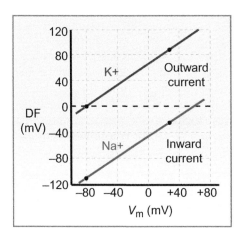

FIGURE 3.5 **The driving force on sodium and potassium ions as a function of membrane potential.** The driving force for the potassium ion is plotted in blue, and the driving force for the sodium ion is plotted in red. The two points on each of these lines represent the physiological range of membrane potentials that most neurons experience (between $-80$ and $+30$ mV). In this physiological range, potassium ions almost always have an outward driving force (positive number on the $y$-axis), and sodium channels always have an inward driving force.

relatively weak inward force on the sodium ion (DF = +30 mV − (+60 mV) = −30 mV). The DF is always defined by the concentration gradient for a given ion, since the concentration gradient determines that ion's equilibrium potential.

## Maintaining Ionic Concentrations

The imbalance of ionic fluxes (small net influx of sodium and large net efflux of potassium) at the resting membrane potential would, in the absence of a counterbalancing mechanism, eventually lead to the dissipation of the normal concentration gradients for sodium and potassium across the cell membrane. In order to counter ionic fluxes that occur at the resting membrane potential, cells express a pump that redistributes sodium and potassium ions. This pump is a large membrane-spanning transporter protein known as the **sodium−potassium ATPase**. For every ATP the pump hydrolyzes, it moves three sodium ions out of the neuron and two potassium ions into the neuron (see Fig. 3.6). In every neuron, this pump runs continuously in the background, and it serves to maintain the sodium and potassium concentration gradients that control the resting membrane potential. Additionally, because this pump moves an unequal number of charges into and out of the neuron (three sodium ions out for every two potassium ions in), it is considered **electrogenic** (that is, it creates a voltage by moving electrical charges). Specifically, activity of the sodium−potassium ATPase makes the inside of the membrane more negative by a few millivolts. See Box 3.2 for a discussion of natural products that target the sodium-potassium ATPase.

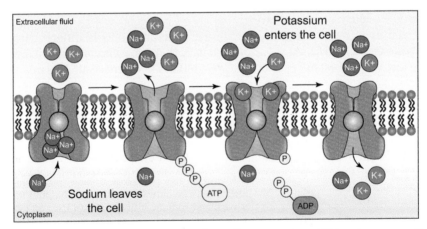

FIGURE 3.6 **The sodium−potassium ATPase.** The sodium−potassium ATPase is a pump that uses the energy provided by cleaving ATP into ADP to move sodium and potassium ions against their concentration gradients. This figure shows the sequence of events (left to right) as the sodium−potassium ATPase pumps sodium out of the cell and potassium into the cell. This pump has three binding sites for sodium and two binding sites for potassium. Thus, with the cleavage of one ATP molecule, three sodium ions are moved out and two potassium ions are moved into the cell.

# BOX 3.2

## DANGERS IN NATURE: PURPLE FOXGLOVE PLANTS, CORAL, AND THE $Na^+/K^+$ ATPase

### *Digitalis purpurea* (Purple foxglove)

Because of its dense rosette of leaves and tall stalks of purple flowers, *Digitalis purpurea* (see Fig. 3.7 left) is a popular plant for gardens across the world. However, *Digitalis* contains the glycosides digoxin and digitoxin, which inhibit the $Na^+/K^+$-ATPase, causing a rise in intracellular sodium, which reverses the direction of the $Na^+/Ca^{2+}$ (high potassium in the blood), low heart rate, and ventricular fibrillation. The effects of overdose occur because the changes in potassium concentration make the potassium reversal potential more positive than normal. This causes persistent inactivation of sodium channels, leading the cell to enter a continuous refractory period (unable to fire action potentials). Without medical intervention, digoxin and digitoxin poisoning can lead to death.

FIGURE 3.7   *Digitalis purpurea* (left) and *Palythoa* coral (right)

exchanger at the resting membrane potential (see Chapter 7). In cardiac muscle cells, high intracellular sodium and calcium ultimately produce a greater than normal contraction of myocardial muscle. *Digitalis* has been used medicinally for patients suffering heart failure since 1785, and digoxin is on the World Health Organization's List of Essential Medicines. However, the strict therapeutic index (ratio of the toxic to the therapeutic doses) for these drugs limits their use, and overdoses cause hyperkalemia

### *Palythoa* Corals

Coral are popular and vibrant additions to tropical aquariums, but handlers of beautiful *Palythoa* corals (see Fig. 3.7 right) should be cautious of contracting one of the most toxic nonprotein poisons in the world, palytoxin. Palytoxin is water-soluble, thermo-stable (unable to be inactivated by boiling temperatures), and can be aerosolized. The danger isn't limited to coral, however, as natural "biomagnification" (the

> **BOX 3.2** *(cont'd)*
>
> increase in toxin concentration via the tissues of organisms higher on the food chain that are resistant to the toxin's virulence) means exposure to palytoxin can occur after ingesting contaminated seafood. Once palytoxin enters the body, it binds strongly to the $Na^+/K^+$ ATPase, with a dissociation constant in the picomolar range. After binding, it transforms the function of this pump to act as a passive nonselective ion channel. This effectively eliminates the concentration gradient for sodium, potassium, and calcium across the plasma membrane, as intracellular contents leak into the blood and affected cells depolarize. Symptoms of palytoxin poisoning include severe vasoconstriction, rhabdomyolysis (skeletal muscle degradation), weakness of the extremities, and heart failure.

## Membrane Conductance and Resistance

The DF on an ion only indicates the magnitude and direction of the force acting on that ion. It does not reflect whether there is any pathway for ion movement (i.e., open ion channels). Even if there is a strong DF on an ion, it can only cross the membrane if such a pathway is available. In biological terms, the ability of charged ions to move across the membrane through a protein pore is referred to as **conductance**. Conductance is an electrical term that indicates the ability of a charged particle (such as an ion) to move from one point to another. Conductance is measured in **siemens** (S), and the magnitude of the conductance is determined by the number of charged particles that can be moved. For example, a channel with high conductance will move a greater number of ions than a small conductance channel.

The inverse of conductance is **resistance**. Resistance is defined as the ability to *prevent* the flow of electrical current, and is measured in Ohms ($\Omega$). In biological terms, the resistance of a cell membrane indicates how well the membrane prevents the flow of ions. A cell membrane with very few open channels has high resistance, whereas a membrane with many open channels has low resistance. Because most neurons have the same number of open channels per unit membrane area when they are at rest, the resistance of a neuronal membrane is inversely proportional to the membrane area of that neuron.

## Ohm's Law

The relationship between resistance, current, and voltage is represented by **Ohm's Law**, which is written as follows:

$$V = IR$$

Ohm's Law can be used to predict how the membrane potential will change in response to current flux across the cell membrane. For example, when charge moves into a neuron there are many pathways for this charge to redistribute back across the membrane (i.e., when there are many open ion channels, which causes a low resistance to charge flux). In this situation, there will be only a relatively small change in the membrane potential in response to the flow of current (ions). In contrast, if there are few open ion channels (thus, a high resistance to charge flux), the change in the membrane potential in response to current flux will be larger. Fig. 3.8 depicts two neurons of different sizes and thus different membrane resistances. The larger cell has more ion channels and therefore a lower membrane resistance, which means that it will have a smaller change in membrane potential for a given amount of charge flux, compared to the smaller neuron, which has fewer ion channels and a higher membrane resistance.

The resistance of a neuron can be measured easily and is often used by experimenters to calculate the impact that electrical charge flux across the cell membrane has on the membrane potential of the neuron. Resistance is an important factor in cellular neurophysiology because changes in membrane potential are triggers for electrical signaling in neurons (e.g., firing an action potential).

In summary, the basic principles of neurophysiology define the mechanisms by which neurons signal within a single cell. This signaling is possible because of the unequal distribution of different ions (e.g., sodium and potassium) across the plasma membrane. This distribution establishes concentration gradients that lead to charge movement (current flow) when pathways for ion flux across the membrane are open. This ionic charge movement creates a current that can change the membrane potential of a cell. The membrane potential of the cell is always governed by a combination of the equilibrium potentials of the ions to which the cell is permeable and the relative permeability of each ion.

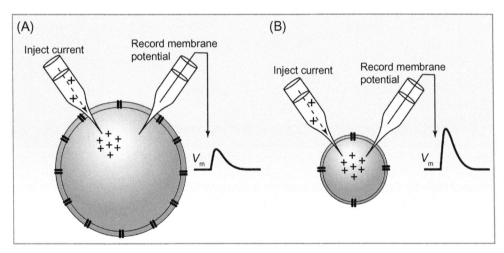

FIGURE 3.8  **The impact of cell size and membrane resistance on changes in membrane potential after current flux.** Because the number of potassium and sodium channels that are open at rest is relatively constant in neuronal membranes, smaller cells (B) have fewer open channels at rest than larger cells (A). As a result, when the same amount of current is injected into both cells, according to Ohm's Law, the smaller cell (B) will experience a larger change in membrane potential than the larger cell (A).

# References

Aller, M.I., Veale, E.L., Linden, A.M., Sandu, C., Schwaninger, M., Evans, L.J., et al., 2005. Modifying the subunit composition of TASK channels alters the modulation of a leak conductance in cerebellar granule neurons. J. Neurosci. 25 (49), 11455–11467. Available from: https://doi.org/10.1523/JNEUROSCI.3153-05.2005.

Crill, W.E., 1996. Persistent sodium current in mammalian central neurons. Annu. Rev. Physiol. 58, 349–362. Available from: https://doi.org/10.1146/annurev.ph.58.030196.002025.

Goldstein, S.A., Bayliss, D.A., Kim, D., Lesage, F., Plant, L.D., Rajan, S., 2005. International Union of Pharmacology. LV. Nomenclature and molecular relationships of two-P potassium channels. Pharmacol. Rev. 57 (4), 527–540. Available from: https://doi.org/10.1124/pr.57.4.12.

Hodgkin, A.L., Katz, B., 1949. The effect of sodium ions on the electrical activity of giant axon of the squid. J. Physiol. 108 (1), 37–77.

Honore, E., 2007. The neuronal background K2P channels: focus on TREK1. Nat. Rev. Neurosci. 8 (4), 251–261. Available from: https://doi.org/10.1038/nrn2117.

<http://www.nytimes.com/2011/06/04/us/04kevorkian.html>, image, Associated Press.

Lu, B., Su, Y., Das, S., Liu, J., Xia, J., Ren, D., 2007. The neuronal channel NALCN contributes resting sodium permeability and is required for normal respiratory rhythm. Cell 129 (2), 371–383. Available from: https://doi.org/10.1016/j.cell.2007.02.041.

Mathie, A., Al-Moubarak, E., Veale, E.L., 2010. Gating of two pore domain potassium channels. J. Physiol. 588 (Pt 17), 3149–3156. Available from: https://doi.org/10.1113/jphysiol.2010.192344.

Poroca, D.R., Pelis, R.M., Chappe, V.M., 2017. ClC Channels and Transporters: Structure, Physiological Functions, and Implications in Human Chloride Channelopathies. Front. Pharmacol. 8, 151. Available from: https://doi.org/10.3389/fphar.2017.00151.

Talley, E.M., Lei, Q., Sirois, J.E., Bayliss, D.A., 2000. TASK-1, a two-pore domain K+ channel, is modulated by multiple neurotransmitters in motoneurons. Neuron 25 (2), 399–410.

Talley, E.M., Solorzano, G., Lei, Q., Kim, D., Bayliss, D.A., 2001. Cns distribution of members of the two-pore-domain (KCNK) potassium channel family. J. Neurosci. 21 (19), 7491–7505.

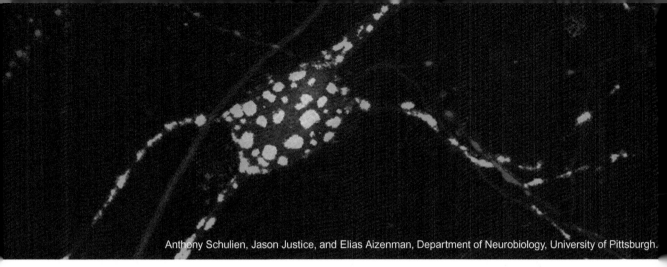

Anthony Schulien, Jason Justice, and Elias Aizenman, Department of Neurobiology, University of Pittsburgh.

# CHAPTER 4

# Ion Channels and Action Potential Generation

## ION CHANNELS

### Gating of Ion Channels

Protein pores in the neuronal plasma membrane that allow ions to flow across the membrane are called **ion channels**. We touched on these channels in Chapter 3. In this chapter, we will cover more details about ion channels that explain how they contribute to neuronal signaling.

Most ion channels can be opened and closed by specific stimuli, and this process is referred to as **channel gating**. In all cases, the stimulus to open an ion channel causes a conformational change in the channel protein's tertiary (three-dimensional) structure. This conformational change creates a path, or pore, through the middle of the channel protein. When the pore is open, ions can move through it, traveling between the intracellular and extracellular sides of the membrane. Critically, each type of ion channel selectively allows only certain types of ions (e.g., calcium, potassium, sodium) to pass through to the other side of the cell membrane. There are several major mechanisms by which these channel protein pores can be induced to open or close (i.e., are gated). Stimuli that can gate ion channels include a ligand molecule binding to either the intracellular or extracellular surface of the channel protein (ligand gating), stretching of the plasma membrane of a neuron (stretch gating), a change in the voltage across the cell membrane (voltage gating), or a combination of ligand and voltage stimuli (see Fig. 4.1). These types of gating mechanisms are described below.

*Ligand-gated ion channels:* Ligand-gated ion channels are opened by the binding of a ligand to either the extracellular or intracellular surface of the channel protein. Channels that open when a ligand binds on their intracellular surface include second-messenger-gated ion channels and calcium-activated potassium channels. These channels sense changes to the intracellular environment that affect the presence of their ligand, and they react to the ligand by opening a protein pore for ion flux. Channels that are gated when a ligand binds on their extracellular surface include ionotropic transmitter receptors (e.g., acetylcholine and glutamate receptors), as well as the resting leak potassium channels (gated by hydrogen ions, or pH) described in Chapter 3.

*Stretch-gated ion channels:* These channels detect membrane stretch by interacting with the intracellular cytoskeleton. As a cell swells, shrinks, or stretches, a series of amino acids in the channel structure are pushed or pulled by the torsion created by this movement, causing a change in channel conformation that opens the ion pore. Channels that are gated

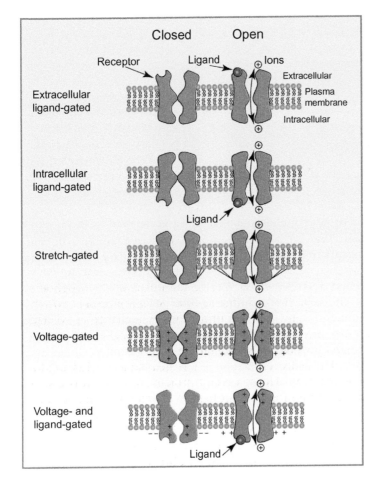

FIGURE 4.1 **General methods of ion channel gating.** Channels can be grouped based on the stimulus or stimuli that open or close them. This includes ion channels that respond to ligands (ligand-gated; top two images), mechanical displacement of the channel when the plasma membrane is stretched (middle image), changes in the voltage across the membrane (fourth image), or a combination of ligand binding and membrane voltage (bottom image).

in response to the stretch of a cell membrane are also referred to as stretch receptors. Stretch receptors can be found in many cell types, including the stereocilia on hair cells of the inner ear and in the membranes of both smooth and skeletal muscle.

*Voltage-gated ion channels*: Channels that open when the voltage across the membrane changes are called voltage-gated ion channels. These include the subtypes of sodium, potassium, and calcium channels that open in response to the depolarization that occurs during an action potential.

Finally, some channels are gated in response to more than one of the three types of stimuli described above. For example, calcium-activated potassium channels are more likely to open both when the membrane is depolarized and when calcium binds to an intracellular binding site. These channels rely on the combined effect of these two influences to create the conformational change that gates (opens) the ion channel pore.

# VOLTAGE-GATED ION CHANNELS

Voltage-gated ion channels are critical for the excitability of neurons, a property that sets neurons apart from most other cells. Because these voltage-gated ion channels are an integral part of how most synapses work, we will cover their structure and function in greater detail here than some of the other types of ion channels.

How can a channel protein detect changes in voltage across a cell membrane? The parts of these proteins that translate changes in membrane voltage to channel gating reside within the **transmembrane** portion of the protein structure (i.e., the portion that is embedded within the neuronal membrane bilayer). Like all proteins, each ion channel is first synthesized as a long string of amino acids, and this is referred to as the **primary protein structure**. Scientists can use a channel's amino acid sequence (which they can derive from the gene that encodes it) to make predictions about its three-dimensional structure based on the properties of each type of amino acid. An amino acid can be categorized as polar, hydrophobic, or charged. Polar amino acids form hydrogen bonds with one another to create the **secondary protein structure** (e.g., $\alpha$-helix or $\beta$-pleated sheet), and the $\alpha$-helical portions of a protein's secondary structure tend to be more stable within a lipid environment such as a cell membrane. As such, amino acids in $\alpha$-helical portions of the protein are likely to be buried in the lipid bilayer or deep within the protein so they avoid contact with the aqueous environments of the intra- and extracellular solutions. Unlike the polar and hydrophobic amino acids, some amino acids are charged. When charged amino acids are present within the protein structure, they can be pulled in one direction or another by a change in voltage across the cell membrane, and they can also form salt bridges (a type of noncovalent molecular interaction) between negatively and positively charged amino acid pairs.

By examining the sequence of polar, hydrophobic, and charged amino acids in the primary protein structure, as well as the predicted secondary protein structure, it is possible to make predictions about the membrane topology of the ion channel protein, such as which parts of the channel are located on the inside face, outside face, or interior of the membrane bilayer (Argos et al., 1982; Kyte and Doolittle, 1982; Noda et al., 1984). These predictions are aided by the use of hydrophobicity plots, which can be used to predict which stretches of the protein structure are more **hydrophobic** (water avoiding) or

**hydrophilic** (water loving). When a long stretch of amino acids is predicted to be hydrophobic, it is assumed to reside within the membrane bilayer, where it is shielded from the aqueous intra- and extracellular environments. Based on the thickness of the plasma membrane ($\sim$3–4 nanometers, or $3-4 \times 10^{-9}$ m), it is estimated that it would take a minimum of 20 amino acids forming a hydrophobic α-helix region to stretch all the way across the plasma membrane. When the amino acid sequences of all voltage-gated ion channels were examined in this manner, it was predicted that they all share a common structural organization (see Fig. 4.2).

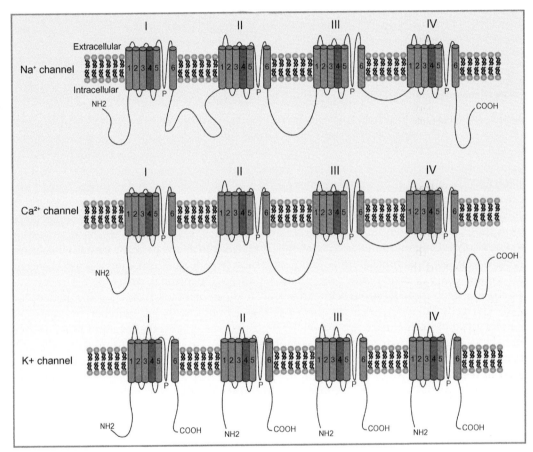

FIGURE 4.2 **The predicted membrane topology of three classes of voltage-gated ion channels.** Voltage-gated sodium (top), calcium (middle), and potassium (bottom) channels all have the same basic membrane topology. Each consists of four domains (Roman numerals I–IV) that wrap around to form a central pore for ion flux (see also Fig. 4.3). In addition, each domain contains six transmembrane segments (numbers S1–S6). The S4 segment in each domain (in *red*) is known as the voltage-sensing segment, because it contains charged amino acids that react to changes in membrane potential. Each domain also contains a "p-loop" (P) between segments 5 and 6 that is thought to form part of the ion pore. The four domains of mammalian sodium and calcium channels are made from one large protein, while each potassium channel domain is constructed from separate protein subunits.

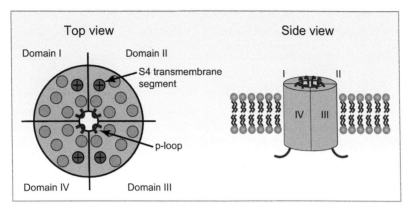

FIGURE 4.3 **The tertiary structure of a voltage-gated ion channel.** (Left) Top-down view of the channel showing the four domains (I–IV) and the six transmembrane segments within each domain. The fourth transmembrane segment (S4) in each domain, is colored red and includes a " + " sign to indicate that this segment contains positively charged amino acids. The top view of the "p-loop" in each domain is shown as purple lines near the pore in the center of the channel. (Right) Side view of the same channel.

Each voltage-gated ion channel protein consists of four repeating domains (I–IV), each of which is composed of six predicted transmembrane segments (S1–S6) connected by intra- and extracellular loops. The predicted loop between segments 5 and 6 is called the "p-loop" (see Fig. 4.2), because it is thought to form the outer vestibule of the pore opening in the channel (see Fig. 4.3). For voltage-gated sodium and calcium channels, this entire structure is created by the folding of a single long string of amino acids. In contrast, each of the four domains of a potassium channel is a separate smaller protein (see Fig. 4.2), and these four proteins come together to form the channel. This configuration allows voltage-gated potassium channels to mix and match single protein domains in order to have diversity in their overall characteristics. The fourth transmembrane segment (S4) in each domain of all voltage-gated ion channels contains a series of four to eight positively charged amino acids at every third amino acid position in the $\alpha$-helix. The six transmembrane segments within each of four domains are thought to be organized into a **tertiary protein structure** (i.e., its folded form) that folds around like staves of a barrel (see Fig. 4.3) to form a pore.

## S4 Segments Act as Voltage Sensors

The positively charged amino acids in the S4 transmembrane segments are the key to sensing membrane voltage. When neurons are at rest, the voltage across the neuronal membrane is about $-65$ mV, which means the inside of the neuron is more negative than the outside. This charge difference causes the positively charged amino acids in the S4 segment to be attracted to the inside of the neuron, and this attraction pulls them toward the inner surface of the neuronal membrane. This pulling force is thought to cause changes in the protein's tertiary structure that reduce the diameter of the pore, closing the channel (see Fig. 4.4).

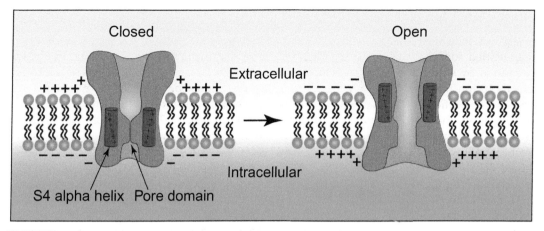

FIGURE 4.4 **The mechanism for voltage gating of an ion channel.** This diagram shows a side view of a voltage-gated ion channel embedded in the plasma membrane with half of the channel structure cut away to reveal the S4 segment (*red*) inside the remaining domains (*orange*). (Left) When the net charge on the inside of the neuron is more negative than the outside (i.e., the membrane potential is negative) the S4 segment is pulled downward by electrical attraction to the inner surface of the membrane. When the S4 segment is pulled inward to this position, the conformation of the rest of the channel changes, closing the pore. (Right) The channel is opened by a depolarization that decreases or reverses the strong negative membrane potential. This depolarization pulls the S4 segment toward the extracellular surface of the membrane, causing a conformational change in the protein structure that opens the channel.

In contrast, when a neuron is depolarized (e.g., during an action potential), the inside surface of the membrane becomes less negative relative to the outside. This causes the positively charged S4 segments to move toward the outer surface of the membrane, pulling on the protein structure to open the pore (see Fig. 4.4). This reaction of the S4 segment to voltage has been described to occur as a "sliding filament" movement, as the S4 α-helix is thought to twist and slide between other transmembrane segments in reaction to voltage changes across the plasma membrane, in what is sometimes called a helical screw motion (Bezanilla, 2008). Since this S4 movement is the trigger for changing the open or closed state of the ion channel, the S4 segment is sometimes referred to as the primary voltage sensor for the ion channel. Because the positively charged amino acids within the S4 segment that traverse the membrane are not hydrophobic, it is hypothesized that when these amino acids pass across the lipid bilayer, the positive charges need to be counterbalanced by association with negative charges. Therefore, it is thought that the positive charges are stabilized within the membrane by salt bridges between negatively charged amino acids on neighboring segments (usually within S2 and S3). Taking these counter charges into account, voltage sensing really involves segments 1–4, so these are sometimes collectively termed the **voltage-sensing domain**. As this voltage-sensing domain changes shape, it also changes the shape of the pore domain of the channel, which includes the S5 and S6 transmembrane segments and the p-loop region. Since membrane voltage does not directly cause the pore to open or close, but instead affects the voltage-sensing domain, which, in turn, affects the conformation of the pore, the effect of membrane voltage on the channel is indirect and is referred to as an **allosteric** effect.

## Voltage-Gated Ion Channel Structure

While predictive methods based on amino acid sequence can shed some light on how a voltage-gated ion channel is folded, the specific tertiary structure of the channel is best determined by studying its crystal structure. The best way to accomplish this is by using a high-resolution microscopy technique called **X-ray crystallography**. In any form of microscopy, the resolution is determined by the wavelength of the radiation used to visualize structures: the smaller the wavelength, the higher the resolution. This is because when light hits something that you are trying to see, it is the diffraction of the light waves around the object that allows you to see it. Therefore, your ability to resolve a small object is determined by the "**diffraction barrier**" of those waves, which is approximately half the wavelength of the radiation (light, X-rays, etc.) you are using to visualize the object. X-rays have a wavelength of 0.1 nm (1 **Angstrom**, represented by the symbol Å), which is about the size of an atom and is the resolution necessary to visualize protein structures. Because they are so small, the diffraction pattern of X-rays as they strike a molecule can be used to reveal the structures of proteins at an atomic level. The diffraction pattern generated by one protein molecule would be too weak to detect, so a large amount of protein must be obtained and crystallized to form a highly ordered array of billions of protein molecules. With this imaging method, X-rays are diffracted by electrons in the crystalized proteins and the diffraction pattern reveals a three-dimensional map of the electrons within the protein structure.

It has been difficult to obtain detailed structural information about large transmembrane proteins because they contain large hydrophobic regions, which makes them difficult to crystallize. For example, it is difficult to crystallize voltage-gated ion channels because each of their four domains contains six hydrophobic transmembrane segments (see Figs. 4.2 and 4.3). Therefore, the first ion channel crystal structures to be obtained were of a bacterial potassium channel that only contains two hydrophobic transmembrane segments per domain (Doyle et al., 1998; Kreusch et al., 1998; Sansom, 1998). Later, improved techniques for crystallizing membrane proteins made it possible to obtain a crystal structure of the mammalian voltage-gated potassium channel at a resolution of 2.4 Angstroms (Long et al., 2005, 2007; see Fig. 4.6). These structures confirmed the membrane topology predicted by hydrophobicity studies and provided detailed information about the molecular surface of channels and the contour of the pore. In 2003, Roderick MacKinnon was awarded the Nobel Prize in Chemistry for his work using X-ray crystallography to determine the structure of potassium channels, an award he shared with Peter Agre, who discovered water channels in cells (see Box 4.1). More recently, the crystal structure of a voltage-gated sodium channel was obtained (Payandeh et al., 2011). Because all voltage-gated ion channels share a similar structural organization, the detailed information provided by studying these potassium and sodium channel crystal structures has allowed scientists to develop more accurate computational models of channel gating for many voltage-gated ion channels.

## Ion Channel Permeation and Selectivity

One aspect of voltage-gated ion channel function that these crystal structures have helped us understand is how ions move through, or **permeate**, the channel pore. Ions such as sodium, potassium, and calcium are charged molecules that are usually hydrated when

## BOX 4.1

## NOBEL PRIZE IN CHEMISTRY, 2003

In 2003, The Nobel Prize in Chemistry was awarded to Roderick MacKinnon (see Fig. 4.5) and Peter Agre *"for discoveries concerning channels in cell membranes."*[1] Peter Agre shared the Nobel Prize as a result of his discovery of water channels in cells (called **aquaporins**). Because MacKinnon's work is more relevant to the topic of this text, we will focus primarily on his discoveries here. The award to Roderick MacKinnon was for his pioneering work to visualize what an ion channel looks like.

To "see" the atomic structure of ion channels, MacKinnon used a technique called X-ray crystallography. This technique utilizes the diffraction pattern that forms when X-rays pass around the protein of interest. But, because most of the X-rays miss atoms in the protein when they pass by, in order to visualize the atoms in the protein, millions of copies of the same protein must be arranged in a single orientation within a crystal. This makes it possible to visualize the diffraction pattern for a given protein because the image is made by X-rays diffracting off all these molecules simultaneously.

FIGURE 4.5  Roderick MacKinnon, 2014.[2]

in the extracellular and cytoplasmic fluids surrounding the openings of channel pores. This hydration is a result of the positively or negatively charged ions attracting water molecules whose spare electrons are attracted to the charged ions (see Fig. 4.7; Conway, 1981). In the presence of water molecules, the ions are said to be **hydrated**. Based on the

## BOX 4.1 (cont'd)

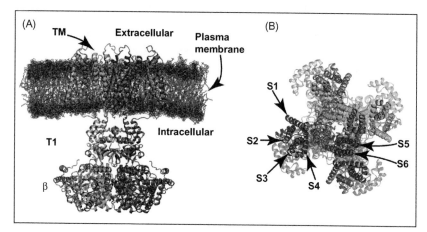

FIGURE 4.6  Diagram of the crystal structure of a potassium channel (Kv1.2). (A) Side view of the channel complex buried in the plasma membrane with the extracellular portion oriented in the upward direction. This view depicts the transmembrane domains (TM), the tetramerization domain (T1; the region that engages in forming the tetramer from four different subunits), and the beta auxiliary subunits (β). Each of the four protein subunits that make up a complete channel is color-coded. (B) Top view of the same channel complex (without the plasma membrane) labeling the six transmembrane segments (S1–S6) within one of the four subunits that make up the channel (*the red color-coded subunit*). The pore can be seen in the middle of this image. *Source: Images generated by Rozita Laghaei using the software VMD and structures deposited in the RCSB protein data bank.*

One of the proteins MacKinnon and his colleagues crystallized was a potassium channel (see Fig. 4.6). The ability to visualize the atomic structure of proteins, both in isolation, and with drugs bound, has significantly advanced our understanding of protein structure and function.

[1] "The Nobel Prize in Chemistry 2003". *Nobelprize.org*. Nobel Media AB 2014. Web. 26 Jun 2018. http://www.nobelprize.org/nobel_prizes/chemistry/laureates/2003/.
[2] Creative Commons image of Roderick MacKinnon: By PotassiumChannel—Own work, CC BY-SA 3.0, https://commons.wikimedia.org/w/index.php?curid = 30730336.

X-ray crystallography view of an ion channel pore, it appeared to researchers that the hydrated radii of ions (sodium = 3.6 Å; potassium = 3.3 Å; calcium = 4.1 Å; chloride = 3.1 Å) were too large to fit through the pore of their respective voltage-gated channels. However, further investigation using X-ray crystallography provided evidence that ions are dehydrated as they pass through ion channel pores and rehydrated on the other side (Zhou et al., 2001). The dehydrated radii of ions (sodium = 0.95 Å; potassium = 1.3 Å; calcium = 1.0 Å; chloride = 1.8 Å) are small enough to feasibly pass through ion channel pores.

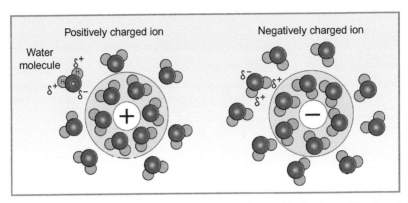

FIGURE 4.7 **The hydration of charged ions in solution.** (Left) A positively charged ion in solution is associated with the negatively charged portion (near the *red oxygen atom*) of the water molecules directly surrounding the ion (inside *blue circle* around the ion). (Right) A negatively charged ion in solution is associated with the positively charged portion (near the *pink hydrogen atoms*) of water molecules directly surrounding the ion.

Voltage-gated ion channels are usually selective for a given type of ion, meaning that they allow only a specific ion species through the pore (for example, voltage-gated potassium channels allow only the flux of potassium, but not sodium or calcium ions). Understanding the role of ion hydration in ion permeability explained how ions could permeate channels, but it did not explain how ion channels are selective for a given ion species. For example, how can potassium channels be selective for potassium ions but not allow the smaller sodium ions to also pass through the pore? That is, how can an ion channel allow a larger ion to pass through its pore but keep a smaller one out? What underlies the remarkable specificity of these channels for a particular ion species? This dilemma was solved when X-ray crystallography revealed the presence of high-affinity binding sites for specific ions within an ion channel pore (Zhou et al., 2001). It is hypothesized that these binding sites are always occupied by the type of ion that permeates the ion channel, and when the channel opens other ions of the same type displace the ones that are already bound inside the channel. This displacement results in a shuttling of ions, driven by electrostatic repulsion (since the permeating ions all have the same charge), as they hop from binding site to binding site and pass through the pore (see Fig. 4.8). In this scenario, sodium does not pass through potassium channels because it does not bind (or at least, does not bind as well as potassium does) to the potassium binding sites within the voltage-gated potassium channel (Armstrong, 2007). This ion hopping process is normally very fast, and allows voltage-gated ion channels to pass a large number of ions over a short period of time (often estimated to be at the limit of diffusion, or about 100 million ions per second).

## Ion Channel Auxiliary Proteins

Voltage-gated ion channels are not only composed of the pore-forming proteins we have discussed thus far. Nonetheless, most investigators focus on the pore-forming proteins of ion channels because they are responsible for most of the common functions of

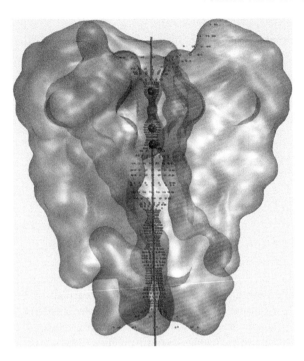

FIGURE 4.8 **Cutaway view of the ion pore in a potassium channel.** The ion channel protein is color-coded to represent areas of charge (*blue* = positive, *white* = neutral, and *yellow* = negative). The potassium ion permeation pathway is outlined in red dots. Potassium ions (*purple spheres*) are shown as they are predicted to pass through the channel pore. *Source: Image generated by Rozita Laghaei using the software VMD and structures deposited in the RCSB protein data bank.*

voltage-gated ion channels, which mainly involves sensing membrane voltage and changing shape to allow ion flux. However, these pore-forming subunits rarely, if ever, function in isolation. Other regulatory proteins, referred to as **auxiliary proteins**, bind to these pore-forming proteins, and together the entire complex makes up the final **quaternary protein structure**.

Auxiliary proteins can have a variety of important functions, including proper transport of newly synthesized ion channels from the Golgi apparatus to the plasma membrane; providing a substrate for transmitter modulation of ion channel function; and modulating some of the voltage-dependent characteristics of the ion channel protein (Isom et al., 1994). Pore-forming subunits are usually named "alpha," while the auxiliary subunits are termed "beta," "gamma," etc. For example, voltage-gated sodium channels have two auxiliary subunits, termed $\beta 1$ and $\beta 2$. Sometimes the auxiliary subunits are covalently linked to the alpha subunit (as is the case for the sodium channel $\beta 2$ subunit), and other times they are noncovalently associated (as is the case for the sodium channel $\beta 1$ subunit). Each voltage-gated potassium channel alpha subunit can have a single noncovalently associated beta subunit, which is located on the intracellular side of the membrane and appears to modulate the activity of the channel (see Fig. 4.6A).

In all voltage-gated channels, the alpha and beta subunits represent families of proteins. There can be many possible combinations of subunits that form a given channel, which allows for a high degree of channel diversity (Coetzee et al, 1999). This diversity is most evident in voltage-gated potassium channels, for which there are more than 30 different alpha subunit types and more than 10 beta subunit types. The different combinations of voltage-gated potassium channels fall into 12 different subfamilies, and the channels within each subfamily share similar characteristics (Gonzalez et al., 2012). Because cells have the ability to express families of channels with different voltage-sensitive and gating characteristics, these channels are able to open under different conditions and to remain open for different lengths of time. This diversity of functions allows potassium channels to be tailored for diverse neuronal functions, including regulating the resting membrane potential, repolarizing the action potential, and modulating transmitter release, as well as for many nonneuronal functions in the body (see Box 4.2 for a description of a natural potassium channel toxin found in a snake venom).

---

BOX 4.2

## DANGERS IN NATURE: BLACK MAMBA SNAKES AND VOLTAGE-GATED POTASSIUM CHANNELS

### *Dendroaspis polylepsis* (Black mamba)

Black mambas (Fig. 4.9) are one of the fastest and most poisonous snakes in the world. A single bite from a black mamba can inject 5–6 times the lethal dose necessary to kill a human, with a nearly 100% fatality rate between 45 minutes and 15 hours unless antivenom treatment is received. Mamba venom is composed primarily of neurotoxins, and, in particular, dendrotoxin, which strongly binds to and blocks potassium channels in motor neurons. This prevents the positively charged potassium from exiting the neuron during an action potential, and the resulting persistent depolarization causes the neuron to fire repetitively. In the periphery, this enhanced release of acetylcholine causes continuous muscle contractions, which lead to convulsions that gradually prevent respiration. Mambas have been known to aggressively chase humans, and can travel up to 12 miles an hour.

FIGURE 4.9  *Dendroaspis polylepsis.*

# IONIC CURRENTS THROUGH VOLTAGE-GATED ION CHANNELS

To study the functions of ion channels, investigators usually use the **voltage-clamp technique**, also called simply "voltage clamp" (see Box 4.3). Voltage clamp is aptly named, since it allows the investigator to hold, or "clamp," the membrane voltage at any level and

---

**BOX 4.3**

## THE VOLTAGE-CLAMP TECHNIQUE

The voltage-clamp technique is an experimental method that allows an experimenter to control (or "command") the desired membrane voltage of the cell. The experimenter uses a set of electronic equipment (referred to here as a voltage-clamp device) to hold the membrane voltage at a desired level (the command voltage) while measuring the current that flows across the cell membrane at that voltage. The voltage-clamp device uses a negative feedback circuit to control the membrane voltage. To do this, the equipment measures the membrane voltage and compares it with the command voltage set by the experimenter. If the measured voltage is different from the command voltage, an error signal is generated and this tells the voltage-clamp device to pass current through an electrode in the neuron in order to correct the error and set the voltage to the command level. This can be accomplished using two microelectrodes inserted into the cell, one to measure voltage and another to pass current (see Fig. 4.10), or using one large-diameter electrode that performs both functions.

A voltage-clamp device functions much like the thermostat that controls the temperature in a home (if your home system could automatically switch between heating and cooling). For example, if you set your home thermostat to 70°F, but the temperature in your home is 65°F, the thermostat will detect this difference and the heater will turn on and heat the home until it reaches the desired temperature. In contrast, if the

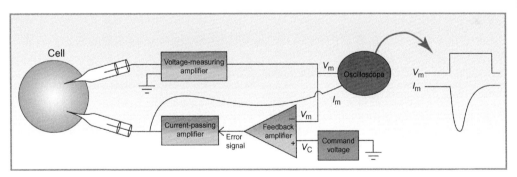

FIGURE 4.10 **The two-electrode voltage-clamp technique.** This diagram depicts the circuit that is used to clamp the voltage of a neuron and measure the current that flows at that membrane voltage.

## BOX 4.3 (cont'd)

temperature in your home is 75°F, then the air conditioning will turn on to cool the home until it reaches the desired 65°F.

If an experimenter is using a voltage-clamp device to control a neuron's membrane voltage, and they set the command voltage to begin at −80 mV and then step to +10 mV, the voltage-clamp device will pass the required current to accomplish this. However, when the membrane potential changes from −80 mV to +10 mV, voltage-gated sodium channels will open, and the sodium flux into the cell will create a current that would normally depolarize the membrane potential. But, the voltage-clamp device will detect this voltage change induced by the sodium flux and pass equal and opposite current to counteract the sodium current, preventing the membrane potential from depolarizing. This compensatory current can be measured by the experimenter and is used to quantify the sodium current that flowed during the step to +10 mV. In this way, not only does a voltage-clamp device hold the voltage at the commanded levels set by the experimenter, but the current that flows to keep the membrane at these voltage levels can be measured. This current, which functions as an indirect measure of the current that flowed during the voltage step, is what is reported in the recorded data during a voltage-clamp experiment.

It is important to understand that the voltage-clamp technique cannot be used to determine which ions are flowing when a current is measured. Only the *net* current flux is measured, and this can be the sum of multiple currents caused by the flow of different types of ions. The net current may even be a mixture of inward and outward currents flowing at the same time. This can be a problem if the experimenter is specifically interested in studying the current of a single type of ion. For example, when recording sodium currents, other types of ionic currents flowing at the same time may mask the sodium currents or be mistaken for them. Therefore, to measure the sodium current selectively, the experimenter must set up the experimental conditions such that only sodium channels can open and/or only sodium current can flow across the membrane. One common method is to use pharmacological agents to block other types of channels. For example, if the experimenter is interested in measuring current through potassium channels, they might block sodium channels with a molecule that only prevents sodium channels from passing current (e.g., tetrodotoxin) but does not affect potassium channels, whereas if they are interested in measuring sodium channel current they might block potassium channels using a pharmacological agent that does not have any effects on sodium channels (e.g., tetraethylammonium). Another method is to replace some of the ions normally present on each side of the membrane with substitutes that will not flow through channels the experimenter does not want to study. For example, an experimenter investigating sodium channels might replace intracellular potassium with cesium ions, which do not permeate potassium channels. Or an experimenter studying potassium channels might replace extracellular sodium with choline ions, which do not permeate sodium channels. In this way, the current through a single type of ion channel can be isolated and studied during an experiment.

simultaneously measure the current that flows across the cell membrane at that membrane voltage. For example, a researcher can clamp the membrane potential at a neuron's resting potential (e.g., −65 mV) and then change ("step") the membrane potential to a more depolarized potential (e.g., +10 mV). This depolarization will cause voltage-gated ion channels to open. Once the channels are open, the driving force on the relevant ions pushes them through the channels. This means the amount of ionic current and the direction of ion movement depends on the driving force on that type of ion at +10 mV (see Chapter 3).

Please note that the equilibrium potentials for any given ion differ slightly according to their intracellular and extracellular concentrations, which vary by species and region of the nervous system. We have selected approximate values here (e.g., $Eq_K = -80$ mV, $Eq_{Na} = +60$ mV), and will use these throughout our discussion for the sake of clarity.

## Voltage-Dependent Activation

The amount of current that flows through voltage-gated ion channels is governed by a balance of two factors. The first is how many channels are open at a specific time, and the second is the driving force on the ions pushing them through each channel. The number of open ion channels is determined by the channel's **voltage-dependence of activation**, while the force pushing ions through the channels is determined by the driving force on the permeant ions. The voltage-dependence of activation is the relationship between membrane voltage and the probability that the channel is open. As described above, the voltage-sensing domain within voltage-gated ion channels includes a charged S4 segment that moves like a piston either up or down, depending on the membrane voltage. The channel opens only when all four S4 segments move up (toward the outside of the membrane). These segments bounce up and down in a stochastic (random) fashion, but the stronger the depolarization, the more time each S4 segment spends in the "up" position. It follows that the stronger the depolarization, the more likely it is at a given moment that all of the S4 segments will be in the "up" position, which is required in order for the channel to open. A channel's voltage-dependence of activation can be plotted as the percentage of ion channels that are open as a function of the membrane voltage (see Figs. 4.11A and 4.12A).

## Current–Voltage Relationships

For a given type of ion channel, it is possible to estimate how much current will flow across the membrane at any given membrane voltage. The voltage-dependence of activation plot (percentage of channels open as a function of voltage) and the driving force plot (force on the ions as a function of voltage) together predict the ion flux that occurs at a given voltage (see Figs. 4.11 and 4.12). For example, imagine an experimenter is measuring the potassium current across a neuronal membrane and has blocked all other voltage-gated ion channels so they will not contribute to the ionic current. At a resting membrane potential of −65 mV, there are almost no potassium channels open (i.e., the percentage open is very near zero). Additionally, at this membrane potential, the driving force on the

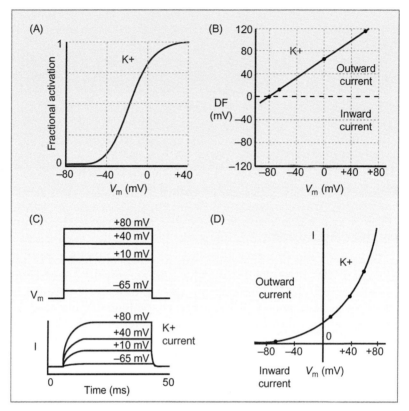

FIGURE 4.11 **The influences that govern the amount of potassium current moving through voltage-gated potassium channels.** (A) A plot of the voltage-dependence of activation for voltage-gated potassium channels. A greater percentage of channels are open at depolarized potentials (note that a fractional activation of 1 means 100% of the channels are open). (B) A plot of the driving force on potassium ions. There is no driving force when the membrane potential is −80 mV. As the membrane potential is depolarized, there is a greater and greater outward force on potassium ions. (C) Sample results from a voltage-clamp experiment designed to measure potassium current. The top graph shows the voltage steps used to change (step) the membrane potential ($V_m$). The bottom graph shows the predicted current in response to these membrane potential steps (by convention, upward deflections of the current indicate outward ion flux). At −65 mV, there is almost no potassium current because very few voltage-gated potassium channels are open (see A) and because there is a very small driving force on potassium ions (see B). However, at +80 mV, there is a large outward potassium current because many channels are open and at this voltage there is a strong driving force. (D) A plot of the peak potassium current measured during voltage steps to a range of voltages (I–V plot). Potassium current increases when stepping to voltages more positive than the resting membrane potential.

potassium ions is very weak because the equilibrium potential for the potassium channel is −80 mV (see Fig. 4.11B), which is fairly close to the resting membrane potential. In this case, even though there is a weak driving force that could push potassium ions out of the cell, there is virtually no potassium current because there are almost no potassium channels open at this voltage.

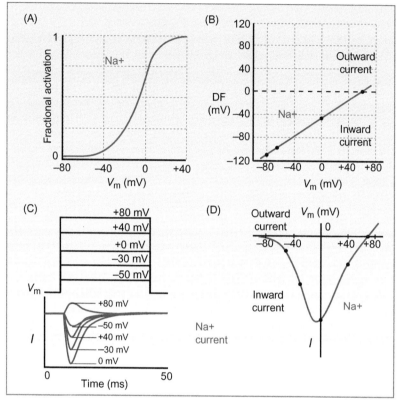

FIGURE 4.12 **The influences that govern the amount of sodium current moving through voltage-gated sodium channels.** (A) A plot of the voltage-dependence of activation of voltage-gated sodium channels. A greater percentage of channels are open at depolarized potentials. (B) A plot of the driving force on sodium ions. At a membrane potential of +60 mV, there is no driving force on sodium ions. As the membrane potential hyperpolarizes, there is a greater and greater inward force on sodium ions across the membrane. (C) Sample results from a voltage clamp experiment designed to measure the sodium current as the membrane is stepped to different voltages. The top graph shows the voltage steps used to change the membrane potential ($V_m$). The bottom graph shows the predicted current in response to these membrane potential steps. At −50 mV, there is a very small sodium current because very few voltage-gated sodium channels are open (see A) and because there is a very large driving force on sodium ions (see B). At +40 mV, there is again a relatively small sodium current because although many sodium channels are open, there is a small driving force. At 0 mV, the sodium current is large because there is a relatively large number of open channels and a relatively strong driving force. (D) A plot of the peak sodium current measured during the voltage steps to a range of voltages (I–V plot). Sodium current increases as the membrane potential is depolarized from −80 mV toward 0 mV, but then starts to get smaller again as the membrane potential approaches the equilibrium potential for the sodium ion (+60 mV).

Using the voltage-clamp technique, an experimenter would record very little current through potassium channels at a membrane potential of −65 mV (see Fig. 4.11C). However, if the experimenter stepped the voltage to +10 mV, the voltage-dependence of activation plot predicts that about 90% of the voltage-gated potassium channels would be open. In addition, there would be a strong driving force that would push potassium ions out of the cell. Therefore, the experimenter would record a large outward potassium current (see Fig. 4.11C, bottom). A step to +40 mV would result in almost all of the potassium channels being open, and there would be an even greater driving force to push potassium out of the cell, leading to an even larger outward current being recorded than when the step was to +10 mV. Finally, with a step to +80 mV, no additional potassium channels would open because virtually all were already open at +40 mV, but the driving force on the potassium ions would be greater. Therefore, the experimenter would record an even larger outward potassium current at +80 mV. For the reasons stated above, the amount of potassium current would continue to increase as the voltage was stepped to more and more depolarized potentials because potassium channels are open and there is a strong driving force.

The relationship between voltage and ionic current is often expressed as an $I-V$ curve, also known as an $I-V$ plot. This type of plot indicates the maximum ionic current ($I$) as a function of the voltage the membrane was stepped to (see Figs. 4.11D and 4.12D). The voltage at which this plot crosses zero on the current axis (i.e., the $y$-axis) indicates the equilibrium potential for the ion in question. By convention, inward current is plotted as a negative value, and outward current is plotted as a positive value.

The $I-V$ curve for voltage-gated potassium channels is relatively straightforward to understand, since both the driving force and the percentage of open channels increase with depolarization. However, while voltage-gated sodium channels have a voltage-dependence of activation relationship that is very similar to that of voltage-gated potassium channels (compare Fig. 4.11A with Fig. 4.12A), currents through sodium channels behave very differently. This is because, whereas the equilibrium potential for potassium is −80 mV (see Fig. 4.11B), sodium has an equilibrium potential of +60 mV (compare Fig. 4.11B with Fig. 4.12B). This difference in equilibrium potentials significantly changes the $I-V$ curve for voltage-gated sodium channels, because a strong depolarization above resting membrane potential is closer to the equilibrium potential for sodium, whereas a similar depolarization is further from the equilibrium potential for potassium. Therefore, the sodium current during a depolarizing voltage step is governed by a balance between opening more sodium channels and reducing the driving force for sodium entry. An understanding of the voltage-gated sodium channel $I-V$ curve is useful to consolidate the principles that guide voltage-gated channel function.

For example, at a resting membrane potential of −65 mV, there are almost no sodium channels open (i.e., the percentage of open channels is very close to zero; see Fig. 4.12A). However, because the equilibrium potential for the voltage-gated sodium channels is +60 mV, the inward driving force on the sodium ion is very strong (see Fig. 4.12B) when the membrane potential is −65 mV. In this case, even though there is a very strong driving force to push sodium ions into the cell, there is virtually no sodium current because there are almost no sodium channels open. Therefore, an experimenter using the voltage clamp technique would record almost no current through sodium channels at −65 mV. However, if the experimenter stepped the voltage to +0 mV, the voltage-dependence of activation plot predicts that about 70% of the voltage-gated sodium channels would be open, and the driving force plot predicts a strong driving force pushing sodium ions into the cell. Therefore, the experimenter would record a large inward sodium current (see Fig. 4.12C). A step to +40 mV would cause almost all of the sodium channels to open, but the driving force on the sodium ions would be very low, since +40 mV is close to the sodium equilibrium potential of +60 mV. This means there would be a small sodium current at this membrane voltage (see Fig. 4.12C). Finally, with a step to +80 mV, no additional sodium channels would be open (they would all already have been open at +40 mV), but the driving force would be now reversed, pushing sodium out of the cell, albeit with a weak force. Therefore, at +80 mV, the experimenter would record a small outward sodium current.

For these reasons, the relationship between voltage and ionic current (I–V curve) for the voltage-gated sodium channel is roughly "U" shaped. Interestingly, this means that the experimenter would record a similar sodium current at −40 mV as at +40 mV. However, as you have seen, the mechanisms that lead to these currents are different. At −40 mV, there are few channels open, but a large inward driving force, while at +40 mV there are many channels open, but a small inward driving force. Unlike the potassium current, the sodium current rapidly moves back toward zero after reaching a peak inward current (compare Fig. 4.11C with Fig. 4.12C). This transient nature of the sodium current is caused by a process called **inactivation**, which will be discussed in detail later in this chapter.

## ACTION POTENTIALS

The role of the nervous system is to control the body by communicating with its neuronal and non-neuronal cells, and an essential step in this communication is the rapid conduction of electrical signals to distant parts of the body. To do this, neurons signal using electric current, which is created by the movement of charged ions across the plasma membrane.

But how is this electric current propagated to distant sites in the body? Some individual neurons, especially motor neurons that have cell bodies in the spinal cord and send signals to your fingers and toes, can have axons that are more than 1 meter long! Yet, when you detect a small object such as a baseball moving quickly in your direction, you can move your hand up to catch that ball within less than one second. There are two physiological processes responsible for such a fast reaction time. The first is the speed of communication between neurons directly at synapses (covered in Chapter 6), and the second is the speed of information transfer along the length of the axonal plasma membrane. The latter is achieved by the conduction of electrical signals, or **action potentials**, within an individual neuron. In the case of motor neurons that innervate the hand or foot, this action potential conduction can occur at speeds of over 100 meters per second (m/s). To understand the mechanisms that underlie this form of cellular communication, we will describe how action potentials are generated and how they move along axons.

As discussed above, when neurons are at rest, the neuronal membrane is most permeable to potassium ions. As a result, the membrane potential (about −65 mV) is close to the potassium equilibrium potential (−80 mV). However, if a neuron is depolarized by 10–15 mV (e.g., by neurotransmitters acting on ligand-gated ion channels whose permeant ions combine to have a mixed equilibrium potential near 0 mV), a small percentage of the available voltage-gated sodium channels react quickly to this depolarization and open (see Fig. 4.13A). Because sodium has an equilibrium potential near +60 mV,

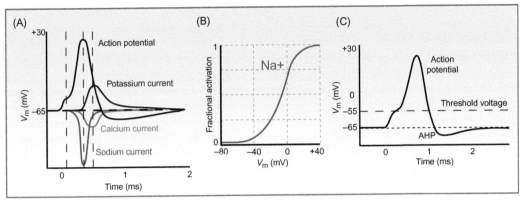

FIGURE 4.13  **The action potential.** (A) The ions that flow across the cell membrane during an action potential are shown as plots (*red* = sodium, *blue* = potassium, and *green* = calcium traces) overlaid onto the action potential waveform (*black trace*). The y-axis labels the membrane voltage for the action potential waveform (no scale given for the ionic currents depicted). By convention, current plotted as negative values (i.e., downward) is inward current (sodium and calcium), while current plotted as positive values (i.e., upward) is outward current (potassium). An action potential is initiated when the membrane potential becomes slightly but sufficiently depolarized, relative to the resting membrane potential (e.g., if ligand-gated ion channels open after transmitter release). (B) The slight depolarization on the rising phase of the action potential activates a small fraction of voltage-gated sodium channels, based on the voltage-dependence of activation for these channels. The opening of these sodium channels further depolarizes the membrane, which recruits the opening of more sodium channels, and this cycle repeats itself in a runaway activation (positive feedback) of more and more sodium channels. (C) The membrane potential at which this runaway activation of sodium channels is initiated is called the threshold voltage. When this positive feedback cycle begins, the rapid depolarization that occurs becomes the rising phase of an action potential. This diagram shows a mild depolarization at the beginning of an action potential that reaches the threshold voltage (*dashed line*). If this initial depolarization reaches threshold, the neuron will fire an action potential. After an action potential occurs, the membrane potential will hyperpolarize below the resting membrane potential (−65 mV, *dotted line*) before returning to the resting membrane potential. This is called the afterhyperpolarization (AHP).

opening sodium channels leads to an influx of positive sodium ions, further depolarizing the membrane, causing more of the available voltage-gated sodium channels to open. In this way, voltage-gated sodium channels trigger a positive feedback loop wherein the opening of some sodium channels depolarizes the membrane further, which in turn recruits even more sodium channels to open, which depolarizes the membrane even further, etc. This positive feedback mechanism leads to the explosive depolarization that is the rising phase of an action potential (see Fig. 4.13A). Action potentials are only triggered by membrane depolarizations that reach the threshold voltage for sodium channel activation. The membrane potential at which this occurs is the threshold voltage for action potential initiation, which is called the **action potential threshold** (see Fig. 4.13). If the depolarization is below threshold, no action potential is generated, but once the membrane potential reaches threshold, local sodium channels are recruited to open because of the positive feedback loop created by depolarization and the opening of sodium channels. The threshold voltage is defined by the sodium voltage-dependence of the activation curve (see Figs. 4.12A and 4.13B).

## Inactivation and Deactivation of Voltage-Gated Ion Channels

As more and more voltage-gated sodium channels are rapidly recruited, the membrane potential quickly reaches about +10 to +30 mV (the exact value depends on the type of neuron). At this voltage, the membrane is more permeable to sodium ($Eq_{Na} = +60$ mV) than to potassium ($Eq_K = -80$ mV). However, this rapid depolarization is short-lived for two reasons. First, very soon after sodium channels open, they rapidly inactivate. **Channel inactivation** occurs when the pore closes despite the continued presence of the stimulus that opened the channel (see Fig. 4.14). This inactivation of sodium channels reduces the sodium current so that it can no longer keep the membrane potential close to the sodium equilibrium potential, allowing the membrane potential to hyperpolarize. Second, voltage-gated potassium channels also open in response to the depolarizing membrane potential. These potassium channels (which are distinct from the resting leak potassium channels that help maintain the resting membrane potential) have a voltage-dependence of activation similar to that of voltage-gated sodium channels, meaning they too are opened by membrane depolarization and are open over roughly the same range of voltages as sodium channels. However, potassium channels open slowly because it takes longer for the channel proteins to change conformation to an open state, relative to sodium channels. The rate of opening is an intrinsic property of a channel protein that reflects how quickly it can change its conformation to the open state. Since the equilibrium potential for potassium is $-80$ mV, an increase in the number of open potassium channels leads to an outward movement of potassium ions, and, since potassium ions are positively charged, this hyperpolarizes the membrane. Thus, both of these events (sodium channel inactivation and the opening of potassium channels) cause the membrane potential to hyperpolarize rapidly toward the resting membrane potential.

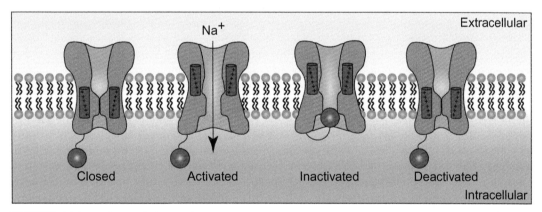

FIGURE 4.14 **Ball and chain or N-type inactivation.** The cytoplasmic surface of a sodium channel protein contains an amino acid sequence that creates a type of "ball and chain" (*red ball* with *black chain*), or "hinged-lid". At rest, this ball is tethered to the channel by the chain (closed). When the membrane depolarizes, a conformational change in the sodium channel protein opens the channel, allowing sodium ions to flow across the membrane (activated), and also creates a binding site for the ball. When the ball binds to the cytoplasmic surface of the open channel, it plugs the ion-conducting pore, inactivating the channel (inactivated). This ball stays bound to the channel pore until the channel deactivates, changing shape back to the closed configuration, at which point the ball is ejected from the pore (deactivated).

In fact, the membrane potential usually hyperpolarizes transiently to a point that is more negative than the resting membrane potential (e.g., −68 mV). This brief hyperpolarization is called the after-hyperpolarization, or **AHP** (see Fig. 4.13C), and it occurs because immediately after the action potential the membrane is temporarily more permeable to potassium than it is when the cell is at rest. The AHP lasts until the voltage-gated potassium channels deactivate. **Channel deactivation** is the closing of an ion channel when the stimulus that opens the channel (here, an increase in membrane voltage) has been removed. For voltage-gated ion channels, both activation and deactivation are caused by conformational changes in the ion channel itself, which are initiated by movement of the voltage-sensing domains of the channel (especially the S4 transmembrane segment in each of the four domains, as described earlier in this chapter). Once potassium channels have deactivated, the membrane potential returns to its resting level.

To summarize this process, an action potential starts when voltage-gated sodium channels open, allowing an inward sodium current that causes a rapid, large depolarization. These sodium channels then inactivate quickly, and potassium channels open, allowing an outward potassium current that repolarizes the neuronal membrane back toward the resting potential. At the end of the action potential, the membrane potential is often temporarily slightly more negative than the resting membrane potential because some potassium channels are still open. These channels eventually close, allowing the membrane to return to its resting potential.

### *Mechanisms of Channel Inactivation*

Channel inactivation occurs when the pore in a channel is closed, even though the stimulus to open is still present. There are two main mechanisms by which channels inactivate. The first is called the "ball and chain" or "hinged lid" mechanism, though it can also be referred to as **"N-type" inactivation**. As the name implies, channels that use this mechanism for inactivation have a group of amino acids on the intracellular surface of the channel that form a "ball or lid" which is tethered to the rest of the channel by a sequence of amino acids called a linker, or "chain" (see Fig. 4.14). The ball of amino acids hanging off the intracellular face of the channel into the cytoplasm has a binding site on the cytoplasmic face of the open pore. This binding site is not present when the pore is closed. However, when the channel opens, the ball or lid very quickly binds to this newly exposed binding site, plugging the pore from the inside of the cell. Voltage-gated sodium channels use this form of inactivation, and it is fast enough to close them very soon after they activate. This inactivation is what limits the time during which voltage-gated sodium channels are open (<0.5 ms) during the rising phase of the action potential. When the membrane potential hyperpolarizes sufficiently to move the S4 segments back to the closed position, the ball or lid is moved out of the channel, allowing the channel to return to the resting state (closed). This process is called **channel deactivation** because the stimulus that opens the channel has now been removed.

The second major type of channel inactivation is called **"C-type" inactivation**, and it is much slower than N-type inactivation. As with N-type channel inactivation, C-type inactivation occurs in the continued presence of the stimulus that opens the channel. For voltage-gated channels, this means that the S4 voltage-sensing segment remains in the activated position. In C-type inactivation, the channel then "C"ollapses or "C"rimps closed

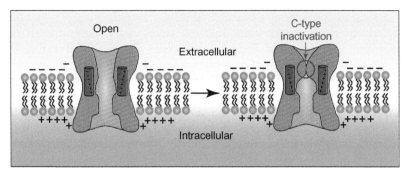

FIGURE 4.15  **C-type inactivation of voltage-gated ion channels.** After a voltage-gated channel opens, if the stimulus that opens it (in this case, depolarization) remains present long enough, the channel pore can close by simply collapsing or crimping shut. This type of inactivation is called C-type inactivation and occurs while the S4 segments are in the open configuration.

using conformational changes among the other transmembrane segments (see Fig. 4.15). Apparently, most voltage-gated channels are not stable in the open configuration for very long, and their pores close by one of these two major inactivation mechanisms.

## Action Potential Refractory Periods

Immediately after the action potential returns to the resting membrane potential, it takes several milliseconds for sodium channels to recover from the inactivation they underwent near the peak of the action potential. Because so many sodium channels are inactivated at the end of an action potential, this delay in recovery from inactivation means that if the membrane is depolarized again within several milliseconds of the first action potential, sodium channels cannot open, which means another action potential cannot be generated. This brief time when a second action potential cannot occur due to sodium channel inactivation is referred to as the **absolute refractory period**. After the absolute refractory period is over, there is another short time period (several more milliseconds) during which only some of the sodium channels have recovered to the resting closed state. During this time, a larger-than-normal depolarization is required to trigger a second action potential. During this time period, the action potential threshold has temporarily increased to a more positive membrane voltage. This period of time is called the **relative refractory period**, because depolarization that is normally sufficient to trigger an action potential cannot, but a larger amount of depolarization can trigger an action potential.

Referring back to Box 3.1, the ball and chain inactivation of sodium channels at depolarized potentials provides a mechanistic explanation for the effects that occur within cardiac muscle cells after a high potassium injection into the blood. The persistent depolarization caused by the change in potassium concentration across the cell membrane leaves the sodium channels in a persistent inactivated state, which means cardiac muscle cells are unable to generate the action potentials that are essential for cardiac muscle contraction. Because of this inactivation mechanism, a hyperpolarized resting membrane potential is required as a starting point for normal electrical signaling in neurons and muscle cells.

## The distributions of sodium and potassium ions across the cell membrane do not change significantly during a single action potential

At first glance, it might seem that the movement of sodium ions into the cell and potassium ions out of the cell during an action potential might dissipate the normal concentration gradient across the cell membrane, but this is not the case. A typical cell has a very large number of potassium ions in its cytoplasm, and the extracellular fluid outside the cell contains an enormous number of sodium ions. It may be obvious that a single action potential will not change the extracellular sodium concentration, but what about the concentrations of sodium and potassium in the cytoplasm? During an action potential, only a small fraction of these ions moves across the membrane, compared to the number of ions that exist in the cytoplasm. To demonstrate this point, we will consider a calculation of these ion numbers for a typical neuron.

First, we will estimate the number of sodium and potassium ions in the cytoplasm of a typical neuron that is 10 μm in diameter. The volume of a sphere is calculated as $4/3 \ (\pi r^3)$, which in this case would equal about 524 μm$^3$. Assuming that the cytoplasmic sodium concentration is 5 mM, and that the cytoplasmic potassium concentration is 130 mM, there are about $1.5 \times 10^9$ total sodium ions in the cytoplasm and $4.1 \times 10^{10}$ total potassium ions in the cytoplasm.

To calculate how many ions move across the cell membrane during an action potential, we will assume that a typical action potential depolarizes the membrane by +100 mV away from the resting potential. The number of charges that must move across the membrane to create this depolarization is given by $Q = C*V$, where $Q$ is the amount of charge moved, $C$ is the capacitance of the membrane the charge is moving across, and $V$ is the change in voltage across the membrane. Typical membrane capacitance is estimated to be about 1 μF/cm$^2$ (or $1 \times 10^{-6}$ F/cm$^2$). Therefore, the total charge moved ($Q$) is equal to $1 \times 10^{-6}$ F/cm$^2$ times 100 mV (0.1 V), which is equal to $1 \times 10^{-7}$ C/cm$^2$. Given that the charge on a monovalent ion is equal to $1.6 \times 10^{-19}$ C, we can convert this to the number of ions moved, which equals $6.25 \times 10^{11}$ ions/cm$^2$ (i.e., $6.25 \times 10^{11}$ ions moved per square centimeter of cell membrane).

A total of $6.25 \times 10^{11}$ ions/cm$^2$ is a very large number of ions, but the size of neurons is not usually measured in centimeters. Given that a typical neuron is 10 μm in diameter, we can convert this number to 6250 ions/μm$^2$. The membrane area of our 10 μm diameter cell would be calculated using $4\pi r^2$, which equals a membrane area of about 314 μm$^2$. Multiplying this membrane area by the number of ions moved per square μm (6250), we obtain an estimate that the +100 mV depolarization that occurs during an action potential involves about two million ($2 \times 10^6$) sodium ions moving into, and about two million potassium ions moving out of, the entire cell. Two million ions also seems like a large number, but compared to what we calculated above for the total number of ions in the cytoplasm ($1.5 \times 10^9$ total sodium ions and $4.1 \times 10^{10}$ total potassium ions), the ion fluxes during a single action potential are a small fraction of the total. Specifically, two million sodium ions flowing into the cell during an action potential add only about 1.3% more sodium ions to the cytoplasm. With two million potassium ions leaving the cell, the cytoplasmic potassium ion concentration only drops by 0.005%.

Performing these calculations drives home the point that the numbers of ions that move across the membrane during one action potential do not have much of an impact on the normal distribution of ions across the membrane. Therefore, the sodium—potassium ATPase described in Chapter 3, is not needed to redistribute ions after a single action potential, but instead is essential on a longer time scale or after a large number of action potentials have occurred in quick succession. Additionally, the sodium—potassium ATPase works on a much slower time scale than the changes in membrane potential that occur during an action potential, so it would not be able to replenish ions quickly enough to keep up with such rapid changes in ionic current.

## Action Potential Propagation

When a neuron initiates an action potential, this electrical signal usually begins at the axon hillock (the first segment of the axon as it leaves the cell body) because there is a high concentration of sodium channels located in this region (see Fig. 4.16). After the membrane potential is depolarized at the axon hillock during an action potential, this depolarization must spread along the axon to the axon terminal in order to trigger neurotransmitter release (see Fig. 4.16). There are two components to this spread of excitation along an axon. The first is the passive spread of electrical current, and the second is the active propagation of the action potential by voltage-gated ion channels.

The passive movement of electrical current along an axon generally occurs after a local change in membrane potential that is not sufficient to bring the neuron to threshold, and

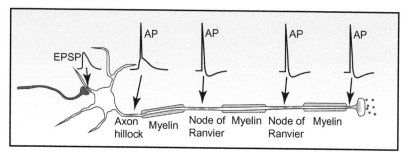

FIGURE 4.16 **Action potentials are usually generated at the axon hillock and then propagated along the axon.** This diagram shows a synapse (*red axon* and nerve terminal) onto the dendrites of a neuron (*white*). The synapse creates an excitatory postsynaptic potential (EPSP) that is generated by ligand-gated ion channels. This EPSP propagates along the dendrites to the cell soma and axon hillock. There is a high concentration of voltage-gated sodium channels (*green*) in the axon hillock, which, when activated by depolarizaton, triggers an action potential (AP). As the action potential generated in the axon hillock travels along the axon, it regenerates an action potential at each "node of Ranvier" on the axon where there is a high concentration of sodium channels (*green*) surrounded by potassium channels (*blue*). The action potential propagates rapidly between the nodes of Ranvier due to the insulating myelin (*orange*), which means it can quickly reach the nerve terminal and lead to neurotransmitter release (*purple dots*). Note that the nodes of Ranvier are not to scale in this image; they are much shorter on real axons (see Fig. 4.18).

thus does not trigger an action potential. If current injected into a cell through an electrode caused a subthreshold depolarization, this depolarization would then spread away from the site of current injection in all directions, and it would decay in amplitude as it spread down the axon due to current that leaks out of the axon.

Current that flows across the membrane at a local site in the neuron creates a depolarization whose size is dependent on Ohm's Law ($V = I^*R$; see Chapter 3). As such, the change in membrane voltage is affected by the resistance of the neuronal membrane. The decrease in amplitude with passive spread down the axon can be described by the equation $V_x = V_0 e^{-x/\lambda}$. In this relationship, $V_x$ is the voltage at a particular distance ($x$) down the axon, $V_0$ is the change in voltage at the current injection site, e is the natural log base ($\sim 2.7$), and $\lambda$ is the **length constant**. The length constant ($\lambda$) is a value that represents how far a change in membrane potential can spread passively down an axon until enough current leaks out of the axon that the depolarization is reduced to $\sim 37\%$ of its initial value. The length constant is governed by several forms of resistance to current flow: resistance of the plasma membrane ($R_m$), resistance of the cytoplasm inside the axon ($R_i$), and resistance of the fluid outside the cell ($R_o$). The specific relationship that describes the length constant is defined as the square root of $R_m/(R_o + R_i)$, meaning that changes in membrane potential will move passively down the axon for a longer distance if the membrane resistance is very high because this greatly reduces passive current flux across the membrane. In contrast, if the resistance to current flow in the axon cytoplasm and/or extracellular fluid is very low, current injected at one site in the neuron will spread quickly and will therefore travel farther from the injection site toward other parts of the cell (i.e., the cell body, axon, or dendrites) before it leaks out.

If current crossing the membrane depolarizes the cell beyond the point at which voltage-gated sodium channel activation occurs (i.e., reaches the action potential threshold), an action potential is generated (see Fig. 4.13). The action potential then spreads passively to the area of membrane adjacent to where it was generated, at a speed that is determined by the length constant. An action potential is such a large depolarization that this passively spreading current activates the sodium channels in the next patch of membrane, and the action potential is regenerated there. Because the length constant governs the speed at which an action potential spreads, action potential conduction velocity is very sensitive to any aspect of the axon anatomy that affects the resistances that govern the length constant ($R_m$, $R_i$, or $R_o$). For example, a common difference between axons is their diameter. Small-diameter axons have a smaller cytoplasmic space, which creates a higher resistance to current flow through the cytoplasm. As such, smaller-diameter axons conduct action potentials more slowly than axons that have a larger diameter.

## Myelin and Nodes of Ranvier

Action potential conduction velocity can be enhanced by indirectly increasing the membrane resistance ($R_m$) with an insulating material that wraps around the axon. This insulating wrap is called **myelin**, and is made from the membrane of glial cells. In the central nervous system **oligodendrocytes** wrap myelin around axons, whereas in the peripheral nervous system **Schwann cells** form the myelin wrap (see Fig. 4.17).

These specialized glial cells wrap around the axons many times with their lipid-rich membrane, which acts as an insulator, increasing $R_m$.

Each glial cell only covers a small length of the axon, with a small gap between myelinated segments. Each gap is called a **node of Ranvier**. In myelinated axons, the action potential can conduct passively along the axon under the sections of myelin sheath, and is only regenerated at the nodes of Ranvier. In general, there are no voltage-gated channels under the myelin, while the nodes of Ranvier have a high concentration of voltage-gated sodium and potassium channels for effective action potential generation (see Figs. 4.16 and 4.18). With this structural organization, myelinated axons can rapidly conduct action potentials over long distances. This form of action potential conduction is called **saltatory conduction**, meaning "conduction that proceeds by leaps," because action potentials effectively "leap" along the axon from node to node. This is in contrast to action potential conduction in unmyelinated axons, which proceeds more slowly, both because such axons have a lower $R_m$ and because action potentials need to be regenerated at each patch of membrane along the entire length of the axon.

Sensory axons that transmit pain and temperature information are unmyelinated and have an action potential conduction velocity of 0.5–2 m/s (or 1–5 miles/h, about as fast as you might walk or run). In contrast, the myelinated axons that connect spinal motor neurons with skeletal muscle can conduct action potentials at 80–120 m/s (or 180–250 miles/h, about as fast as a race car drives). Interestingly, the length of each myelin segment appears to vary between different types of axons and may be regulated to optimize action potential timing within systems where such fine-tuning is essential for information processing (e.g., the auditory system; see Ford et al., 2015).

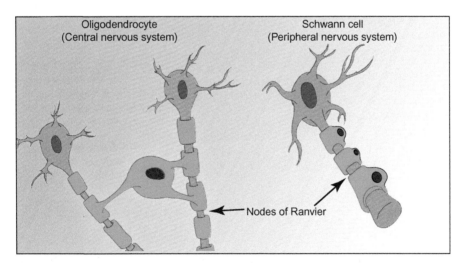

FIGURE 4.17 **Glial cells that form myelin on axons.** (Left) In the central nervous system, oligodendrocytes (blue) form a myelin sheath around axons. One oligodendrocyte can contribute myelin to more than one axon. (Right) In the peripheral nervous system, Schwann cells form myelin on axons, and one Schwann cell wraps a portion of only one axon (the diagram on the right shows three separate glial cells, each creating one myelin segment). Spaces between myelin on an axon are called nodes of Ranvier.

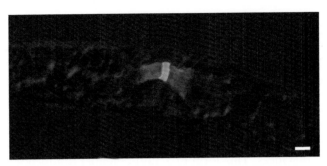

FIGURE 4.18 **Voltage-gated sodium and potassium channels are concentrated at the nodes of Ranvier of myelinated axons.** Immunohistochemical staining of one node of Ranvier in a myelinated rat sciatic nerve axon. The green label identifies the position of voltage-gated sodium channels. The red label is a myelin-specific protein that is found at the connection between the ends of each myelin segment and the axon (Caspr1). The blue label identifies voltage-gated potassium channels (Kv1.2). The dark gray shapes on either side of this node of Ranvier are outlines of the myelin sheath visualized using contrast optics. Scale bar = 2 μm. *Source: Image provided by Matthew Rasband and Peter Shrager, University of Rochester Medical Center.*

# References

Argos, P., Rao, J.K., Hargrave, P.A., 1982. Structural prediction of membrane-bound proteins. Eur. J. Biochem. 128 (2–3), 565–575.

Armstrong, C.M., 2007. Life among the axons. Annu. Rev. Physiol. 69, 1–18. Available from: https://doi.org/10.1146/annurev.physiol.69.120205.124448.

Bezanilla, F., 2008. How membrane proteins sense voltage. Nat. Rev. Mol. Cell Biol. 9 (4), 323–332. Available from: https://doi.org/10.1038/nrm2376.

Coetzee, W.A., Amarillo, Y., Chiu, J., Chow, A., Lau, D., McCormack, T., et al., 1999. Molecular diversity of K+ channels. Ann. N. Y. Acad. Sci. 868, 233–285.

Conway, B.E., 1981. Ionic Hydration in Chemistry and Biophysics. Elsevier Scientific Pub. Co.: distributors for the U.S. and Canada, Elsevier/North-Holland, Amsterdam; New York.

Doyle, D.A., Morais Cabral, J., Pfuetzner, R.A., Kuo, A., Gulbis, J.M., Cohen, S.L., et al., 1998. The structure of the potassium channel: molecular basis of K+ conduction and selectivity. Science 280 (5360), 69–77.

Ford, M.C., Alexandrova, O., Cossell, L., Stange-Marten, A., Sinclair, J., Kopp-Scheinpflug, C., et al., 2015. Tuning of Ranvier node and internode properties in myelinated axons to adjust action potential timing. Nat. Commun. 6, 8073. Available from: https://doi.org/10.1038/ncomms9073.

Gonzalez, C., Baez-Nieto, D., Valencia, I., Oyarzun, I., Rojas, P., Naranjo, D., et al., 2012. K(+) channels: function-structural overview. Compr Physiol 2 (3), 2087–2149. Available from: https://doi.org/10.1002/cphy.c110047.

Isom, L.L., De Jongh, K.S., Catterall, W.A., 1994. Auxiliary subunits of voltage-gated ion channels. Neuron 12 (6), 1183–1194.

Kreusch, A., Pfaffinger, P.J., Stevens, C.F., Choe, S., 1998. Crystal structure of the tetramerization domain of the Shaker potassium channel. Nature 392 (6679), 945–948. Available from: https://doi.org/10.1038/31978.

Kyte, J., Doolittle, R.F., 1982. A simple method for displaying the hydropathic character of a protein. J. Mol. Biol. 157 (1), 105–132.

Long, S.B., Campbell, E.B., Mackinnon, R., 2005. Crystal structure of a mammalian voltage-dependent Shaker family K+ channel. Science 309 (5736), 897–903. Available from: https://doi.org/10.1126/science.1116269.

Long, S.B., Tao, X., Campbell, E.B., MacKinnon, R., 2007. Atomic structure of a voltage-dependent K+ channel in a lipid membrane-like environment. Nature 450 (7168), 376–382. Available from: https://doi.org/10.1038/nature06265.

Noda, M., Shimizu, S., Tanabe, T., Takai, T., Kayano, T., Ikeda, T., et al., 1984. Primary structure of Electrophorus electricus sodium channel deduced from cDNA sequence. Nature 312 (5990), 121–127.

Payandeh, J., Scheuer, T., Zheng, N., Catterall, W.A., 2011. The crystal structure of a voltage-gated sodium channel. Nature 475 (7356), 353–358. Available from: https://doi.org/10.1038/nature10238.

Sansom, M.S., 1998. Ion channels: a first view of K+ channels in atomic glory. Curr. Biol. 8 (13), R450–R452.

Zhou, Y., Morais-Cabral, J.H., Kaufman, A., MacKinnon, R., 2001. Chemistry of ion coordination and hydration revealed by a K+ channel–Fab complex at 2.0 Å resolution. Nature 414, 43–48.

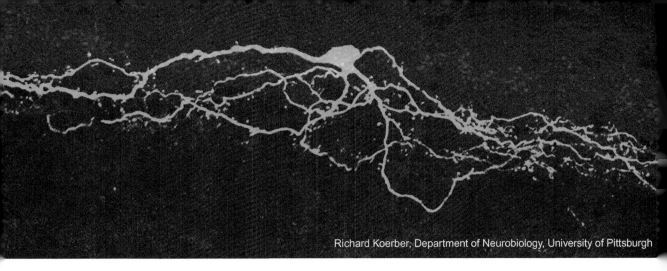

Richard Koerber, Department of Neurobiology, University of Pittsburgh

CHAPTER 5

# Electrical Synapses

## HISTORY OF ELECTRICAL SYNAPSES

In the early 1900s, evidence was mounting that there was a gap (which we now know to be the synaptic cleft) between neurons in the brain at specialized sites of contact. This evidence set the stage for investigations into how neurons communicate across this gap. The debate that followed was primarily between **neuropharmacologists**, who provided evidence of chemical communication, and **electrophysiologists**, who provided evidence of electrical communication.

Early electrophysiologists arguing in favor of electrical communication at synapses were comfortable with the concept of chemical communication in the body, but considered it to be relegated to the endocrine system, where chemically mediated events occurred on the time scale of seconds or minutes. After all, when a person reacts to a ball that is thrown at them unexpectedly, they sense the object with their eyes, relay that information to the brain, make a decision about what to do, and then send a motor command to muscles to either move or raise a hand to catch or deflect the object. All of these events occur in less than a second! Therefore, electrophysiologists reasoned that the nervous system must work too quickly for communication between neurons to be mediated by the diffusion of chemicals across a gap between cells.

This debate has been characterized as the "war of the soups and sparks" (reviewed by Valenstein, 2005), with those who argued for chemical communication in the "soups" camp and those in favor of electrical communication in the "sparks" camp. In this chapter, we will focus primarily on the evidence supporting electrical communication and the discovery of electrical synapses.

## Early Evidence in Favor of Electrical Communication in the Nervous System

In 1791, Luigi Galvani, an Italian physician and physicist, discovered that the muscles in a frog leg would contract when a spark of static electricity made contact with the nerve innervating the muscles. He also showed that when two different metals were used (one in contact with the spinal cord, and the other in contact with muscle), an electric charge was generated that caused the muscles to contract (see Fig. 5.1). Galvani hypothesized that a force called "animal electricity" activated frog muscles to contract and that this electricity was generated within the body. In honor of his discovery, this effect was called **Galvanism**, a historical term that today we might call the study of electrophysiology, at least in the context of this textbook. At the time, Galvani did not know that this electricity was carried by ions. Instead, he speculated that it was carried by an electrical fluid which was delivered to the muscles by the nerves that innervate them.

One of Galvani's contemporaries, Alessandro Volta, did not agree with the concept of an "animal electric fluid," and instead speculated that this effect was the result of a "direct current of electricity produced by chemical action" (which is currently considered a modern definition of "Galvanism" in the field of physics). In 1800, Volta created an "electric pile" by stacking plates made of two different metals (zinc and copper) in contact with a salt solution, effectively creating a source of electricity (see Fig. 5.2). This electric pile was a precursor to the modern battery. Volta used this approach to argue that Galvani had created an electric current in his experiments by using two different metals to contact the spinal cord and muscles in a frog. He advertised his new invention as an "artificial electric organ" and emphasized that animal tissue was not required to create the

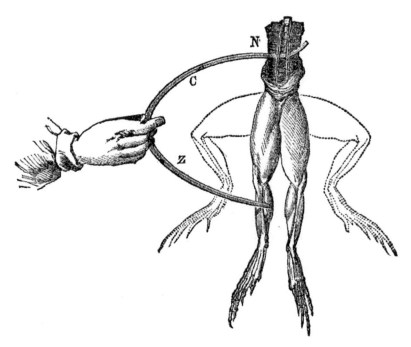

FIGURE 5.1 Galvani's drawing of an experiment in which two different metal probes touch between the body cavity of a frog and the muscles in the legs, causing the leg muscles to twitch. *Source: Wiki public domain; From David Ames Wells, 1859, The Science of Common Things: A Familiar Explanation of the First Principles of Physical Science. For Schools, Families, and Young Students. Publisher Ivison, Phinney, Blakeman, 323 pages; p. 290.*

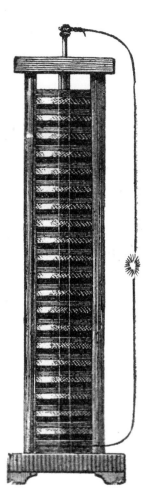

FIGURE 5.2 Diagram of an electric or Voltaic pile created by Alessandro Volta in 1800. *Source: Wiki public domain; From the book "Opfindelsernes Bog" 1878 by André Lütken.*

electricity. We now know that the electricity that causes muscles to contract *is* generated by the body, but Volta was correct that this electricity is produced by chemical action—specifically, the movement of ions. Volta's invention and interpretation of his data are recognized by naming the unit of electrical potential the "Volt."

## Discovery of Electrical Synapses

In the 1950s, two research groups independently discovered evidence of direct electrical coupling between neurons. Watanabe (1958) discovered that electrical synapses exist between neurons in the cardiac ganglion of lobsters, and Furshpan and Potter (1959) identified electrical synapses in the giant motor synapse of the crayfish. Further evidence of direct electrical coupling between neurons accumulated with reports in fish (Bennett et al., 1963; Robertson, 1963; Kriebel et al., 1969) and birds (Martin and Pilar, 1963; see Nagy et al., 2018 for review). Initially, scientists studying mammals assumed that these "primitive" electrical synapses would not be present in mammalian brains. They reasoned that direct electrical coupling between neurons in lower animal species was consistent with the more "primitive" functions of their nervous systems.

For example, the crayfish giant motor synapse is part of the circuit that triggers a rapid flip of the tail (the "tail flip"), which is essential for this animal's escape response. Scientists observed that the crayfish tail flip was mediated by electrical synapses, and they reasoned that electrical synapses were sufficient for this reflex because it is a primitive behavior not present in mammals, and that the electrical synapses were critical for the immense speed of the escape response. Furthermore, most of the lower species that had been shown to use electrical synapses between some of their specialized neurons were cold-blooded animals (for example, fish and crayfish), which need to be able to function in the cold environmental temperatures of a stream, lake, or ocean. Scientists reasoned that the biochemical processes underlying chemical transmitter release would be more sensitive to temperature and thus too slow to be effective in a cold environment. The consensus among scientists in these early days was that invertebrates and lower vertebrates seemed to use electrical synapses in varied environmental temperatures when very high speed of communication was essential.

It was also found that some electrical synapses in lower animal nervous systems facilitate the precise synchronous activation of a neuronal network. Carew and Kandel (1977) studied the ink discharge of a sea slug (*Aplysia californica*), which is a defensive response that is initiated when *Aplysia* are disturbed and is used to keep predators away. These researchers demonstrated that when the tail of the animal is touched, sensory neurons excite the motor neurons that innervate an ink gland. These motor neurons are electrically connected (coupled) to one another, which ensures that they all fire the same robust burst of action potentials together. This direct electrical connection serves to synchronize the activation of these motor neurons, producing a robust discharge of ink from the gland (see Fig. 5.3).

FIGURE 5.3 **The defensive response of the sea slug *Aplysia californica*.** When threatened, the sea slug releases a plume of ink. This ink discharge is triggered by the synchronous activation of electrically coupled motor neurons. *Source: Adapted from Barsby, T., 2006. Drug discovery and sea hares: bigger is better. Trends Biotechnol. 24, 1–3 (Barsby, 2006).*

The assumption that electrical synapses only mediated relatively simple behaviors in primitive animal nervous systems gave way to the realization that electrical synapses do exist in mammalian nervous tissue (Baker and Llinas, 1971; reviewed in Nagy and Dermietzel, 2000). It has subsequently been shown that electrical synapses play important roles in central nervous system function. In fact they are found in every part of the mammalian central nervous system and are produced by many, if not all, mammalian neurons during development. Given their importance in the nervous system, we will first describe the characteristics of electrical synapses and then discuss some of their functions in the mammalian nervous system.

## STRUCTURE AND PHYSIOLOGICAL CHARACTERISTICS OF ELECTRICAL SYNAPSES

As anatomists began visualizing the specialized connections between neurons at the ends of axons, in addition to the synaptic specializations described in Chapter 3, they documented other types of connections between cells. One such connection is called a **gap junction** (see Fig. 5.4). It became clear that these gap junctions are the sites of electrical communication between neurons at electrical synapses. When viewed with an electron microscope, gap junctions have a very characteristic appearance. First, the membranes of the two connected cells are separated by a gap of only about 3–4 nm. This is much narrower than the normal gap between cells outside of the gap junction (e.g., at the synaptic cleft this gap is about 30 nm wide, and the gap between neurons in non-synaptic regions is about 20 nm). Furthermore, electron micrographs reveal electron-dense material inside both cells at the site of the electrical synapses, which is thought to represent a collection of scaffolding and modulatory proteins associated with the gap junction, akin to the pre- and postsynaptic densities that are present at chemical synapses.

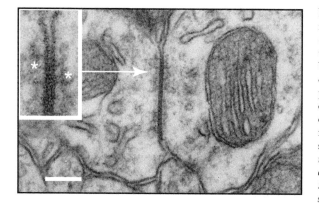

FIGURE 5.4 Electron micrograph of a neuronal gap junction (*arrow*) between two dendrites in the mouse neocortex. Gap junctions are characterized by a short distance between the plasma membranes of the two connected cells. Scale bar = 200 nm. Arrow points to the gap junction in the image that is enlarged in the inset. This enlargement more clearly reveals the presence of electron-dense material (*asterisks*) in the cytoplasm on each side of the gap junction. *Source: From Atlas of ultrastructural neurocytology.* <http://synapseweb.clm.utexas.edu/atlas>; J. Spacek contributor. SynapseWeb, Kristen, M. Harris, P.I. <http://synapseweb.clm.utexas.edu/>.

## Gap Junction Structure

A gap junction is a specialized formation of intercellular channels, or **protein pores**, that form a complex with regulatory proteins. Each of these channels is a multimeric protein made up of two "hemichannels," which are called **connexons** in vertebrates. The two cells separated by the gap junction each contribute one connexon per channel. When a connexon from one cell aligns with an opposing connexon from the other cell, they form a complete channel (pore) that spans the gap junction, connecting the two cells (Goodenough and Paul, 2009).

In mammals, each connexon is a hexamer composed of six **connexin** (Cx) protein subunits (see Fig. 5.5). New connexins are produced in the cell body and are transported to the site of gap junction formation, which can be in the cell body or other locations within neurons. They are then inserted into the plasma membrane and aligned with the connexins of opposing hemichannels synthesized by the coupled neighboring cell. Adhesion and scaffolding proteins are thought to aid in the clustering and organization of gap junction proteins. Intracellular scaffolding proteins have binding sites for regulatory proteins, including kinases and other intracellular signaling systems (see Miller and Pereda, 2017 for review).

The connexin protein subunits that form the gap junction protein pores are expressed from a multigene protein family (Dermietzel et al., 2000; Rozental et al., 2000). The mammalian genome codes for about 20 different connexin proteins. In the adult nervous system, specific types of connexins are expressed in neurons, while others appear to be present in glia (Nagy et al., 2004). A cell can assemble a **homomeric connexon** from six

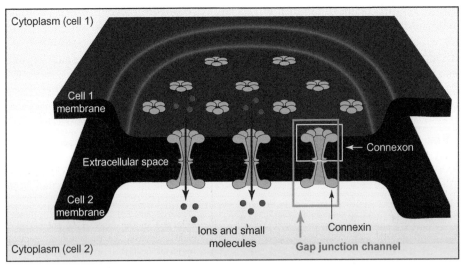

FIGURE 5.5  **Diagram of a gap junction.** Six connexin subunits (*light blue*) come together to form one connexon (*yellow box*), and many connexons are embedded in the synaptic membrane of both cells (1 and 2) at the gap junction. One connexon from each cell comes together to form a complete gap junction channel (green box) with a pore in its center. The gap junction channels can pass ions and small molecules from the cytoplasm of one cell into the cytoplasm of another.

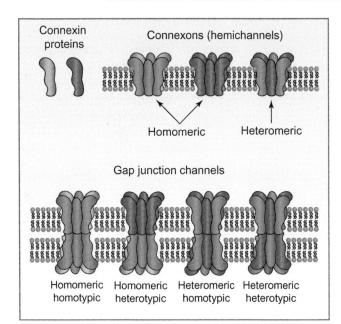

FIGURE 5.6 **The formation of connexon channels from connexin protein subunits.** Each cell assembles a connexon hemichannel from six individual connexin proteins (top). Connexons can be assembled using six proteins of the same subtype (homomeric), or a mix of different proteins (heteromeric). Then, two hemichannels come together (one from each cell that will be electrically coupled) to form the final gap junction channel (bottom). These gap junction channels can be mixed and matched with connexons from each cell to form a variety of gap junction channel types.

connexins of the same type, or it can assemble a **heteromeric connexon** by combining different connexin types. In the same vein, two connexons of the same type can align to form a homotypic channel, while two connexons of different types can align to form a heterotypic channel (see Fig. 5.6).

## Physiological Characteristics of Electrical Synapses

An important physical characteristic of gap junctions is the size of the pore because this constrains which molecules can pass between cells at electrical synapses. There are two ways to characterize molecules with respect to their ability to pass through a gap junction pore.

The first is the size of the molecule as measured by its atomic mass. The protein pores that form gap junctions at electrical synapses are estimated to be large enough to pass anything with an atomic mass of less than about 1000 Da (1 kDa), though the pore size varies substantially by the types of connexins that make up the channels. The **dalton** is the standard unit of mass used to describe quantities on the atomic scale. One dalton is defined as 1/12th of the mass of a single carbon-12 atom. A hydrogen atom has a mass of about one dalton, and inorganic ions that might pass through a gap junction pore at an electrical synapse range in mass from 22–40 Da (e.g., sodium, potassium, chloride, and calcium ions). When estimated based on mass, this gap junction pore size appears plenty large enough to allow naturally occurring inorganic ions to pass through.

However, ions are charged, and thus hydrated (surrounded by water) when dissolved in the cell cytoplasm (see Chapter 2). With this in mind, an ion has a hydrated diameter that also has to be taken into consideration. Therefore, a second characteristic to consider when evaluating which ions can pass through a gap junction is the actual size of the pore, which is the diameter of the space the molecules must pass through. Based on the sizes of molecules that can pass through a gap junction, it has been estimated that the pore diameter is about 10–20 Å, depending on the particular connexin protein isoforms that form the pore (Beblo and Veenstra, 1997; Wang and Veenstra, 1997; Veenstra, 2001; Gong and Nicholson, 2001; Hormuzdi et al., 2004). It is hypothesized that monovalent ions passing through a gap junction pore are partially hydrated, with water already in the pore aiding in the passage of the charged ion through the pore. Hydrated ions have diameters of about 3–5 Å, so pore size does not generally limit the flux of ions through gap junctions.

Because gap junction pores are so large, inorganic ions are not the only molecules that can pass through them. In fact, gap junctions are known to allow important cellular signaling molecules to pass between cells through these channels. These signaling molecules include cyclic nucleotides (cAMP and cGMP, roughly 550 Da in size; Lawrence et al., 1978) and inositol 1,4,5-trisphosphate (IP3; roughly 420 Da in size; Sáez et al., 1989; Sanderson, 1995). The cellular effects of the above molecules in synaptic transmission are covered in Chapter 11 and Chapter 12.

In an experimental setting, the transfer of electrical information across gap junctions is most often measured by making intracellular microelectrode recordings between two electrically coupled neurons (see Fig. 5.7). When ions pass through gap junctions, they

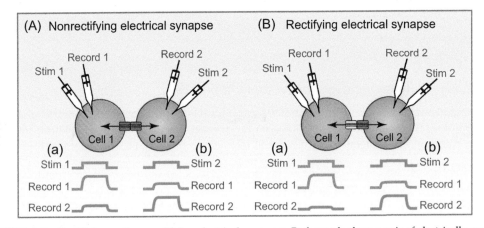

FIGURE 5.7  **Rectifying and nonrectifying electrical synapses.** Both panels show a pair of electrically-coupled cells with a recording and stimulating electrode in each. A. When a depolarizing current (Stim 1) is injected into Cell 1, this cell depolarizes (Record 1 in column a) and also depolarizes Cell 2, though to a lesser degree (Record 2 in in column a). Injecting the same amount of current into Cell 2 (Stim 2) depolarizes Cell 2 (Record 2 in column b) and also depolarizes Cell 1, though to a lesser degree (Record 1 in column b). However, the amount by which the current is reduced as it goes through the gap junction is the same in both directions (compare Record 2 in column a with Record 1 in column b). This is a nonrectifying electrical synapse. B. In contrast, at a rectifying electrical synapse, when a depolarizing current (Stim 1) is injected into Cell 1, this depolarizes Cell 2 (Record 2 in column a) substantially less than when the same amount of depolarizing current (Stim 2) is injected into Cell 2 and depolarizes Cell 1 (compare Record 2 in column a with Record 1 in column b).

change the distribution of charges across the membrane, creating a change in membrane potential (see Fig. 5.7), and this exchange of ions is usually bidirectional. Most gap junction channels are nonrectifying, which means ionic current going from one side of the gap junction to the coupled neuron on the other can flow equally well in either direction. However, a minority of connexon channels are **rectifying**, which means they pass current more preferentially in one direction than in the other (see Marder, 2009; Fig. 5.7). Rectifying gap junctions are usually formed when the connexon hemichannels contributed by each cell are of different connexin protein subtypes (heterotypic channels; see Fig. 5.6). In contrast, when electrical synapses pass current equally in both directions, the connexon protein hemichannels contributed by each cell to form the complete channel are both of the same protein type (homotypic channels; see Fig. 5.6).

When current passes from one neuron across a gap junction to a second neuron, it can change the membrane potential of the second neuron in a number of ways. A single gap junction can transmit a signal that is depolarizing or hyperpolarizing, according to whatever voltage changes occur in the cells connected by the gap junction. That is, a depolarizing change in a neuron's membrane potential will in turn depolarize an electrically coupled neuron, while a hyperpolarizing change in a neuron will hyperpolarize a neuron coupled to it. Additionally, changes in membrane potential transmitted by gap junctions are analog (continuously variable), meaning that they do not require action potentials and do not occur in the discrete amounts called quanta that are characteristic of vesicular release of neurotransmitter at chemical synapses (see Chapter 6). Note that while most electrical synapses are bidirectional, we will discuss how one neuron affects a second, coupled neuron using the terms "presynaptic" and "postsynaptic." In this case, the "presynaptic cell" is the one propagating an action potential or other change in membrane potential, and the "postsynaptic cell" is the one responding to the current that flows from the presynaptic cell through the gap junction (see Fig. 5.8).

When an action potential (or other membrane depolarization) reaches a gap junction on the presynaptic side of the synapse, ions flow through gap junction channels, creating a current flux across the membrane that depolarizes the postsynaptic cell. The size of that postsynaptic depolarization is governed in part by the number of connexon channels at the gap junction because this determines how many ions flow into the postsynaptic cell at a given voltage, and therefore the size of the resulting ionic current. Thus, the amount of current ($I$) flowing across a gap junction is proportional to $V$ (the voltage across the gap junction) and inversely proportional to $R$, the total resistance of all of the gap junction channels, as described by Ohm's Law ($V = IR$). The above is also true for presynaptic membrane hyperpolarizations. Remember that gap junctions can relay any change in membrane potential from the presynaptic cell to the postsynaptic cell, whether the change is depolarizing or hyperpolarizing. Given these properties, electrical synapses are strongest when they contain many gap junction channels and the postsynaptic cell is smaller than the presynaptic cell. However, in most cases a presynaptic action potential will only depolarize the postsynaptic cell slightly, causing a subthreshold depolarization, rather than generating an action potential (see Fig. 5.9).

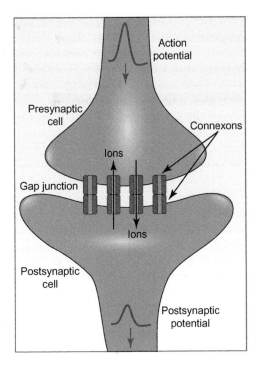

FIGURE 5.8 **The transfer of an electrical potential from a presynaptic cell to a postsynaptic cell through connexon channels.** An action potential in the presynaptic cell (*upper red trace*) results in a smaller depolarization in the postsynaptic cell (*lower red trace*). The ions that pass through gap junctions result in current flux that affects membrane potential.

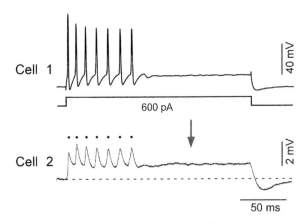

FIGURE 5.9 Conduction of action potential signals in one cell through a gap junction to a second cell. When a depolarizing current injection (600 pA square step) is given to cell 1 via an electrode, that cell fires action potentials (*top trace*). The depolarizations resulting from each of the action potentials in cell 1 cause small "spikelets" in cell 2 (indicated by the dot above each spikelet). These spikelets ride on top of a small, constant depolarization (*arrow*) that is also seen in cell 1 in response to the current injection. Note the difference in the voltage scale bar when comparing top and bottom traces. *Source: Adapted from Curti, S., Hoge, G., Nagy, J.I., Pereda, A.E., 2012. Synergy between electrical coupling and membrane properties promotes strong synchronization of neurons of the mesencephalic trigeminal nucleus. J. Neurosci. 32, 4341–4359 (Curti et al., 2012).*

Simultaneous electrical recordings from both pre- and postsynaptic cells has shown that there is little synaptic delay at an electrical synapse (see Fig. 5.10). That is, the time between the onset of a change in membrane potential in one cell and the concomitant change in an electrically coupled cell is very short (~0.1 ms). While some chemical synapses in the mammalian brain can have very short synaptic delays at mammalian body temperature (e.g., ~0.15 ms; Sabatini and Regehr, 1999), at other chemical synapses, the delay between the arrival of an action potential at the axon terminal and the onset of a change in voltage in the postsynaptic neuron can be around 1–2 ms (see Chapter 6).

Another method for studying electrical synapses takes advantage of the fact that gap junctions allow the passage of small molecules. If the experimenter injects a small molecule into one cell, the molecule will move through gap junctions between that cell and any surrounding cells to which it is electrically coupled. In this type of experiment, a researcher can fill a microelectrode with a dye or other label that is small enough to travel through the type of gap junction being investigated, and inject the dye into a single cell. The dye will slowly diffuse out of the electrode into the cell and then cross through gap junction channels into surrounding electrically coupled cells.

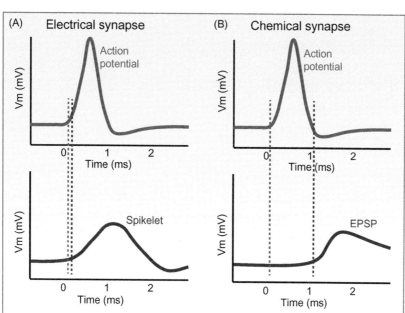

FIGURE 5.10  **Synaptic delays at electrical and chemical synapses.** Synaptic transmission through a gap junction has essentially no synaptic delay (~0.1 ms), compared to a typical chemical synapse, which can have up to a 1–2 ms synaptic delay (see Chapters 6–8). (A) An action potential in the presynaptic neuron (top) leads, with little to no synaptic delay, to a small "spikelet" (spike-like depolarization) in the postsynaptic neuron when an electrical synapse connects these two cells. (B) An action potential in the presynaptic neuron leads, with about a 1 ms synaptic delay, to an excitatory postsynaptic potential (EPSP) in the postsynaptic neuron when a chemical synapse connects these two cells.

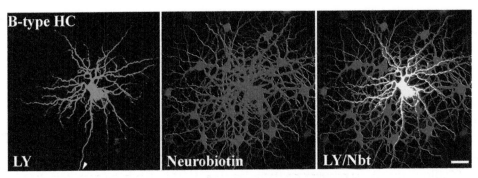

FIGURE 5.11  **Fluorescent dyes can reveal groups of electrically coupled neurons.** In this experiment, type B horizontal cells (B-type HC) in the rabbit retina were loaded with fluorescent dyes to show how many of these cells are connected to a single central cell. A single type B horizontal cell was loaded with two dyes simultaneously—one that cannot pass through the gap junctions between horizontal cells (LY; *green*) and one that can pass through gap junctions in these cells (neurobiotin; *red*). The image on the left shows the full extent of the injected cell using the green dye that does not pass through gap junctions. The image in the center panel shows the red dye that moved from the injected cell into neighboring electrically coupled cells. The image on the right is an overlay of the left and center images (LY/Nbt). Yellow indicates locations where both the red and the green dye were found, which, in this case, was only in the injected cell and its proximal processes. Scale bar = 25 μm. LY, Lucifer yellow dye; Nbt, neurobiotin dye. *Source: Adapted with permission from O'Brien, J.J., Li, W., Pan, F., Keung, J., O'Brien, J., Massey, S.C., 2006. Coupling between A-type horizontal cells is mediated by connexin 50 gap junctions in the rabbit retina. J. Neurosci. 26, 11624–11636 (O'Brien et al., (2006).*

When the injected cell and the surrounding tissue are viewed under a microscope, it is possible to see all of the cells that form gap junctions with the injected one (see Fig. 5.11).

# ROLES OF ELECTRICAL SYNAPSES

Electrical synapses are found in a range of areas of the nervous system, and they are more prevalent during the development of the nervous system than in adulthood. Regardless of the age and/or area of the nervous system, they serve a wide variety of functions, including synchronizing the activity of networks of neurons that are interconnected by gap junctions (see Connors, 2017 for review). We discuss several examples of this capability, including during development and in behaviors in the fully mature nervous system.

## The Role of Electrical Synapses in the Developing Mammalian Nervous System

As the mammalian nervous system develops, gap junctions appear very early and are widespread. Before chemical synapses have formed or matured, electrical synapses are often present and contribute to the proper development of the nervous system (see Hormuzdi et al., 2004 for review). In fact, during early development, electrical synapses are more common than they are in the adult nervous system. Several specific developmental events, including neurulation, cellular differentiation, neuronal migration, axon pathfinding, and the formation of neuronal circuits, are facilitated by the exchange of cellular signals through gap junctions. When these developmental events conclude, the gap junctions that facilitated them are often downregulated.

## *Electrical Synapses in the Development of the Neuromuscular Synapse*

The establishment of chemical communication between nerve and muscle cells is preceded by transient gap junctions between these cells and the exchange of chemical messengers through the connexon channels (Fischbach, 1972; Allen and Warner, 1991; Hall and Sanes, 1993). It is hypothesized that these early electrical connections allow the exchange of chemical messengers that may facilitate the formation of the neuromuscular junction. After this early period of synapse formation, the presence of gap junctions between dendrites of motor neurons serves to synchronize their activity patterns (Fig. 5.12). This synchronization of electrical activity between motor neurons facilitates the establishment of early exuberant connections, while the subsequent loss of these gap junctions permits a form of Hebbian competition that leads to synapse elimination and aids in the maturation of the chemical synapses in this system (see Box 5.1). For Hebbian plasticity to govern synapse elimination at the neuromuscular junction, there needs to be a mismatch in activity between competing motor axons. When investigators studied the activity patterns of motor neurons in the spinal cord during development, they found that all motor neurons are active early in development, and that this synchronous activity is facilitated by the presence of electrical synapses between the motor neurons (Fulton et al., 1980; Walton and Navarrete, 1991; Chang et al., 1999). Even modestly strong electrical

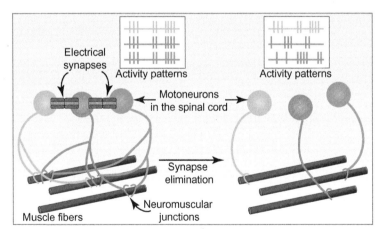

FIGURE 5.12 **Gap junctions between spinal motor neurons regulate competition during the period before and during synapse elimination.** (Left) Electrical synapses between dendrites of developing motor neurons synchronize their activity patterns (inset; activity patterns), preventing activity-dependent competition at neuromuscular junctions. (Right) Later in development, gap junctions have disappeared, desynchronizing motor neuron electrical activity (inset; activity patterns), allowing activity-dependent competition to eliminate excess neuromuscular synapses onto muscle cells.

## BOX 5.1

## SYNAPSE ELIMINATION AT THE NEUROMUSCULAR JUNCTION

Initially, motor neurons make **exuberant synapses** with peripheral muscle cells, meaning that they form more synapses than will be required in the mature tissue. As a result, developing muscle cells are innervated by more than one motor neuron axon, a phenomenon termed **multiple innervation**. This developmental overabundance of neuromuscular synapse formation ensures that the entire peripheral musculature receives adequate innervation. Subsequently, this overabundance of synapses needs to be pruned and adjusted to achieve the one-to-one motor axon innervation of each muscle cell that is characteristic of the adult pattern of connections. This period of adjustment is called **synapse elimination**, and it requires an activity-dependent competition between motor axons to determine which one will persist on each muscle cell (see Fig. 5.12). The process of synapse elimination has been monitored by repeated **in vivo** observations of the same neuromuscular synapses over a 1-week period in the mouse (Fig. 5.13). These observations indicate that one of the motor inputs gradually takes over most if not all of the postsynaptic area previously occupied by a competitor. The motor axon that loses the competition retracts from the neuromuscular junction.

The experimental evidence for this activity-dependent competition comes from studies in which experimenters used tetrodotoxin (TTX; a sodium channel blocker; see Box 6.2) to block action potential activity in motor axons, and found that synapse elimination was delayed until after the block was removed (Thompson et al., 1979).

Scientists investigating the mechanisms underlying activity-dependent synapse elimination found evidence that synapses became stronger when the pre- and post-synaptic cells were active simultaneously, but were weakened when the pre- and post-synaptic cellular activity was asynchronous. This modulation is a form of Hebbian plasticity (see Box 5.2). It was proposed that activity-dependent competition at the developing neuromuscular junction was governed by the generation of local "protection" and "elimination" molecular signals within the synaptic cleft (see Fig. 5.14). The hypothesis was that an active nerve terminal would generate both a protection and an elimination signal. Thus, it would be supported by the protection signal, but not be sensitive to harm from the elimination signal. In contrast, nearby inactive nerve terminals would be affected by the elimination signal, but not benefit from the protection signal.

In the search for molecules that might mediate these signals, one attractive candidate is brain-derived neurotrophic factor (BDNF). The BDNF hypothesis of synapse elimination (see Fig. 5.15) is based on the hypothesis that calcium flux through active nicotinic acetylcholine receptors might trigger the postsynaptic muscle cell secretion of proBDNF, a precursor of BDNF. Since at this stage of development multiple axon terminals synapse onto the same muscle cell, the proBDNF secreted by that cell can reach all of these terminals, not just the active presynaptic terminal that triggered the secretion. However, the active nerve

## BOX 5.1 (cont'd)

terminal is hypothesized to secrete an enzyme (matrix metalloprotease; MMP) into the local synaptic cleft that can cleave the proBDNF into mature BDNF. The inactive nerve terminals are affected by the proBDNF secretion, but do not secrete the enzyme that cleaves it into BDNF. proBDNF and BDNF preferentially bind to different presynaptic motor nerve terminal receptors; proBDNF binds to a receptor called p75, and BDNF binds to a receptor called TrkB. Therefore, active nerve terminals that secrete MMP receive protective BDNF signals through TrkB receptors, while inactive nerve terminals receive proBDNF signals through p75 receptors (a hypothesized elimination signal). This BDNF hypothesis is proposed as one candidate explanation for Hebbian plasticity at the neuromuscular junction during synapse elimination. Validation of this hypothesis in synapse elimination will require further study.

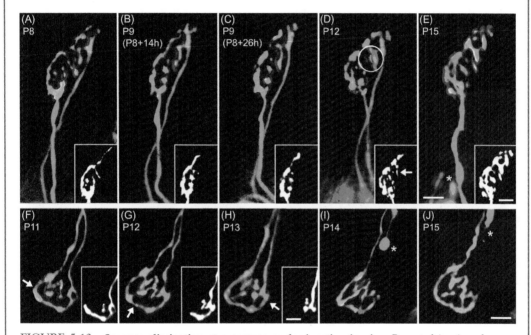

FIGURE 5.13 **Synapse elimination at a neuromuscular junction in vivo.** Repeated in vivo observations from a transgenic mouse expressing a cyan fluorescent protein in one motor neuron, and a green fluorescent protein in another motor neuron, that both innervate a single neuromuscular junction. Over about 1 week (P = postnatal day), one of the two motor axons "wins" the synapse elimination competition, and the other retracts from the neuromuscular junction. A–E and F–J represent two examples, with insets representing the neuromuscular junction territory occupied by the cyan-labeled nerve terminal over the course of several days. Scale bars = 10 μm. *Source: Adapted from Walsh, M.K., Lichtman, J.W., 2003. In vivo time-lapse imaging of synaptic takeover associated with naturally occurring synapse elimination. Neuron 37, 67–73 (Walsh and Lichtman, 2003).*

## BOX 5.1 (cont'd)

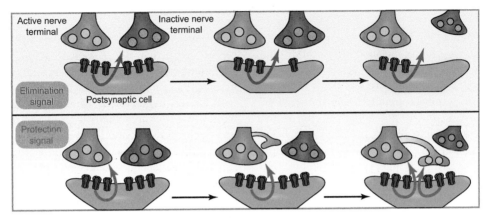

FIGURE 5.14 **Protection and elimination molecular signals that might control synapse elimination at the neuromuscular junction.** The proposed mechanisms depicted in the top and bottom rows of this series of drawings are hypothesized to occur at the same time. In the top row, an active nerve terminal (*green*) is generating an elimination signal (red arrow) that only affects inactive nerve terminals in the vicinity. The bottom row shows that the active nerve terminal is producing a protection signal (green arrow) that only active terminals can benefit from. Thus, this signal provides support for the active synapse and allows the active nerve terminal to take over areas of the synapse left unoccupied by inactive nerve terminals that have been eliminated.

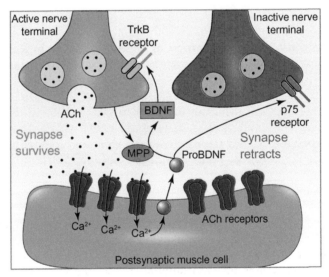

FIGURE 5.15 **The BDNF hypothesis of synapse elimination.** This diagram represents an elaboration of the process of synapse elimination in Fig. 5.14, in which elimination and protection signals are hypothesized to be present. This image depicts the hypothesis that active nerve terminals release the enzyme MMP and induce the postsynaptic cell to release proBDNF. Then, MMP near the active nerve terminal converts proBDNF into BDNF. At the active nerve terminal, the BDNF acts on presynaptic TrkB receptors, which creates a support signal for this axon terminal. In contrast, at the inactive nerve terminal, the proBDNF acts on presynaptic p75 receptors, which creates a retraction signal for this axon terminal.

> BOX 5.2
>
> ## HEBBIAN PLASTICITY
>
> **Hebbian plasticity** is a form of synaptic plasticity named after Donald O. Hebb (1904–1985), a Canadian psychologist who developed theories about cellular mechanisms for learning in the nervous system (see Fig. 5.16). Since he attempted to explain psychological findings using cellular neuroscience mechanisms, he is often referred to as the "father of **neuropsychology**" (Milner, 1993).
>
> In Hebb's most famous work, "The Organization of Behavior: A Neuropsychological Theory," he outlined a theory to explain how neural circuits are formed and modified as a result of experience. In this book, Hebb stated:
>
> "When an axon of cell A is near enough to excite a cell B and repeatedly or persistently takes part in firing it, some growth process or metabolic change takes place in one or both cells such that A's efficiency, as one of the cells firing B, is increased."
>
> Hebb's explanation of his theory is often summarized as "cells that fire together, wire together." This theory predicts that there is an exchange of signaling molecules between neurons that synapse with one another and are active together, and that this exchange leads to the support and strengthening of the chemical synaptic connection between the two cells.
>
> Hebb's theory is popular because it conceptualizes a cellular mechanism for experience-dependent synaptic plasticity, and this theory is sometimes applied to learning in nervous system circuits (see Chapter 14 for more details).

synapses can lead to considerable synchrony of action potentials in a network of motor neurons. This early synchronous activity is thought to allow a single muscle fiber to receive innervation from multiple axons during the period of synapse formation, since there is no activity mismatch to create activity-dependent competition. However, after synapse formation, the electrical synapses between motor neurons in the spinal cord are eliminated (Walton and Navarrete, 1991; Chang et al., 1999), which allows motor neurons to be activated independently, facilitating activity-dependent competition at the neuromuscular junction (see Fig. 5.12; Personius and Balice-Gordon, 2001).

### Electrical Synapses in the Development of Cortical Synapses

The formation of circuits within the cerebral cortex is a large and complex developmental challenge. Millions of developing neurons need to make appropriate connections with an even greater number of axons and dendrites, and all of this specific connectivity must occur in the complex and changing environment of the brain. To accomplish this task, many central nervous system circuits employ gap junctions to assist in proper chemical synapse development. For many types of chemical synapses to develop properly, electrical activity (both spontaneous and triggered by sensory input) is critical. This early electrical activity is important not just for chemical synapse formation, but also for cell migration, neuronal differentiation, the determination of neurotransmitter type, and synaptic plasticity (Rakic and Komuro, 1995; Katz and Crowley, 2002; Spitzer et al., 2004).

As chemical synapses form in the developing cortex, activity-dependent and/or chemical-messenger-mediated plasticity are thought to be critical for strengthening some connections and weakening others, as appropriate synaptic partners are chosen and chemical synapses mature to form the circuits that will be important for cortical processes in the adult nervous system (Shatz, 1990). Circuits that develop within the cortex often organize into small clusters or columns, which will be activated by spatially and functionally related synaptic inputs during adulthood. Gap junction connections between all of the developing neurons in a cluster or column can mediate the temporal coordination of electrical activity and the exchange of chemical messengers for the entire neuron population (Connors et al., 1983; Dudek et al., 1986; Christie et al., 1989). Electrical coupling can ensure that chemical synapses onto this group of neurons will be encouraged to develop in the correct pattern, based on Hebbian plasticity rules (see Fig. 5.16).

In addition to the synchronous activity of the entire group of developing neurons, gap junctions also allow for the exchange of intracellular chemical messengers that are hypothesized to aid in chemical synapse formation. Interestingly, these developing cortical columns have been shown to exhibit large, synchronous increases in cytosolic calcium. The calcium waves that propagate through these neurons are mediated by the passage of the second messenger IP3 through gap junction connections between cells, and this IP3 triggers the release of calcium ions from intracellular stores within each neuron (Yuste et al., 1995; Kandler and Katz, 1995; see Chapter 12 for discussion of IP3 as a second messenger). It is believed that the passage of IP3 between these developing cortical neurons is

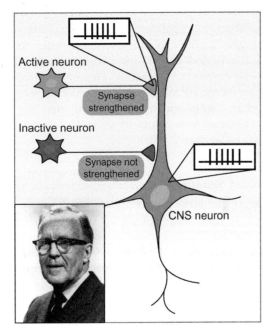

FIGURE 5.16 Diagram of Hebbian plasticity as it can be applied to a CNS neuron. The neuron (*orange*) receives two inputs: an active input (*green*) that excites the CNS neuron shown (depicted by the burst of action potentials; vertical black lines shown in the *white boxes*), and an inactive input (*red*). Hebbian plasticity rules would predict that the active input is strengthened. Inset: Photograph of Donald O. Hebb (University of British Columbia Archives, [UBC 5.1/3771]).

more important than the passage of electrical signals, because the calcium waves propagate through more neurons in a cluster than the electrical signals do. Furthermore, the biochemical changes induced by the calcium waves are thought to be important for chemical synapse formation and maturation. Collectively, these events are thought to contribute to the formation of appropriate functional networks of the mature cortex.

In the rodent cerebral cortex, electrical synapses form in early postnatal life and persist for the first 2 weeks after birth (see Fig. 5.17). During this time, chemical synapses are

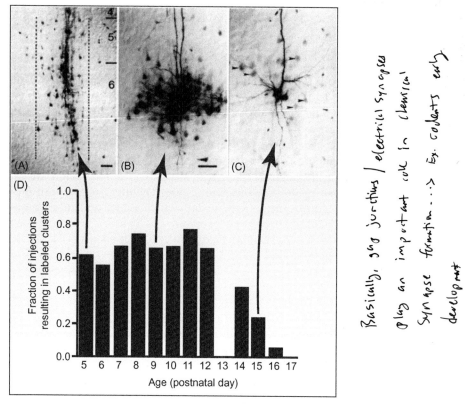

FIGURE 5.17 **The transient presence of electrical coupling between cortical neurons during development.** In this experiment, researchers injected neurobiotin molecules into a single neuron and measured its spread to neighboring neurons. Neurobiotin is a cellular label that is small enough to pass through many types of gap junctions. Panels A–C show the spread of neurobiotin from a single injected neuron in the cortex to neurons that are electrically coupled to it. Panel D plots the percentage of injected neurons from which dye spread to neighboring cells. A high percentage of neurons are electrically coupled with other neurons at young postnatal ages (5–12 days after birth), but this number decreases dramatically after postnatal day 13. Furthermore, the pattern of coupled neurons changes from a columnar appearance at very young postnatal ages (days 5–6) to a more clustered arrangement around the injected cell soma at older ages. *Source: Adapted from Peinado, A., Yuste, R., Katz, L.C., 1993. Extensive dye coupling between rat neocortical neurons during the period of circuit formation. Neuron 10, 103–114 (Peinado et al., 1993).*

forming, as evidenced by increases in dendrite and axon length, spine density on dendrites, and the presence of chemical synaptic markers (Miller, 1988; Killackey et al., 1995). The formation of chemical synapses coincides with increases in the sensitivity of neurons to chemical transmitters (due to the clustering of receptors for chemical transmitters), and increases in spontaneous and evoked chemical transmitter release (Armstrong-James and Fox, 1988; Schlaggar and O'Leary, 1994; Stern et al., 2001). Therefore, while chemical synapses are still immature, electrical synapses coordinate synchronous activity among neuron populations. After the second postnatal week, chemical synapses have matured sufficiently and electrical synapses are lost (see Fig. 5.17). In this context, electrical synapses play a critical but transient role in synapse formation.

## Roles of Electrical Synapses in the Adult Mammalian Nervous System

As stated earlier in the chapter, electrical synapses are abundant during nervous system development and, while they are not as prevalent in the adult nervous system, they do persist in some regions and contribute significantly to the functions of these areas. One approach to evaluating the importance of electrical synapses in the adult is to selectively remove them and observe how the nervous system functions without them. Once a specific connexin gene has been identified, it is possible to create a knockout mouse that lacks the corresponding connexin protein. Because it is expressed widely in the nervous system, the connexin gene Cx36 has been a popular target for knockout experiments. Mice lacking Cx36 have significant deficits in the visual system and inferior olive, and they perform poorly in behavioral tasks that depend on memory function (Bennett and Zukin, 2004; Connors and Long, 2004; Hormuzdi et al., 2004; LeBeau et al., 2003; Sohl et al., 2004, 2005; Frisch et al., 2005).

These experiments lead to the conclusion that electrical synapses are critical for a number of functions in the adult nervous system. The synchronization of neuronal activity has been proposed as a primary role for electrical synapses (Bennett and Zukin, 2004; Connors and Long, 2004; Connors, 2017). Evidence suggests that gap junctions between inhibitory neurons in the cerebral cortex and thalamus are likely responsible for synchronizing networks of neurons so that they fire action potentials nearly simultaneously, at particular frequencies, in a repeated pattern. These patterns of activity may be responsible for a variety of cognitive functions, including learning and memory.

Electrical synapses are also essential for the transmission and modulation of rod and cone signals in the retina. For example, rod photoreceptors in the retina activate a subtype of amacrine cells called "AII." Light activation of rods triggers a different subtype of amacrine cells (A18) to release dopamine, which modulates gap junction channels on AII cells and closes them. Depending on the level of light, the retina alters the electrical coupling between these AII amacrine cells to balance the needs for sensitivity and spatial resolution in our vision (see Fig. 5.18; Hormuzdi et al., 2004; Bloomfield and Volgyi, 2009)

The examples above (synchronous brain activity patterns in the neocortex and coupling between retinal amacrine cells) are just two of many known functions for gap junctions in the adult nervous system. Other brain areas that utilize gap junctions to mediate particular

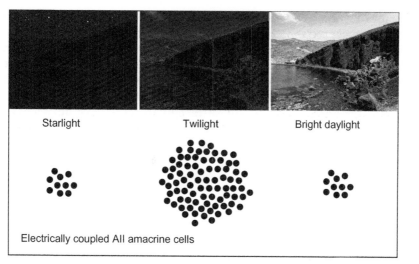

FIGURE 5.18  **Electrical coupling between amacrine cells changes according to the level of background illumination.** In the bottom panels, the number of dots indicates the numbers of electrically coupled amacrine cells for the light level indicated. In dark-adapted retinas, only a small number of AII amacrine cells are electrically coupled to one another (left panel). When the retina is adapted to see in conditions of low light, such as at twilight, retinas have the greatest degree of amacrine cell electrical coupling (center panel). In strong daylight, retinas adapt by reducing amacrine cell electrical coupling.

functions include the olfactory bulb, where gap junctions synchronize mitral cell activity (Friedman and Strowbridge, 2003); the suprachiasmatic nucleus, where gap junctions are required for normal circadian activity (Long et al., 2005; Tsuji et al., 2016); the hypothalamus, where electrical synapses allow synchronized burst firing that triggers pulsatile oxytocin release (Hatton et al., 1984); the inferior olive, where synchronous activation of Purkinje cells allows for the temporal precision of movement controlled by the cerebellum (Lang et al., 2014); the brainstem, where electrical synapses between neurons that control respiration modulate respiratory frequency (Bou-Flores and Berger, 2001); and the trigeminal nucleus, where glial coupling is regulated by sensory inputs that are important in generating chewing movements by synchronous activation of a locomotor central pattern generator (Condamine et al., 2017). The common theme of these examples is that gap junctions allow groups of cells to coordinate their function, leading to synchronous electrical activity and/or the exchange of small signaling molecules.

## ELECTRICAL SYNAPSE PLASTICITY

We usually think of gap junctions as a consistently open pathway for small-molecule or ion movement between cells. Indeed, one of the distinguishing features often cited when comparing electrical and chemical synapses is that electrical synapses are not as "plastic," or sensitive to modulation. While it is true that more mechanisms for modulation of chemical synapses have been described, it is clear that under certain conditions, gap junction channels at electrical synapses can be induced to close, which modulates the amount of flow through the gap junction pore.

In fact, gap junction channels in mammalian neurons can be modulated by a variety of neurotransmitters and biochemical factors (see for review: O'Brien, 2014; Haas et al., 2016; Miller and Pereda, 2017). One of the most potent modulators of gap junctions is intracellular calcium, which acts as an intracellular messenger in cells and can be released from intracellular storage organelles (e.g., endoplasmic reticulum) by the second messenger IP3 (see Chapter 12 for more details). For example, in the developing somatosensory cortex (7–10 days after birth in the rat), the neurotransmitter serotonin can close gap junction channels between neurons by acting on a metabotropic serotonin receptor (Rorig and Sutor, 1996). Serotonin receptors in this system activate a second messenger cascade in the neuron that generates IP3, which in turn triggers the release of calcium from the endoplasmic reticulum. The serotonin-triggered elevation in cytoplasmic calcium then closes gap junction channels, uncoupling these developing cortical neurons from one another.

The gap junctions in these same developing cortical neurons can also be closed by exposure to the monoamine neurotransmitters norepinephrine and dopamine (Rorig et al., 1995; Rorig and Sutor, 1996). The receptors for these monoamine transmitters do not couple to the generation of IP3, but instead increase cAMP production in cortical neurons. The second messenger cAMP activates protein kinase A (PKA), which can phosphorylate gap junction proteins and induce them to close.

Protein phosphorylation is in fact a common mechanism for regulating gap junctions, and can have opposing effects on gap junction coupling, depending on the kinase that is activated. The dynamic, light-activated control of AII retinal amacrine cell coupling is a good example of bidirectional control by different kinases and cell signaling pathways (Vaney, 1994; Kothmann et al., 2012; O'Brien, 2014). In the retina, light acting on rod and cone photoreceptors activates both horizontal cells and bipolar cells. Bipolar cells in turn synapse onto amacrine cells and ganglion cells (see Fig. 5.19), and amacrine cells also form synapses with one another. In this circuit, the amacrine cells are inhibitory interneurons that modulate the signal transmitted to ganglion cells (Masland, 2012). There are many subtypes of amacrine cells, and the AII amacrine cells have both D1 dopamine receptors and NMDA glutamate receptors. They receive glutamate from bipolar cells and dopamine from A18 amacrine cells (see Fig. 5.21). Interestingly, the electrical coupling between AII neurons is increased by NMDA receptor activation, but decreased by dopamine receptor activation (see Fig. 5.20), and may underlie the effects of light levels on coupling between retinal amacrine cells described above (Bloomfield and Volgyi, 2009). The balance between glutamate acting on NMDA receptors and dopamine acting on D1 receptors regulates the level of gap junction coupling among this amacrine cell population.

NMDA glutamate receptors (Kothmann et al., 2012) and D1 dopamine receptors (Hampson et al., 1992) mediate opposite effects on gap junction channels because each receptor activates a different subtype of kinase in the AII amacrine cell. NMDA receptors are ligand-gated ion channels that primarily conduct sodium and potassium, but also allow calcium ions to flow through the channel into the cell cytoplasm (see Chapter 11 for

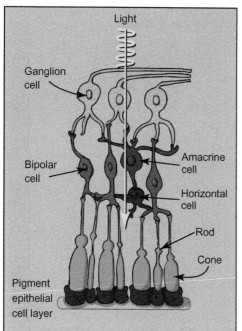

FIGURE 5.19 **The laminar organization of the retina.** Ganglion cells are on the surface of the retina and relay visual information to higher brain centers. Horizontal cells, bipolar cells, and amacrine cells relay information from photoreceptors to the ganglion cells. Light penetrates the other layers before reaching the rod and cone photoreceptor cells.

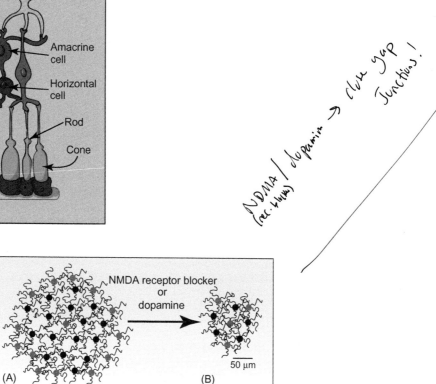

FIGURE 5.20 **Coupling between amacrine cells can be controlled by NMDA glutamate receptors and dopamine.** Diagram of top views of flattened retinas showing coupling among AII amacrine cells in the retina. In these diagrams, the red cell in the middle indicates the one that was injected with a low-molecular-weight tracer molecule. The surrounding cells colored black are strongly coupled, while the cells colored gray are coupled more weakly. (A) Under control conditions there is constant glutamate release that promotes extensive coupling between AII amacrine cells. (B) When an NMDA receptor blocker or dopamine is applied, the number of coupled cells decreases, reducing the number of cells into which the marker molecules can flow.

more details on ligand-gated ion channels). This calcium influx binds to calmodulin and activates a calcium-calmodulin-dependent protein kinase (CaMKII), which phosphorylates connexin 36 (Cx36) in AII amacrine cells. CAMKII phosphorylates different sites on Cx36 than other kinases do, and when it phosphorylates these sites, the connexon pore opens (see Fig. 5.21). In contrast, D1 dopamine receptors are metabotropic receptors that couple

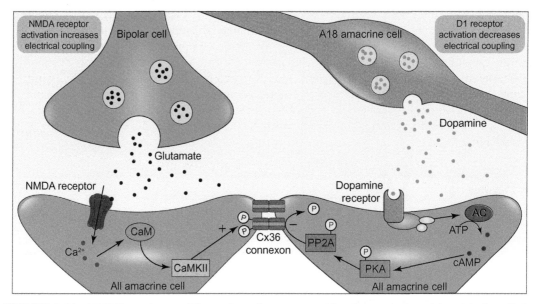

FIGURE 5.21  NMDA receptors on AII amacrine cells are activated by glutamate released from bipolar cells in response to light (left). This NMDA receptor activation leads to an influx of calcium, which activates CaMKII. CaMKII then phosphorylates Cx36 channels, which increases coupling between amacrine cells. When light intensities are higher (right side), A18 type amacrine cells release dopamine, which activates D1 dopamine receptors on AII amacrine cells. These D1 receptors activate adenylyl cyclase, generating cAMP, which increases PKA activity. PKA phosphorylates the phosphatase PP2A, which dephosphorylates Cx36 channels, leading to a decrease in coupling. This biphasic modulation of electrical coupling allows AII amacrine cells to modulate visual perception and acuity when the retina is exposed to different levels of light.

to the cAMP signaling pathway in the cytoplasm (see Chapter 12 for more details). When cAMP is increased within AII amacrine cells, it activates a cAMP-dependent protein kinase (PKA), which phosphorylates the phosphatase PP2A, which removes phosphate groups from Cx36. This dephosphorylation of Cx36 leads to closing of the connexon pore (see Fig. 5.21). By tapping into two different signaling cascades, glutamate and dopamine can have opposite effects on gap junction function in AII amacrine cells.

The above examples are only a subset of several cases of gap junction plasticity mediated by various neurotransmitter systems in a wide range of neuronal circuits (see O'Brien, 2014 for review of others). It is apparent that electrical synapse function can be modulated and controlled by cells, although in a more limited manner than is possible at chemical synapses.

In summary, electrical synapses have a number of interesting properties that distinguish them from chemical synapses. First, they do not use neurotransmitters, as chemical synapses do, but instead rely on the direct exchange of ions (and other small molecules) through gap junction channels to allow one neuron to influence one or more other neurons. Second, a change in the membrane potential of a neuron can cause a parallel, if smaller, change in the membrane potential of a neuron to which it is electrically coupled.

Finally, whereas the speed with which chemical synapses release and respond to neurotransmitters can vary somewhat, from approximately 0.2–2 ms, electrical synapses are always extremely fast, requiring about 0.1 ms for a signal in one neuron to reach a coupled neuron. One of the main roles electrical synapses play in areas of the nervous system as diverse as the retina and the neocortex is to synchronize the activity of many neurons within a network. However, chemical synapses have remarkable diversity and versatility not found in electrical synapses. In subsequent chapters we will explore the properties of chemical synapses and discuss the many ways they contribute to the function of the nervous system.

# References

Allen, F., Warner, A., 1991. Gap junctional communication during neuromuscular junction formation. Neuron 6, 101–111.
Armstrong-James, M., Fox, K., 1988. The physiology of developing cortical neurons. Cereb. Cortex 7, 237–272.
Baker, R., Llinas, R., 1971. Electrotonic coupling between neurones in the rat mesencephalic nucleus. J. Physiol. 212, 45–63.
Barsby, T., 2006. Drug discovery and sea hares: bigger is better. Trends Biotechnol. 24, 1–3.
Beblo, D.A., Veenstra, R.D., 1997. Monovalent cation permeation through the connexin40 gap junction channel. Cs, Rb, K, Na, Li, TEA, TMA, TBA, and effects of anions Br, Cl, F, acetate, aspartate, glutamate, and NO3. J. Gen. Physiol. 109 (4), 509–522.
Bennett, M.V., Zukin, R.S., 2004. Electrical coupling and neuronal synchronization in the Mammalian brain. Neuron 41, 495–511.
Bennett, M.V.L., Aljure, E., Nakajima, Y., Pappas, G.D., 1963. Electrotonic junctions between teleost spinal neurons: electrophysiology and ultrastructure. Science 141, 262–264.
Bloomfield, S.A., Volgyi, B., 2009. The diverse functional roles and regulation of neuronal gap junctions in the retina. Nat. Rev. Neurosci. 10, 495–506.
Bou-Flores, C., Berger, A.J., 2001. Gap junctions and inhibitory synapses modulate inspiratory motoneuron synchronization. J. Neurophysiol. 85, 1543–1551.
Carew, T.J., Kandel, E.R., 1977. Inking in Aplysia californica. I. Neural circuit of an all-or-none behavioral response. J. Neurophysiol. 40, 692–707.
Chang, Q., Gonzalez, M., Pinter, M.J., Balice-Gordon, R.J., 1999. Gap junctional coupling and patterns of connexin expression among neonatal rat lumbar spinal motor neurons. J. Neurosci. 19, 10813–10828.
Christie, M.J., Williams, J.T., North, R.A., 1989. Electrical coupling synchronizes subthreshold activity in locus coeruleus neurons in vitro from neonatal rats. J. Neurosci. 9, 3584–3589.
Condamine, S., Lavoie, R., Verdier, D., Kolta, A., 2017. Functional rhythmogenic domains defined by astrocytic networks in the trigeminal main sensory nucleus. Glia 66, 311–326.
Connors, B.W., 2017. Synchrony and so much more: diverse roles for electrical synapses in neural circuits. Dev. Neurobiol. 77, 610–624.
Connors, B.W., Long, M.A., 2004. Electrical synapses in the mammalian brain. Annu. Rev. Neurosci. 27, 393–418.
Connors, B.W., Benardo, L.S., Prince, D.A., 1983. Coupling between neurons of the developing rat neocortex. J. Neurosci. 3, 773–782.
Curti, S., Hoge, G., Nagy, J.I., Pereda, A.E., 2012. Synergy between electrical coupling and membrane properties promotes strong synchronization of neurons of the mesencephalic trigeminal nucleus. J. Neurosci. 32, 4341–4359.
Dermietzel, R., Kremer, M., Paputsoglu, G., Stang, A., Skerrett, I.M., Gomes, D., et al., 2000. Molecular and functional diversity of neural connexins in the retina. J. Neurosci. 20, 8331–8343.
Dudek, F.E., Snow, R.W., Taylor, C.P., 1986. Role of electrical interactions in synchronization of epileptiform bursts. Adv. Neurol. 44, 593–617.
Fischbach, G.D., 1972. Synapse formation between dissociated nerve and muscle cells in low density cell cultures. Dev. Biol. 28, 407–429.

Friedman, D., Strowbridge, B.W., 2003. Both electrical and chemical synapses mediate fast network oscillations in the olfactory bulb. J. Neurophysiol. 89, 2601–2610.

Frisch, C., De Souza-Silva, M.A., Sohl, G., Guldenagel, M., Willecke, K., Huston, J.P., et al., 2005. Stimulus complexity dependent memory impairment and changes in motor performance after deletion of the neuronal gap junction protein connexin36 in mice. Behav. Brain. Res. 157, 177–185.

Fulton, B.P., Miledi, R., Takahashi, T., 1980. Electrical synapses between motoneurons in the spinal cord of the newborn rat. Proc. R. Soc. Lond. B. Biol. Sci. 208, 115–120.

Furshpan, E.J., Potter, D.D., 1959. Transmission at the giant motor synapses of the crayfish. J. Physiol. 145, 289–325.

Gong, X.Q., Nicholson, B.J., 2001. Size selectivity between gap junction channels composed of different connexins. Cell Commun. Adhes. 8 (4-6), 187–192.

Goodenough, D.A., Paul, D.L., 2009. Gap junctions. Cold Spring Harb. Perspect. Biol. 1 (1), a002576.

Haas, J.S., Greenwald, C.M., Pereda, A.E., 2016. Activity-dependent plasticity of electrical synapses: increasing evidence for its presence and functional roles in the mammalian brain. BMC. Cell. Biol. 17 (Suppl 1), 14.

Hall, Z.W., Sanes, J.R., 1993. Synaptic structure and development: the neuromuscular junction. Cell 72, 99–121. Suppl.

Hampson, E.C., Vaney, D.I., Weiler, R., 1992. Dopaminergic modulation of gap junction permeability between amacrine cells in mammalian retina. J. Neurosci. 12, 4911–4922.

Hatton, G.I., Perlmutter, L.S., Salm, A.K., Tweedle, C.D., 1984. Dynamic neuronal-glial interactions in hypothalamus and pituitary: implications for control of hormone synthesis and release. Peptides 5 (Suppl 1), 121–138.

Hormuzdi, S.G., Filippov, M.A., Mitropoulou, G., Monyer, H., Bruzzone, R., 2004. Electrical synapses: a dynamic signaling system that shapes the activity of neuronal networks. Biochim. Biophys. Acta 1662, 113–137.

Kandler, K., Katz, L.C., 1995. Neuronal coupling and uncoupling in the developing nervous system. Curr. Opin. Neurobiol. 5, 98–105.

Katz, L.C., Crowley, J.C., 2002. Development of cortical circuits: lessons from ocular dominance columns. Nat. Rev. Neurosci. 3, 34–42.

Killackey, H.P., Rhoades, R.W., Bennett-Clarke, C.A., 1995. The formation of a cortical somatotopic map. Trends Neurosci. 18, 402–407.

Kothmann, W.W., Trexler, E.B., Whitaker, C.M., Li, W., Massey, S.C., O'Brien, J., 2012. Nonsynaptic NMDA receptors mediate activity-dependent plasticity of gap junctional coupling in the AII amacrine cell network. J. Neurosci. 32, 6747–6759.

Kriebel, M.E., Bennett, M.V., Waxman, S.G., Pappas, G.D., 1969. Oculomotor neurons in fish: electrotonic coupling and multiple sites of impulse initiation. Science 166, 520–524.

Lang, E.J., Tang, T., Suh, C.Y., Xiao, J., Kotsurovskyy, Y., Blenkinsop, T.A., et al., 2014. Modulation of Purkinje cell complex spike waveform by synchrony levels in the olivocerebellar system. Front. Syst. Neurosci. 8, 210.

Lawrence, T.S., Beers, W.H., Gilula, N.B., 1978. Transmission of hormonal stimulation by cell-to-cell communication. Nature 272 (5653), 501–506.

LeBeau, F.E., Traub, R.D., Monyer, H., Whittington, M.A., Buhl, E.H., 2003. The role of electrical signaling via gap junctions in the generation of fast network oscillations. Brain Res. Bull. 62, 3–13.

Long, M.A., Jutras, M.J., Connors, B.W., Burwell, R.D., 2005. Electrical synapses coordinate activity in the suprachiasmatic nucleus. Nat. Neurosci. 8, 61–66.

Marder, E., 2009. Electrical synapses: rectification demystified. Curr. Biol. 19, R34–R35.

Martin, A.R., Pilar, G., 1963. Transmission through the Ciliary Ganglion of the Chick. J. Physiol. 168, 464–475.

Masland, R.H., 2012. The tasks of amacrine cells. Vis. Neurosci. 29, 3–9.

Miller, M., 1988. Development of projection and local circuit neurons in neocortex. Cereb. Cortex 7, 133–175.

Miller, A.C., Pereda, A.E., 2017. The electrical synapse: molecular complexities at the gap and beyond. Dev. Neurobiol. 77, 562–574.

Milner, P.M., 1993. The mind and Donald O. Hebb. Sci. Am. 268, 124–129.

Nagy, J.I., Dermietzel, R., 2000. Gap junctions and connexins in the mammalian central nervous system. Adv. Mol. Cell Biol. 30, 323–396.

Nagy, J.I., Dudek, F.E., Rash, J.E., 2004. Update on connexins and gap junctions in neurons and glia in the mammalian nervous system. Brain Res. Brain Res. Rev. 47, 191–215.

Nagy, J.I., Pereda, A.E., Rash, J.E., 2018. Electrical synapses in mammalian CNS: past eras, present focus and future directions. Biochim Biophys Acta 1860, 102–123.

O'Brien, J., 2014. The ever-changing electrical synapse. Curr. Opin. Neurobiol. 29, 64–72.

O'Brien, J.J., Li, W., Pan, F., Keung, J., O'Brien, J., Massey, S.C., 2006. Coupling between A-type horizontal cells is mediated by connexin 50 gap junctions in the rabbit retina. J. Neurosci. 26, 11624–11636.

Peinado, A., Yuste, R., Katz, L.C., 1993. Extensive dye coupling between rat neocortical neurons during the period of circuit formation. Neuron 10, 103–114.

Personius, K.E., Balice-Gordon, R.J., 2001. Loss of correlated motor neuron activity during synaptic competition at developing neuromuscular synapses. Neuron 31, 395–408.

Rakic, P., Komuro, H., 1995. The role of receptor/channel activity in neuronal cell migration. J. Neurobiol. 26, 299–315.

Robertson, J.D., 1963. The occurrence of a subunit pattern in the unit membranes of club endings in Mauthner cell synapses in goldfish brains. J. Cell Biol. 19, 201–221.

Rorig, B., Sutor, B., 1996. Regulation of gap junction coupling in the developing neocortex. Mol. Neurobiol. 12, 225–249.

Rorig, B., Klausa, G., Sutor, B., 1995. Dye coupling between pyramidal neurons in developing rat prefrontal and frontal cortex is reduced by protein kinase A activation and dopamine. J. Neurosci. 15, 7386–7400.

Rozental, R., Giaume, C., Spray, D.C., 2000. Gap junctions in the nervous system. Brain Res. Brain Res. Rev. 32, 11–15.

Sabatini, B.L., Regehr, W.G., 1999. Timing of synaptic transmission. Annu. Rev. Physiol. 61, 521–542.

Sáez, J.C., Connor, J.A., Spray, D.C., Bennett, M.V., 1989. Hepatocyte gap junctions are permeable to the second messenger, inositol 1,4,5-trisphosphate, and to calcium ions. Proc. Natl. Acad. Sci. USA 86 (8), 2708–2712.

Sanderson, M.J., 1995. Intercellular calcium waves mediated by inositol trisphosphate. Ciba Found. Symp. 188, 175–189.

Schlaggar, B.L., O'Leary, D.D., 1994. Early development of the somatotopic map and barrel patterning in rat somatosensory cortex. J. Comp. Neurol. 346, 80–96.

Shatz, C.J., 1990. Impulse activity and the patterning of connections during CNS development. Neuron 5, 745–756.

Sohl, G., Odermatt, B., Maxeiner, S., Degen, J., Willecke, K., 2004. New insights into the expression and function of neural connexins with transgenic mouse mutants. Brain Res. Brain Res. Rev. 47, 245–259.

Sohl, G., Maxeiner, S., Willecke, K., 2005. Expression and functions of neuronal gap junctions. Nat. Rev. Neurosci. 6, 191–200.

Spitzer, N.C., Root, C.M., Borodinsky, L.N., 2004. Orchestrating neuronal differentiation: patterns of Ca2C spikes specify transmitter choice. Trends Neurosci. 27, 415–421.

Stern, E.A., Maravall, M., Svoboda, K., 2001. Rapid development and plasticity of layer 2/3 maps in rat barrel cortex in vivo. Neuron 31, 305–315.

Thompson, W., Kuffler, D.P., Jansen, J.K., 1979. The effect of prolonged, reversible block of nerve impulses on the elimination of polyneuronal innervation of new-born rat skeletal muscle fibers. Neuroscience 4, 271–281.

Tsuji, T., Tsuji, C., Ludwig, M., Leng, G., 2016. The rat suprachiasmatic nucleus: the master clock ticks at 30 Hz. J. Physiol. 594, 3629–3650.

Valenstein, E.S., 2005. The War of the Soups and the Sparks: The Discovery of Neurotransmitters and the Dispute Over How Nerves Communicate. Columbia University Press, New York.

Vaney, D.I., 1994. Patterns of neuronal coupling in the retina. Prog. Retinal Eye Res. 13 (1), 301–355.

Veenstra, R.D., 2001. Determining ionic permeabilities of gap junction channels. In: R. Bruzzone, C. Giaume (Eds.), Connexin Methods and Protocols. Methods in Molecular Biology, vol. 154. Humana Press, pp. 293–311.

Walsh, M.K., Lichtman, J.W., 2003. In vivo time-lapse imaging of synaptic takeover associated with naturally occurring synapse elimination. Neuron 37, 67–73.

Walton, K.D., Navarrete, R., 1991. Postnatal changes in motoneurone electrotonic coupling studied in the in vitro rat lumbar spinal cord. J. Physiol. 433, 283–305.

Wang, H.Z., Veenstra, R.D., 1997. Monovalent ion selectivity sequences of the rat connexin43 gap junction channel. J. Gen. Physiol. 109 (4), 491–507.

Watanabe, A., 1958. The interaction of electrical activity among neurons of lobster cardiac ganglion. Jpn. J. Physiol. 8 (4), 305–318.

Yuste, R., Nelson, D.A., Rubin, W.W., Katz, L.C., 1995. Neuronal domains in developing neocortex: mechanisms of coactivation. Neuron 14, 7–17.

# PART II

# REGULATION OF CHEMICAL TRANSMITTER RELEASE

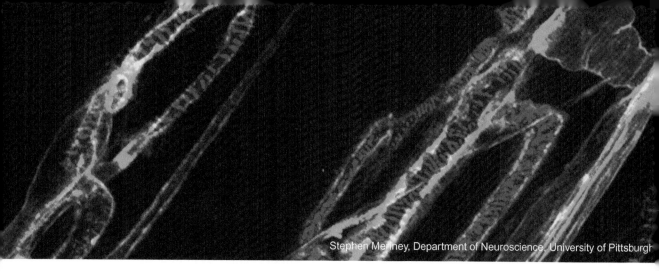

# CHAPTER 6

# Function of Chemical Synapses and the Quantal Theory of Transmitter Release

The nervous system uses chemicals to communicate between cells at most synapses. When an action potential (an electrical signal) reaches the nerve ending, it must be converted into the release of chemicals that can cross the synaptic cleft. This process is termed **"excitation-secretion coupling."** The action potential-induced depolarization of the membrane triggers specific biochemical events that liberate chemicals from the nerve ending (the details of these biochemical events will be discussed in detail in Chapter 8). The released chemicals can then diffuse across the synapse and bind to selective receptors on a postsynaptic cell. Finally, these receptors must transduce the binding of the chemical ligands into an electrical and/or chemical signal within the postsynaptic cell. The effect of this complex process is to convert an electrical signal in the presynaptic cell (the action potential) into a chemical signal (the release of chemicals from the nerve ending), which then crosses the synaptic cleft and initiates a signal in the postsynaptic cell.

## COSTS AND ADVANTAGES OF CHEMICAL COMMUNICATION

Chemical communication involves a complex molecular process that is slower than direct electrical communication. Nevertheless, these molecular processes are highly conserved across many species and are ubiquitous throughout the nervous system. What are the advantages of using chemical transmitter release for communication between cells, and what are the neurobiological costs (see Table 6.1)?

In contrast to electrical synapses (discussed in Chapter 5), chemical synapses rely on the conversion of a presynaptic action potential into a biochemical trigger for the release of chemical neurotransmitters. A large number of intracellular proteins must work together to (1) prepare chemicals for release; (2) provide a voltage-dependent trigger for the release of chemical neurotransmitters so their release can be controlled by action potentials; and (3) move these chemicals across the presynaptic membrane into the synaptic cleft.

The complex molecular steps required for the release of neurotransmitters make chemical transmission a relatively inefficient mechanism of cellular communication. As a result, the probability of neurotransmitter release following a presynaptic action potential can be low at many synapses, especially in the central nervous system. The process of transmitter release is also susceptible to disruption: biochemical changes within the cell, genetic mutations or changes in gene expression of critical proteins, and/or the pharmacological action of drugs can interfere with any of the individual steps required for release, disrupting the entire process. Even when successful, chemical transmission is slower than electrical transmission, often requiring up to 2 ms to complete, compared with ~0.1 ms for electrical synapses. Finally, the release of chemical transmitters requires energy in the form of ATP. In light of these drawbacks, chemical transmission may seem too inefficient, unpredictable, and biologically "expensive" to be a dependable method of communication in a nervous system that must control such complex and important functions as coordinated locomotion, sensing the outside world, controlling critical organs (e.g., heart, lungs), and underlying cognitive processes such as learning, memory, and thinking through complex problems.

TABLE 6.1  Costs and Advantages of Chemical Transmitter Release

| **Costs of Chemical Communication** |
| --- |
| Susceptibility to disruption and disease |
| Relatively low speed of communication |
| Energy use required |
| Low efficiency |
| **Advantages of Chemical Communication** |
| Synaptic plasticity |
| Capable of excitation or inhibition |
| Communication can be amplified |
| Diversity in types of signals communicated |

However, the advantages of chemical communication far outweigh the costs. This is in part because many important functions of the nervous system, especially learning and memory, require cells to change how they communicate in response to past experiences. This capability is called **synaptic plasticity**. Chemical synapses are critical to the function of the nervous system in part because they are capable of a wide array of plasticity mechanisms (see Chapter 14). Another advantage to chemical communication is that a small, brief chemical signal can be amplified in size and duration so its effects can alter the activity of many neurons and, in some cases, persist for a long time compared to the milliseconds-long duration of signals through electrical synapses. Additionally, whereas communication via electrical synapses is limited to small depolarizing or hyperpolarizing effects or to molecules small enough to pass through the gap junctions, chemical transmission can cause a remarkable diversity of signals. This includes excitation or inhibition (depolarizing or hyperpolarizing membrane potential, respectively), but there are also many transmitters that modulate excitatory or inhibitory communication, either enhancing or reducing it, and, due to the diversity of neurotransmitter receptors, such modulation can occur via many physiological mechanisms.

## ELECTRICAL FOOTPRINTS OF CHEMICAL TRANSMITTER RELEASE

To study and understand chemical transmitter release at synapses, investigators use techniques that allow them to "listen in" on such communication. In the 1950s and 1960s, several investigators began using the **microelectrode technique** (see Box 6.1, Fig. 6.1) to record the postsynaptic "footprints" of chemical transmitter release (Fatt and Katz, 1952; Katz and Miledi, 1965). The groundbreaking data that Sir Bernard Katz and colleagues collected using this approach significantly advanced our understanding of synaptic transmission. Subsequent work by Ulf von Euler and Julius Axelrod showed that chemical transmitters are stored and released from nerve terminals, and that they are broken down soon after release, which terminates their action. Katz, Euler and Axelrod were jointly awarded the 1970 Nobel Prize in Physiology or Medicine "for their discoveries concerning the humoral transmitters in the nerve terminals and the mechanism for their storage, release and inactivation" (nobelprize.org).

Sir Bernard Katz and colleagues used the microelectrode technique to evaluate the mechanisms responsible for the release of chemicals, especially the role of calcium ions. During such recordings, they observed a small depolarization of membrane potential on the rising edge of action potentials recorded from the muscle cell at the neuromuscular junction (see Fig. 6.2). This depolarization occurred about 1 ms after the presynaptic axon was stimulated with an electrical pulse, and was only recorded when the microelectrode tip was placed in the muscle cell very close to where the nerve terminal made contact with the muscle cell. The nerve terminal at motor synapses has a plate-like structure, and is referred to as the "endplate." Since the small depolarizing membrane potentials they observed were only recorded near the nerve terminal, the researchers termed these small events **endplate potentials** (EPPs).

More recently, researchers showed that an EPP spreads along the muscle cell from the site of nerve contact. They were able to do this by using specific components of the venom

## BOX 6.1

## THE MICROELECTRODE TECHNIQUE AND THE RECORDING OF SYNAPTIC EVENTS

The microelectrode technique is a method that can be used to measure the charge difference between the inside and outside of a cell (i.e., the cell's membrane voltage). This technique uses two **electrodes** (electrical conductors), one placed in the solution bathing the outside of the cell, and another placed inside the cell (see Fig. 6.1). In order to be able to place an electrode inside a cell, investigators use a glass microelectrode with a very sharp tip (diameter of 0.1–0.3 μm) that is filled with conducting saline solution (often 3 M potassium chloride) and contains a fine wire. Microelectrodes are made from glass capillary tubes (diameter ~1 mm) that are heated in the middle. When the glass is hot enough to soften, the two ends of the capillary tube are pulled apart rapidly, dividing the tube into two sharp-tipped halves, each of which can be used as a microelectrode. The tip is so small that it can be gently inserted into a cell without causing significant damage. An electrical amplifier compares the intracellular and extracellular charges recorded by the electrodes, and the difference between these charges is the membrane potential of the cell.

The microelectrode technique was used to measure how the release of the chemical transmitter acetylcholine onto muscle cells affected the membrane potential of the muscle cells. Since postsynaptic receptors for acetylcholine are not saturated at the neuromuscular junction (meaning that not all of them are activated when neurotransmitter is released in response to an action potential), changes in muscle cell membrane potential could be used to calculate the amount of acetylcholine released from the motor neuron.

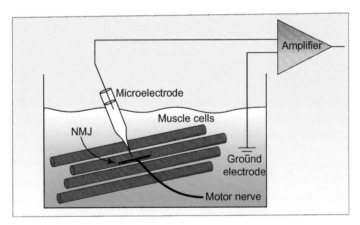

FIGURE 6.1  **Diagram of a microelectrode recording an electrical signal from a postsynaptic muscle cell.** A microelectrode is inserted into a muscle fiber (*red*) in close proximity to the motor nerve ending that forms a neuromuscular junction synapse (NMJ). When the microelectrode is inside the muscle cell, measurements can be made to compare the charge inside the muscle cell with the charge in the solution surrounding the muscle cells, which tells the experimenter the membrane potential of the muscle cell. Small changes in the membrane potential that occur in response to synaptic transmission are then amplified and recorded.

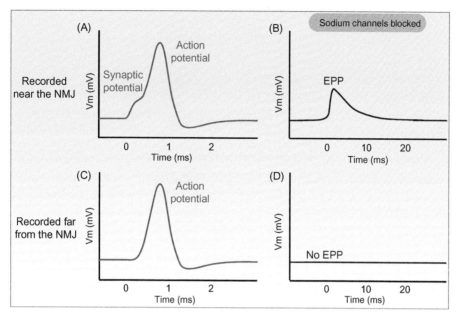

FIGURE 6.2 **Recordings of action potentials and endplate potentials (EPPs) at locations that are either near (A and B) or far from (C and D) the NMJ on a muscle cell.** Under physiological conditions (*red traces*), the action potential has a synaptic potential on the rising edge when recorded near the NMJ (A), but not when recorded far from the NMJ (C). If sodium channels are selectively blocked in muscle cells (using μ-conotoxin; see Box 6.2), the action potential is blocked in the postsynaptic recordings from these cells (B and D). Under these conditions, an endplate potential (EPP) is recorded near the NMJ (B), but decays with distance such that there is no noticeable EPP when recorded far from the NMJ (D). Note the difference in the time scales between the action potential (*red*) and endplate potential traces (*blue*).

of South Pacific cone snails (conotoxins; see Box 6.2). A family of these cone snail toxins, the μ-conotoxins, block sodium channels (Yanagawa et al., 1987), and one of these selectively blocks the subtype of sodium channel that generates action potentials in muscles, but not the subtype that is responsible for action potentials in nerves. Using this muscle-specific μ-conotoxin, experimenters selectively blocked action potentials in muscle cells without blocking presynaptic action potentials that trigger acetylcholine release from motor neurons. This was necessary for these experiments because EPPs at the neuromuscular junction typically depolarize the membrane enough to trigger an action potential in the muscle cell, which masks the amplitude of the EPP as well as its rise and decay times. Eliminating this muscle cell action potential revealed the EPP (see Fig. 6.2B), which at the frog neuromuscular junction ranges from about 30–50 mV in amplitude. Using μ-conotoxin as a tool, researchers measured EPPs from a postsynaptic muscle cell at different distances from the endplate and measured how EPP magnitude changed with distance from the endplate. As the recording electrode was placed further and further from the endplate, the EPPs became smaller and had slower rising and falling phases (see Fig. 6.4). These changes reflected the passive spread of a local EPP from the site of nerve contact along the length of a muscle cell (see Chapter 13, for more discussion of passive properties of cell membranes).

## BOX 6.2

## NATURAL MOLECULES AS EXPERIMENTAL TOOLS

Advances in neuroscience research are often limited by the availability of experimental tools. An unexpected source of such tools is venoms and toxins evolved by organisms as mechanisms of defense and predation. Components of these venoms and toxins are very effective at paralyzing prey by selectively targeting subtypes of ion channels and/or neurotransmitter receptors. Below we discuss some of the most common toxins used in neuroscience research and the plants and animals from which they are derived (see Fig. 6.3). Additionally, we describe how natural fluorescent molecules produced by a jellyfish were harnessed to be used as tools to label proteins.

**Conotoxins.** Fish-hunting cone snails inhabit the tidal pools of the South Pacific and hunt fish using venom-filled harpoons which they shoot into prey from their "proboscis" (an elongated appendage). The venom contains hundreds of specific peptide toxins, many of which are selective for ion channels or receptors. These include a variety of μ-conotoxins that selectively inhibit voltage-gated sodium channel subtypes, ω-conotoxins that inhibit voltage-gated calcium channel subtypes, and α-contoxins that inhibit specific subtypes of nicotinic receptors. In addition to their use as experimental tools (Terlau and Olivera, 2004), one of the μ-conotoxins has been FDA-approved for use in treating chronic pain (Prialt; Safavi-Hemami et al., 2019; see Box 7.2).

**Agatoxin.** The funnel web spider (*Agelenopsis aperta*) produces venom that contains several proteins known as agatoxins (Adams, 2004; Pringos et al., 2011). The ω-agatoxin IVA protein is a selective antagonist at the Cav2.1 (P/Q-type) voltage-gated calcium channel (the type that is commonly used at synapses to bring calcium across the plasma membrane and trigger chemical neurotransmitter release).

**Scorpion toxins.** There are two toxins used to study the large-conductance calcium-activated potassium channel (Bergeron and Bingham, 2012). Iberiotoxin (IbTX), which blocks these channels, is a component of the venom of the Eastern Indian red scorpion (*Buthus tamulus*). The "deathstalker" scorpion (*Leiurus quinquestriatus hebraeus*) produces a toxin component called charybdotoxin that also selectively blocks large-conductance calcium-activated potassium channels.

**Tetrodotoxin.** The pufferfish has an endosymbiotic (a form of symbiosis in which one organism lives inside the other) bacterium in its body that produces tetrodotoxin, a toxin that is selective for voltage-gated sodium channels. In fact, this bacterium can be found in a variety of marine and terrestrial species in which tetrodotoxin is also found, including the blue-ringed octopus (Bane et al., 2014; Magarlamov et al., 2017). Despite the presence of this dangerous toxin, pufferfish ("fugu") is a delicacy in Japan. Specially trained chefs purposely leave some tetrodotoxin in the fish they prepare so the consumer can experience the desired side-effects of the meal: a tingling sensation in their mouth and a sense of euphoria.

**Bungarotoxin.** The *Bungarus* genus of snakes, also called Kraits, produce a venom that contains several types of bungarotoxin. One of these toxins, α-bungarotoxin, binds selectively to some types of nicotinic acetylcholine receptors and inhibits their activation by acetylcholine (Chang, 1999; Utkin, 2013). This toxin is valuable not only as an experimental tool to inhibit acetylcholine receptors, but can also be labeled and used to visualize the location of these receptors.

BOX 6.2 (cont'd)

FIGURE 6.3 Toxins used in neuroscience research are found in a range of plants and animals. Clockwise from top left: *Pionoconus magus* (H. Zell—Own work, CC BY-SA 3.0) from which μ-conotoxins (blockers of voltage gated sodium channels) are derived, *Buthus tamulus* (Shantanu Kuveskar CC BY-SA 4.0) from which iberiotoxin (a calcium-activated potassium channel blocker) is derived, *Chondrodendron tomentosum* (Krzysztof Ziarnek, Kenraiz CC BY-SA 4.0) from which curare (an acetylcholine receptor blocker) is derived, *Bungar candi* (Wibowo Djatmiko (Wie146) CC BY-SA 3.0) from which bungarotoxin (an acetylcholine receptor blocker) is derived, black widow spider (Shenrich91 CC BY-SA 3.0) from which latrotoxin (which causes synaptic vesicle fusion at synapses) is derived, European bee-wolf wasp (Alvesgaspar—Own work, CC BY-SA 3.0) from which philanthotoxin (a glutamate receptor blocker) is derived; center: *Agelenopsis aperta* (Mfield, Matthew Field, http://www.photography.mattfield.com—Own work, GFDL 1.2) from which Agatoxin (a calcium channel blocker) is derived.

**Curare**. A South American vine plant (*Chondrodendron tomentosum*) produces an alkaloid that is known as tubocurarine, or curare. Local Indian tribes prepare poisons for the tips of their arrows from this plant. Curare is now known to be a selective inhibitor of the nicotinic acetylcholine receptor (Bowman, 2006)

## BOX 6.2 (cont'd)

**Philanthotoxin.** A component of the venom of the European bee-wolf wasp (a wasp that hunts bees), philanthotoxin, is an antagonist at several ionotropic receptor channels (including nicotinic acetylcholine receptors and some types of ionotropic glutamate receptors). Philanthotoxins are polyamines which immediately but reversibly paralyze their prey. δ-Philanthotoxin ("PhTX-433"), is the venom component most commonly used in laboratories, and it is usually used in experiments as a blocker of some types of ionotropic glutamate receptors (Piek, 1982).

**Latrotoxin.** The venom of the black widow spider contains a toxin called α-latrotoxin. This is a large protein (120 kDa) that triggers synaptic vesicle exocytosis from presynaptic nerve terminals in the absence of action potential activity (Sudhof, 2001). To accomplish this, α-latrotoxin binds to two distinct families of neuronal cell-surface receptors, neurexins and latrophilins, that are expressed on nerve terminals. After binding to these proteins, α-latrotoxin is thought to insert into the presynaptic plasma membrane to stimulate exocytosis (the detailed mechanisms that cause this are still under study).

**Botulinum and tetanus toxins.** Anaerobic rod-shaped, spore-forming bacteria of the genus *Clostridium* synthesize and release protein toxins called clostridial neurotoxins. These include seven distinct toxins from *Clostridium botulinum* and one toxin from *Clostridium tetani* that are considered some of the most dangerous naturally occurring toxins (Binz et al., 2010). Because these toxins selectively target SNARE proteins in the nerve terminal, they are useful tools to study exocytosis (see Chapter 8). In addition, one of the botulinum toxins (Botulinum toxin A; BOTOX) is used cosmetically to smooth facial wrinkles (by blocking transmitter release, relaxing facial muscles).

**Green fluorescent protein.** Some organisms have evolved fluorescent molecules (perhaps used to attract prey) that, like the toxins listed above, have been harnessed for research purposes. The jellyfish *Aequorea victoria* makes a protein called aequorin, which releases blue light after binding calcium. This blue light is then absorbed by another protein these organisms make, green fluorescent protein (GFP), which in turn gives off a green light (Zimmer, 2009). Three scientists received the 2008 Nobel Prize in Chemistry for isolating GFP from jellyfish (Osamu Shimomura), expressing it in other organisms (Martin Chalfie), and mutating it to create fluorescent proteins in a variety of colors that can be expressed in cells as experimental markers (Roger Tsien).

## SPONTANEOUS RELEASE OF SINGLE NEUROTRANSMITTER VESICLES

When Fatt and Katz made microelectrode recordings very close to the endplate in the absence of a presynaptic action potential (that is, without stimulating the axon synapsing onto the muscle cell), they observed very small, spontaneous depolarizations in the

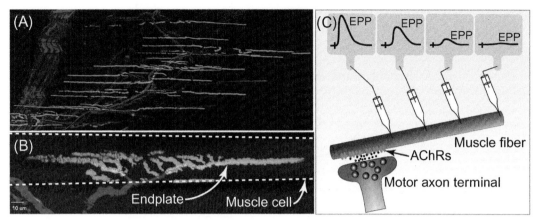

FIGURE 6.4 **The frog neuromuscular "endplate" and the recording of synaptic potentials from associated muscle cells.** (A) Low-magnification view of a frog skeletal muscle (in this case, the cutaneous pectoris) stained with two fluorescent dyes. The red fluorescent dye identifies the protein tubulin, a cytoskeletal protein in motor axons. The green fluorescent dye identifies postsynaptic acetylcholine receptors (using Alexa-488-labeled α-bungarotoxin, a component of snake venom that selectively binds to these proteins; see Box 6.2). Postsynaptic acetylcholine receptors (AChRs) are densely localized only at synapses, so their labeling identifies neuromuscular junctions, which are termed "endplates" because they take the form of long, green-stained lines, or "plates" on the muscle. (B) A high-magnification view of a single neuromuscular junction, showing a branch of the motor nerve (*red*) innervating a single muscle cell (outlined using a dotted line). The red-stained axon terminates in a complex, branched, green-stained "endplate" structure that represents one neuromuscular synapse. (C) Simplified diagram of the postsynaptic endplate potentials (EPPs) that can be recorded after action potential stimulation from the frog muscle at different distances from the neuromuscular junction. When microelectrodes are inserted very close to the endplate, the resulting EPP is large and has a fast rise time from the baseline membrane potential to the peak of the EPP. With distance from the neuromuscular junction, the recorded EPP becomes smaller and its time course becomes slower. *Source: Images in (A) and (B) were obtained by Blake Vuocolo from the laboratory of Stephen D. Meriney, Department of Neuroscience, University of Pittsburgh (unpublished).*

membrane potential of the frog muscle cell. These depolarizations had roughly the same duration as an EPP, but were much smaller in amplitude, averaging only 0.5–1 mV, compared to 30–50 mV for an EPP (see Fig. 6.5). Furthermore, these small depolarizations could not be recorded if the microelectrode was inserted in the muscle cell outside of the endplate region. Fatt and Katz termed these spontaneous depolarizations **miniature endplate potentials** (mEPPs).

In order to determine what caused EPPs and mEPPs, researchers performed several experiments. First, applying a selective acetylcholine receptor blocker (e.g., curare, a naturally occurring toxin from a South American vine; see Box 6.2) to the neuromuscular junction blocked both EPPs and mEPPs. Second, blocking the extracellular enzyme acetylcholinesterase (AChE), which normally rapidly breaks down acetylcholine released from the presynaptic neuron, lengthened the duration of EPPs and mEPPs and slightly increased their amplitude. Finally, when experimenters cut the motor nerve and waited 2 weeks for the axon and nerve terminal to degenerate, both EPPs and mEPPs disappeared from the muscle cells. The conclusions from these experiments were that (1) a large amount of the chemical transmitter acetylcholine is released from the nerve

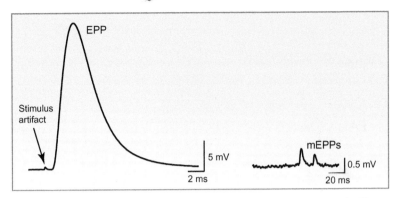

FIGURE 6.5  **Synaptic potentials recorded from a mouse neuromuscular junction.** (Left) An endplate potential recorded following suprathreshold stimulation (i.e., strong enough to generate an action potential) of the motor nerve. The EPP shown is the average of 20 individual EPPs recorded when the nerve was stimulated at low frequency (0.2 Hz). (Right) In the same muscle cell, spontaneous mEPPs, such as the two pictured here, were recorded from the muscle cell even when no stimulus was given to the motor nerve. These mEPPs occurred randomly at a frequency of about 3–4 per second. Note the difference in both the scale bars for magnitude (mV) and time (ms) between the left and right recordings. *Source: The data shown here are recordings from the laboratory of Stephen D. Meriney, Department of Neuroscience, University of Pittsburgh (unpublished).*

terminal in response to stimulation of the nerve (to generate the EPP), and (2) mEPPs are caused by acetylcholine that is spontaneously released from the nerve terminal even when the nerve is not stimulated.

These observations led researchers to ask exactly how mEPPs occur. In particular, they wondered whether mEPPs are caused by the spontaneous leakage of single molecules of acetylcholine from the nerve terminal. To answer this question, they applied different concentrations of acetylcholine from a small tube onto the endplate region of a muscle cell and recorded changes in the membrane potential of that cell. If individual molecules of acetylcholine were responsible for one mEPP, then adding extra acetylcholine should increase mEPP frequency, due to the increased number of acetylcholine molecules contacting the endplate region and triggering mEPPs. However, experimenters found that low concentrations of acetylcholine applied to the neuromuscular junction caused no detectable change in either the membrane potential of the muscle cell or the size and frequency of spontaneous mEPPs. In contrast, higher concentrations of acetylcholine caused a depolarization of the muscle cell, and when experimenters gradually increased the amount of acetylcholine they applied, this depolarization also increased gradually and could reach the size of a spontaneous mEPP. Based on these observations, the researchers concluded that individual mEPPs could not be caused by a single molecule of acetylcholine, but instead must be the result of the spontaneous but infrequent release of relatively large amounts of acetylcholine from the nerve terminal (though not as large as the amounts released in response to stimulation). Somehow, the nerve terminal was releasing these bursts of acetylcholine randomly and in the absence of nerve stimulation.

Another important observation was that there was little variation in the amplitude of mEPPs. Using patch clamp techniques (a variant of the voltage-clamp technique) to measure the ion flux through single acetylcholine receptors, later investigators estimated that a mEPP occurred when ~2000 acetylcholine receptors were activated (opened) in the muscle cell (Wathey et al., 1979). It was determined that these ~2000 acetylcholine receptors were activated when the nerve terminal synchronously released about 5000–10,000 molecules of acetylcholine (Kuffler and Yoshikami, 1975).

# THE QUANTAL THEORY OF CHEMICAL TRANSMITTER RELEASE

The next critical question to answer was: If mEPPs were caused by the synchronous release of a fixed number (or package) of acetylcholine molecules, what caused an EPP? Could an EPP be made up of many packages of acetylcholine of a fixed size all being released from the nerve terminal at the same time? If this were the case, the amplitude of an EPP should always be a multiple of the amplitude of the fixed-size mEPPs.

In their earliest attempts to compare the EPP size with the mEPP size, experimenters tried stimulating the motor nerve at the frog neuromuscular junction repeatedly at a low frequency (0.2 Hz) to record individual EPPs. However, this method was subject to an experimental complication in studying EPP size, namely that the current through open ion channels depends on the membrane potential.

Why was this the case, and why did it pose a problem for quantifying EPP size and comparing it to mEPP size? The acetylcholine receptor is a ligand-gated ion channel that passes roughly equal amounts of sodium and potassium (plus a very small amount of calcium). Therefore, this channel has an equilibrium potential that is roughly halfway between the equilibrium potentials for sodium (+60 mV) and potassium (−80 mV), or about −10 mV. This means the net inward driving force on ions decreases as the membrane potential approaches the channel's equilibrium potential. Consequently, there is a greater net flux of ions through the acetylcholine receptor channel when the muscle cell is at the resting membrane potential (about −80 mV in muscle cells) than there is at the peak of the EPP (about −30 mV). Fig. 6.6 depicts the acetylcholine receptor channel current–voltage relationship (I–V plot; Fig. 6.6A) and the potassium and sodium currents moving through acetylcholine receptor channels at different membrane potentials (Fig. 6.6B). For mixed ion channels such as the acetylcholine receptor channel, the equilibrium potential is the voltage at which the collective flux of permeable ions in each direction is equal, resulting in no net charge movement (even though there can be unequal amounts of sodium and potassium moving through the channel).

Because the driving force on ions moving through acetylcholine channels varies widely as a function of membrane voltage, the opening of an acetylcholine receptor causes a larger change in the membrane potential when the cell is at rest (−80 mV) than it does at the peak of the EPP (about −30 mV). This means that in the low-frequency stimulation experiments mentioned above, the current through each channel was low enough at the peak of the EPP that differences in the number of open acetylcholine receptor channels were difficult to detect. Consequently, the recorded EPPs did not vary significantly in amplitude; in fact, they varied in amplitude by an amount that was much smaller than an mEPP. For this reason, investigators needed a way to study action potential-triggered transmitter release independent of these large differences in driving force. Otherwise, their

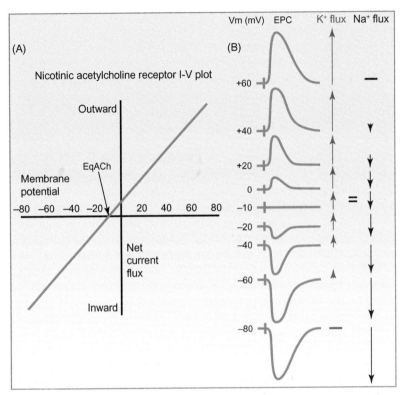

FIGURE 6.6 Current−voltage relationship, diagram of the endplate current (EPC) size measured at different membrane potentials, and the underlying sodium ($Na^+$) and potassium ($K^+$) currents through the acetylcholine receptor channel at the listed membrane potentials. (A) Nicotinic acetylcholine receptor channel current−voltage relationship. (B) Changes in ion flux through the nicotinic acetylcholine receptor channel at given voltages. At +60 mV, the EPC is a large outward current because there is a large outward $K^+$ current and no net $Na^+$ movement through the acetylcholine receptor channel at the $Na^+$ equilibrium potential. At −10 mV, the EPC is not observable because there is equal net flux of $Na^+$ into and $K^+$ out of the muscle cell [this is the equilibrium potential (EqACh) shown in (A)]. At −80 mV, the EPC is a large outward current because there is no net $K^+$ movement at the $K^+$ equilibrium potential and there is a large inward $Na^+$ current. At membrane potential values between these mentioned above, there is a different mixture of net $K^+$ and $Na^+$ current, as determined by the driving force on each ion.

measurements of EPP size would be confounded by the changing membrane potential during the measurement.

Sir Bernard Katz and colleagues discovered that reducing the concentration of calcium in the extracellular solution greatly reduced the amount of transmitter released following an action potential. The same effect could be achieved by keeping extracellular calcium at 2 mM (the normal physiological concentration) while increasing the concentration of magnesium, because at high extracellular concentrations, magnesium ions compete with calcium ions for flux through the presynaptic voltage-gated calcium channels.

When Katz and colleagues reduced extracellular calcium from 2 to 0.5 mM, the size of the EPP decreased substantially, to only a few millivolts in amplitude. This was advantageous because the peaks of the EPPs were much closer to the resting membrane potential, so the effect of membrane voltage on ion flux through the acetylcholine channels was minimized. In these low-calcium experiments, the magnitude of the presynaptic action potential-evoked EPPs varied in amplitude by fixed amounts that were multiples of the size of a mEPP. Furthermore, when the researchers reduced extracellular calcium even further (e.g., to 0.2 mM), motor nerve stimulation occasionally failed to cause any EPP at all. When an EPP did occur in this low-calcium condition, it was sometimes the same size as a mEPP. At other times the EPP magnitude was two or three times the size of a mEPP (see Fig. 6.7). These observations were critical because they demonstrated that when the extracellular calcium concentration was low, EPP size varied in multiples of the size of a mEPP (reviewed by Augustine and Kasai, 2007).

We now know that the neurotransmitter molecules responsible for a mEPP are contained within the synaptic vesicles that populate the presynaptic nerve terminal. Each synaptic vesicle contains roughly the same amount of neurotransmitter, so the release of one vesicle results in a mEPP. EPPs are caused by the simultaneous release of multiple

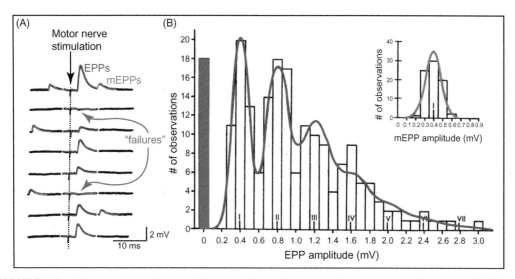

FIGURE 6.7  **Measurements of transmitter release from the mammalian neuromuscular junction.** (A) When calcium in the extracellular bathing solution is reduced from 2 to 1 mM, and magnesium chloride is increased from 1 to 6 mM, the size of action-potential-evoked EPPs (*red events*) decreases such that their amplitude fluctuates by a multiple of the amplitude of a spontaneous mEPP (*blue events*). Occasionally, nerve stimulation does not evoke an EPP and results in a "failure" (*dark green*). (B) The plot of hundreds of EPP amplitudes reveals a "peaky" histogram that can be fit using a Poisson distribution (*red line*). This Poisson distribution predicts the number of times the nerve stimulation would fail to evoke an EPP at all (vertical dark green bar at "0"), and the number of times an EPP would be the same size as a mEPP, or double, triple, or quadruple the size of a mEPP. The mEPP amplitude distribution is shown as an inset (upper right, fit by a *blue line*). *Source: Adapted from (A) Liley, A.W., 1956. The quantal components of the mammalian end-plate potential. J. Physiol. 133, 571–587; (B) Boyd, I.A., Martin, A.R. 1956a. Spontaneous subthreshold activity at mammalian neural muscular junctions. J. Physiol. 132, 61–73.*

vesicles, and are integer multiples of the size of an mEPP. Researchers referred to the consistent amount of neurotransmitter contained in a single synaptic vesicle as a **quantum** (plural: quanta).

To understand these experimental results, Katz and colleagues applied concepts from probability and statistics to their data in the following way: If a nerve terminal has a fixed number of distinct transmitter release sites ($n$), each with one synaptic vesicle that is ready to fuse to the neuronal membrane and release transmitter, the amount of transmitter released depends on the probability ($p$) that release will occur at a given site. Vesicle release is a binomial process, analogous to flipping coins. When you flip a coin, it has a 50% probability ($p$) of landing with heads facing up. Therefore, if you flip 100 coins ($n$), the predicted number of times you should observe heads is $n \times p$ ($100 \times 0.5$), or 50. In reality, since this is a random process, the actual number of times the coin lands as heads will fluctuate around the predicted value of 50.

In the case of a nerve terminal, the probability that a single vesicle will fuse with the plasma membrane and release transmitter is usually much lower than 0.5, and may even be closer to 0.05. Therefore, if a neuromuscular synapse has 100 release sites, each with one vesicle ready to release transmitter, binomial theory predicts that the number of vesicles that will release their contents should be 5 ($100 \times 0.05$). Of course, this is a random process, so the amount of transmitter released after any given stimulus will fluctuate around this mean value, just as in the coin toss example above. Conversely, if the probability that release occurs from one site is equal to 0.05, then the probability that no release will occur from that site is $1 - p$, or 0.95. If there are 100 release sites, the probability that none of them will release neurotransmitter is $(0.95)^{100}$, which equals 0.00592 (or about 0.6% of the time). In this way, binomial theory predicts the number of times a nerve terminal is likely to release different numbers of neurotransmitter-containing vesicles, based on the probability that each vesicle will or will not be released after each nerve stimulation. It also predicts how many times there is likely to be no neurotransmitter release from any of the release sites despite stimulation of the nerve.

It is difficult to independently measure the number of neurotransmitter release sites at a synapse or the probability that a site will release a vesicle in response to an action potential. Instead, experimenters typically measure the total amount of neurotransmitter released from a synapse after any given stimulus by recording postsynaptic potentials (mEPPs and EPPs). If we assume that the probability of release is much smaller than the number of release sites (which is true for synapses where the probability of neurotransmitter release from a single release site approaches, but is not equal to, zero), then we can use a simplified approach that utilizes a Poisson distribution to describe the data. This approach makes it possible to predict the rates at which random events are likely to occur. Experimenters used a Poisson distribution to fit the experimental data they recorded from a neuromuscular junction when it was bathed in a low-calcium extracellular solution (see Fig. 6.7). Taking into account the inherent variability in electrophysiological recordings, this Poisson fitting method closely matched the rates at which transmitter release events were observed. Each time the motor nerve was stimulated, the result was *either* a "failure" (no EPP at all) *or* an EPP that was a multiple of the size of a single mEPP. Similar results were observed at both frog and mammalian neuromuscular junctions (Fatt and Katz, 1952; Del Castillo and Katz, 1954; Boyd and Martin, 1956a,b; Liley, 1956).

The conclusions drawn from analyses of these experiments led to the formulation of the **quantal theory of chemical neurotransmitter release**, which states that:

1. Chemical neurotransmitter is only released in multimolecular "packages," termed quanta, each containing thousands of neurotransmitter molecules.
2. These packages of neurotransmitter are sometimes released spontaneously in the absence of action potentials, because the probability of release in the absence of action potentials is slightly greater than zero.
3. The release of one quantum is independent of the release of other quanta, and release of a quantum occurs in a random manner, with low statistical probability.
4. Presynaptic action potential activity causes the influx of calcium ions, and this intracellular calcium significantly increases the probability of quantal release.

In Chapter 8, we will cover the mechanisms by which action potential-triggered calcium entry increases the probability that these vesicles fuse with the presynaptic plasma membrane and release a quantum of neurotransmitter, as well as the mechanisms by which synaptic vesicles occasionally fuse with the neuronal membrane spontaneously, releasing neurotransmitter and causing a mEPP.

Neuromuscular synapses are so large, and have so many transmitter release sites, that under normal physiological conditions each action potential triggers the release of hundreds of quanta. For example, one frog neuromuscular synapse typically contains about 21,000 single vesicle release sites and releases about 350 quanta in response to a single action potential (Laghaei et al., 2018). The quantal theory of transmitter release is thought to apply not only to neuromuscular synapses, but to all chemical synapses in the nervous system. Different types of synapses can release different numbers of quanta, depending on the size of the synapse. For example, single bouton-like synapses in the central nervous system may only contain 10–20 single vesicle release sites and release between 1 and 5 quanta following each action potential. Other types of synapses have such a low probability of neurotransmitter release that they might release only 0–2 quanta per action potential.

## QUANTAL ANALYSIS OF CHEMICAL TRANSMITTER RELEASE AT THE NEUROMUSCULAR JUNCTION

In addition to providing a conceptual mechanism for the release of chemical transmitters, quantal theory also opened the door for quantitative analyses of the neurotransmitter release process itself. Using the quantal theory framework, scientists can not only understand how chemicals are released from nerve terminals, but also gain insight into ways in which neurotransmitter release can be modulated. For example, if the amount of neurotransmitter released at a synapse changes in response to an organism's experience, disease, or the effects of a drug, quantal analysis allows a researcher to investigate whether the changes observed are the result of alterations on the presynaptic or postsynaptic side of the synapse. It also provides clues about the specific mechanisms for these changes. Thus, the results of quantal analysis can guide a more detailed investigation into the cellular mechanisms responsible for synaptic plasticity (see Chapter 14).

How can the quantal theory of neurotransmitter release be used to generate a quantitative analysis of synaptic vesicle release characteristics? The underlying assumption of the quantal theory is that neurotransmitter is only released in quanta. Therefore, one straightforward quantitative approach to studying synapses is to ask how many quanta are released after an action potential. Because spontaneous postsynaptic depolarizations (e.g., mEPPs at the neuromuscular junction) are caused by the release of a single quantum, the magnitude of this depolarization can be used to estimate how many of these quanta are released at a given synapse in response to an action potential. This value is known as the **quantal content** (QC) of the synapse, and is defined as the number of synaptic vesicles that fuse with the plasma membrane during a presynaptic action potential. The QC can be calculated by dividing the amplitude of the average EPP by the amplitude of the average mEPP.

Analyses of mEPPs and EPPs assume that there is little impact of driving force on the accuracy of the amplitudes of these potentials. More accurate measurements of these values can be obtained using voltage clamp measurements (see Box 4.3) of synaptic current during spontaneous and action potential-triggered transmitter release. The spontaneous miniature endplate currents (mEPCs) and action potential-triggered endplate currents (EPCs) recorded using the voltage clamp technique are not influenced by driving force since the membrane potential is held constant during these recordings of current through ligand-gated channels (transmitter receptors) in the membrane. Using the microelectrode recording technique or the voltage clamp technique, investigators can quantify the number of vesicles (quanta) released during an action potential (quantal content) and can use this as a measure of the amount of transmitter released.

The quantal theory of neurotransmitter release can be used to make additional inferences about the mechanisms of vesicular release. For example, experimenters are often interested in determining how synapses might have changed following intense activity, disease, or exposure to a drug. If mEPPs and EPPs from the synapse of interest can be measured both before and after such an event (e.g., before and after drug application), it is possible to use quantal analysis to predict which aspects of synaptic function might have changed.

Based on quantal theory, the measured impact of neurotransmitter released from one nerve terminal can be described as follows:

The magnitude ($m$) of a measured event in the postsynaptic cell is equal to the product of the number ($n$) of release sites times the probability ($p$) of release at each site times the size of a spontaneous mEPP (quantum; $q$), or: $\mathbf{m = n \times p \times q}$.

In the application of quantal theory, it is assumed that all release sites have the same probability of transmitter release and that the size of the measured mEPP is the same no matter from which release site it occurs. Later we will discuss whether these assumptions are likely to be true at all synapses, but for the purposes of this discussion we will assume they are true.

Let's look at a case in which an experimenter found that the application of a drug to a neuromuscular synapse increased the size of the EPP within seconds, but did not change mEPP size (see Fig. 6.8A). Since the drug effect occurred within seconds, too quickly for the nerve terminal to have assembled the proteins required to create new release sites, there was likely *not* a change in the number of transmitter release sites ($n$), as these take longer to form or be eliminated. Therefore, according to quantal theory ($m = n \times p \times q$), either $p$ or $q$ must have changed. Since $q$ (the size of a spontaneous mEPP) also did not change, $p$ (the probability of release at each site) must have increased when the drug was applied to the synapse. Based on these results, the experimenter would conclude that the drug acted on the presynaptic cell to increase the probability that an action potential would trigger the release of transmitter, but did not have an effect on the postsynaptic cell.

In a different experiment (see Fig. 6.8B), the researcher applied a drug which decreased the size of both the mEPP and the EPP, without changing the ratio of their sizes (i.e., the quantal content, average EPP/average mEPP). Because the size of the mEPP decreased, these data indicated that there was a change in $q$ in response to the drug. That is, the impact of the release of one quantum of neurotransmitter on the postsynaptic cell decreased, which could have occurred for one of two reasons. First, the drug could have caused a reduction in the amount of transmitter loaded into a single synaptic vesicle. However, this type of

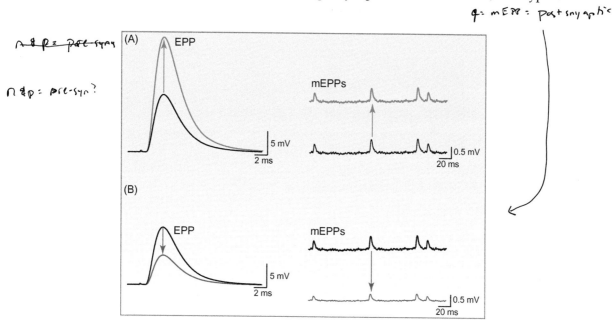

FIGURE 6.8 Two examples of changes in transmitter release that can be studied using quantal analysis. (A) In this diagram, the control recordings (*black traces*) of EPPs and mEPPs are altered by a drug (*green traces*) that increases the EPP amplitude, but has no effect on the mEPPs (note that the mEPP traces are offset for clarity; there is no change in membrane potential between the control and drug recordings). If the effect on EPP amplitude occurs too quickly for there to be a change in the number of release sites ($n$), then one would conclude that the drug changes the probability of release ($p$). (B) In a different experiment, the control recordings (*black traces*) are altered by a different drug such that both the amplitude of the EPP and mEPPs are reduced (*red traces*). Since the size of the measured mEPPs is reduced but the ratio of EPP/mEPP sizes has not changed, one would conclude that the drug reduces the size of a quantum ($q$; note that the mEPP traces are offset for clarity; there is no change in membrane potential).

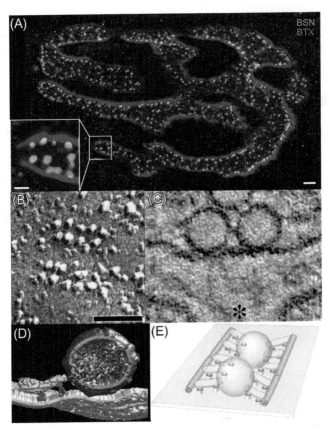

FIGURE 6.9  Quantification of the number of active zones and vesicles at each active zone within a neuromuscular synapse. (A) Fluorescent light microscopy view of a mouse neuromuscular junction stained for two proteins [postsynaptic acetylcholine receptors in red, using Alexa 594-α-bungarotoxin (BTX), and presynaptic active zones in green, using an antibody to the protein bassoon (BSN)]. Active zones can be identified as small "dot-like" structures within the nerve terminal regions of the image (scale bar = 2 μm). The inset shows an enlarged view of one part of this nerve terminal (scale bar = 0.5 μm). (B) Freeze-fracture electron micrograph of a mouse active zone. The large "bumps" are thought to be transmembrane proteins associated with this site of vesicle fusion, and are hypothesized to include voltage-gated calcium channels. Scale bar = 100 nm. (C) Transmission electron micrograph of docked vesicles at a single active zone in the mouse neuromuscular junction. The two circular structures represent synaptic vesicles that are docked at the active zone (note close apposition to the plasma membrane). The synaptic cleft is shown by the "*". (D) Electron microscope tomography representation of a single synaptic vesicle (*blue-purple sphere*) docked at an active zone in the mouse neuromuscular junction by unidentified proteins (*yellow and red*). This image was created by imaging the same area with the electron microscope from different angles and recreating a three-dimensional representation of the structure. (E) Artist's rendering of the mouse active zone that contains two docked synaptic vesicles (*blue spheres*). Undefined components of the active zone structure include elements (*orange and yellow*) that appear to coordinate the position of the synaptic vesicles and green "pegs" that are interpreted to either be connections to transmembrane proteins in the active zone, or the transmembrane proteins themselves. This drawing is based on electron microscope tomography data (as shown in D). *Source: Adapted from (A) Laghaei, R., Ma, J., Tarr, T.B., Homan, A.E., Kelly, L., Tilvawala, M.S., et al., 2018. Transmitter release site organization can predict synaptic function at the neuromuscular junction. J. Neurophysiol. 119, 1340−1355; (B) Fukuoka, T., Engel, A.G., Lang, B., Newsom-Davis, J., Prior, C., Wray, D.W., 1987. Lambert-Eaton myasthenic syndrome: I. Early morphological effects of IgG on the presynaptic membrane active zones. Ann. Neurol. 22, 193−199; (C−E) Nagwaney, S., Harlow, M.L., Jung, J.H., Szule, J.A., Ress, D., Xu, J., et al., 2009. Macromolecular connections of active zone material to docked synaptic vesicles and presynaptic membrane at neuromuscular junctions of mouse. J. Comp. Neurol. 513, 457−468.*

change has not typically been observed at synapses, so is unlikely to have occurred. Second, the drug could have caused a change in the number or function of postsynaptic transmitter receptors, meaning that the effect of a single quantum of neurotransmitter on the postsynaptic cell was reduced because fewer receptors were available or able to respond to the quantum of transmitter. This second scenario is more commonly observed as a mechanism by which a drug can cause a change in response to transmitter release.

Changes in the number of neurotransmitter release sites ($n$) can be quantified using anatomical markers that identify such sites. For example, at the neuromuscular junction experimenters can use antibodies directed against proteins found at neurotransmitter release sites, which are called **active zones**. The antibodies can be attached to fluorescent markers that show where active zones are located at a single synapse (see Fig. 6.9). The number of active zones can be counted by using light microscopy to determine the number of sites at which transmitter-containing vesicles are collected. But without an electron microscopic study, it is not possible to determine how many single vesicle release sites are contained within each active zone, which is perhaps the most appropriate measure of "$n$" (see Fig. 6.9).

Using these quantal analysis methods, it is possible to predict whether observed changes in transmitter release are pre- or postsynaptic, which makes quantal analysis a powerful tool for studying the mechanisms that underlie changes in neurotransmitter release and postsynaptic responses, including changes that result from activity, drugs, neuromodulators, aging, or disease.

## QUANTAL ANALYSIS OF CHEMICAL TRANSMITTER RELEASE AT CENTRAL SYNAPSES

The quantal theory of chemical transmitter release is thought to apply to all synapses, across all species, as a basic mechanism for chemical release from neurons. In fact, the control of vesicle release by calcium and the proteins involved in this process are some of the most conserved mechanisms in nature (see Chapter 8). Nevertheless, it is sometimes complicated and difficult to apply quantal analysis to the study of transmitter release at many central nervous system synapses.

Unlike muscle cells, which each receive a single synaptic input at the neuromuscular junction, most neurons in the central nervous system have multiple synaptic inputs. In fact, central neurons often each receive thousands of synaptic inputs, which are distributed over a complex neuronal architecture that includes the soma and extensive dendritic branches. Further, most people who study synapses onto neurons in the central nervous system use a recording technique (microelectrode or voltage clamp) that records voltages or currents at the soma, not the axons or dendrites. Since synapses can be located anywhere from the soma to distant dendritic branches, the distance between the site of recording (soma) and the site of neurotransmitter receptors receiving synaptic input can be quite variable.

In our earlier discussions of microelectrode recordings that Katz and colleagues made at the neuromuscular junction, we saw that if the recording electrode was positioned outside the synaptic endplate region on the muscle, postsynaptic potentials were difficult to record. This is because these potentials decayed in size when they propagated passively from the endplate region to more distant regions of the postsynaptic muscle cell (see Chapter 13). Similarly, synaptic potentials generated at a distance from the soma decay in size as they spread passively from the dendrites to a recording site in the soma (see Fig. 6.10). Accurate measurement is further complicated by the fact that dendrites in the central nervous system contain voltage-gated ion channels that open in response to very small depolarizations (e.g., the depolarization created by a single synaptic excitatory postsynaptic potential). The active ionic conductances on dendrites can affect the membrane potential, altering the size of the synaptic potential as it travels to the cell soma (see Chapter 13, for more details). As a result, the magnitudes of spontaneous synaptic potentials at the soma are variable in size, so the size of the postsynaptic potential recorded at the soma is not always a reflection of where it originated on the dendrites.

These factors can be significant complications when trying to perform quantal analysis at a synapse. When recording synaptic input to a neuron in the central nervous system, an experimenter can record **excitatory postsynaptic potentials** (**EPSPs**; these are essentially the same as EPPs, but they are called "EPSPs" at central synapses since these synapses are not endplates), as well as spontaneous **miniature excitatory postsynaptic potentials** (**mEPSPs**). Suppose an experimenter stimulates a synaptic input to a central neuron and measures the postsynaptic response at the soma. Given that the mEPSPs they record will vary in size for the reasons explained above, how can the experimenter determine the size of one quantum—that is, the size of the mEPSPs that originate from the particular input they are stimulating?

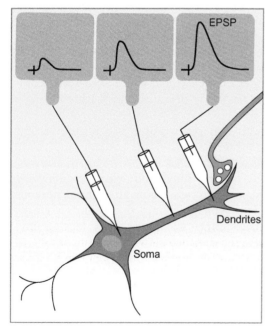

FIGURE 6.10 **Diagram demonstrating the expected decay in excitatory postsynaptic potentials (EPSPs) as they propagate from synapses onto distal dendrites to the soma of a neuron.** In this diagram, the size of the EPSP is shown as recorded by electrodes placed at different locations in the dendritic tree or in the soma.

Several clever experimental strategies have been employed to distinguish between mEPSPs originating from different synaptic inputs. One of these takes advantage of the fact that **hypertonic** saline (obtained by adding 300–500 mM sucrose to the extracellular solution bathing the cell) dramatically increases the frequency of mEPSPs. It is not known why this occurs, though it may be due to changes in vesicle fusion in response to stretching of the cell membrane. When hypertonic solution is applied to a very small region on a dendrite where nerve stimulation induces EPSPs, it causes a burst of mEPSPs from that specific synaptic region. This tells experimenters how big a mEPSP originating from that specific site will be when it is recorded at the soma, and they can use this information to apply quantal analysis.

Another experimental technique used to determine the impact of one quantum on the postsynaptic membrane potential is to replace the calcium ions in the extracellular solution bathing the cell with strontium ions. In solution, strontium is a divalent cation that is similar enough to calcium in its properties that it can pass through voltage-gated calcium channels at synapses. Once inside the cytoplasm of the nerve terminal, strontium is less efficient at triggering synaptic vesicle fusion, and substantially increases the time period over which vesicles are released in response to an action potential. The result is that EPSPs triggered by action potentials, which typically result from the near-synchronous release of many quanta, are instead spread out in time (see Fig. 6.11). As a result, the strontium triggers a dispersed burst of individual quanta from only the stimulated inputs. Now, when the response to a stimulus occurs, the EPSP has effectively been broken down into its component quanta, which can be seen in experimental recordings as independent events separated from one another in time.

Another issue that complicates quantal analysis at synapses in the central nervous system is a key assumption of quantal theory: that all transmitter release sites have the same

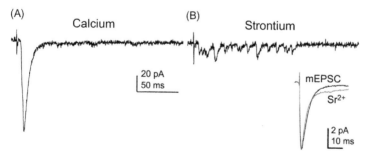

FIGURE 6.11  **Excitatory postsynaptic currents (EPSCs) recorded from the somatosensory cortex in the presence of extracellular calcium (A) or strontium (B).** (A) In normal extracellular calcium, an action potential evokes a fast, synchronous release of transmitter, resulting in an EPSC in the postsynaptic cell. (B) When the same experiment is performed with strontium in the extracellular fluid, action potentials evoke a desynchronized burst of single events that are dispersed in time. The inset (lower right) shows a comparison of the amplitude and time course of a single event evoked in strontium, compared to a mEPSC. The similarity of these two provides evidence that events evoked in strontium are single quantal events, just like mEPSCs. *Source: Adapted from Bender, K.J., Allen, C.B., Bender, V.A., Feldman, D.E., 2006. Synaptic basis for whisker deprivation-induced synaptic depression in rat somatosensory cortex. J. Neurosci. 26, 4155–4165.*

probability of releasing transmitter. At the neuromuscular junction, this assumption has historically been accepted as likely to be true, as there was no evidence to the contrary (although recent work has challenged this assumption, even at the neuromuscular junction; Peled and Isacoff, 2011; Tabares et al., 2007; Wyatt and Balice-Gordon, 2008; Gaffield et al., 2009; Melom et al., 2013). However, in the central nervous system, evidence has existed for some time that some synaptic inputs onto a postsynaptic neuron have different probabilities of release than other inputs onto the same postsynaptic neuron. Experimenters who use quantal analysis at central nervous system synapses nevertheless generally assume that the small group of inputs that are stimulated in their experiments all have the same probability of release, since they originate from the same bundle of stimulated axons. Whether or not this is true is a caveat that must be acknowledged when interpreting results from these experiments.

Lastly, since quantal analysis depends on measuring the size of synaptic events very accurately, it is assumed that the postsynaptic membrane has sufficient receptors to respond to *all* of the neurotransmitter molecules that are released. This is certainly the case at the neuromuscular junction, where the postsynaptic muscle membrane has such a high density of receptors that they are packed side-by-side into the membrane at about 16,000 per square micron (Unwin, 2013). However, there is debate about the number of postsynaptic receptors at central synapses. Some researchers argue that at a subset of central synapses the release of a single quantum of neurotransmitter can saturate most available receptors (Frerking and Wilson, 1996), which means it is possible that not all released neurotransmitter molecules will impact the postsynaptic cell because all of the receptors may already have bound neurotransmitter. If released neurotransmitter does saturate most available receptors, then it is impossible to accurately measure the size of multiquantal events because the recorded postsynaptic response is always of the same or of similar magnitude. In this case, it is impossible to apply quantal analysis to electrical recordings from the postsynaptic cell.

Despite the complications described above, quantal analysis of chemical neurotransmitter release is a powerful and popular technique used to study synapses. It is more commonly applied to the neuromuscular junction, but there are also many investigators who successfully navigate these complications to apply this approach to the study of central nervous system synapses. For example, based on quantal analysis, **long-term potentiation** in the hippocampus has been argued to involve both pre- and postsynaptic changes (see Fig. 6.12; Lisman and Harris, 1993; MacDougall and Fine, 2014)

## OPTICAL QUANTAL ANALYSIS

Thanks to the development of optical reporters that show when and where a single synaptic vesicle fuses with the plasma membrane, investigators can now visualize the number of vesicles that fuse within a nerve terminal without the need to record using microelectrodes (Oertner, 2002; Balaji and Ryan, 2007; Peled and Isacoff, 2011; Melom et al., 2013). In one approach, the investigator adds a molecular fluorescent reporter of vesicle fusion to each presynaptic vesicle to visualize vesicle fusion directly

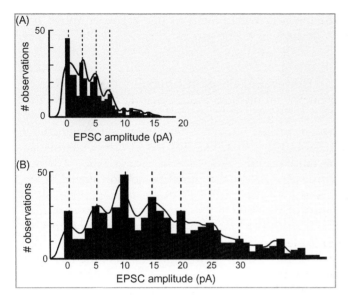

FIGURE 6.12 **Quantal analysis of action-potential-evoked EPSCs in area CA1 of the hippocampus.** (A) Plot of evoked EPSC amplitudes under control conditions. The bar at 0 pA represents failures to detect an EPSC. The dotted lines indicate the peak amplitudes in the distribution that were fit to a Poisson function (hypothesized single quantal EPSC amplitudes). (B) After induction of long-term potentiation (see Chapter 14), there is an increase in the size of single quantum ($q$; the position on the x-axis at which peaks occur, as represented by dotted lines), an increase in the probability of observing events of larger amplitude (the peaks with the largest number of observations occur at larger amplitudes), and a reduction in the number of failures (size of the peak at 0 pA), both of which indicate a change in "$p$". *Source: Adapted from Lisman, J.E., Harris, K.M., 1993. Quantal analysis and synaptic anatomy—integrating two views of hippocampal plasticity. Trends Neurosci. 16, 141–147.*

(see Chapter 8). In another approach, the investigator adds a reporter of postsynaptic receptor function that detects the impact of transmitter molecules released from a single vesicle. These approaches reveal additional details of the transmitter release process, and in some cases have shown that individual release sites can be specialized for different forms of transmitter release (e.g., mEPPs and EPPs can sometimes originate from separate release sites; see Fig. 6.13; Peled and Isacoff, 2011; Melom et al., 2013). As these optical reporters continue to be used in future experiments, many additional details in the regulation of chemical transmitter release will undoubtedly be revealed.

# SUMMARY

Observations of evoked and spontaneous neurotransmitter release by early neurophysiologists led to the understanding that at chemical synapses, neurotransmitter is released in fixed amounts called quanta that correspond to the release of neurotransmitter molecules contained in one synaptic vesicle from the presynaptic terminal. These findings established a framework for understanding this process of vesicle release called quantal theory. With some simplifying assumptions, quantal theory can be used to predict the probability of vesicle release under a given set of conditions, and how release probability is affected by the presence of drugs and/or changes in synaptic structure or physiology.

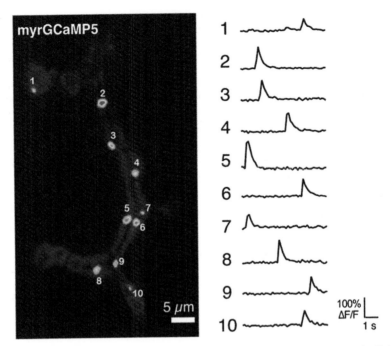

FIGURE 6.13  Optical quantal analysis at the neuromuscular junction of *Drosophila*. (Left) Fluorescent signals (generated by a myristoylated GCaMP5 calcium sensor; myrGCaMP5) collected over a 10-s time period reveal the location within the nerve terminal from which spontaneous mEPPs originated. The neuromuscular junction is outlined in a faint green stain, and the location of each synaptic vesicle release site is shown as a bright fluorescent spot (numbered). (Right) Time course of fluorescent change at each of the numbered locations shown on the left. These types of studies have revealed that each release site at the *Drosophila* neuromuscular junction has an independent probability of releasing a synaptic vesicle, and that this probability is heterogeneous at locations across this synapse. Similar optical recordings of individual synaptic vesicle fusion events after nerve stimulation also indicate that the single vesicle release events that make up an EPP originate in a heterogeneous manner from a subset of active zones. *Source: Adapted from Melom, J.E., Akbergenova, Y., Gavornik, J.P., Littleton, J.T., 2013. Spontaneous and evoked release are independently regulated at individual active zones. J. Neurosci. 33, 17253–17263.*

# References

Adams, M.E., 2004. Agatoxins: ion channel specific toxins from the American funnel web spider, Agelenopsis aperta. Toxicon 43, 509–525.

Augustine, G.J., Kasai, H., 2007. Bernard Katz, quantal transmitter release and the foundations of presynaptic physiology. J. Physiol. 578, 623–625.

Balaji, J., Ryan, T.A., 2007. Single-vesicle imaging reveals that synaptic vesicle exocytosis and endocytosis are coupled by a single stochastic mode. Proc. Natl. Acad. Sci. USA 104, 20576–20581.

Bane, V., Lehane, M., Dikshit, M., O'Riordan, A., Furey, A., 2014. Tetrodotoxin: chemistry, toxicity, source, distribution and detection. Toxins (Basel) 6, 693–755.

Bender, K.J., Allen, C.B., Bender, V.A., Feldman, D.E., 2006. Synaptic basis for whisker deprivation-induced synaptic depression in rat somatosensory cortex. J. Neurosci. 26, 4155–4165.

Bergeron, Z.L., Bingham, J.P., 2012. Scorpion toxins specific for potassium (K+) channels: a historical overview of peptide bioengineering. Toxins (Basel) 4, 1082–1119.

# References

Binz, T., Sikorra, S., Mahrhold, S., 2010. Clostridial neurotoxins: mechanism of SNARE cleavage and outlook on potential substrate specificity reengineering. Toxins (Basel) 2, 665–682.

Bowman, W.C., 2006. Neuromuscular block. Br. J. Pharmacol. 147 (Suppl. 1), S277–S286.

Boyd, I.A., Martin, A.R., 1956a. Spontaneous subthreshold activity at mammalian neural muscular junctions. J. Physiol. 132, 61–73.

Boyd, I.A., Martin, A.R., 1956b. The end-plate potential in mammalian muscle. J. Physiol. 132, 74–91.

Chang, C.C., 1999. Looking back on the discovery of alpha-bungarotoxin. J. Biomed. Sci. 6, 368–375.

Del Castillo, J., Katz, B., 1954. Quantal components of the end-plate potential. J. Physiol. 124, 560–573.

Fatt, P., Katz, B., 1952. Spontaneous subthreshold activity at motor nerve endings. J. Physiol. 117, 109–128.

Frerking, M., Wilson, M., 1996. Saturation of postsynaptic receptors at central synapses? Curr. Opin. Neurobiol. 6, 395–403.

Fukuoka, T., Engel, A.G., Lang, B., Newsom-Davis, J., Prior, C., Wray, D.W., 1987. Lambert-Eaton myasthenic syndrome: I. Early morphological effects of IgG on the presynaptic membrane active zones. Ann. Neurol. 22, 193–199.

Gaffield, M.A., Tabares, L., Betz, W.J., 2009. The spatial pattern of exocytosis and post-exocytic mobility of synaptopHluorin in mouse motor nerve terminals. J. Physiol. 587, 1187–1200.

Katz, B., Miledi, R., 1965. Propagation of electric activity in motor nerve terminals. Proc. R. Soc. Lond. B. Biol. Sci. 161, 453–482.

Kuffler, S.W., Yoshikami, D., 1975. The number of transmitter molecules in a quantum: an estimate from iontophoretic application of acetylcholine at the neuromuscular synapse. J. Physiol. 251, 465–482.

Laghaei, R., Ma, J., Tarr, T.B., Homan, A.E., Kelly, L., Tilvawala, M.S., et al., 2018. Transmitter release site organization can predict synaptic function at the neuromuscular junction. J. Neurophysiol. 119, 1340–1355.

Liley, A.W., 1956. The quantal components of the mammalian end-plate potential. J. Physiol. 133, 571–587.

Lisman, J.E., Harris, K.M., 1993. Quantal analysis and synaptic anatomy—integrating two views of hippocampal plasticity. Trends Neurosci. 16, 141–147.

MacDougall, M.J., Fine, A., 2014. The expression of long-term potentiation: reconciling the preists and the postivists. Philos. Trans. R. Soc. Lond. B. Biol. Sci. 369, 20130135.

Magarlamov, T.Y., Melnikova, D.I., Chernyshev, A.V., 2017. Tetrodotoxin-producing bacteria: detection, distribution and migration of the toxin in aquatic systems. Toxins (Basel) 9, 166.

Melom, J.E., Akbergenova, Y., Gavornik, J.P., Littleton, J.T., 2013. Spontaneous and evoked release are independently regulated at individual active zones. J. Neurosci. 33, 17253–17263.

Nagwaney, S., Harlow, M.L., Jung, J.H., Szule, J.A., Ress, D., Xu, J., et al., 2009. Macromolecular connections of active zone material to docked synaptic vesicles and presynaptic membrane at neuromuscular junctions of mouse. J. Comp. Neurol. 513, 457–468.

Oertner, T.G., 2002. Functional imaging of single synapses in brain slices. Exp. Physiol. 87, 733–736.

Peled, E.S., Isacoff, E.Y., 2011. Optical quantal analysis of synaptic transmission in wild-type and rab3-mutant Drosophila motor axons. Nat. Neurosci. 14, 519–526.

Piek, T., 1982. delta-Philanthotoxin, a semi-irreversible blocker of ion-channels. Comp. Biochem. Physiol. C 72, 311–315.

Pringos, E., Vignes, M., Martinez, J., Rolland, V., 2011. Peptide neurotoxins that affect voltage-gated calcium channels: a close-up on omega-agatoxins. Toxins (Basel) 3, 17–42.

Safavi-Hemami, H., Brogan, S.E., Olivera, B.M., 2019. Pain therapeutics from cone snail venoms: from Ziconotide to novel non-opioid pathways. J. Proteomics 190, 12–20.

Sudhof, T.C., 2001. Alpha-Latrotoxin and its receptors: neurexins and CIRL/latrophilins. Annu. Rev. Neurosci. 24, 933–962.

Tabares, L., Ruiz, R., Linares-Clemente, P., Gaffield, M.A., Alvarez de Toledo, G., Fernandez-Chacon, R., et al., 2007. Monitoring synaptic function at the neuromuscular junction of a mouse expressing synaptopHluorin. J. Neurosci. 27, 5422–5430.

Terlau, H., Olivera, B.M., 2004. Conus venoms: a rich source of novel ion channel-targeted peptides. Physiol. Rev. 84, 41–68.

Unwin, N., 2013. Nicotinic acetylcholine receptor and the structural basis of neuromuscular transmission: insights from Torpedo postsynaptic membranes. Q. Rev. Biophys. 46, 283–322.

Utkin, Y.N., 2013. Three-finger toxins, a deadly weapon of elapid venom—milestones of discovery. Toxicon 62, 50–55.

Wathey, J.C., Nass, M.M., Lester, H.A., 1979. Numerical reconstruction of the quantal event at nicotinic synapses. Biophys. J. 27, 145–164.

Wyatt, R.M., Balice-Gordon, R.J., 2008. Heterogeneity in synaptic vesicle release at neuromuscular synapses of mice expressing synaptopHluorin. J. Neurosci. 28, 325–335.

Yanagawa, Y., Abe, T., Satake, M., 1987. Mu-conotoxins share a common binding site with tetrodotoxin/saxitoxin on eel electroplax Na channels. J. Neurosci. 7, 1498–1502.

Zimmer, M., 2009. GFP: from jellyfish to the Nobel prize and beyond. Chem. Soc. Rev. 38, 2823–2832.

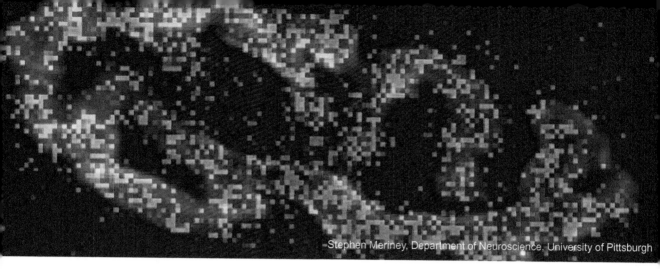

Stephen Meriney, Department of Neuroscience, University of Pittsburgh

CHAPTER 7

# Calcium Homeostasis, Calcium Channels, and Transmitter Release

## CALCIUM AS A TRIGGER FOR NEUROTRANSMITTER RELEASE

Once the debate between the Reticular Theory and the Neuron Doctrine had been settled, it was clear that the nervous system was not made up of a continuous network of interconnected cells, but rather that neurons were separate entities, suggesting there must be a way for them to communicate with one another. In Chapter 5, we described the mechanisms by which neurons communicate at electrical synapses, and in Chapter 6, we introduced the quantal theory of chemical neurotransmitter release. But what are the mechanisms that neurons use to communicate with each other at chemical synapses? In Chapters 7–9 we will discuss the major mechanisms that govern the release of molecules that mediate chemical synaptic transmission. We will begin with a discussion of the ion that triggers chemical neurotransmitter release: calcium.

### The Distribution of Calcium Ions Across the Cell Membrane

Calcium is a divalent cation, and the fact that it is charged means it can cause an electrical current when it crosses the plasma membrane. However, calcium is more important to cells as a biochemical signaling ion. Although calcium is the most abundant mineral in the body, most calcium is bound up in **hydroxyapatite**, a mineral that makes up most of the composition of bones. The remaining calcium is distributed throughout the extracellular fluid, serum, and body tissues. In the blood, calcium is maintained in a narrow range between 8.5 and 10.5 mg/dL (Goldstein, 1990), but about 50% of this is bound to proteins and other molecules. The other half is in the form of free ionic calcium (∼4.4 mg/dL), which the body tightly regulates to a concentration of about 1.1–1.4 mM. The ability to measure and maintain this type of equilibrium in the face of external changes is called **homeostasis**. The body's calcium homeostasis mechanisms include calcium-sensing receptors on cells in the parathyroid gland and thyroid gland that regulate the release of hormones (calcitonin and parathyroid hormone). These hormones act on bones, kidneys, and the small intestine to regulate calcium homeostasis in the body.

The brain is insulated from the rest of the body by the **blood–brain barrier** (see Box 7.1). This semipermeable capillary barrier allows the circulatory system to vascularize the brain, but tightly regulates which molecules can pass between the blood and the brain tissue (Daneman and Prat, 2015). As such, the environment within the extracellular spaces around brain cells is different than the environment around the cells in other parts of the body. The blood–brain barrier contains transmembrane pumps that assist in regulating the calcium concentration in the **cerebrospinal fluid**, helping to prevent blood changes in calcium from having a strong effect on brain calcium levels (Keep et al., 1999). In addition, cerebrospinal fluid contains many fewer total proteins than blood does, which means there are fewer proteins available that would normally bind calcium ions. As a result, the free ionic calcium concentration in the cerebrospinal fluid is close to 70%–75% of the total calcium, making the free ionic calcium concentration in the nervous system

---

BOX 7.1

## THE BLOOD–BRAIN BARRIER

All neurons in the central nervous system are located very close to **capillaries** (the smallest blood vessels in the body), which provide required nutrients and remove waste. In general, substances can diffuse relatively freely across capillary walls. However, blood capillaries in the central nervous system contain a "**blood–brain barrier**" that limits the diffusion of substances in and out of the capillaries, only allowing the diffusion of water, gases, and some lipid-soluble molecules, while regulating the passage of glucose, amino acids, and other molecules by selective transporters.

The blood–brain barrier is a cellular barrier between the circulatory system and the central nervous system that is located along the walls of the vascular capillaries that run through the central nervous system (see Fig. 7.1). The blood–brain barrier is formed by **endothelial cells** (cells that line the interior surface of blood vessels) that have tight junctions between them, which effectively restrict the diffusion of substances and cells across the capillary wall. The blood–brain barrier is also supported by **pericytes**, contractile cells that regulate blood flow in capillaries and communicate with endothelial cells to regulate the

## BOX 7.1 (cont'd)

permeability of the endothelial barrier. Finally, astrocytes extend protoplasmic processes called **endfeet** that ensheath capillaries in the brain and spinal cord, further reducing the permeability of these capillaries. Astrocytes are critical for the development and maintenance of the blood–brain barrier because they secrete substances that support the function of the endothelial cells and pericytes.

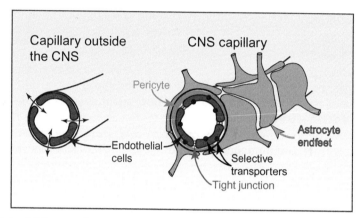

FIGURE 7.1 **Differences between a general blood vessel capillary (capillary outside the CNS) and a capillary within the central nervous system (CNS capillary).** Capillaries outside the CNS allow the diffusion of many substances across the capillary wall, while CNS capillaries are ensheathed by astrocyte endfeet that contribute to the blood–brain barrier, protecting the central nervous system by limiting the movement of substances from the circulatory system into the central nervous system.

about 1.5–2.0 mM (Jones and Keep, 1987; Egelman and Montague, 1999; Jimerson et al., 1980; Somjen, 2004).

It is common for ion concentrations inside and outside neurons to be in the millimolar range (see Chapter 2 for a discussion of the distributions of sodium, potassium, and chloride), so a 1.5–2 mM calcium concentration in the nervous system extracellular space is not unusual. However, the cytoplasmic concentration of calcium inside neurons is very low, typically 0.00005–0.0001 mM, or 50–100 nM (for comparison, the cytoplasmic concentration of sodium is about 15 mM, potassium is about 140 mM, and chloride is about 10 mM). As a result, the difference in calcium concentration across the plasma membrane is roughly five orders of magnitude (approximately 20,000–40,000:1). This calcium concentration gradient creates a very large driving force for calcium entry near resting membrane potential (see Chapter 2 for a detailed discussion of driving force), which means that when calcium channels open near resting membrane potential in the plasma membrane, a large amount of calcium flows in through the channels and there is a sharp rise in intracellular calcium in the vicinity of the open channels. Because the cytoplasmic calcium concentration is normally low, but it can undergo a large, rapid local increase when a calcium channel opens, calcium serves as a temporally and spatially precise biochemical trigger within cells.

## Cellular Mechanisms Used to Maintain the Very Low Intracellular Calcium Concentration

Cells use many mechanisms to maintain a very low intracellular free calcium ion concentration, including plasma membrane transporters, intracellular storage organelles, and protein buffers. These mechanisms work in concert to restore cytoplasmic calcium to a range of 50–100 nM following any calcium influx into the cell.

### Presynaptic Calcium Ion Plasma Membrane Transporters

There are two plasma membrane transporters that extrude calcium from the cytoplasm after it rises above resting levels (Brini and Carafoli, 2011). Moving calcium out of the cell requires energy because even when intracellular calcium rises significantly above resting levels (e.g., to 10 μM), the calcium concentration outside of the cell is always much higher (1.5–2 mM). Therefore, pumps must use a source of energy to move calcium against this strong calcium concentration gradient.

The first transporter, or pump, is the **plasma membrane calcium–magnesium ATPase** (PMCA). This transporter is part of a large family of ion and lipid transporters called "P-ATPases," named for their ability to phosphorylate themselves using ATP (Palmgren and Nissen, 2011).

The PMCA pump contains a high-affinity calcium-binding site on its cytoplasmic surface. **Binding affinity** refers to the strength of a binding interaction between a single molecule and its binding partner. We typically measure the strength of binding by the **equilibrium dissociation constant ($K_D$)**. $K_D$ is calculated as the ratio of the unbinding rate ($K_{off}$) divided by the binding rate ($K_{on}$), and $K_D$ and affinity are inversely related. Thus, the smaller the $K_D$ value, the higher the binding affinity of a molecule for its binding partner. The PMCA pump also associates with a type of **calmodulin** (which is short for "calcium-modulated protein," a family of common mediators of calcium action in cells) in the cytoplasm. When intracellular calcium levels rise, the calcium activates calmodulin, which binds to the PMCA pump and increases the affinity of the pump's calcium-binding site.

The PMCA has a $K_D$ of about 300–500 nM (Carafoli, 1991; Mangialavori et al., 2010), which means that when the cytoplasmic calcium concentration is elevated from 50–100 nm to 300–500 nM, there is a 50% likelihood that a calcium ion will occupy the binding site on the PMCA at any given moment. Therefore, when calcium rises above resting levels, the high-affinity calcium-binding site on the cytoplasmic surface of the PMCA binds a calcium ion, triggering the auto-phosphorylation of the PMCA using ATP (with magnesium as a cofactor). This phosphorylation changes the configuration of the PMCA, repositioning the calcium-binding site such that the bound calcium ion is translocated to the other side of the membrane (see Fig. 7.2). In this extracellular-facing configuration, the

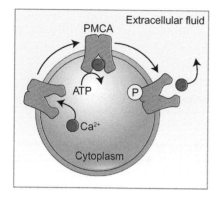

FIGURE 7.2 **The plasma membrane calcium–magnesium ATPase.** The plasma membrane calcium–magnesium ATPase (PMCA) binds calcium in the cytoplasm when the concentration of calcium is elevated to about 500 nM. When calcium binds, ATP is hydrolyzed, which leads to the autophosphorylation of the PMCA pump. Once phosphorylated, the pump changes configuration and translocates the bound calcium ion to the extracellular surface of the cell, where it is released into the extracellular fluid.

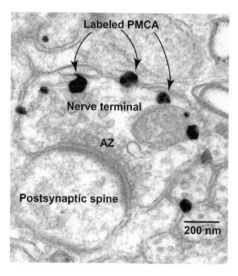

FIGURE 7.3 **The distribution of PMCA pumps at nerve terminals in the molecular layer of the cerebellum.** PMCA pumps were identified using a gold-labeled antibody (seen as dark patches here). These pumps are located away from the active zone (i.e., neurotransmitter release sites), often at the back of the nerve terminal. *Source: Adapted from Burette, A., Weinberg, R.J., 2007. Perisynaptic organization of plasma membrane calcium pumps in cerebellar cortex. J. Comp. Neurol. 500, 1127–1135.*

PMCA calcium-binding site now has a low affinity for calcium, so it releases the calcium ion into the extracellular space. The entire process is reset by phosphatases that cleave the phosphate off the PMCA, causing it to revert back to its original configuration, ready for another round of calcium transport. The PMCA transports one calcium ion out of the cell for every ATP that is hydrolyzed, and is considered to be a high-affinity, **low-capacity** (slow-working) pump.

It is hypothesized that at some CNS nerve terminals, PMCA pumps are not located at neurotransmitter release sites (Burette and Weinberg, 2007; see Fig. 7.3). In these terminals, PCMA pumps are not likely to be directly coupled to sites of calcium influx (i.e., within the active zone), but rather appear to be strategically positioned to only pump calcium after it has accumulated within the entire nerve terminal. This is because these PMCA pumps respond to volume-averaged increases in the nerve terminal's calcium ion concentration, as opposed to the local increases in calcium concentration that occur in the immediate area around an open calcium channel. These volume-averaged increases are not likely to occur after a single action potential, and instead occur predominantly after bursts of action potential activity. In contrast, work on a peripheral parasympathetic ganglion showed that PMCA is located very close to the active zone in these neurons (Juhaszova et al., 2000), where it may contribute to the local active zone control of intracellular calcium concentration. Furthermore, because PMCA is located close to the active zone, it may be in a position to pump calcium back into the synaptic cleft, which would reduce changes in the local extracellular calcium ion concentration after calcium enters the nerve terminal at active zones during action potential activity.

The second pump that contributes to the removal of excess calcium from the cytoplasm is the **sodium−calcium exchanger** (NCX), which belongs to a large family of calcium/cation exchangers (Brini and Carafoli, 2011). As the name implies, the NCX pump exchanges calcium ions for sodium ions across the plasma membrane, moving one calcium ion out of the cell for every three sodium ions that it moves into the cell (see Fig. 7.4). This pump also needs to transport calcium ions against the strong calcium concentration gradient, but it does not utilize ATP as an energy source for this process. Instead, the NCX takes advantage of the strong sodium gradient that exists across the cell membrane. Moving sodium ions down their concentration gradient provides the energy needed to move calcium ions against their concentration gradient.

Because the NCX pump moves two positive charges (because calcium is a divalent cation) out of the cell for every three positive charges (because sodium is a monovalent cation), it creates a net +1 positive charge movement into the cell for every cycle of the pump. If the NCX pump were more frequently active, this **electrogenic** action would slightly depolarize the cell. However, the NCX pump is not frequently active because it has a lower calcium-binding affinity than the PMCA transporter does; its estimated calcium-binding affinity is in the range of 700 nm−1 μM (Lee et al., 2007; Blaustein et al., 1991). Despite this low affinity, the NCX pump works at a higher capacity than the PMCA, meaning that it pumps calcium faster while it is active. However, since the calcium concentration in the nerve terminal rarely reaches levels high enough to activate the NCX pump, it is thought to only be active briefly following high-frequency action potential activity. NCX is generally thought to be located far from the active zone, on nerve terminal membrane that is not close to neurotransmitter release sites (Juhaszova et al., 2000). These characteristics make the NCX pump well-suited for facilitating rapid recovery from brief high calcium signals in the nerve terminal.

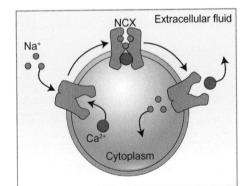

FIGURE 7.4 **The NCX transporter uses the sodium gradient as the energy to move calcium against its concentration gradient.** Three sodium ions move into the cell down the sodium concentration gradient in order to pump one calcium ion out of the cell.

## Presynaptic Cellular Organelles That Buffer Calcium

In addition to plasma membrane pumps, nerve terminals have organelles that assist in calcium homeostasis. One such organelle is the **endoplasmic reticulum**. The endoplasmic reticulum is typically known more for its functions in the cell body, but it also exists in axons and nerve terminals, where it contributes to calcium homeostasis (de Juan-Sanz et al., 2017) by serving as an intracellular calcium storage organelle. The resting calcium concentration inside the endoplasmic reticulum is estimated to be about 150–500 µM, varying significantly between cell types and with cell activity (Yu and Hinkle, 2000; de Juan-Sanz et al., 2017).

The endoplasmic reticulum contains multiple protein complexes that contribute to calcium homeostasis (see Fig. 7.5). The first is the sarco-endoplasmic reticulum $Ca^{2+}$-ATPase (SERCA). SERCA is a pump that works in exactly the same way as the PMCA pump described previously, except that it is located on the ER. There, it binds to calcium ions in the cytoplasm and pumps them into the lumen of the endoplasmic reticulum.

The second is the ryanodine receptor, which mediates calcium-induced calcium release (a positive-feedback mechanism) from the endoplasmic reticulum. The ryanodine receptor is a type of ion channel that is opened by cytoplasmic calcium, allowing more calcium to pass from the endoplasmic reticulum into the cytoplasm. At synapses, the

[SERCA = PMCA]

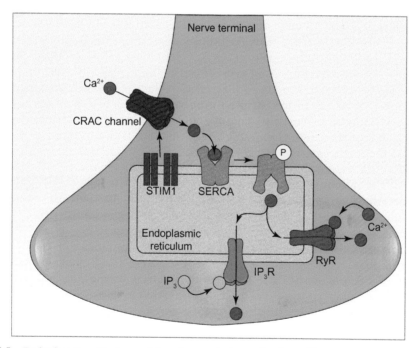

FIGURE 7.5  **Endoplasmic reticulum calcium homeostasis mechanisms.** The protein STIM1 links the endoplasmic reticulum to a plasma membrane CRAC channel to bring calcium into the lumen of the ER using a SERCA pump, and two ligand-gated channels can release calcium from the ER ($IP_3R$ and RyR). These mechanisms are described in detail in the text.

ryanodine receptor appears to be involved in activity-dependent changes in synaptic transmission, since calcium entry during action potential activity can trigger calcium release from the endoplasmic reticulum. The calcium released from the endoplasmic reticulum then contributes to calcium-triggered neurotransmitter release during subsequent action potential activity (see, e.g., Unni et al., 2004; Futagi and Kitano, 2015; Johenning et al., 2015).

The third protein is the inositol trisphosphate receptor ($IP_3R$). This protein is a ligand-gated ion channel that is activated by the second messenger $IP_3$. When activated by $IP_3$, the $IP_3R$ ligand-gated ion channel opens to permit calcium release from the endoplasmic reticulum. This receptor is involved in a large number of modulatory events at synapses, often triggered by G-protein-coupled receptors (see Chapter 12 for more details).

Finally, proteins called the STIM1 and store-operated calcium release-activated calcium (CRAC) channels can sense the calcium concentration within the ER and replenish it when it becomes too low. STIM1 senses calcium on the portion of the protein that is inside the ER lumen. If a STIM1 protein detects a low calcium concentration in the endoplasmic reticulum, it associates with other STIM1 proteins to create a multi-protein complex that binds to the Orai protein (Prakriya, 2013), which forms the pore of the CRAC channel. The binding of the STIM1 complex to the CRAC channel induces the channel to open and conduct calcium from the extracellular fluid into the cell cytoplasm (Hogan and Rao, 2015). This occurs within local domains positioned very close to the endoplasmic reticulum where SERCA pumps this calcium into the endoplasmic reticulum.

Mitochondria are another important type of organelle involved in calcium homeostasis (Kwon et al., 2016). Mitochondria are well-known for their role in cell metabolism and the generation of ATP, which neurons require for energy-dependent events, including synaptic transmission. However, they also take up and store large amounts of calcium after high-frequency stimulation (Kirichok et al., 2004), making them an important contributor to synaptic calcium homeostasis.

A mitochondrion has an inner and outer membrane that separate it into two aqueous compartments, the intermembrane space and the matrix. The mitochondrial matrix is a major cellular calcium storage pool and can be used to dynamically control cellular calcium concentration. Most of the calcium within the mitochondrial matrix is in the form of an insoluble calcium phosphate, and the remaining calcium is free, though the percentage of bound or free calcium can change. Only the free calcium ions move across the mitochondrial membranes.

Movement of calcium across the outer mitochondrial membrane is mediated by a voltage-dependent anion channel (VDAC). This relatively nonselective channel allows calcium to move into the mitochondrial intermembrane space (Shoshan-Barmatz et al., 2017), whereas more selective processes control calcium entry across the inner membrane (in and out of the matrix). There are two major mechanisms for the movement of calcium across the inner membrane (see Fig. 7.6). First, the inner membrane contains a **mitochondrial calcium uniporter (MCU)**. The MCU is a channel that mediates the influx of calcium down its concentration gradient into the matrix. Second, the inner membrane also contains a **mitochondrial sodium−calcium exchanger (mNCX)** that is similar to the NCX on the

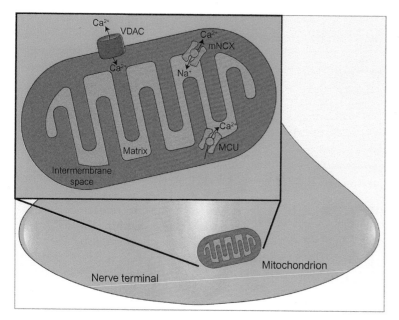

FIGURE 7.6 Mechanisms of calcium homeostasis found in mitochondria. Mitochondria move calcium across the outer membrane using an anion channel (VDAC), and use two major mechanisms to move calcium across the inner membrane (MCU and mNCX). The function of these proteins is described in more detail in the text.

plasma membrane, which was discussed earlier (Palty and Sekler, 2012). The mNCX uses the sodium ion gradient as the energy to move calcium out of the matrix against its concentration gradient. Using these two mechanisms, mitochondria can rapidly buffer large and sudden increases in calcium. Mitochondria do not seem to be engaged in the calcium homeostasis process following small calcium influx events, such as those that result from a single action potential; however, they are recruited during high-frequency activity that causes a large influx of calcium (Kim et al., 2005).

### *Nerve Terminals Contain Endogenous Calcium Buffer Proteins*

Neurons express a large number of proteins that bind calcium ions (Schwaller, 2010). These calcium-binding proteins can exist within the cytoplasm at concentrations ranging from 10 μM to several millimolar. Some types of neurons can be classified by the predominant calcium-binding protein they express. Calcium-binding proteins that have been used to classify neurons include **calretinin**, **calbindin**, and **parvalbumin**. In the context of synaptic function, these proteins act as mediators of calcium-triggered events in neurons and/or as buffers that quickly reduce the concentration of free calcium ions in the cytoplasm after a rapid calcium influx into the presynaptic terminal. Calcium-binding proteins can often influence synaptic function and mediate some forms of synaptic plasticity (Pongs et al., 1993; Schwaller et al., 2002; Caillard et al., 2000; see Chapter 14).

In summary, nerve terminals have a variety of mechanisms that work together to tightly control the concentration of calcium within the cytoplasm (see Fig. 7.7). These mechanisms maintain a resting intracellular concentration of around 50–100 nM and respond to the rapid calcium influx caused by action potentials, limiting the spread of calcium ions within the cytoplasm and quickly returning the calcium concentration to resting levels.

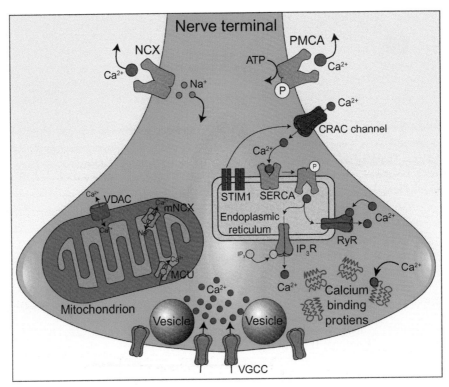

FIGURE 7.7 **Summary of calcium buffering and handling mechanisms in the nerve terminal.** This diagram provides a summary of the variety of mechanisms that control calcium in the nerve terminal. These include calcium entry through the voltage-gated calcium channels (VGCC), which are positioned next to synaptic vesicles that are ready for neurotransmitter release; calcium-binding proteins; and pumps in the plasma membrane (NCX and PMCA). Also depicted is a mitochondrion that moves calcium across its outer membrane using an anion channel (VDAC), and uses two major mechanisms to move calcium across its inner membrane (MCU and mNCX). Finally, the protein STIM1 links the endoplasmic reticulum to a plasma membrane CRAC channel to bring calcium into the lumen of the ER using a SERCA pump, and two ligand-gated channels can release calcium from the ER (IP$_3$R and RyR). These mechanisms are described in detail in the text.

# CONTROL OF NEUROTRANSMITTER RELEASE BY CALCIUM IONS

Early researchers studying chemical neurotransmitter release found that the concentration of extracellular calcium strongly influenced the amount of neurotransmitter released at a synapse (Katz and Miledi, 1965). When the concentration of extracellular calcium was experimentally decreased from 2 mM to 0.2 mM, the probability of action potential-evoked neurotransmitter release also decreased (as discussed in Chapter 6). These results imply that an action potential opens a transient pathway for calcium entry, which triggers chemical neurotransmitter release. During the brief depolarization caused by an action potential, the concentration of calcium outside the nerve terminal is proportional to the influx of calcium moving down its concentration gradient. This interpretation is supported by an experiment in which researchers removed all calcium from the extracellular solution, and then added controlled amounts of calcium via a pipette positioned very close to the synapse being studied. When extracellular calcium was absent, there was no action potential-evoked **endplate potential (EPP)**, but when calcium was expelled from the pipette onto the synapse during stimulation of the nerve, an EPP could be measured in the postsynaptic muscle cell (see Fig. 7.8). This experiment demonstrated that calcium is necessary for chemical neurotransmitter release.

However, these experiments provided only indirect evidence of the effects of cytoplasmic calcium ion concentration on the probability of chemical neurotransmitter release. In this section we will explore the relationship between calcium and neurotransmitter release further, and will provide experimental evidence that calcium ions are both necessary and sufficient for chemical neurotransmitter release. To more directly explore the relationship between intracellular calcium concentration and neurotransmitter release, it was useful for researchers to increase calcium concentration within the nerve terminals directly [without needing to use voltage-gated calcium channels (VGCC) to bring it in from the extracellular space]. At the squid giant synapse, experimenters accomplished this by using a microelectrode to inject calcium ions directly into the nerve terminal, which produced a transient increase in neurotransmitter release (Miledi, 1973). However, most nerve terminals are too small to permit insertion of a microelectrode to directly inject calcium ions. In addition, the calcium ion buffering and handling

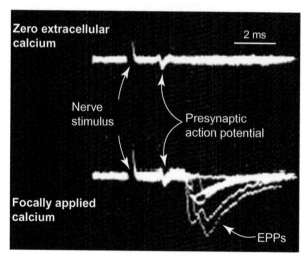

FIGURE 7.8 **Electrical response to nerve stimulation and focally applied calcium at a neuromuscular junction.** Using extracellular electrodes positioned on top of the neuromuscular junction. The nerve stimulus was detected as an electrical "artifact," the presynaptic action potential was detected as a local extracellular current due to the sodium influx, and the postsynaptic EPPs were detected as large downward currents. In the absence of extracellular calcium, stimulation of the presynaptic nerve resulted in a presynaptic action potential, but no postsynaptic endplate potential (top). However, when calcium was applied focally to the nerve terminal during stimulation of the nerve, both a presynaptic action potential and a postsynaptic endplate potential could be measured (bottom). *Source: Adapted from Katz, B., Miledi, R., 1965. The effect of calcium on acetylcholine release from motor nerve terminals. Proc. R. Soc. Lond. B. Biol. Sci. 161, 496–503.*

mechanisms described above quickly return the calcium concentration in the nerve terminal to a resting level after free calcium ions are injected. Therefore, injecting calcium into the nerve terminal is both an impractical and an inefficient method of increasing its cytoplasmic calcium concentration. For this reason, researchers developed other experimental techniques to persistently increase calcium concentration specifically in axon terminals, and we will discuss three of these.

One effective experimental approach was to use a **calcium ionophore** (a chemical that creates a pathway for calcium to cross cell membranes) to make the nerve terminal plasma membrane persistently permeable to calcium ions (Kita and Van der Kloot, 1974, 1976). At the frog neuromuscular junction, application of a calcium ionophore increased both the frequency of spontaneous neurotransmitter release (mEPPs) and the magnitude of action potential-evoked neurotransmitter release (EPPs). A similar technique used **calcium liposomes**, experimentally fabricated phospholipid vesicles that contain calcium ions. When these liposomes are added to the solution outside nerve terminals, they fuse with the plasma membrane and deposit their contents into the cytoplasm of the nerve terminal. The calcium liposome technique was used at the frog neuromuscular junction to demonstrate that the addition of calcium ions directly to the nerve terminal cytoplasm could increase the frequency of mEPPs and the magnitude of EPPs (Rahamimoff et al., 1978).

Later, a molecule known as **caged calcium** was developed. This molecule has a "cage" structure that can hold a calcium ion inside it, which means the calcium ion is not free to affect neurotransmitter release. Caged calcium can be introduced to a nerve terminal either by injecting it via a microelectrode, or by backfilling the presynaptic terminal via the cut end of a nerve. When the caged calcium molecule is exposed to ultraviolet (UV) light, it breaks apart and releases the calcium ion into the cytoplasm in the presynaptic terminal, and the freed calcium ion is then able to participate in triggering neurotransmitter release (see Fig. 7.9; Kaplan and Ellis-Davies, 1988). Caged calcium became a powerful tool that gave synaptic researchers more precise control over the concentration of calcium in nerve terminals. By applying a brief flash of UV light focused on the active zone of the nerve terminal, an experimenter could produce a transient rise in intraterminal calcium, the magnitude of which could be regulated by varying the intensity and duration

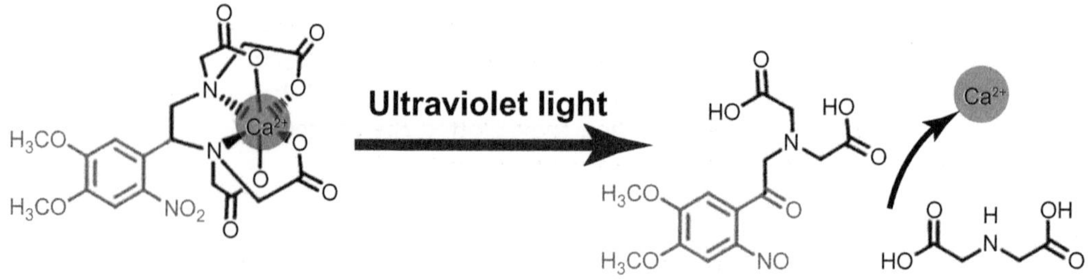

FIGURE 7.9 **A caged calcium molecule allows experimenters to liberate free calcium ions in the nerve terminal in response to UV illumination.** One type of caged calcium molecule, DM-nitrophen (left), binds calcium with high affinity. Calcium-bound DM-nitrophen can be injected into large nerve terminals. Upon exposure to ultraviolet light, the "cage" is cleaved into two fragments, liberating a free calcium ion into the cytoplasm (right).

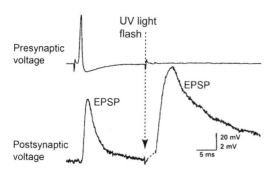

FIGURE 7.10 Voltage recordings from the presynaptic nerve terminal (top) and postsynaptic cell (bottom) at the squid giant synapse after loading the nerve terminal with the caged calcium molecule DM-nitrophen. A presynaptic action potential (top trace, left) evokes an EPSP (bottom trace, left) that is smaller than the EPSP evoked by a large flash of UV light (bottom trace, right). *Source: Adapted from Delaney, K.R., Zucker, R.S., 1990. Calcium released by photolysis of DM-nitrophen stimulates transmitter release at squid giant synapse. J. Physiol. 426, 473–498.*

of the UV light flash (Zucker, 1993). This method was used at the squid giant synapse to demonstrate that a rise in intraterminal calcium was sufficient to trigger chemical neurotransmitter release in the absence of a presynaptic action potential (see Fig. 7.10; Delaney and Zucker, 1990; Zucker, 1993).

Taken together, the experiments described above demonstrate that calcium is capable of triggering chemical neurotransmitter release in the absence of a presynaptic action potential. This is an important finding because it shows that neurotransmitter release is not dependent on presynaptic voltage-gated ion channels (sodium, potassium, or calcium), or depolarization of the axon terminal. Instead, voltage-gated ion channels simply participate in physiological mechanisms that increase the concentration of calcium within a nerve terminal, and it is the increase in calcium concentration that actually triggers neurotransmitter release. The sodium and potassium channels in neurons are still critical for neurotransmitter release in a physiological context, since they are required for the generation of action potentials, which open VGCCs in the nerve terminal. However, voltage-gated ion channels can be bypassed by experimentally elevating the calcium concentration in the nerve terminal by other means.

## The Nonlinear Relationship Between Calcium and Neurotransmitter Release

Although we have presented experimental evidence that calcium ion entry into the nerve terminal increases the probability of chemical neurotransmitter release, we have not fully explored the detailed relationship between calcium and neurotransmitter release (the **calcium-release relationship**). Researchers first examined this relationship at the frog neuromuscular junction by varying the concentration of extracellular calcium (Jenkinson, 1957; Dodge and Rahamimoff, 1967). When they plotted the concentration of extracellular calcium against the magnitude of neurotransmitter released (measured as EPP size), they found that the calcium-release relationship was nonlinear (see Fig. 7.11). For example, small changes in extracellular calcium concentration led to much larger changes in neurotransmitter release (measured as changes in EPP size). When plotted on a log scale, the relationship had a slope of about 4. Remarkably, every synapse that has been studied (across all species and across all neuron types) displays a similar calcium-release relationship (with estimated slopes of 3–5). These results show that there is a roughly fourth-order relationship between calcium and neurotransmitter release.

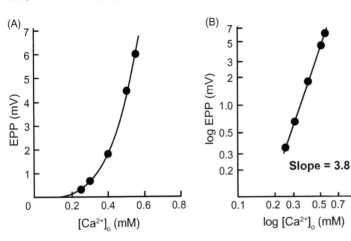

FIGURE 7.11 **Transmitter release as a function of the concentration of extracellular calcium.** Transmitter release was quantified by measuring the amplitudes of EPPs recorded at the frog neuromuscular junction as the concentration of extracellular calcium ($[Ca^{2+}]_o$) was varied. (A) When plotted using linear scales, it can be seen that there is a nonlinear relationship between calcium and neurotransmitter release. (B) When plotted on a log-log scale, the relationship is linear with a slope of 3.8.

What is responsible for this fourth-order calcium-release relationship, and what does it tell us about calcium-triggered chemical neurotransmitter release? The most straightforward interpretation of the nonlinear relationship between calcium concentration and neurotransmitter release is that the fusion of one synaptic vesicle is triggered by the action of more than one calcium ion, since if one calcium ion could trigger the fusion of one synaptic vesicle, changing the concentration of calcium would be predicted to change neurotransmitter release in a linear fashion. Initially, it was proposed that perhaps there were four calcium-binding sites on synaptic vesicles, all of which had to be occupied to trigger a vesicle to fuse with the plasma membrane. In addition, it was thought that the binding of one calcium ion might change the binding affinity for other calcium ions, suggesting there was a set of cooperative binding reactions. However, this early theory was hard to reconcile with subsequent biochemical and proteomic studies of synaptic vesicle proteins. One such protein, synaptotagmin, was hypothesized to be the calcium sensor for neurotransmitter release. Synaptotagmin can bind multiple calcium ions (two to five), and there are 8–15 synaptotagmin proteins attached to each synaptic vesicle (see Chapter 8). These data indicate that there are far more than just four calcium-binding sites on vesicles (there may be 16–75). Current hypotheses propose that multiple synaptotagmin proteins participate in calcium-triggered synaptic vesicle fusion, but the specific number of calcium ions that need to bind to one synaptotagmin protein, and how many of these proteins need to be calcium-bound to cause a synaptic vesicle to fuse, is currently under study. A computational modeling study that is consistent with biochemical and proteomic data argues that a fourth-order calcium-release relationship can be achieved if there are more synaptotagmin proteins per vesicle than are needed to bind calcium to trigger vesicle fusion (Dittrich et al., 2013). Validation of this model and determination of the specific details of this reaction will require further study.

## Where Are Calcium Channels Located Within the Nerve Terminal?

Scientists studying calcium's role in neurotransmitter release initially did not know where the calcium sources needed to be located within the nerve terminal in order to trigger vesicle release. Were the calcium channels close to the vesicles or far away? This spatial relationship is important to understand because it influences how far calcium ions need to diffuse to reach synaptic vesicles, which in turn affects how much calcium must enter the presynaptic terminal to trigger release.

Some synaptic vesicles in the nerve terminal are "docked," meaning they are ready to be released as soon as calcium is available, and they are found near the plasma membrane. To determine the spatial relationship between VGCCs and docked vesicles, investigators injected two different types of calcium buffers into the nerve terminal—EGTA (a relatively slowly binding calcium buffer) and BAPTA (a faster acting calcium buffer)—and compared their effects (Adler et al., 1991). The idea was that if calcium channels were very close to docked synaptic vesicles, there would not be much time between calcium entry and neurotransmitter release, because the calcium would not need to diffuse far from where it entered through calcium channels in order to trigger neurotransmitter release. Therefore, a calcium buffer would not have much time to interfere with calcium-triggered neurotransmitter release. In contrast, if calcium channels were far from docked synaptic vesicles, an injected buffer would have more time to bind the calcium ions before they could reach docked vesicles, which means an injected buffer with slow binding kinetics could impede neurotransmitter release. When the effects of the slow-acting EGTA buffer and the faster-acting BAPTA buffer were compared at the squid giant synapse, EGTA (the slower buffer) had no effect on vesicle release, but BAPTA (the faster buffer) substantially reduced neurotransmitter release (see Fig. 7.12). Because only the faster calcium buffer was able to compete with vesicles

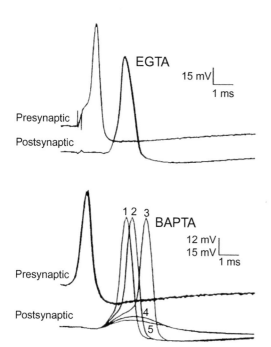

FIGURE 7.12 **The effects of calcium buffers on neurotransmitter release at the squid giant synapse presynaptic nerve terminal.** In these experiments, the presynaptic nerve terminal was stimulated to generate a presynaptic action potential. This was followed after a 1–2-ms delay by a postsynaptic action potential because the normal amount of neurotransmitter release at this synapse was strong enough to bring the postsynaptic cell to the action potential threshold. In the top panel, EGTA, a slow calcium buffer, was injected into the nerve terminal, but the postsynaptic action potential was unchanged, showing that EGTA had no effect on chemical neurotransmitter release. In the bottom panel, the injection of BAPTA, which buffers calcium faster than EGTA, gradually blocked neurotransmitter release such that the resulting depolarization gradually became below threshold with time after the injection, and no longer caused a postsynaptic action potential. The five numbered postsynaptic traces shown were recorded successively as BAPTA diffused into the nerve terminal to gradually block neurotransmitter release. *Source: Adapted from Adler, E.M., Augustine, G.J., Duffy, S.N., Charlton, M.P., 1991. Alien intracellular calcium chelators attenuate neurotransmitter release at the squid giant synapse. J. Neurosci. 11, 1496–1507.*

for calcium binding, investigators concluded that the calcium channels that provide the source for calcium-triggered neurotransmitter release must be very close to the calcium-binding sites on vesicles. Based on the speed of BAPTA binding, it was hypothesized that calcium must bind to synaptic vesicle calcium-binding sites in less than about 200 μs after entry into the nerve terminal.

The conclusion that calcium channels are positioned very close to neurotransmitter release sites was later supported by optical experiments that used calcium-sensitive florescent dyes injected into nerve terminals. These dyes change their fluorescent properties when they bind calcium ions, which allowed researchers to localize calcium entry sites by observing where changes in the fluorescence of the dyes occurred. This technique was first applied to the squid giant synapse (Smith et al., 1993) and revealed that calcium entry during nerve stimulation did not occur in all areas of the presynaptic nerve terminal, but was restricted to the region that was in close contact with the postsynaptic cell (see Fig. 7.13).

Using a similar imaging technique at the frog neuromuscular junction, other researchers confirmed that calcium entry occurred near the site of neurotransmitter release (Wachman et al., 2004; Luo et al., 2011). The advantage of using the neuromuscular synapse to study localized sites of calcium entry is that the neurotransmitter release sites are organized into parallel rows along the length of the nerve terminal. These rows can be visualized using a fluorescent label for postsynaptic acetylcholine receptors, which are positioned directly opposite presynaptic active zone neurotransmitter release sites (see Fig. 7.14). Using this experimental preparation, investigators also noticed that calcium entry did not flood the neurotransmitter release sites uniformly. This finding led to the conclusions that at this synapse, (1) there are relatively few VGCC within each release site, and (2) each calcium channel has a relatively low probability of opening following a single action potential (Luo et al., 2011). If these data are representative of the mechanisms that control most synapses (and it has been argued that this might be true; Tarr et al., 2013), it appears that active zones have few VGCC, these channels are closely positioned adjacent to docked synaptic vesicles, and each channel has a relatively low probability of opening during an action potential (about 20%–30% of the time).

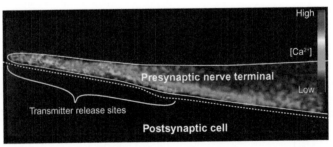

FIGURE 7.13  Calcium-sensitive dye fluorescence in the squid giant synapse nerve terminal after presynaptic stimulation. When the presynaptic nerve terminal was stimulated at a high frequency, the calcium-sensitive dye detected the largest calcium entry (*red color*) along the edge of the nerve terminal that was adjacent to the postsynaptic cell. *Source: Adapted from Smith et al. (1993).*

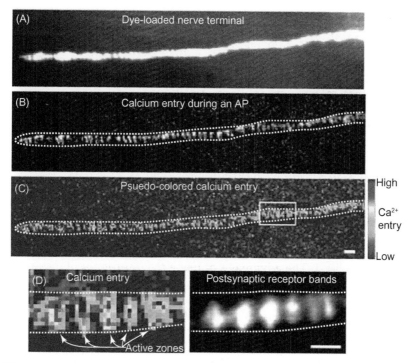

FIGURE 7.14  **Imaging calcium at the frog neuromuscular junction.** (A) Calcium dye loaded into the nerve terminal. In panels (B–D) the nerve terminal is indicated using a dotted line. (B) When the presynaptic neuron is stimulated, evoking an action potential, the presynaptic calcium-induced increase in fluorescence is restricted to defined locations along the length of the synapse. (C) The intensity of the fluorescence change is converted to a pseudo-color scale to more clearly represent sites of high calcium concentrations. (D) (Left) Enlargement of the area in C shown by the white box, demonstrating restricted sites of calcium entry within active zone (arrows). (Right) The same region stained with a label for the bands of postsynaptic receptors, which are predictive of the location of presynaptic active zones. Comparing the left and right panels, it is clear that calcium entry is restricted to active zone regions of the frog nerve terminal. Scale bars = 2 μm. *Source: Adapted from Wachman et al. (2004).*

The hypothesis that presynaptic calcium channels are positioned only at neurotransmitter release sites was also supported by direct labeling of these channels. Using a fluorescence-tagged toxin (ω-conotoxin GVIA, see Box 7.2) that binds selectively to the type of VGCC located at the frog neuromuscular junction, investigators showed that these labeled toxins bound only at presynaptic active zones (see Fig. 7.15). At the frog neuromuscular junction, this labeling is found in bands located at the active zones of this nerve terminal (see Chapter 8 for more details).

## Cytoplasmic Calcium Microdomains

What concentration of calcium normally triggers neurotransmitter release? To answer this question, researchers manipulated the concentration of calcium in the nerve

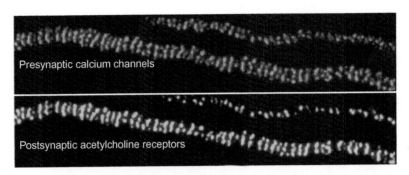

FIGURE 7.15  **A frog neuromuscular junction labeled with two fluorescent toxins.** These images show the alignment of presynaptic voltage-gated calcium channels with postsynaptic acetylcholine receptors (which are predictive of the location of presynaptic active zones). In this image, the nerve terminal is stained with ω-conotoxin GVIA (*orange*) to localize presynaptic calcium channels, and with α-bungarotoxin (*green*) to localize postsynaptic receptors. The top and bottom panels of this figure represent the same nerve terminal imaged first in orange (top) and then in green (bottom). A comparison of these two panels reveals that the calcium channel label (*orange*) is perfectly aligned with the acetylcholine receptor label (*green*). *Source: Adapted from Robitaille, R., Adler, E.M., Charlton, M.P., 1990. Strategic location of calcium channels at transmitter release sites of frog neuromuscular synapses. Neuron 5, 773–779.*

terminal (using the caged calcium approach to liberate specific amounts of calcium, in combination with calcium imaging to calculate the resulting calcium concentration) and measured the amount of neurotransmitter release that occurred at different calcium concentrations (Heidelberger et al., 1994; Schneggenburger and Neher, 2000). Based on their results, they estimated that neurotransmitter release is normally triggered when calcium rises to 10–100 μM in the nerve terminal. The next question researchers asked was how much of the spatial environment within the nerve terminal would be exposed to a calcium concentration of 10–100 μM after a single action potential stimulation. To answer this question, they needed to consider the buffering and handling of calcium ions within the nerve terminal and the specific localization of VGCCs very close to the neurotransmitter release sites within the nerve terminal. By using calcium-sensitive dyes with a low affinity for calcium, such that they only fluoresce when calcium rises to the relevant 10–100 μM range, the researchers demonstrated that calcium only reaches this concentration in the immediate vicinity of open calcium channels (see Fig. 7.17; Llinas et al., 1992). Thus, after VGCCs open, the concentration of calcium reaches the micromolar range only in a very localized volume of the nerve terminal cytoplasm that is just inside the pore of open calcium channels. The local cytoplasmic volume where calcium reaches 10–100 μM around a cluster of calcium channels is called a **microdomain** (see Fig. 7.18), and the volume around the pore of a single open calcium channel is called a **nanodomain**. Because this high concentration of calcium is required to trigger vesicle fusion and neurotransmitter release, it is hypothesized that synaptic vesicles must be positioned within a microdomain in order to be triggered to fuse with the plasma membrane.

## BOX 7.2

## DRUGS FROM NATURE: CONE SNAIL TOXINS AND CHRONIC PAIN TREATMENT

### Conus magus

The conotoxins derived from predatory *Conus magus* snails (see Fig. 7.16) have been extensively used by scientists to investigate calcium channel inhibition, but this toxin was also approved in 2004 as a therapy for chronic pain management under the name ziconotide (brand name Prialt). Conotoxins block N-type VGCCs, resulting in reduced release of nociceptive transmitters in the dorsal horn of the spinal cord, including glutamate, calcitonin gene-related peptide, and substance P. Tolerance and dependence seem to be minimal in patients receiving chronic ziconotide treatment, but the necessary administration method of this

FIGURE 7.16  *Conus magus*. Richard Parker (CC BY 2.0).

### BOX 7.2 (cont'd)

drug is extremely invasive and requires intrathecal lumbar puncture, colloquially known as a spinal tap. Chronic ziconotide treatment is usually achieved by the placement of an infusion pump in the spinal cord, and, while invasive, provides powerful antinociceptive benefits for patients experiencing chronic pain and who are resistant to or unable to maintain opioid-based treatments.

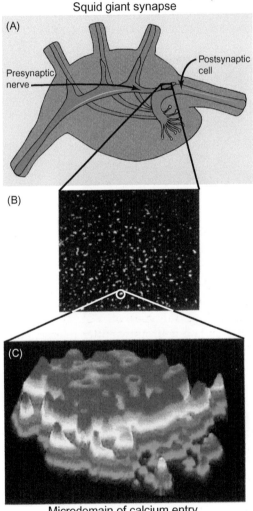

FIGURE 7.17 **Imaging calcium entry during action potential activity using a low-affinity calcium-sensitive dye.** (A) The presynaptic axon (*blue*) of the squid giant synapse was filled with n-aeqourin-j, a calcium-sensitive dye that only fluoresces when the calcium concentration reaches 100 μM. (B) After action potential activity, the small spots show where increases in fluorescence can be detected in the presynaptic axon. (C) An enlargement of one such localized calcium spot, psuedocolored for fluorescence intensity, represents a single microdomain of calcium entry. *Source: Adapted from Llinas, R., Sugimori, M., Silver, R.B., 1992. Microdomains of high calcium concentration in a presynaptic terminal. Science 256, 677–679.*

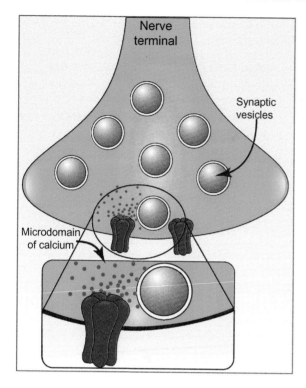

FIGURE 7.18 **Calcium microdomains.** A calcium microdomain is the small, localized volume of the nerve terminal cytoplasm near the mouth of open calcium channels, within which the concentration of calcium ions (red dots) reaches the micromolar range after voltage-gated channels open. Only synaptic vesicles that are in the active zone are exposed to this concentration of calcium.

# VOLTAGE-GATED CALCIUM CHANNELS IN NERVE TERMINALS

As described above, calcium ions enter the presynaptic nerve terminal through VGCCs in response to action potentials. Chapter 3, described the basic function of voltage-gated channels, but in this chapter we will cover more specific characteristics of VGCCs that explain the role they play in nerve terminals, including subtypes of VGCCs that are expressed in nerve terminals, how their properties regulate neurotransmitter release, and how these channels respond to action potential stimuli to create the calcium ion trigger for neurotransmitter release.

## The Structure of Voltage-Gated Calcium Channels

VGCCs primarily consist of an α-1 subunit that is made up of a single long string of amino acids with a repeating structure reminiscent of all voltage-gated channels (see Chapter 3). The α-1 subunit is made up of four domains (I–IV), each consisting of six transmembrane segments (S1–S6), with S4 containing charged amino acids that allow the protein to sense voltage across the membrane (see Fig. 7.19). Although all voltage-gated channels share this general structure, VGCCs are a diverse group of proteins that include three major families (Cav1, Cav2, and Cav3) with specific properties. One of the main differences between these families is their sensitivity to membrane voltage. Cav3 channels are considered low-voltage-activated channels since they open near the resting membrane potential, while Cav1 and Cav2 channels are considered high-voltage-activated channels since they open at relatively depolarized membrane potentials.

Different calcium channel types are selectively expressed in different organs in the body and across different parts of a single neuron. Only the high-voltage-activated type calcium channels are effective sensors of action potential activity (or any very strong depolarization), and therefore capable of functioning as the activity-dependent trigger for neurotransmitter release. If low-voltage-activated calcium channels were expressed in axon terminals, they would mediate frequent and robust calcium ion entry at voltages near the

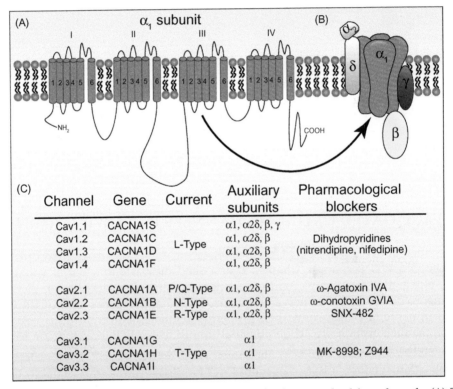

FIGURE 7.19 **The structural organization and diversity of voltage-gated calcium channels.** (A) The predicted distribution of amino acids across the plasma membrane for all voltage-gated calcium channels. (B) Assembly of the α-1 subunit of a voltage-gated calcium channel (*orange*), in combination with auxiliary subunits [β (*yellow*), α2-δ (*blue*), and γ (*red*)] that can associate with the α-1 subunit. (C) Table listing the family members for the Cav1, Cav2, and Cav3 calcium channel types, including their gene names, common terms to describe the calcium current they create, subunits that assemble to make up each channel, and the selective pharmacological blockers used to study each type.

resting membrane potential, which means calcium entry would not be specifically related to action potentials. Therefore, low-voltage-activated calcium channels (Cav3) are generally expressed in the dendrites and cell bodies of neurons, but not at axon terminals (McKay et al., 2006). In the dendrites and cell body, Cav3 channel types contribute to postsynaptic integration of voltage changes in dendrites and the cell body (see Chapter 13 for more details).

The Cav1 family of high-voltage-gated calcium channels is a diverse group of channels that are rarely associated with action potential-triggered neurotransmitter release at synapses. Instead, the Cav1 family of calcium channels is primarily expressed in cardiac tissue, skeletal muscle, and endocrine tissue, though some members of this family are also present in the nervous system (Lipscombe et al., 2004). In endocrine cells, such as pancreatic beta cells, Cav1 calcium channels provide the calcium ion trigger for hormone secretion (Ashcroft et al., 1994; Wiser et al., 1999). Within neurons, Cav1 calcium channels are best known for their role in regulating gene expression, cell survival, and some forms of plasticity in the nervous system. Nonetheless, in some sensory organs, Cav1 calcium channels are localized to neurotransmitter release sites (e.g., in hair cells and some parts of the retina; Fuchs, 1996; Heidelberger and Matthews, 1992).

The Cav2 family of channels regulates synaptic transmission at most synapses for two main reasons. First, like other high-voltage-gated channels, Cav2 channels will only open in response to a strong depolarization such as an action potential. Second, Cav2 channels contain a binding site for proteins that regulate synaptic vesicle docking and fusion at the nerve terminal active zone. This binding site is made up of a sequence of amino acids in the cytoplasmic linker between Cav2 domains II and III, and it is called the "Synaptic Protein Interaction" or "SynPrInt" site. The presence of the SynPrInt site on Cav2-type calcium channels ensures that they can be tightly associated with the neurotransmitter release machinery, which is critical because of the very tightly controlled calcium ion buffering and handling within the nerve terminal that keeps cytoplasmic calcium levels low under most conditions. If VGCCs were not tightly associated with neurotransmitter release sites (active zones) via the SynPrInt site, calcium influx through these channels would not efficiently trigger chemical neurotransmitter release.

Among the Cav2 family of channels, Cav2.1 (P/Q-type channels) and Cav2.2 (N-type channels) are the most common at synapses. Researchers were able to determine which of these Cav2 channel types are used at a given synapse by making electrophysiological recordings of synaptic currents or potentials after the addition of the selective blockers for each calcium channel type (see Figs. 7.19, 7.20, and Box 7.2). Some synapses appear to use only one Cav2 family member (e.g., the mouse NMJ employs the P/Q-type channel, Cav 2.1), while other synapses use two or more Cav2 family members (e.g., auditory brainstem synapses use P/Q-, N-, and R-type channels, Cav2.1, 2.2, and 2.3; Wheeler et al., 1994). The subtypes of Cav2 channels expressed at synapses can be developmentally regulated (Iwasaki et al., 2000).

There are several advantages to having more than one type of calcium channel at synapses. These include that each type of calcium channel opens at slightly different voltages; that each type of calcium channel can inactivate at different rates; and that each type of calcium channel is sensitive to different forms of modulation (see below). As a result, a

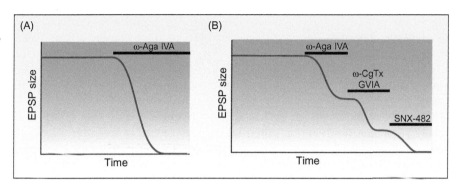

FIGURE 7.20 **Application of selective calcium channel blockers can reveal which calcium channels are present at synapses.** In these experiments, neurotransmitter release is represented by EPSP size as a function of time during the application of compounds that block specific voltage-gated calcium channels. (A) Neurotransmitter release at some synapses can be completely eliminated by a blocker for a single type of calcium channel. In this diagram, EPSPs are completely blocked by ω-Agatoxin IVA (ω-Aga IVA), a selective blocker of P/Q-type calcium channels (Cav2.1), indicating that Cav2.1 is the only type of voltage-gated calcium channel that is tightly associated with the neurotransmitter release site at this particular synapse. (B) Neurotransmitter release at some synapses is only partially blocked by a selective blocker of a single type of calcium channel, but can be completely blocked using a combination of blockers for several types of calcium channels. In this example, a combination of ω-AgaIVA, ω-CgTX GVIA, and SNX-482 is required to block all action potential-evoked neurotransmitter release, indicating that this synapse employs P/Q- N-, and R-type calcium channels (Cav2.1, Cav2.2, and Cav2.3) to regulate neurotransmitter release.

synapse that employs more than one type of calcium channel has more ways in which it can be fine-tuned to control the magnitude of neurotransmitter released in response to an action potential. Having varying types of calcium channels gives a synapse more potential for plasticity, which is an essential feature for synapses, especially in the central nervous system where fine-tuning synaptic transmission is important for higher brain functions (see Chapter 14).

Given that calcium ions trigger synaptic vesicle release, it makes sense that VGCCs are a common target for modulation. VGCCs can be heavily modulated by G-protein-coupled receptors or protein kinases (see Chapter 12), and by synaptic proteins that regulate vesicle fusion (see Chapter 8). Variability in the amino acid sequences of different calcium channel types results in differences in how these can be modulated.

## Calcium Entry Into a Presynaptic Terminal During an Action Potential

Examining the activation of calcium currents during a step depolarization in a voltage clamp experiment (see Fig. 7.21) shows that VGCCs are much slower to open than voltage-gated sodium channels. Before examining calcium current activation during an action potential, it is instructive to first examine calcium current activation during longer voltage steps. Several important points about calcium channel function can be garnered from these data. First, depolarization steps to voltages that are likely to be at the peak of an action potential (+10 to +30 mV) have the potential to activate 60%–90% of

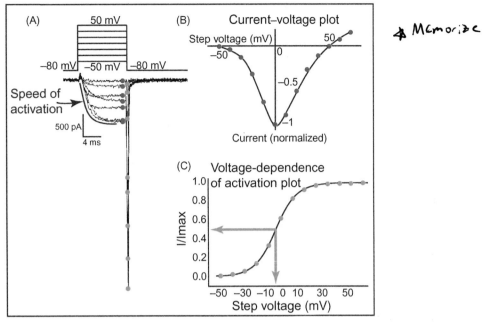

FIGURE 7.21 **Calcium current activated by depolarizing voltage steps.** (A) A family of calcium currents activated by a series of voltage steps from −80 mV to a range of depolarizing voltages, which are indicated above. The magnitude of the peak calcium current during the voltage steps is shown by the red symbols and plotted in (B). The magnitude of the tail current generated after each voltage step upon repolarization back to −80 mV is shown by the green symbols and plotted in (C). The blue curve is fit to the trace with the largest current, and this can be used to calculate the speed of calcium current activation. (B) Current−voltage plot for calcium currents, measured as described in (A). The current gets larger as a greater number of channels are activated with larger voltage steps, but then gets smaller as the voltage steps approach the equilibrium potential for calcium (+50 mV under these recording conditions). (C) Plot of the voltage-dependence of calcium current activation, as described in (A). The current at each voltage is expressed as a ratio of the maximum current in the series (I/Imax). The voltage that activates 50% of calcium current (about −5 mV) is shown by the green arrows. *Source: Adapted from Pattillo et al. (1999).*

channels (see data plotted in green in Fig. 7.21C), but only if that depolarization is held long enough for all these channels to activate. Second, calcium channels take several milliseconds to open (see the speed-of-activation curve in blue in Fig. 7.21A). Therefore, an action potential that depolarizes to +20 mV for less than 1 ms is not likely to activate most of the calcium channels in a nerve terminal, simply because the action potential depolarization is not long enough for most calcium channels to activate.

In addition, the driving force on calcium is an important factor in calcium entry during an action potential. During an action potential, the rapid but brief membrane depolarization to about +20 mV activates a portion of the calcium channels in the nerve terminal, but there is a weak inward driving force for calcium entry at +20 mV. However, the action potential only stays at peak voltage for a short time before sodium channels inactivate and potassium channels open, quickly repolarizing the membrane back to resting membrane potential. This causes the driving force for calcium to increase dramatically,

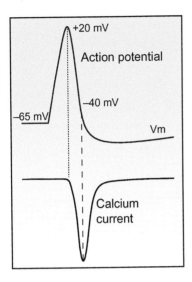

FIGURE 7.22 **Calcium current during an action potential.** The top panel shows a representative action potential waveform. The bottom panel shows the calcium current during that action potential. At the peak of the action potential (dotted line), there is little or no calcium entry since calcium channels are slow to open and there is little calcium driving force at +20 mV. As the action potential repolarizes, the calcium channels that opened during the depolarization are still open, but there is a much larger driving force when the membrane potential repolarizes to near resting levels. Therefore, the peak of the calcium current usually occurs during the falling phase of the action potential (dashed line).

since the membrane potential is now far from the calcium equilibrium potential. Calcium channels are slow to close, which means that while the action potential is rapidly repolarizing, the calcium channels that were opened by the depolarizing phase of the action potential remain open for some time. The increased driving force during the repolarization step, combined with the calcium channels that remain open, causes a sharp rise in calcium influx called a calcium **tail current**. (This type of current can be observed in electrophysiology experiments when a cell is abruptly repolarized after depolarizing steps that activate calcium current; see Fig. 7.21A.) The tail current that occurs during the action potential repolarization step is large in amplitude but brief in duration, because the repolarization of the membrane eventually causes the calcium channels to close (see Fig. 7.22).

## The Role of Potassium Channels in Shaping Calcium Entry During an Action Potential

The speed of the repolarization phase of the action potential is critical to creating the calcium tail current that shapes calcium entry during an action potential. Because the opening of potassium channels affects the rate of repolarization during an action potential, any change in potassium channel activation during the action potential has an indirect effect on calcium entry and neurotransmitter release (see Sabatini and Regehr, 1997).

In fact, there are two families of potassium channels that can contribute to action potential repolarization: voltage-gated and calcium-activated potassium channels. Voltage-gated potassium channels open near the peak of the action potential and, in combination with sodium channel inactivation, end the depolarizing phase of the action potential and begin the rapid repolarization phase. Therefore, if voltage-gated potassium channels are modulated or pharmacologically blocked, the action potential shape is changed starting at the

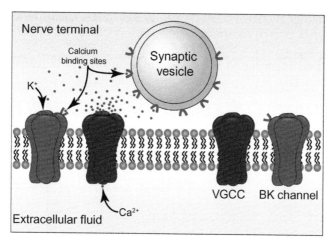

FIGURE 7.23  Calcium-binding sites on synaptic vesicle proteins and calcium-activated potassium channels are positioned to sense the calcium influx within the microdomain of calcium entry through voltage-gated calcium channels at synapses. Transmitter release sites include both voltage-gated calcium channels (*purple*) and calcium-activated potassium channels (BK channel; *red*). Calcium that enters the nerve terminal in a microdomain around open calcium channels binds to docked synaptic vesicles to trigger vesicle fusion with the plasma membrane. At the same time, calcium binds to calcium-activated potassium channels to alter the shape of the presynaptic action potential.

peak. If more voltage-gated potassium channels open, the action potential repolarizes faster and does not last as long. This decreases the time during which the action potential is above a voltage sufficient to activate calcium channels, which means that fewer calcium channels open, leading to a smaller calcium current. In contrast, if fewer voltage-gated potassium channels open, repolarization is slower and the action potential depolarization lasts longer. The longer depolarization opens more calcium channels, leading to a larger calcium current. In fact, a blocker of voltage-gated potassium channels has been approved for treatment of the neuromuscular disease Lambert-Eaton myasthenic syndrome (see Box 7.3).

Calcium-activated potassium channels are activated synergistically by a combination of depolarization and intracellular calcium binding (Fig. 7.23; Berkefeld et al., 2010). Because most calcium entry occurs late in the action potential, these calcium-activated potassium channels often open relatively late in the repolarization phase of the action potential. Therefore, when these channels are modulated or pharmacologically blocked, they often alter the late phases of repolarization and the **after-hyperpolarization**. Changes to these phases of the action potential predominately affect the speed with which the action potential returns to a voltage where there is a substantial calcium driving force. As such, calcium-activated potassium channels may not affect how many calcium channels open during an action potential, but instead may affect the size of the calcium tail current generated upon repolarization. Depending on when the action potential is altered during the repolarization phase, changes in the balance between driving force and calcium channel closure can cause the net calcium current during the action potential to either increase or decrease. In synapses, calcium-activated potassium channels are clustered at active zones along with VGCCs (Robitaille et al., 1993; see Fig. 7.24), because each calcium-activated potassium channel needs to be very close to open calcium channels in order to bind calcium ions before the calcium buffering and handling mechanisms take over and lower the intracellular calcium concentration back to resting levels.

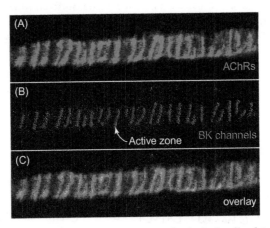

FIGURE 7.24 **Calcium-activated potassium channels are selectively localized to active zone regions of the frog motor nerve terminal.** (A) Postsynaptic acetylcholine receptors (AChRs) are distributed in bands along the length of the frog neuromuscular junction and are positioned immediately opposite to presynaptic active zones. In this image, the postsynaptic receptors are visualized using a green-fluorescent label on α-bungarotoxin. (B) Calcium-activated potassium channels (BK channels) are visualized using a toxin labeled with a red fluorescent dye (IBTX-Dylight-650). (C) An overlay of the two labels in (A) and (B) demonstrates that calcium-activated potassium channels are selectively expressed and colocalized in active zone regions of the neuromuscular junction. *Source: Vuocolo, Bingham, and Meriney, unpublished observations.*

BOX 7.3

## DISEASE FOCUS: LAMBERT–EATON MYASTHENIC SYNDROME

Lambert–Eaton myasthenic syndrome (LEMS) is an autoimmune disease in which autoantibodies (antibodies produced by the body against a part of its own tissues) attack presynaptic calcium channels in the active zone of the neuromuscular junction. This autoimmune attack reduces the number of presynaptic calcium channels, resulting in less calcium entry during a presynaptic action potential, and therefore less neurotransmitter release from the synapse (Tarr et al., 2015; see Fig. 7.25). This reduction in neurotransmitter

## BOX 7.3 (cont'd)

release causes muscle weakness, since the transmitter-induced postsynaptic depolarization (EPP) may be reduced such that it is below threshold for muscle contraction. LEMS patients report weakness primarily in the upper legs, hips, upper arms, and shoulders, which can make walking and self-care difficult. It is not clear why LEMS preferentially attacks these muscle groups.

Roughly 50% of LEMS patients have, or will develop, small-cell lung cancer (SCLC). SCLC is a cancer of neuroendocrine origin that is often associated with heavy, life-long cigarette smoking. SCLC tumor cells express the same types of VGCCs as the neuromuscular junction. The immune system produces antibodies against these calcium channels to attack the lung tumor, but the same antibodies also attack the neuromuscular junction calcium channels. The remaining 50% of LEMS patients are "idiopathic," meaning they have no known associated disease and produce antibodies against presynaptic calcium channels for no known reason. Regardless of the cause, the resulting disease is the same.

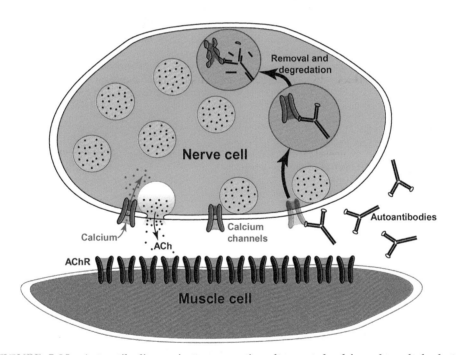

FIGURE 7.25 **Autoantibodies against presynaptic voltage-gated calcium channels leads to LEMS.** Diagram depicting the antibody attack on the nerve terminal that leads to the internalization and destruction of presynaptic voltage-gated calcium channels in LEMS. *Source: Adapted from MDA.org/.*

## BOX 7.3 (cont'd)

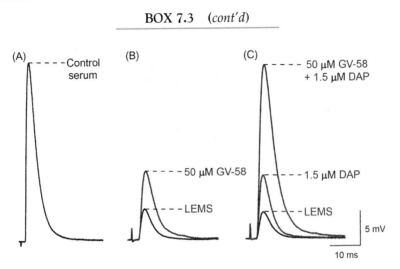

FIGURE 7.26 **Endplate potentials (EPPs) recorded from the mouse neuromuscular junction after a single action potential stimulation in control and LEMS conditions.** (A) Control EPPs are very large since these synapses normally release a large amount of the neurotransmitter acetylcholine. (B) In LEMS model mice (created by daily injection of LEMS patient serum into mice) neuromuscular synapses become very weak and do not release very much neurotransmitter (resulting in the small EPP labeled "LEMS"). However, this weak synapse can release about twice as much neurotransmitter when exposed to GV-58. (C) At a mouse synapse that has been weakened by LEMS patient serum, DAP increases EPP size by about twofold. However, when DAP and GV-58 are applied together, the amount of neurotransmitter released by the synapse recovers completely, resulting in an EPP of the same size as in the healthy control synapse. *Source: Adapted from Tarr et al. (2014).*

Treatment for LEMS can include immunosuppressants to weaken the autoimmune attack, but these drugs usually have side effects that make them less likely to be used. Most LEMS patients are treated using drugs that only reduce their symptoms and do not address the cause of the disease. Since patients have fewer calcium channels on their nerve terminals, drugs that can increase calcium entry during an action potential through the remaining calcium channels have been shown to be effective. For example, the most common treatment is a potassium channel blocker (3,4-diaminopyridine; DAP), which slightly prolongs the action potential duration, increasing the number of calcium channels that open and therefore increasing presynaptic calcium influx. The increase in calcium influx in turn increases neurotransmitter release and partially relieves symptoms. However, DAP can cause dangerous side effects at high doses that broaden the action potential too much. For example, it can act on potassium channels in the central nervous system, increasing the risk of seizures. Therefore, DAP cannot be administered to patients at a dose that would fully restore muscle strength.

> **BOX 7.3** (cont'd)
>
> A second potential treatment option is currently under development, but to date has only been studied in animals. This new treatment, a compound called GV-58, is a voltage-gated calcium channel gating modifier that holds calcium channels open for longer than normal. This increases calcium influx through the remaining calcium channels at a LEMS synapse after they have been opened by an action potential. Interestingly, GV-58 is more effective during longer depolarizations of the nerve terminal. Therefore, when DAP and GV-58 are administered simultaneously, they can work together to create a stronger effect on action potential-evoked EPPs than when either is used alone (Tarr et al., 2014; see Fig. 7.26).
>
> Further information about LEMS can be found on the Muscular Dystrophy Association website: https://www.mda.org/disease/lambert-eaton-myasthenic-syndrome.

## References

Adler, E.M., Augustine, G.J., Duffy, S.N., Charlton, M.P., 1991. Alien intracellular calcium chelators attenuate neurotransmitter release at the squid giant synapse. J. Neurosci. 11, 1496–1507.

Ashcroft, F.M., Proks, P., Smith, P.A., Ammala, C., Bokvist, K., Rorsman, P., 1994. Stimulus-secretion coupling in pancreatic beta cells. J. Cell. Biochem. 55, Suppl:54–Suppl:65.

Berkefeld, H., Fakler, B., Schulte, U., 2010. $Ca^{2+}$-activated $K^+$ channels: from protein complexes to function. Physiol. Rev. 90, 1437–1459.

Blaustein, M.P., Goldman, W.F., Fontana, G., Krueger, B.K., Santiago, E.M., Steele, T.D., et al., 1991. Physiological roles of the sodium-calcium exchanger in nerve and muscle. Ann. N. Y. Acad. Sci. 639, 254–274.

Brini, M., Carafoli, E., 2011. The plasma membrane $Ca(2)+$ ATPase and the plasma membrane sodium calcium exchanger cooperate in the regulation of cell calcium. Cold Spring Harb. Perspect. Biol. 3.

Burette, A., Weinberg, R.J., 2007. Perisynaptic organization of plasma membrane calcium pumps in cerebellar cortex. J. Comp. Neurol. 500, 1127–1135.

Caillard, O., Moreno, H., Schwaller, B., Llano, I., Celio, M.R., Marty, A., 2000. Role of the calcium-binding protein parvalbumin in short-term synaptic plasticity. Proc. Natl. Acad. Sci. USA 97, 13372–13377.

Carafoli, E., 1991. The calcium pumping ATPase of the plasma membrane. Annu. Rev. Physiol. 53, 531–547.

Daneman, R., Prat, A., 2015. The blood-brain barrier. Cold Spring Harb. Perspect. Biol. 7, a020412.

de Juan-Sanz, J., Holt, G.T., Schreiter, E.R., de Juan, F., Kim, D.S., Ryan, T.A., 2017. Axonal endoplasmic reticulum $Ca(2+)$ content controls release probability in CNS nerve terminals. Neuron 93, 867–881 e866.

Delaney, K.R., Zucker, R.S., 1990. Calcium released by photolysis of DM-nitrophen stimulates transmitter release at squid giant synapse. J. Physiol. 426, 473–498.

Dittrich, M., Pattillo, J.M., King, J.D., Cho, S., Stiles, J.R., Meriney, S.D., 2013. An excess-calcium-binding-site model predicts neurotransmitter release at the neuromuscular junction. Biophys. J. 104 (12), 2751–2763.

Dodge Jr., F.A., Rahamimoff, R., 1967. Co-operative action a calcium ions in transmitter release at the neuromuscular junction. J. Physiol. 193, 419–432.

Egelman, D.M., Montague, P.R., 1999. Calcium dynamics in the extracellular space of mammalian neural tissue. Biophys. J. 76, 1856–1867.

Fuchs, P.A., 1996. Synaptic transmission at vertebrate hair cells. Curr. Opin. Neurobiol. 6, 514–519.

Futagi, D., Kitano, K., 2015. Ryanodine-receptor-driven intracellular calcium dynamics underlying spatial association of synaptic plasticity. J. Comput. Neurosci. 39, 329–347.

Goldstein, D.A., 1990. Serum Calcium. In: Walker, H.K., Hall, W.D., Hurst, J.W. (Eds.), Clinical Methods: The History, Physical, and Laboratory Examinations, 3rd edition. Butterworths, Boston, 1990. Chapter 143.

Heidelberger, R., Matthews, G., 1992. Calcium influx and calcium current in single synaptic terminals of goldfish retinal bipolar neurons. J. Physiol. 447, 235–256.

Heidelberger, R., Heinemann, C., Neher, E., Matthews, G., 1994. Calcium dependence of the rate of exocytosis in a synaptic terminal. Nature 371, 513–515.

Hogan, P.G., Rao, A., 2015. Store-operated calcium entry: mechanisms and modulation. Biochem. Biophys. Res. Commun. 460, 40–49.

Iwasaki, S., Momiyama, A., Uchitel, O.D., Takahashi, T., 2000. Developmental changes in calcium channel types mediating central synaptic transmission. J. Neurosci. 20, 59–65.

Jenkinson, D.H., 1957. The nature of the antagonism between calcium and magnesium ions at the neuromuscular junction. J. Physiol. 138 (3), 434–444.

Jimerson, D.C., Wood, J.H., Post, R.M., 1980. Cerebrospinal fluid calcium. In: Wood, J.H. (Ed.), Neurobiology of Cerebrospinal Fluid 1. Springer US, Boston, MA, pp. 743–749.

Johenning, F.W., Theis, A.K., Pannasch, U., Ruckl, M., Rudiger, S., Schmitz, D., 2015. Ryanodine receptor activation induces long-term plasticity of spine calcium dynamics. PLoS Biol. 13, e1002181.

Jones, H.C., Keep, R.F., 1987. The control of potassium concentration in the cerebrospinal fluid and brain interstitial fluid of developing rats. J. Physiol. 383, 441–453.

Juhaszova, M., Church, P., Blaustein, M.P., Stanley, E.F., 2000. Location of calcium transporters at presynaptic terminals. Eur. J. Neurosci. 12, 839–846.

Kaplan, J.H., Ellis-Davies, G.C., 1988. Photolabile chelators for the rapid photorelease of divalent cations. Proc. Natl. Acad. Sci. USA 85, 6571–6575.

Katz, B., Miledi, R., 1965. The effect of calcium on acetylcholine release from motor nerve terminals. Proc. R. Soc. Lond. B. Biol. Sci. 161, 496–503.

Keep, R.F., Ulanski 2nd, L.J., Xiang, J., Ennis, S.R., Lorris Betz, A., 1999. Blood-brain barrier mechanisms involved in brain calcium and potassium homeostasis. Brain Res. 815, 200–205.

Kim, M.H., Korogod, N., Schneggenburger, R., Ho, W.K., Lee, S.H., 2005. Interplay between Na+/Ca2+ exchangers and mitochondria in Ca2+ clearance at the calyx of Held. J. Neurosci. 25, 6057–6065.

Kirichok, Y., Krapivinsky, G., Clapham, D.E., 2004. The mitochondrial calcium uniporter is a highly selective ion channel. Nature 427, 360–364.

Kita, H., Van der Kloot, W., 1974. Calcium ionophore X-537A increases spontaneous and phasic quantal release of acetylcholine at frog neuromuscular junction. Nature 250, 658–660.

Kita, H., Van Der Kloot, W., 1976. Effects of the ionophore X-537A on acetylcholine release at the frog neuromuscular junction. J. Physiol. 259, 177–198.

Kwon, S.K., Hirabayashi, Y., Polleux, F., 2016. Organelle-specific sensors for monitoring Ca(2+) dynamics in neurons. Front. Synaptic Neurosci. 8, 29.

Lee, S.H., Kim, M.H., Lee, J.Y., Lee, S.H., Lee, D., Park, K.H., et al., 2007. Na+/Ca2+ exchange and Ca2+ homeostasis in axon terminals of mammalian central neurons. Ann. N. Y. Acad. Sci. 1099, 396–412.

Lipscombe, D., Helton, T.D., Xu, W., 2004. L-type calcium channels: the low down. J. Neurophysiol. 92, 2633–2641.

Llinas, R., Sugimori, M., Silver, R.B., 1992. Microdomains of high calcium concentration in a presynaptic terminal. Science 256, 677–679.

Luo, F., Dittrich, M., Stiles, J.R., Meriney, S.D., 2011. Single-pixel optical fluctuation analysis of calcium channel function in active zones of motor nerve terminals. J. Neurosci. 31, 11268–11281.

Mangialavori, I., Ferreira-Gomes, M., Pignataro, M.F., Strehler, E.E., Rossi, J.P., 2010. Determination of the dissociation constants for Ca2+ and calmodulin from the plasma membrane Ca2+ pump by a lipid probe that senses membrane domain changes. J. Biol. Chem. 285, 123–130.

McKay, B.E., McRory, J.E., Molineux, M.L., Hamid, J., Snutch, T.P., Zamponi, G.W., et al., 2006. Ca(V)3 T-type calcium channel isoforms differentially distribute to somatic and dendritic compartments in rat central neurons. Eur. J. Neurosci. 24, 2581–2594.

Miledi, R., 1973. Transmitter release induced by injection of calcium ions into nerve terminals. Proc. R. Soc. Lond. B Biol. Sci. 183 (1073), 421–425.

Palmgren, M.G., Nissen, P., 2011. P-type ATPases. Annu. Rev. Biophys. 40, 243–266.

Palty, R., Sekler, I., 2012. The mitochondrial Na(+)/Ca(2+) exchanger. Cell Calcium. 52, 9–15.

Pattillo, J.M., Artim, D.E., Simples Jr., J.E., Meriney, S.D., 1999. Variations in onset of action potential broadening: effects on calcium current studied in chick ciliary ganglion neurones. J. Physiol 514 (Pt 3), 719–728.

Pongs, O., Lindemeier, J., Zhu, X.R., Theil, T., Engelkamp, D., Krah-Jentgens, I., et al., 1993. Frequenin—a novel calcium-binding protein that modulates synaptic efficacy in the Drosophila nervous system. Neuron 11, 15–28.

Prakriya, M., 2013. Store-operated Orai channels: structure and function. Curr. Top Membr. 71, 1–32.

Rahamimoff, R., Meiri, H., Erulkar, S.D., Barenholz, Y., 1978. Changes in transmitter release induced by ion-containing liposomes. Proc. Natl. Acad. Sci. USA 75, 5214–5216.

Robitaille, R., Adler, E.M., Charlton, M.P., 1990. Strategic location of calcium channels at transmitter release sites of frog neuromuscular synapses. Neuron 5, 773–779.

Robitaille, R., Garcia, M.L., Kaczorowski, G.J., Charlton, M.P., 1993. Functional colocalization of calcium and calcium-gated potassium channels in control of transmitter release. Neuron 11 (4), 645–655.

Sabatini, B.L., Regehr, W.G., 1997. Control of neurotransmitter release by presynaptic waveform at the granule cell to Purkinje cell synapse. J. Neurosci. 17 (10), 3425–3435.

Schneggenburger, R., Neher, E., 2000. Intracellular calcium dependence of transmitter release rates at a fast central synapse. Nature 406 (6798), 889–893.

Schwaller, B., 2010. Cytosolic Ca2+ buffers. Cold Spring Harb. Perspect. Biol. 2, a004051.

Schwaller, B., Meyer, M., Schiffmann, S., 2002. 'New' functions for 'old' proteins: the role of the calcium-binding proteins calbindin D-28k, calretinin and parvalbumin, in cerebellar physiology. Studies with knockout mice. Cerebellum 1, 241–258.

Shoshan-Barmatz, V., Krelin, Y., Chen, Q., 2017. VDAC1 as a Player in Mitochondria-Mediated Apoptosis and Target for Modulating Apoptosis. Curr. Med. Chem. 24 (40), 4435–4446.

Smith, S.J., Buchanan, J., Osses, L.R., Charlton, M.P., Augustine, G.J., 1993. The spatial distribution of calcium signals in squid presynaptic terminals. J. Physiol 472, 573–593.

Somjen, G.G., 2004. Ions in the Brain: Normal Function, Seizures, and Stroke. Oxford University Press, New York.

Tarr, T.B., Dittrich, M., Meriney, S.D., 2013. Are unreliable release mechanisms conserved from NMJ to CNS? Trends Neurosci. 36 (1), 14–22.

Tarr, T.B., Lacomis, D., Reddel, S.W., Liang, M., Valdomir, G., Frasso, M., Wipf, P., Meriney, S.D., 2014. Complete reversal of Lambert-Eaton myasthenic syndrome synaptic impairment by the combined use of a K+ channel blocker and a Ca2+ channel agonist. J. Physiol. 592 (16), 3687–3696.

Tarr, T.B., Wipf, P., Meriney, S.D., 2015. Synaptic Pathophysiology and Treatment of Lambert-Eaton Myasthenic Syndrome. Mol. Neurobiol. 52 (1), 456–463.

Unni, V.K., Zakharenko, S.S., Zablow, L., DeCostanzo, A.J., Siegelbaum, S.A., 2004. Calcium release from presynaptic ryanodine-sensitive stores is required for long-term depression at hippocampal CA3-CA3 pyramidal neuron synapses. J. Neurosci. 24, 9612–9622.

Wachman, E.S., Poage, R.E., Stiles, J.R., Farkas, D.L., Meriney, S.D., 2004. Spatial Distribution of Calcium Entry Evoked by Single Action Potentials within the Presynaptic Active Zone. J. Neurosci. 24 (12), 2877-2785.

Wheeler, D.B., Randall, A., Tsien, R.W., 1994. Roles of N-type and Q-type Ca2+ channels in supporting hippocampal synaptic transmission. Science. 264 (5155), 107–111.

Wiser, O., Trus, M., Hernández, A., Renström, E., Barg, S., Rorsman, P., Atlas, D., 1999. The voltage sensitive Lc-type Ca2+ channel is functionally coupled to the exocytotic machinery. Proc. Natl. Acad. Sci. USA 96 (1), 248–253.

Yu, R., Hinkle, P.M., 2000. Rapid turnover of calcium in the endoplasmic reticulum during signaling. Studies with cameleon calcium indicators. J. Biol. Chem. 275, 23648–23653.

Zucker, R.S., 1993. The calcium concentration clamp: spikes and reversible pulses using the photolabile chelator DM-nitrophen. Cell. Calcium. 14 (2), 87–100.

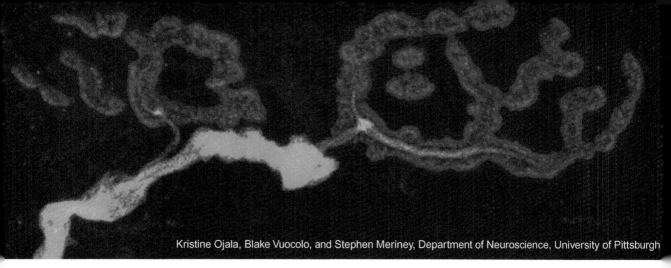

Kristine Ojala, Blake Vuocolo, and Stephen Meriney, Department of Neuroscience, University of Pittsburgh

CHAPTER 8

# Cellular and Molecular Mechanisms of Exocytosis

The main focus of this chapter is to address how action potential-induced calcium entry into the nerve terminal triggers vesicle fusion and the release of chemical transmitter molecules into the synaptic cleft. First, however, we will detail the experimental evidence that transmitter release occurs by fusion of synaptic vesicles with the neuronal plasma membrane. Why is vesicle fusion currently accepted as the mechanism for quantal transmitter release, and could there be an alternative (see Tauc, 1997)?

## DISCOVERY OF THE MECHANISMS OF NEUROTRANSMITTER RELEASE

Early pharmacologists determined that nerve terminals contain high concentrations of chemical transmitters, and Chapter 7, detailed how action potentials cause neurotransmitter release by activating voltage-gated calcium channels in the active zone region of nerve terminals. But what is the molecular basis for calcium-triggered neurotransmitter release? It is important to keep in mind that the synaptic delay between a presynaptic action potential and a resulting chemical transmitter response in a postsynaptic cell can be 1 ms or shorter. About half of that 1 ms is needed to open voltage-gated calcium channels and bring calcium into the nerve terminal during the repolarizing phase of the presynaptic

# 156    8. CELLULAR AND MOLECULAR MECHANISMS OF EXOCYTOSIS

action potential. The remaining 0.5 ms includes not only the release of neurotransmitter, but its diffusion across the cleft, its binding to postsynaptic receptors, and the activation of those receptors to cause a postsynaptic potential. Therefore, the mechanism that leads to calcium-triggered chemical transmitter release must act in less than 0.5 ms.

## Is Neurotransmitter Released Through a Channel in the Presynaptic Membrane?

How is it possible for calcium to trigger the release of neurotransmitter from the nerve terminal in less than 0.5 ms? We will consider several hypotheses that were proposed and tested by scientists studying neurotransmitter release.

The nerve terminal cytoplasm is the site of synthesis for some transmitters (acetylcholine, GABA, and the amines; see Chapters 16–18). The concentration of acetylcholine in the cytoplasm of cholinergic nerve terminals is relatively high (Morel et al., 1978), so could acetylcholine be released through a calcium-gated acetylcholine-conducting channel in the presynaptic membrane? If such a channel existed, it could certainly open very quickly (based on the speed of activation of calcium-activated potassium channels), and it could conduct a large number of acetylcholine molecules each time it opened (ion channels can often conduct a large number of ions).

Nevertheless, multiple pieces of experimental evidence fail to support the presence of an acetylcholine-conducting channel in the presynaptic membrane. First, when the driving force on acetylcholine was changed by experimentally altering the acetylcholine concentration gradient across the membrane, no change in the size of a single quantum release event was detected (Poulain et al., 1986), as would have been expected if the driving force on acetylcholine were pushing it through channels. Second, when the driving force on acetylcholine was changed by altering the presynaptic membrane potential, there was still no change in the size of quantal events (Young and Chow, 1987). Third, acetylcholine molecules are positively charged, so if an acetylcholine channel did exist it should be possible to measure current through that channel during neurotransmitter release. Young and Chow (1987) looked for such a current by patch-clamping the presynaptic membrane of a cultured frog motor neuron while simultaneously monitoring transmitter release from a postsynaptic muscle cell, but could not detect any channel openings during neurotransmitter release events (see Fig. 8.1).

All of the above evidence argues against an acetylcholine-conducting channel in the nerve terminal membrane as an explanation for quantal transmitter release. Consistent with these data, no channels have been discovered that can conduct acetylcholine molecules.

## Experimental Evidence Supporting Synaptic Vesicle Fusion With Plasma Membrane as the Mechanism for Quantal Transmitter Release

It took more than 20 years after Sir Bernard Katz and colleagues proposed the quantal theory for chemical transmitter release (see Chapter 6) for scientists to discover

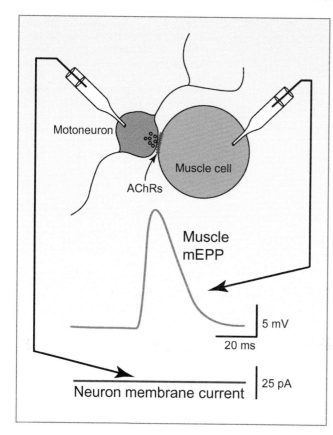

FIGURE 8.1 **Simultaneous recording of presynaptic membrane current and postsynaptic membrane potential during quantal release of transmitter.** Patch-clamp recordings from a presynaptic cell body in a cultured neuron (blue) showed no membrane current during spontaneous quantal transmitter release events (mEPP) that were recorded from a postsynaptic muscle cell (green). *Source: Adapted from Young, S.H., Chow, I., 1987. Quantal release of transmitter is not associated with channel opening on the neuronal membrane. Science 238, 1712–1713.*

experimental methods that could prove synaptic vesicles were the mechanism for the quantal release of transmitter.

First, a new method was developed for rapidly freezing tissue in preparation for electron microscopy. Traditional electron microscopy requires chemical fixatives that crosslink proteins in the tissue to preserve fine structure. The problem with using this approach to study the process of neurotransmitter release at synapses is that neurotransmitter release is very fast, and chemical fixatives are very slow to fix tissues. As a result, traditional images of synapses showed many synaptic vesicles in nerve terminals, but failed to capture a vesicle actually fusing with plasma membrane.

To solve this problem, laboratories perfected the technique of first stimulating the nerve of a frog neuromuscular junction preparation, then freezing the preparation milliseconds later by dropping it onto a block of copper that had been cooled with liquid helium (Heuser et al., 1976). In order to increase the number of vesicles released after stimulation, and thus increase their chances of observing evidence of vesicle fusion events, researchers applied a high concentration of a potassium channel blocker to broaden the presynaptic action potential and significantly increase the calcium influx that triggers transmitter release. When this approach was used for thin-section electron microscopy, evidence of

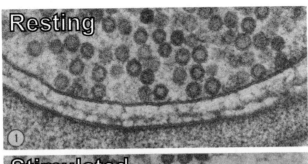
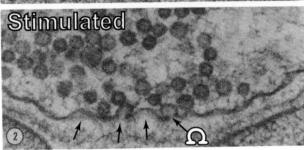

FIGURE 8.2 Thin-section electron microscopic evidence that synaptic vesicles fuse with plasma membrane during transmitter release. In the top panel, the nerve was not stimulated (resting) before the preparation was frozen, and no synaptic vesicles can be observed fusing with plasma membrane. In the bottom panel, the nerve was stimulated 3.5 ms before the preparation was rapidly frozen. In this case, numerous synaptic vesicles were caught in the act of fusing with the plasma membrane (arrows). In cross-section, these vesicle fusion events have the shape of the Greek letter omega ($\Omega$; see white outline of representative event), and so they are sometimes referred to as "omega profiles". *Source: Adapted from Heuser, J.E., Reese, T.S., 1981. Structural changes after transmitter release at the frog neuromuscular junction. J. Cell Biol. 88, 564–580.*

vesicles fusing with the presynaptic plasma membrane could be observed several milliseconds after nerve stimulation (see Fig. 8.2). Taking into account the time required for action potential conduction down the axon, this timing aligned perfectly with when chemical transmitter release occurred. On the other hand, if the nerve was not stimulated, no vesicles were observed fusing with the plasma membrane (Heuser et al., 1976, 1979; Heuser and Reese, 1981; Torri-Tarelli et al., 1985). Related approaches have been applied to other neuronal preparations and have yielded similar results (Watanabe, 2016).

Another method, called **freeze-fracture electron microscopy**, was developed to observe these fusion events along the length of the entire active zone in the frog neuromuscular junction (see Fig. 8.3). The same rapid freezing process was used as in the experiments described above, but the tissue was not sectioned; instead, a knife blade was used to crack the frozen tissue. When frozen tissue cracks, it often breaks right between the bilayers of the cell, like opening up the two halves of a sandwich (see Fig. 8.3B and C). Experimenters coated the cracked surfaces of the presynaptic membrane bilayer with metal to create a foil replica, which they viewed using an electron microscope to reveal the shapes of the proteins in the membrane. In the absence of nerve stimulation, the transmembrane proteins within the active zone could be seen arranged in a double row (see Fig. 8.3D). However, when the tissue was rapidly frozen within milliseconds after nerve stimulation, the normal double row of active zone proteins was disrupted by indentations ("pits") that indicated the location of vesicle membrane fusing with the plasma membrane (see Fig. 8.3E). The timing of the appearance of "pits" a few milliseconds after nerve stimulation provided strong evidence that synaptic vesicles fused with plasma membrane during action potential activity. However, when the tissue was frozen 50 ms after nerve stimulation the pits were not observed and the double row of active zone proteins observed when

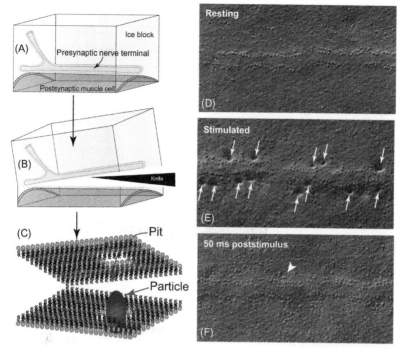

FIGURE 8.3  Freeze-fracture technique and images of the frog neuromuscular junction reveal evidence of vesicle fusion events. (A) Tissue is prepared for freeze-fracture by dropping a synaptic preparation floating in saline onto a chilled copper block, freezing the synapse in a block of ice. (B) The frozen tissue is cracked with a knife and fractures between halves of a cell bilayer. (C) Diagram of a cell bilayer that has been split open by freeze-fracture. Transmembrane proteins remain associated with one half of the bilayer and appear as particles. Where they previously extended through the other half of the bilayer, a hole or pit is left behind. A fusing synaptic vesicle can also leave a large "pit" in the fractured membrane (not diagramed). (D) Freeze-fracture image of one half of the fractured bilayer from a resting frog active zone. Two double rows of large intramembraneous particles define the active zone region. These particles are thought to include the presynaptic voltage-gated calcium channels and calcium-activated potassium channels. (E) When the tissue was frozen rapidly 3.5 ms after nerve stimulation in the presence of a high concentration of a potassium channel blocker, numerous "pits" were observed (arrows). These pits are interpreted as locations of synaptic vesicle fusion. (F) When the tissue was frozen rapidly 50 ms after nerve stimulation, the "pits" were no longer present, but the normal double row of active zone particles was disrupted and clusters of particles had appeared (arrowhead).

the cell was at rest was disrupted, perhaps indicating changes in the plasma membrane that persisted in the aftermath of vesicle fusion (see Fig. 8.3F).

A final method that has provided evidence of synaptic vesicle fusion with the plasma membrane during transmitter release is based on measuring the capacitance of the nerve terminal membrane (Neher and Marty, 1982). Electrically, a neuron can be represented as a circuit that includes a battery (the membrane potential), a resistor (the resistance across the membrane determined by the number of open channels), and a capacitor, or separator of charge (the plasma membrane). The plasma membrane is a good capacitor (see Chapter 3), and the larger the surface area of the membrane, the larger its capacitance.

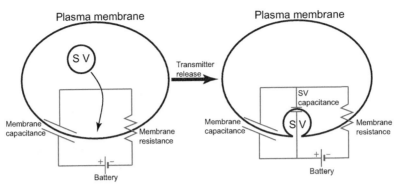

FIGURE 8.4  **Electrical representation of a neuron before and after synaptic vesicle fusion.** (Left) A representation of a neuron as an electrical circuit. (Right) Changes in the circuit when a synaptic vesicle (SV) fuses with the plasma membrane. When the plasma membrane from the synaptic vesicle joins the cell membrane, it increases the total surface area of the nerve terminal. Electrically, the addition of synaptic vesicle membrane adds a capacitor in parallel, increasing plasma membrane capacitance.

If the membrane around synaptic vesicles fuses with the plasma membrane during neurotransmitter release, then vesicle release should be accompanied by an increase in the membrane area of the nerve terminal. Since capacitance is proportional to membrane surface area, vesicle fusion should therefore result in an increase in capacitance (see Fig. 8.4).

Measuring capacitance in nerve terminals is very tricky and requires overcoming two major obstacles. First, the experimenter must find a nerve terminal that is large enough and accessible for patch clamp electrophysiology techniques. Second, the experimenter must be able to detect the relatively small capacitance increase that occurs when the membrane from a small synaptic vesicle (40–50 nm in diameter) is added to the nerve terminal plasma membrane.

Before this experimental approach was attempted in neurons, it was used with endocrine cells, which release hormones from very large vesicles (or granules). For example, in adenal chromaffin cells, catecholamines are released from 300 to 400-nm granules that fuse with plasma membrane. The resulting capacitance increase, or "step jump," is large enough to be easily detected (see Fig. 8.5). Using a combination of electrical detection methods for catecholamine release (amperometry; see Box 8.1) and measurements of capacitance, experimenters showed that each catecholamine release event was accompanied by a step increase in capacitance (see Fig. 8.5).

In nerve terminals, synaptic vesicles are much smaller (only 40–50 nm in diameter), so experimenters have usually focused on measuring the capacitance increase that occurs following the action potential-evoked synchronous release of large numbers of vesicles at giant synapses (Kim and von Gersdorff, 2010). In other cases, however, experimenters have sought to measure the capacitance jump from the fusion of a single synaptic vesicle. Because all electrical measurements, including these capacitance measurements, have some electrical noise associated with them, it is difficult to detect an event (such as single vesicle fusion) whose signal is small enough to get lost in the noise. To increase the signal-to-noise ratio of their recordings, experimenters recorded 100,000–400,000 mEPSCs

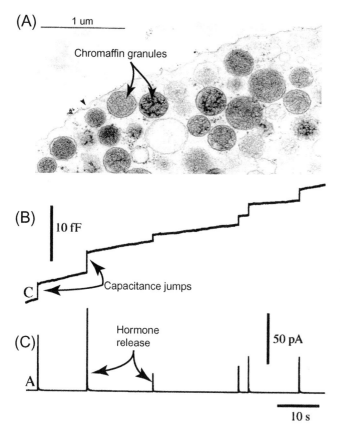

FIGURE 8.5 Measurements of membrane capacitance combined with cell-attached patch amperometry to detect catecholamine release by endocrine cells. (A) Electron micrograph of an adrenal chromaffin cell containing large, chromaffin granules. (B) Measurements of large step increases in capacitance. (C) The step increases in capacitance shown in (B) correspond to catecholamine release events detected using amperometry. *Source: Adapted from Plattner, H., Artalejo, A.R., Neher, E., 1997. Ultrastructural organization of bovine chromaffin cell cortex-analysis by cryofixation and morphometry of aspects pertinent to exocytosis. J. Cell Biol. 139, 1709–1717. Gong, L.W., Hafez, I., Alvarez de Toledo, G., Lindau, M., 2003. Secretory vesicles membrane area is regulated in tandem with quantal size in chromaffin cells. J. Neurosci. 23, 7917–7921.*

---

BOX 8.1

## TECHNIQUE FOCUS: AMPEROMETRY

**Amperometry** is an electrochemical detection method that takes advantage of an oxidization or reduction reaction between the chemical of interest and a carbon fiber that is placed in close proximity to the site of chemical release and has a fixed electrical potential applied across it. Voltammetry is a subclass of amperometry in which the current is measured by varying the electrical potential that is applied to the carbon fiber electrode (sometimes called "fast-scan cyclic voltammetry"). When neurotransmitter is released, neurotransmitter molecules come into contact with the carbon fiber. The resulting oxidation or reduction reaction transfers electrons to or from the carbon fiber, which creates a current that can be detected (Neher and Marty, 1982; Bucher and Wightman, 2015). The conventional amperometry approach can be combined with patch clamp techniques to allow the simultaneous detection of capacitance and neurotransmitter release using a technique called "patch amperometry" (Dernick et al., 2007).

## BOX 8.1  (cont'd)

Importantly, the amperometry/voltammetry technique has traditionally been limited to detecting transmitters that can undergo an oxidation or reduction reaction. Catecholamines are the easiest chemical transmitters to study using this approach, because they are oxidized when they come in contact with the carbon fiber electrode (see Fig. 8.6).

It is possible to use amperometry to detect other chemical transmitters, but this requires more complex carbon fiber electrodes that are coated with enzymes which react with the chemical transmitter of interest and mediate the formation, via an oxidase, of an electroactive product (often $H_2O_2$). This enzymatic approach has been applied to the study of the neurotransmitters glutamate and acetylcholine (Oldenziel and Westerink, 2005; Sarter et al., 2009).

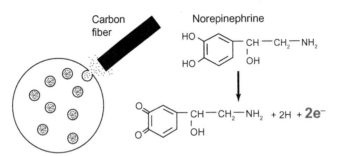

FIGURE 8.6  **Amperometric detection of catecholamines.** (Left) A carbon fiber is positioned near a cell that is releasing catecholamines and an electrical potential is applied. (Right) The catecholamine norepinephrine is oxidized upon contact with the carbon fiber and generates two electrons, which cause a current that can be measured.

---

from a single synapse and averaged the associated capacitance measurements (Sun et al., 2002). In the average of the recordings, the random noise "canceled out," making the averaged capacitance signal easier to distinguish. This method was used at the giant synapse in the auditory brainstem (the calyx of Held) to reveal the average increase in capacitance caused when a single synaptic vesicle fuses with plasma membrane (see Fig. 8.7), as well as the average time course of recovery of the vesicle membrane via endocytosis (see Chapter 9).

Taken together, "omega profiles" in thin sections, "large pits" in freeze-fracture images from rapidly frozen nerve terminals just after action potential stimulation, and capacitance increases associated with chemical transmitter release are all strong evidence that quantal transmitter release is associated with fusion of synaptic vesicles with the plasma membrane.

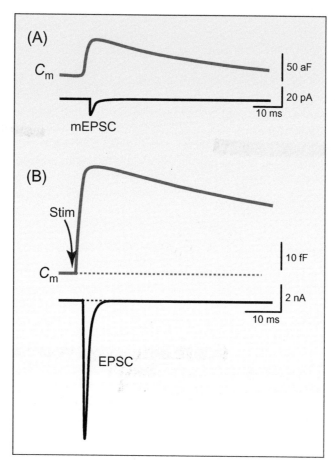

FIGURE 8.7  **Diagram of capacitance measurements during miniature and evoked synaptic vesicle release at a giant synapse in the auditory brainstem of a mouse (the calyx of Held).** (A) Diagram of the average capacitance trace during spontaneous vesicle fusion. In order to be able to resolve such small capacitance events above the noise in such recordings, experimenters were required to average more than 300,000 spontaneous miniature synaptic currents (mEPSC) along with the corresponding average capacitance trace ($C_m$) from these mEPSCs. Each mEPSC creates an average capacitance increase of about 50 aF ($10^{-18}$ F), and this capacitance decayed with a time constant of about 50 ms (an estimate of how long the vesicle stays in the plasma membrane before endocytosis). (B) Following a brief stimulation of the nerve terminal, a large EPSC could be measured in the postsynaptic cell. Under these conditions, a large capacitance jump was measured in the presynaptic nerve terminal ($C_m$) that decayed with a time constant of about 100 ms. If a calcium channel blocker was added (e.g., cadmium), there was no measured EPSC and the capacitance jump was also blocked (dotted lines). Note difference in scales between (A) and (B).

# BIOCHEMICAL MECHANIMS OF CALCIUM-TRIGGERED SYNAPTIC VESICLE FUSION

Previously, we defined the active zone as a region of the nerve terminal where synaptic vesicles fuse with the plasma membrane to release transmitter into the synaptic cleft.

Which proteins are collected in this active zone region of the nerve terminal, how do they coordinate the preparation of synaptic vesicles for fusion, and how does calcium influx trigger vesicle fusion in less than 0.5 ms after an action potential?

Initially, scientists reasoned that calcium influx into the nerve terminal might trigger a biochemical cascade that leads to vesicle fusion. One of the first hypotheses to be tested was that a calcium-dependent phosphorylation event leads to synaptic vesicle fusion. However, the exploration of this and other calcium-dependent biochemical events instead revealed mechanisms for synaptic vesicle storage and movement, also known as **vesicle trafficking**, within the nerve terminal (see Chapter 9), but did not lead to the identification of the molecular mechanisms for synaptic vesicle fusion. The conceptual breakthrough that led to more successful experiments only came when neuroscientists began to pay attention to cell biologists' investigations of the Golgi apparatus.

## Study of Vesicle Fusion

Cell biologists found yeast cells, which divide rapidly and are amenable to genetic mutation, to be an efficient system in which to identify the mechanisms cells use to transport proteins from the endoplasmic reticulum, through the Golgi apparatus, and ultimately to different areas of the cell (Schekman, 1982, 1985; Waters et al., 1991; Rothman and Orci, 1992). Importantly, cell biologists used these cells as a model system to study the roles different proteins play in **membrane trafficking**, the process by which vesicles are fused with the plasma membrane in order to insert transmembrane proteins into the membrane or release secretory proteins from the cell. As scientists interested in mammalian membrane trafficking evaluated cell biologists' findings in yeast cells, it became clear that vesicle fusion mechanisms used by yeast cells are conserved at mammalian nerve terminals. When neuroscientists looked for **homologs** of yeast proteins in the presynaptic terminals of mammalian synapses, they found that the same family of proteins that carries out vesicle fusion in yeast cells is expressed in the active zone of the synapse. These similarities led to the conclusion that the mechanisms cells evolved for **membrane fusion** and **fission** are used in a variety of other cellular functions. Neuroscientists reasoned that since neurons are just another type of cell, and synaptic vesicle fusion is very similar to the type of vesicle fusion that occurs within the Golgi apparatus, neurons would not have needed to evolve a separate mechanism for vesicle fusion at the nerve terminal (Ferro-Novick and Jahn, 1994).

There is a major difference between how vesicles fuse with membranes in the Golgi apparatus and how they fuse at the nerve terminal: neurons use calcium influx as the trigger for vesicle fusion at synapses, but in the Golgi apparatus, membrane fusion occurs in the absence of a calcium trigger. Neuroscientists therefore hypothesized that synapses use the same basic framework for vesicle fusion as in other parts of the cell, but possess an additional calcium-dependent regulatory mechanism to control fusion at synapses. This insight dramatically advanced our understanding of the calcium-dependent mechanisms of neurotransmitter release at synapses, and was the foundation for rapid scientific advances in the field. In 2013, the Nobel Prize for Physiology or Medicine was awarded to three investigators who were pioneers in this broad area of research (see Box 8.2).

## BOX 8.2

## 2013 NOBEL PRIZE IN PHYSIOLOGY OR MEDICINE FOR THE DISCOVERY OF THE MECHANISMS OF VESICLE TRAFFICKING AND FUSION IN CELLS

The ability of a cell to move proteins or chemicals around within cells (incuding between organelles, and to secrete them across the cell membrane) is essential for many functions in the body. In the nervous system, this includes the mechanisms that regulate synaptic vesicle fusion with the plasma membrane.

The research that led to the discovery of mechanisms that cells use to traffic and fuse vesicles with plasma membrane occurred between three laboratories, starting with work in yeast cells. In the 1970s, Randy Schekman used yeast cells to study vesicle movement between the Golgi apparatus and the plasma membrane. Schekman took advantage of the experimental ability to create genetic mutants in yeast cells, identifying malfunctions in the movement of vesicles between different compartments of the cell that were caused by different genes. This was followed in the 1980s and 1990s by work in James Rothman's laboratory that showed how these vesicles fuse with the plasma membrane, identifying critical proteins in the process. Thomas Sufhoff examined these processes at synapses in neurons and discovered the specializations that are required within the active zone of the synapse to prepare synaptic vesicles for fusion, including the critical calcium-sensitive steps that allow synapses to control this process (Fig. 8.8).

For more details, see: "The Nobel Prize in Physiology or Medicine 2013". Nobelprize.org. Nobel Media AB 2014. Web. 6 Jan 2018. <http://www.nobelprize.org/nobel_prizes/medicine/laureates/2013/>

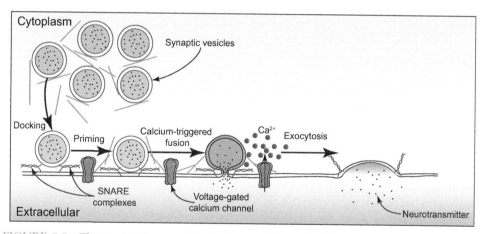

FIGURE 8.8  The regulated process by which synaptic vesicles fuse with the plasma membrane. Work by Schekman, Rothman, and Sudhof led to the discovery of the protein machinery that controls vesicle docking, priming, and fusion (including the SNARE complexes). Sudhof, in particular, identified calcium-dependent mechanisms for regulating this process at nerve terminals.

## Proteins Involved in Calcium-Triggered Vesicle Release

With the yeast research as a guide, proteins essential for calcium-triggered neurotransmitter release were identified in the active zone and on synaptic vesicles. As investigators began to characterize all of the proteins that were found at synapses, it became clear that the synapse is a very crowded place, with hundreds of proteins concentrated in this very small region of the neuron (see Fig. 8.9). It is not easy to arrange all of these proteins into a mechanistic scheme that characterizes synaptic function (especially since most of these proteins are modulatory and may only serve to mediate subtle adjustments to neurotransmitter release). Instead, we will incorporate a subset of the most important proteins into a more manageable model of the main events in calcium-triggered neurotransmitter release, including interacting elements from the plasma membrane, synaptic vesicle membrane, and the cytoplasm.

In particular, proteomic studies of synaptic vesicles isolated from the central nervous system revealed that their membranes contain a variety of proteins (see Fig. 8.10; Takamori et al., 2006). We will discuss some of these proteins—including synaptobrevin (sometimes called VAMP), the most abundant synaptic vesicle protein (see Fig. 8.10C;

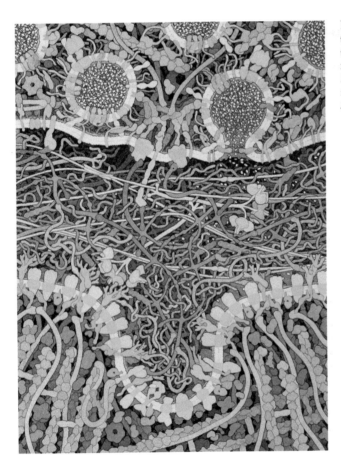

FIGURE 8.9 Artist's rendering of the tangle of diverse proteins that are concentrated around sites of chemical transmitter release at the neuromuscular junction. *Source: From Goodsell, D.S., 2009. The neuromuscular synapse. Biochem. Mol. Biol. Educ. 37, 204–210.*

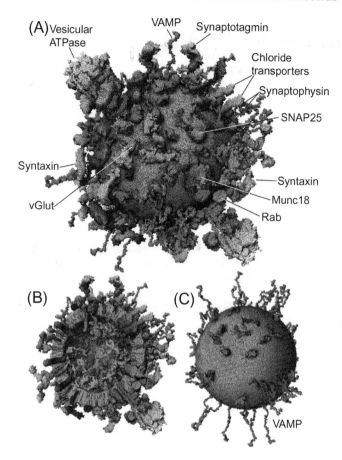

FIGURE 8.10 Artist's rendering of an average synaptic vesicle and the proteins that populate the membrane. (A) Representation of a synaptic vesicle that has been labeled to identify the major proteins on synaptic vesicles. (B) Cross-section of a synaptic vesicle which reveals the major synaptic release proteins and the extent to which these proteins protrude into the cytoplasm and lumen of the vesicle. (C) Representation of a synaptic vesicle showing a single type of vesicle release protein called VAMP (or synaptobrevin), which is the most abundant protein on synaptic vesicles. *Source: Adapted from Takamori, S., et al., 2006. Molecular anatomy of a trafficking organelle. Cell 127, 831–846.*

Takamori et al., 2006); synaptotagmin; and Rab—as part of our hypothesized mechanistic model of calcium-triggered transmitter release. Other vesicular proteins will be discussed when we cover neurotransmitter packaging into synaptic vesicles (vesicular-ATPase, and other transporters; see Chapters 16–18). Some of the proteins shown in Fig. 8.10 have either unclear functions or subtle modulatory functions that we will not discuss here.

The speed at which vesicle fusion occurs is a critical constraint of any schematic description of the mechanisms that underlie calcium-triggered transmitter release. After calcium entry, vesicles must fuse with the plasma membrane within several hundred microseconds, and it is essential to take this time limit into account when proposing a molecular mechanism for vesicle fusion. As we will describe, in order to achieve the necessary speed of fusion, synaptic vesicles must be prepared to fuse far in advance of the action potential that actually triggers fusion. This preparation process leaves synaptic vesicles on the verge of fusion with the plasma membrane, just waiting for the calcium signal that completes the fusion process. We will also describe a slower recovery step that occurs after synaptic vesicle fusion with the plasma membrane, before the vesicle membrane can be selectively pulled back into the nerve terminal (endocytosis; see Chapter 9).

# 8. CELLULAR AND MOLECULAR MECHANISMS OF EXOCYTOSIS

The hypothesized basic mechanism of calcium-triggered vesicle fusion is centered around a set of four proteins. Three are responsible for "docking" the synaptic vesicle to the plasma membrane: syntaxin, SNAP25, and synaptobrevin/VAMP. The assembly of these three proteins is sometimes called the **CORE complex**. The fourth protein is a calcium sensor, synaptotagmin, which contains two calcium-binding domains called **C2 domains** (C2A and C2B).

As stated above, the CORE complex of proteins "dock" synaptic vesicles to the plasma membrane. Syntaxin is an **integral membrane protein** that is attached to the plasma membrane of the nerve terminal and has one alpha helix chain that extends into the cytoplasm. SNAP25 has two alpha helix chains in the cytoplasm and is anchored to the plasma membrane by **palmitoyl side chains** (long fatty acid chains linked to a series of cysteines in the amino acid sequence of the protein that bury in the plasma membrane) (see Fig. 8.11). Synaptobrevin/VAMP is an integral vesicle membrane protein, with one alpha helix in the neuronal cytoplasm.

Fig. 8.11 illustrates the general hypothesis for how the CORE complex proteins work together to facilitate calcium-triggered synaptic vesicle fusion with plasma membrane. The CORE complex forms by the coiling together of the alpha-helices of syntaxin, SNAP25, and synaptobrevin/VAMP. These three CORE complex proteins are sometimes called **"SNARE" proteins** based on the discovery that when coiled together, they form a binding site for other proteins that regulate their coiling (the word SNARE is a conjunction of "Soluble NSF-Attachment protein REceptor").

## Experimental Evidence That the CORE Complex Is Critical for Transmitter Release

Perhaps the best evidence that a particular protein is important in neurotransmitter release is the observation that removing or experimentally altering that protein affects neurotransmitter release. The most definitive proof of the critical role that SNARE proteins play in transmitter release comes from scientific experiments using the clostridial neurotoxins responsible for tetanus and botulism (see Box 8.3). Tetanus toxin and the seven botulinum toxins are selective proteases that cleave SNARE proteins at specific locations on each protein (see Fig. 8.12). Each of these toxins consists of two polypeptide chains, the "heavy" chain and the "light" chain. Tetanus and botulinum toxins gain selective entry into synapses by binding with their heavy chain to membrane proteins on the inside (luminal) surface of vesicular membranes, which are accessible from outside the cell after exocytosis (see Fig. 8.13). During synaptic vesicle endocytosis, the toxins are engulfed by the vesicle and carried into the nerve terminal, where the light chains are cleaved from the heavy chains and ejected into the cytoplasm to act as selective proteases for SNARE proteins.

When botulinum toxin or tetanus toxin is applied to synapses, neurotransmitter release (both spontaneous and action potential-evoked) is completely blocked. The synaptic dysfunction caused by any of these toxins is strong evidence that the SNARE proteins are

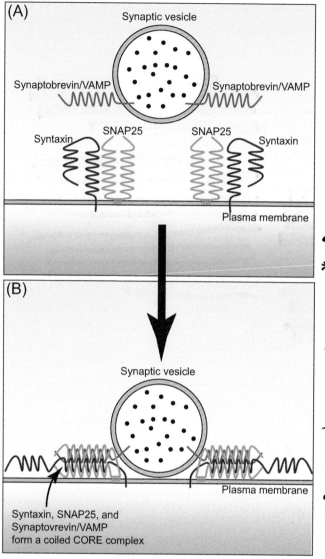

FIGURE 8.11 **CORE complex proteins are associated with the attachment of a synaptic vesicle in close proximity to the plasma membrane.** (A) Three major proteins that are essential to calcium-triggered synaptic vesicle fusion. Synaptobrevin (or VAMP) is an integral vesicle membrane protein; syntaxin is an integral plasma membrane protein; and SNAP25 is anchored to the plasma membrane by palmitoylation. (B) In preparation for synaptic vesicle fusion with the plasma membrane, synaptobrevin/VAMP, syntaxin, and SNAP25 form a CORE complex by coiling together their alpha-helical regions.

absolutely essential for synaptic vesicle fusion with the plasma membrane (Niemann et al., 1994; Pantano and Montecucco, 2014).

With syntaxin, SNAP25, and synaptobrevin/VAMP identified as the critical CORE complex of proteins that mediate the fusion of synaptic vesicles with the plasma membrane, we will now expand our discussion to the molecular processes that prepare vesicles for fusion, regulate the formation of the CORE complex, and control the vesicle fusion reaction.

## BOX 8.3

## DISEASE FOCUS: BOTULISM AND TETANUS

Tetanus and botulism are well-known, potentially fatal diseases. Both are caused by potent neurotoxins that are synthesized by Gram-positive anaerobic bacteria (*Clostridium tetani* and *Clostridium botulinum*). There are seven types of botulinum toxin (designated by letters A–G) and one type of tetanus toxin. Botulinum toxins are among the most deadly toxins known, so powerful that a single gram of toxin could kill more than 1 million people if evenly distributed in an active form (Simpson, 1989; Ting and Freiman, 2004).

Tetanus is usually caused by a laceration from a contaminated object that introduces environmental bacterial spores to the wound, but it is rare in countries that require vaccination. Symptoms of tetanus include spasms and stiffness in jaw muscles (which is why tetanus is often referred to as "lockjaw"), neck muscles, or abdominal muscles, as well as difficulty swallowing and painful body spasms.

Botulism is usually caused by exposure to botulinum bacteria in spoiled food. This disease is rare in developed countries, but can be a problem when hygiene during food production is not well controlled. Symptoms of botulism include difficulty swallowing, dry mouth, facial weakness, blurred vision, drooping eyelids, trouble breathing, nausea, vomiting, and abdominal cramps.

Even though both diseases are caused by toxins that target the SNARE proteins, they are characterized by a different set of clinical symptoms because of the general distribution of the toxins after they enter the body. Tetanus toxin moves easily from infected motor neurons near its site of entry into the spinal cord, where it poisons the synapses of spinal cord circuits. The resulting uncoordinated spinal motor output can create the spastic paralysis characteristic of this disease. Conversely, botulinum toxin usually enters the body via the gastrointestinal tract, and thus has widespread access to many peripheral neuron types that innervate the gut. However, botulinum toxin does not infect the spinal cord as efficiently as tetanus toxin, so it primarily targets the peripheral nervous system.

Despite their history as major disease-causing toxins, one variant of botulinum toxin (serotype A; BoNT/A) has been shown to be useful in cosmetics and as a treatment for a variety of medical conditions. Cosmetically, BoNT/A can be injected locally to smooth out facial lines and wrinkles. Clinically, BoNT/A has been shown to be effective at treating blepharospasm and strabismus (both caused by hyperactive extraocular muscles), dystonias, spasticity, tremors, vocal disorders, cerebral palsy in children, gastrointestinal disorders, tension and migraine headaches, and some pain syndromes (Ting and Freiman, 2004; Montecucco and Molgo, 2005; Davletov et al., 2005; Bhidayasiri and Truong, 2005; Dressler, 2012; Dutta et al., 2016). However, the beneficial effects of these treatments can be short-lived because motor nerve terminals recover from the toxin attack by re-synthesizing the proteins that are cleaved to reestablish previous function after about 3 months (Lam, 2003).

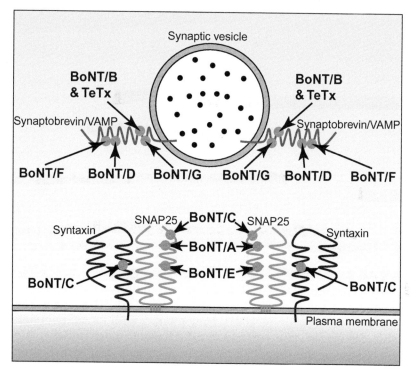

FIGURE 8.12 **Selective targeting of particular SNARE proteins by each serotype of botulinum toxin (BoNT/A-G) or by tetanus toxin (TeTx).** The pink dots represent cleavage sites on each SNARE protein for the labeled toxin. These toxins can only cleave SNARE proteins when they are not coiled with one another.

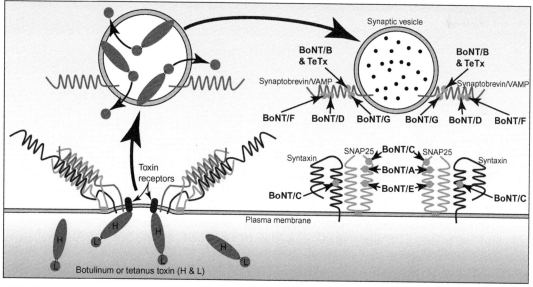

FIGURE 8.13 **Diagram of botulinum toxin and tetanus toxin entry into nerve terminals.** Toxins (*pink*) consist of heavy (H) and light (L) chains. Heavy chains recognize the vesicle lumen portion of synaptic vesicle proteins (*blue ovals*) that are exposed during exocytosis. Upon endocytosis, the toxins are engulfed into the nerve terminal. Once inside synaptic vesicles after endocytosis, the light chain of these toxins is moved out of the vesicle into the cytoplasm where selective proteolytic cleavage of SNARE proteins occurs.

# How Do Synaptic Vesicles Move to the Correct Location in the Nerve Terminal Prior to Release?

Synaptic vesicles are initially formed in the Golgi apparatus, where proteins critical for their function are synthesized and inserted into the plasma membrane. After budding from the Golgi, these vesicles use motor proteins called kinesins to travel down the axon along a cytoskeleton pathway made up of microtubules and actin, until they arrive at the nerve terminal (Pfeffer, 1999; Cai et al., 2007).

Once in the nerve terminal, small clear synaptic vesicles are reused many times for transmitter release, and thus are recycled within the nerve terminal without the immediate need for the neuron to make more vesicles. To move synaptic vesicles within the nerve terminal environment, tethering proteins interact to chaperone (aid in the movement of) vesicles between storage pools, recycling pools, and the docking sites for transmitter release (see Chapter 9, for more information on these synaptic vesicle pools).

Vesicle docking is facilitated by a family of proteins attached to synaptic vesicles, called Rab proteins. Rab proteins are GTPases that use GTP hydrolysis to switch between active and inactive configurations, allowing them to associate and disassociate from other proteins as needed (Zerial and McBride, 2001). They interact with PRA1, Rabphillin, RIM, and other proteins in the active zone environment to actively move vesicles into the proper location for tethering and docking at the active zone. For example, Rab proteins interact with cytoskeletal cytomatrix proteins that are very long and stretch like tufts of hair from the active zone into the nearby nerve terminal cytoplasm. Two of these proteins, piccolo and bassoon, are hypothesized to assist in tethering synaptic vesicles in the vicinity of the active zone by providing a scaffold for PRA1, Rabphillin, and RIM to chaperone the vesicles into position (see Fig. 8.14) for association with SNARE proteins. Regulators of SNARE protein interaction (munc-13 and munc-18; see below) then dock the vesicles in the active zone. Piccolo and bassoon are sometimes called "peripheral organizers" because of their role in aiding organizing this region of the synapse, but with no major role in causing vesicle fusion.

To provide evidence of the hypothesized function of piccolo and bassoon, investigators used a powerful experimental approach to test the role of a protein in any cellular process. It is possible to remove, or "knock out," the gene in an experimental animal (e.g., a mouse or *Drosophila*) that codes for a specific protein of interest. With the candidate protein missing, an experimenter can evaluate how the system functions without that protein. Mukherjee et al. (2010) genetically knocked out (removed completely) or knocked down (reduced the expression levels of) piccolo and bassoon proteins. Synapses deficient in piccolo and/or bassoon had fewer synaptic vesicles concentrated at the nerve terminal and fewer docked synaptic vesicles at the active zone. However, there were few or no differences in the magnitude of neurotransmitter release after low-frequency stimulation (which was insufficient to deplete the remaining docked vesicles), indicating that the loss of piccolo and/or bassoon impairs vesicle movement to the active zone but not fusion. These data support the hypothesis that piccolo and bassoon are peripheral organizers of synaptic vesicles that serve as cytomatrix pathways for vesicle tethering to the active zone. Note that they are likely not the *only* such organizers; the fact that some vesicles were still moved to active zones in piccolo/bassoon knockout synapses suggests the existence of

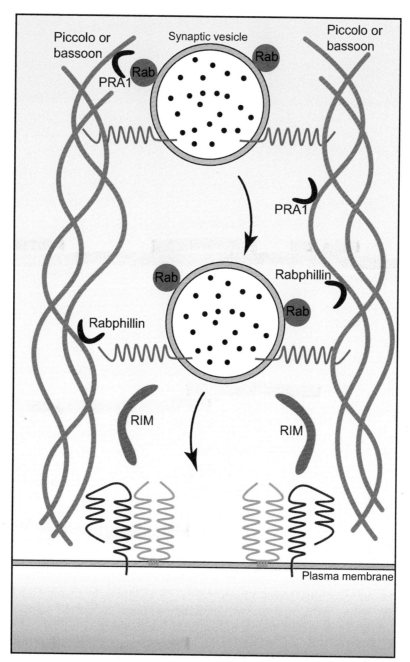

FIGURE 8.14 **Cytoskeletal cytomatrix proteins assist in moving synaptic vesicles into a vesicle docking site in the nerve terminal.** Piccolo and bassoon are long proteins that interact with a variety of chaperone partners (PRA1, Rabphillin, and RIM) and Rab proteins on synaptic vesicles. These proteins move vesicles along a directed path and position them so that the SNARE proteins can bind together, tethering vesicles to the neuronal plasma membrane.

parallel mechanisms for vesicle replenishment, involving different proteins. One of the known limitations of knockout/knockdown experiments is that cells often have multiple redundant pathways for carrying out a given function, so that even when one pathway is experimentally disrupted, the others will compensate for its loss, which can sometimes make it difficult to interpret the results of these experiments.

The proteins presented above, which have important roles in synaptic vesicle transport and tethering, are of course only a subset of the many proteins that participate in these functions. Those interested in a more comprehensive review of many of the other proteins can consult Pfeffer (1999) and Cai et al. (2007).

## How Are SNARE Proteins Directed to Coil Together Properly to "Dock" a Synaptic Vesicle to the Plasma Membrane?

Synaptic vesicle fusion with the plasma membrane requires the SNARE proteins to selectively and tightly bind to one another with proper alignment and stoichiometry. The CORE complex must include one alpha helix from synaptobrevin/VAMP, one from syntaxin, and two from SNAP25 (see Fig. 8.11). If these proteins were free to form complexes with one another without any regulation, many incorrect protein associations might occur and vesicles might be bound up together or with other organelles. For example, SNARE proteins are also present on endoplasmic reticulum, Golgi, and transport vesicles that contain cellular cargo other than chemical transmitter. Therefore, CORE complex formation in the active zone needs to be tightly controlled to direct appropriate interactions. Accordingly, the proteins Munc-13 and Munc-18 (sometimes called "SM" proteins, referring to the family of "Sec/Munc" proteins) serve as critical regulators of CORE complex formation. What is the mechanism by which Munc-13 and Munc-18 regulate CORE complex formation? Munc-13 may serve as an intermediate chaperone for synaptic vesicles, bridging the gap between the so-called "peripheral organizers" of synaptic vesicle tethering (piccolo and bassoon and their associated interactions with Rab and RIM proteins) and the CORE complex proteins that dock synaptic vesicles. RIM proteins contain separate Munc13- and Rab-binding sites, so that all three proteins can link together as synaptic vesicles are moved toward a docking site in the active zone (Dulubova et al., 2005). Meanwhile, at the awaiting vesicle docking site, Munc-18 binds to syntaxin and holds it in a "folded" configuration that prevents it from forming the CORE complex with the other SNARE proteins; for this reason, Munc-18 is sometimes called a "SNARE protector" (Fig. 8.15; Hata et al., 1993). When the Rab/RIM/Munc13 complex chaperones the synaptic vesicle into the docking site, Munc-13 and Munc-18 are thought to interact in a manner that dissociates Munc-18 from syntaxin, allowing syntaxin to unfold so that the CORE complex can coil together and tether the vesicle to the docking site (Shen et al., 2007; see Fig. 8.15). In support of this mechanistic model, formation of the CORE complex has been shown to be dependent on the chaperoning of syntaxin from the "folded" to the "open" configuration (Dulubova et al., 1999).

Some evidence for the importance of these protein interactions in neurotransmitter release comes from experiments in the large calyx of Held synapse in the auditory brainstem, in which researchers used point mutations to selectively disrupt the

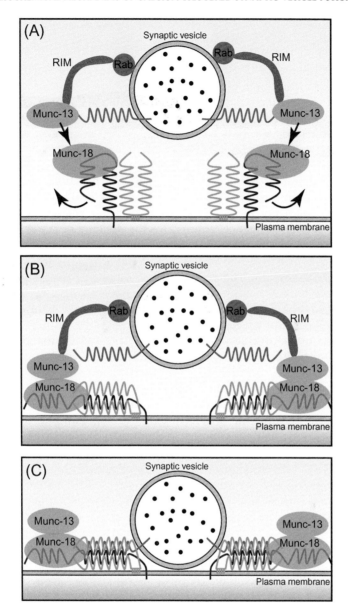

FIGURE 8.15 **Proteins that facilitate synaptic vesicle docking of SNARE proteins.** (A) Synaptic vesicles must be moved into position in a regulated manner. Rab, RIM, and Munc-13 may form a three-protein complex that regulates synaptic vesicle docking. When no vesicle is present at the docking site, Munc-18 binds to syntaxin in a folded configuration that prevents CORE complex formation. (B) When a vesicle is chaperoned into the area of the docking site, the Rab/RIM/Munc-13 complex interacts with Munc-18 to dissociate it from syntaxin, allowing syntaxin to unfold. (C) In its newly unfolded configuration, syntaxin can now coil together with synaptobrevin/VAMP and SNAP25 to form the CORE complex.

Rab3/RIM/Munc13 protein–protein interaction sites. When a mutation impaired the ability of these three proteins to bind together, the pool of docked vesicles available for neurotransmitter release (sometimes called the "readily releasable pool of vesicles"; Dulubova et al., 2005; also see Chapter 9) was smaller. These data suggest that the interaction of Rab3, RIM, and Munc13 is important for chaperoning vesicles into the active zone region of the nerve terminal. In addition, when Munc-13 was mutated in a way that prevented it from interacting with Munc-18, action potential-triggered neurotransmitter release was completely blocked, suggesting that Munc-13-mediated removal of Munc-18 from syntaxin is critical for vesicle docking at release sites to proceed (Aravamudan et al., 1999; Augustin et al., 1999; Richmond et al., 1999; Varoqueaux et al., 2002). Similar results were observed after mutating RIM, such that it did not function, indicating RIM has a similarly important role in vesicle docking. However, mutating Rab to interfere with its function had only a relatively minor effect on transmitter release, suggesting that Rab facilitates these interactions but is not required for the docking of synaptic vesicles.

The physiological relevance of the SNARE-regulating function of Munc-18 has been supported by studies in neurons (Shen et al., 2015). If Munc-18 is mutated in such a way that it cannot be displaced from syntaxin, all transmitter release is blocked (both spontaneous and action potential-evoked). In addition, when Munc-13 and Munc-18 were absent from the nerve terminal, docked vesicles were unstable and fell back into an undocked configuration very rapidly (He et al., 2017). This implies that the Munc-18 rearrangement of syntaxin configuration is absolutely essential for synaptic vesicles to dock to neurotransmitter release sites (Weimer et al., 2003; Toonen and Verhage, 2007; Shen et al., 2015).

In summary, Rab proteins on the outside of synaptic vesicles interact with the RIM protein to facilitate the movement of synaptic vesicles to a docking site at the active zone. At the unoccupied vesicle docking site, Munc-18 binds to syntaxin and holds it in a folded conformation that prevents SNARE complex formation. The Rab/RIM/Munc-13 complex chaperones a vesicle into the docking site and interacts with Munc-18 to allow syntaxin to unfold. Once in the unfolded configuration, syntaxin can now coil together with synaptobrevin/VAMP and SNAP25 to form the CORE complex, which tethers the vesicle to the plasma membrane and functions as the scaffold that all other synaptic vesicle fusion reactions revolve around.

### Coupling of Voltage-Gated Calcium Channels to the Active Zone

As we discussed in Chapter 7, action-potential-mediated calcium influx needs to occur near calcium sensors in order to trigger neurotransmitter release. This is why voltage-gated calcium channels must be closely associated with the synaptic vesicle docking site. Voltage-gated calcium channels that are coupled to transmitter release sites have an amino acid sequence called a "synaptic protein interaction site," or SynPrInt site (also discussed in Chapter 7). The SynPrInt site is a string of amino acids on one cytoplasmic loop of the calcium channel that can bind to the folded configuration of syntaxin, as well as SNAP25 and synaptotagmin. At the unoccupied vesicle docking site, voltage-gated calcium channels are bound to the folded configuration of syntaxin (see Fig. 8.16). When syntaxin

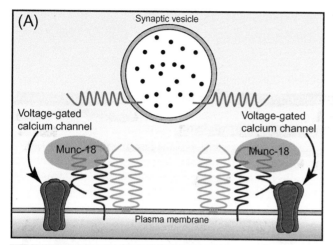

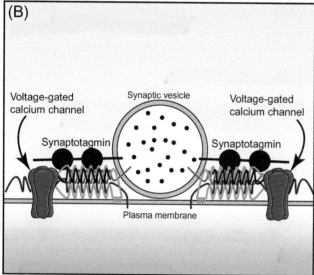

FIGURE 8.16 Voltage-gated calcium channel association with the active zone vesicle docking site. (A) In the absence of a docked synaptic vesicle, voltage-gated calcium channels bind to the folded configuration of syntaxin (held in this folded configuration by Munc-18; see Fig. 8.15). (B) After vesicle docking and priming, calcium channels are thought to bind to SNAP25 and synaptotagmin (colored and drawn as in Figs. 8.15 above and 8.17 below).

unfolds as part of the vesicle docking process, the calcium channels are thought to dissociate from the unfolded syntaxin and form new associations with SNAP25 and synaptotagmin. These associations keep the voltage-gated calcium channel in close proximity to a docked vesicle that is being prepared for fusion.

## How Do SNARE Proteins of the CORE Complex Work With Synaptotagmin (the Calcium Sensor) to Regulate Calcium-Triggered Transmitter Release?

When a CORE complex is unregulated, the coiling of the three SNARE proteins continues slowly all the way to the ends that insert into vesicle membrane (synaptobrevin/

VAMP) and plasma membrane (syntaxin and SNAP25). In the endoplasmic reticulum and Golgi apparatus, where there is no calcium control over SNARE coiling, this coiling slowly leads vesicles to fuse with the plasma membrane. However, at nerve terminals, two additional proteins join with the CORE complex to regulate the extent to which the SNARE proteins complete their coiling together (Sudhof and Rothman, 2009; Trimbuch and Rosenmund, 2016). In nerve terminals, the cytoplasmic protein complexin coils together with the CORE complex, and it is hypothesized that complexin assists in preventing the SNARE proteins from completing their coiling (see Fig. 8.17). The vesicle membrane protein synaptotagmin also binds to the CORE complex and is hypothesized to perform two independent functions. First, synaptotagmin assists with complexin in preventing CORE complex-mediated vesicle fusion. Second, synaptotagmin contains two calcium-binding domains, called C2 domains, that act as the primary calcium sensor for transmitter release (see Fig. 8.17).

The inclusion of complexin and synaptotagmin in the CORE complex allows the vesicle to be held in a fusion-ready state that is waiting for calcium ion entry to trigger fusion. This fusion-ready state is sometimes called a **primed** synaptic vesicle. This primed state is thought to leave synaptic vesicles on the verge of vesicle fusion, waiting for calcium entry. In fact, spontaneous neurotransmitter release (miniature synaptic events) may occasionally occur by the random escape of a synaptic vesicle from this primed state, even without calcium binding to synaptotagmin. Miniature synaptic events may also occur by two other mechanisms. First, cytoplasmic resting calcium ions may bind to synaptotagmin, though this is rare because synaptotagmin has low-affinity calcium-binding sites and the resting cytoplasmic calcium concentration is low. Second, voltage-gated calcium channels occasionally open at the resting membrane potential, even though it is very unlikely (but not impossible) for the types of voltage-gated calcium channels that exist at synapses to open at such a hyperpolarized potential. The contribution of each of these mechanisms to spontaneous transmitter release appears to vary by synapse type.

After an action potential reaches the presynaptic nerve terminal, some fraction of voltage-gated calcium channels open. A single action potential depolarization opens between 10% and 90% of calcium channels, depending on the calcium channel type and the duration of the action potential. After calcium ion entry, these ions bind to a variety of calcium-binding proteins in the nerve terminal, including synaptotagmin.

Calcium ions binding to synaptotagmin neutralize the negative charge on the C2 domains, allowing them to insert into plasma membrane and/or vesicle membrane and create conformational changes in the synaptotagmin protein that rapidly release the hold that complexin and synaptotagmin have on the CORE complex. In addition to releasing the hold, the calcium-binding-induced conformational changes in synaptotagmin rapidly increase the probability that the vesicle membrane will fuse to the plasma membrane (see Fig. 8.18). This calcium-triggering event converts what would be a slow, uncontrolled vesicle fusion in the absence of complexin and synaptotagmin, into a rapid fusion event that is synchronized across many primed vesicles in a synapse based on the timing of action-potential-mediated calcium entry. Synchronous calcium-triggered vesicle fusion creates the evoked synaptic potential that can be measured at synapses following a presynaptic action potential.

II. REGULATION OF CHEMICAL TRANSMITTER RELEASE

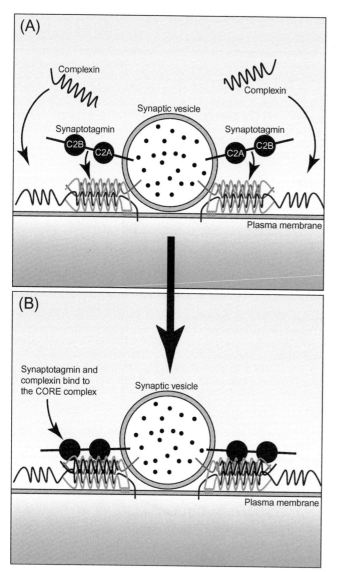

FIGURE 8.17 **Priming of docked synaptic vesicles.** (A) After the CORE complex forms, allowing a vesicle to become docked (A), the cytoplasmic protein complexin and the vesicle membrane protein synaptotagmin bind to the coiled SNARE proteins of the CORE complex (B). The CORE complex holds the vesicle in a fusion-ready (primed) state, but complexin and synaptotagmin prevent actual fusion with the neuronal membrane.

## Evidence That Synaptotagmin Is the Calcium Sensor at Active Zones

Why is it thought that synaptotagmin is the calcium sensor for fast synchronous neurotransmitter release? A series of experimental results that support this hypothesis are outlined here:

1. There are many proteins in the nerve terminal that bind calcium (including Rabphillin, RIM, Munc-13, and Doc2), but the protein that triggers fast synchronous neurotransmitter release after an action potential should bind calcium with low affinity

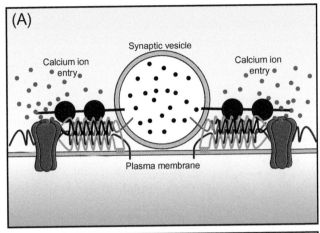

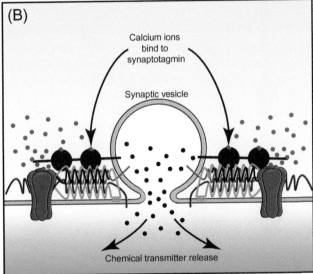

FIGURE 8.18 Calcium-triggered synaptic vesicle fusion. (A) Voltage-gated calcium channels that are positioned in the active zone open when an action potential depolarizes the nerve terminal. Calcium ions (red dots) enter the nerve terminal cytoplasm near the mouths of open channels. (B) Calcium ions then bind to synaptotagmin, rapidly increasing the probability that synaptic vesicles will fuse with plasma membrane.

(in the micromolar range, specifically 10–100 μM; see Chapter 7), to minimize the chance that it will be activated by cytoplasmic calcium in the absence of an action potential. Synaptotagmin has two calcium-binding domains (C2A and C2B) that both bind calcium in the micromolar range (Brose et al., 1992). Most other calcium-binding proteins in the nerve terminal have higher-affinity calcium-binding sites.

2. Because calcium only reaches these micromolar concentrations near the mouth of an open calcium channel after an action potential, the protein that serves as the trigger for action potential-evoked transmitter release should be positioned in close proximity to

voltage-gated calcium channels in the active zone. As a vesicle membrane protein that binds to the CORE complex after vesicle docking, synaptotagmin is closely positioned near voltage-gated calcium channels (Leveque et al., 1994).

3. The protein that is the calcium trigger for transmitter release should, when calcium-bound, engage in interactions that increase the probability of vesicle fusion with the plasma membrane. When synaptotagmin binds calcium, the bound calcium ions screen the negative charges on the C2 domains, rendering these regions of the proteins lipophilic. As a result, calcium-bound C2 domains of synaptotagmin can bind to cell membranes (Davis et al., 1999). It is still debated whether these domains bind to vesicle membrane or to the plasma membrane, and it is also possible that C2A binds to one and C2B to the other. Regardless of the specific sites of synaptotagmin action, it is hypothesized that membrane binding pulls the vesicle membrane closer to the plasma membrane, increasing the probability and speed of vesicle fusion to the plasma membrane. It is not yet known how C2 domain binding brings vesicle membrane and plasma membrane together, although several specific mechanistic hypotheses have been proposed (Chang et al., 2018; Gundersen and Umbach, 2013).

4. The gene knockout technique was used to evaluate the role of synaptotagmin at hippocampal synapses (Geppert et al., 1994; Nishiki and Augustine, 2004). Synaptotagmin knockout mice homozygous for the missing gene die at birth because they cannot breathe well enough, so experimenters could not study synaptotagmin knockout synapses in vivo. They instead made hippocampal cell cultures from control (wild-type) and embryonic synaptotagmin knockout mice, permitted these neuronal cultures to form synapses and mature their synaptic contacts for 2 weeks in vitro, and compared synaptic transmission in control and synaptotagmin knockout synapses using whole-cell patch clamp techniques.

    In hippocampal cultures, control synapses had large, fast synchronous postsynaptic currents (many quantal events release at the same short time delay after a presynaptic stimulus), followed by a small burst of **asynchronous neurotransmitter release** (single quantal events that occurred at variable delay after a presynaptic stimulation). Synaptotagmin knockout synapses showed asynchronous neurotransmitter release, but not fast synchronous neurotransmitter release (Geppert et al., 1994; see Fig. 8.19). These results support the hypothesis that synaptotagmin is the calcium sensor for fast synchronous neurotransmitter release. The fact that asynchronous neurotransmitter release was not eliminated by knocking out synaptotagmin implies the existence of other calcium-sensitive proteins (see #1 above) that can increase the probability of neurotransmitter release after nerve stimulation, but not in a fast, efficient manner like synaptotagmin does.

5. Using a model organism in which it is easier to make specific mutations to synaptotagmin (the fruit fly, *Drosophila*), experimenters evaluated the role of synaptotagmin at neuromuscular synapses (Littleton et al., 1993; 1994; DiAntonio and Schwarz, 1994). One experimental measure that can be used to evaluate the effects of mutations of synaptotagmin is to calculate the calcium-release relationship (see Chapter 7). At all synapses, this relationship normally has a slope when plotted on a log scale of about 3–5. At control (wild-type) *Drosophila* neuromuscular junctions, experimenters calculated a calcium-release relationship slope of 3.5 (see Fig. 8.20). A

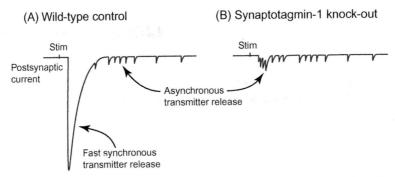

FIGURE 8.19  **Recordings obtained from control (wild-type) hippocampal synapses in culture (A) and after knock out of the synaptotagmin-1 protein (B).** In control synapses, stimulation of the presynaptic nerve (Stim) resulted in a fast, synchronous postsynaptic current, followed by a short burst of asynchronous transmitter release. After knocking out synaptotagmin-1, the fast synchronous transmitter release was eliminated, leaving only the asynchronous transmitter release.

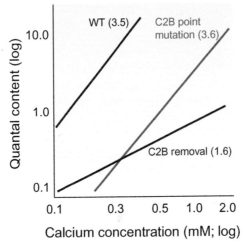

FIGURE 8.20  **Diagram of the calcium-release relationship plot at the *Drosophila* neuromuscular junction in control synapses and those in which synaptotagmin has been mutated.** Control synapses (*WT*) have a calcium-release relationship with a slope of 3.5 (*black line*). A single point mutation in the C2B calcium-binding domain maintains the same relationship (slope = 3.6; *red line*), but shifts the sensitivity of neurotransmitter release to calcium to the right, meaning that more calcium is required to trigger the same amount of transmitter release. A mutation that removes the entire C2B calcium-binding domain from synaptotagmin reduces calcium-triggered transitter release, and changes the slope of the calcium-release relationship (slope = 1.6; *blue line*).

point mutation (a change in one amino acid) in the C2B domain of synaptotagmin did not change the slope of the calcium-release relationship, but the sensitivity of neurotransmitter release to calcium became much weaker than at control synapses, with more calcium being required to trigger the same amount of neurotransmitter release. When the C2B domain was removed entirely, the slope of the calcium-release relationship became significantly more shallow (slope = 1.6), and significantly more calcium was required to trigger a normal magnitude of neurotransmitter release (see Fig. 8.20). These data provide strong evidence that synaptotagmin is the calcium sensor for calcium-triggered transmitter release, since mutations that affect its calcium-binding functions can cause striking changes in the calcium-release relationship.

Taken together, the above studies strongly support the hypothesis that synaptotagmin is the primary calcium sensor for fast calcium-triggered vesicle release. However, there is also evidence that other calcium-binding proteins in the nerve terminal can act as secondary calcium sensors that increase the probability of neurotransmitter release (Sun et al., 2007), though they may be slower-acting and/or have a higher affinity for calcium than synaptotagmin. The identity of these secondary calcium sensors is still debated, but they have been proposed to selectively trigger slower asynchronous neurotransmitter release (for example, Doc2 or synatotagmin-7; Wen et al., 2010; Yao et al., 2011), spontaneous neurotransmitter release (Doc2; Groffen et al., 2010), and/or to increase the probability of neurotransmitter release during short-term plasticity (synaptotagmin-7; Jackman et al., 2016; see Chapter 14).

Multiple isoforms of synaptotagmin are expressed in the nervous system (Sudhof, 2002; Chen and Jonas, 2017). In fact, there are 17 synaptotagmin genes in mammals. Interestingly, only eight of these bind calcium. Four of the synaptotagmin subtypes are hypothesized to be expressed on the vesicle membrane and function as fast calcium sensors for transmitter release (Xu et al., 2007), but one of these may reside in the plasma membrane instead of the vesicle membrane (Li et al., 2017) and play a role in spontaneous transmitter release, asynchronous release triggered by action potentials, and/or short-term synaptic plasticity. synaptotagmin subtypes may also be expressed on the postsynaptic side of the synapse and participate in trafficking receptors into the plasma membrane (Wu et al., 2017). The functions of many subtypes of synaptotagmin have not yet been identified.

## Recovery and Disassembly of the SNARE Protein CORE Complex After Synaptic Vesicle Fusion

Following synaptic vesicle fusion, the vesicle membrane is selectively pulled back into the cytoplasm by a process called **endocytosis** (see Chapter 9) to be re-used. In order to be recycled as a functional synaptic vesicle, the patch of membrane to be endocytosed must include only proteins that are part of vesicular membrane, not proteins that are found in the plasma membrane. Therefore, prior to endocytosis, the CORE complex needs to be uncoiled to separate synaptobrevin/VAMP from syntaxin and SNAP25. Because the SNARE proteins of the CORE complex are so stably and tightly coiled during synaptic vesicle docking and fusion, chaperone proteins and an expenditure of energy are required to uncoil them (Sollner et al., 1993; Zhao et al., 2015).

SNARE uncoiling is orchestrated by the cytoplasmic chaperone proteins NSF and αSNAP. The hypothesized model for this SNARE disassembly begins with αSNAP proteins binding around the coiled CORE complex. NSF then joins the αSNAP-CORE complex, and together they form a "cap" on the SNARE proteins (see Fig. 8.21B). The NSF-αSNAP "cap" uses the energy provided by ATP hydrolysis to explosively uncoil the CORE complex (Ryu et al., 2015; see Fig. 8.21C).

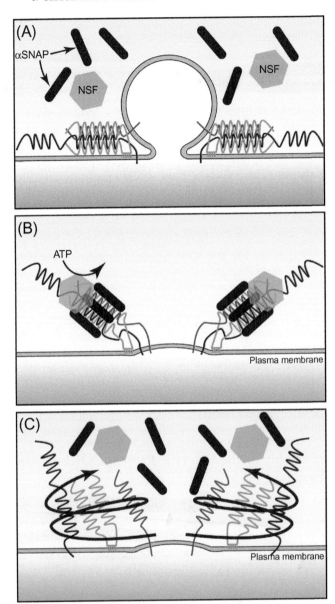

FIGURE 8.21  **CORE complex uncoiling after the process of vesicle fusion is complete.** (A) Cytoplasmic αSNAP and NSF assist in CORE complex uncoiling. (B) αSNAP recognizes the fully coiled SNARE complex after vesicle fusion is complete, and several of these proteins bind around the SNARE complex. Cytoplasmic NSF binds to the αSNAP-SNARE complex to form a "cap" on the SNARE complex. (C) NSF uses the energy of ATP to twist the "cap" and rapidly uncoil the SNARE complex.

# References

Aravamudan, B., Fergestad, T., Davis, W.S., Rodesch, C.K., Broadie, K., 1999. Drosophila UNC-13 is essential for synaptic transmission. Nat. Neurosci. 2, 965–971.

Augustin, I., Rosenmund, C., Sudhof, T.C., Brose, N., 1999. Munc13-1 is essential for fusion competence of glutamatergic synaptic vesicles. Nature 400, 457–461.

Bhidayasiri, R., Truong, D.D., 2005. Expanding use of botulinum toxin. J. Neurol. Sci. 235, 1–9.

Brose, N., Petrenko, A.G., Sudhof, T.C., Jahn, R., 1992. Synaptotagmin: a calcium sensor on the synaptic vesicle surface. Science 256, 1021–1025.

Bucher, E.S., Wightman, R.M., 2015. Electrochemical analysis of neurotransmitters. Annu. Rev. Anal. Chem. (Palo Alto Calif) 8, 239–261.

Cai, Q., Pan, P.Y., Sheng, Z.H., 2007. Syntabulin-kinesin-1 family member 5B-mediated axonal transport contributes to activity-dependent presynaptic assembly. J. Neurosci. 27, 7284–7296.

Chang, S., Trimbuch, T., Rosenmund, C., 2018. Synaptotagmin-1 drives synchronous $Ca(2+)$-triggered fusion by C2B-domain-mediated synaptic-vesicle-membrane attachment. Nat. Neurosci. 21, 33–40.

Chen, C., Jonas, P., 2017. Synaptotagmins: that's why so many. Neuron 94, 694–696.

Davis, A.F., Bai, J., Fasshauer, D., Wolowick, M.J., Lewis, J.L., Chapman, E.R., 1999. Kinetics of synaptotagmin responses to $Ca2+$ and assembly with the core SNARE complex onto membranes. Neuron 24, 363–376.

Davletov, B., Bajohrs, M., Binz, T., 2005. Beyond BOTOX: advantages and limitations of individual botulinum neurotoxins. Trends Neurosci. 28, 446–452.

Dernick, G., de Toledo, G.A., Lindau, M., 2007. The patch amperometry technique: design of a method to study exocytosis of single vesicles. In: Michael, A.C., Borland, L.M. (Eds.), Electrochemical Methods for Neuroscience. CRC Press/Taylor & Francis, Boca Raton, FL.

DiAntonio, A., Schwarz, T.L., 1994. The effect on synaptic physiology of synaptotagmin mutations in Drosophila. Neuron 12 (4), 909–920.

Dressler, D., 2012. Clinical applications of botulinum toxin. Curr. Opin. Microbiol. 15, 325–336.

Dulubova, I., Sugita, S., Hill, S., Hosaka, M., Fernandez, I., Sudhof, T.C., et al., 1999. A conformational switch in syntaxin during exocytosis: role of munc18. EMBO J. 18, 4372–4382.

Dulubova, I., Lou, X., Lu, J., Huryeva, I., Alam, A., Schneggenburger, R., et al., 2005. A Munc13/RIM/Rab3 tripartite complex: from priming to plasticity? EMBO J. 24, 2839–2850.

Dutta, S.R., Passi, D., Singh, M., Singh, P., Sharma, S., Sharma, A., 2016. Botulinum toxin the poison that heals: a brief review. Natl. J. Maxillofac. Surg. 7, 10–16.

Ferro-Novick, S., Jahn, R., 1994. Vesicle fusion from yeast to man. Nature 370, 191–193.

Geppert, M., Goda, Y., Hammer, R.E., Li, C., Rosahl, T.W., Stevens, C.F., et al., 1994. Synaptotagmin I: a major $Ca2+$ sensor for transmitter release at a central synapse. Cell 79, 717–727.

Goodsell, D.S., 2009. The neuromuscular synapse. Biochem. Mol. Biol. Educ. 37, 204–210.

Gong, L.W., Hafez, I., Alvarez de Toledo, G., Lindau, M., 2003. Secretory vesicles membrane area is regulated in tandem with quantal size in chromaffin cells. J. Neurosci. 23, 7917–7921.

Groffen, A.J., Martens, S., Diez Arazola, R., Cornelisse, L.N., Lozovaya, N., de Jong, A.P., et al., 2010. Doc2b is a high-affinity $Ca2+$ sensor for spontaneous neurotransmitter release. Science 327, 1614–1618.

Gundersen, C.B., Umbach, J.A., 2013. Synaptotagmins 1 and 2 as mediators of rapid exocytosis at nerve terminals: the dyad hypothesis. J. Theor. Biol. 332, 149–160.

Hata, Y., Slaughter, C.A., Sudhof, T.C., 1993. Synaptic vesicle fusion complex contains unc-18 homologue bound to syntaxin. Nature 366, 347–351.

He, E., Wierda, K., van Westen, R., Broeke, J.H., Toonen, R.F., Cornelisse, L.N., et al., 2017. Munc13-1 and Munc18-1 together prevent NSF-dependent de-priming of synaptic vesicles. Nat. Commun. 8, 15915.

Heuser, J.E., Reese, T.S., 1981. Structural changes after transmitter release at the frog neuromuscular junction. J. Cell Biol. 88, 564–580.

Heuser, J.E., Reese, T.S., Landis, D.M., 1976. Preservation of synaptic structure by rapid freezing. Cold Spring Harb. Symp. Quant. Biol. 40, 17–24.

Heuser, J.E., Reese, T.S., Dennis, M.J., Jan, Y., Jan, L., Evans, L., 1979. Synaptic vesicle exocytosis captured by quick freezing and correlated with quantal transmitter release. J. Cell Biol. 81, 275–300.

Jackman, S.L., Turecek, J., Belinsky, J.E., Regehr, W.G., 2016. The calcium sensor synaptotagmin 7 is required for synaptic facilitation. Nature 529 (7584), 88–91.

Kim, M.H., von Gersdorff, H., 2010. Extending the realm of membrane capacitance measurements to nerve terminals with complex morphologies. J. Physiol. 588, 2011–2012.
Lam, S.M., 2003. The basic science of botulinum toxin. Facial Plast. Surg. Clin. North Am. 11, 431–438.
Leveque, C., el Far, O., Martin-Moutot, N., Sato, K., Kato, R., Takahashi, M., et al., 1994. Purification of the N-type calcium channel associated with syntaxin and synaptotagmin. A complex implicated in synaptic vesicle exocytosis. J. Biol. Chem. 269, 6306–6312.
Li, Y.C., Chanaday, N.L., Xu, W., Kavalali, E.T., 2017. synaptotagmin-1- and synaptotagmin-7-dependent fusion mechanisms target synaptic vesicles to kinetically distinct endocytic pathways. Neuron. 93 (3), 616–631.
Littleton, J.T., Stern, M., Schulze, K., Perin, M., Bellen, H.J., 1993. Mutational analysis of Drosophila synaptotagmin demonstrates its essential role in Ca(2 + )-activated neurotransmitter release. Cell 74, 1125–1134.
Littleton, J.T., Stern, M., Perin, M., Bellen, H.J., 1994. Calcium dependence of neurotransmitter release and rate of spontaneous vesicle fusions are altered in Drosophila synaptotagmin mutants. Proc. Natl. Acad. Sci. USA 91, 10888–10892.
Montecucco, C., Molgo, J., 2005. Botulinal neurotoxins: revival of an old killer. Curr. Opin. Pharmacol. 5, 274–279.
Morel, N., Israel, M., Manaranche, R., 1978. Determination of ACh concentration in torpedo synaptosomes. J. Neurochem. 30, 1553–1557.
Mukherjee, K., Yang, X., Gerber, S.H., Kwon, H.B., Ho, A., Castillo, P.E., Liu, X., Südhof, T.C., 2010. Piccolo and bassoon maintain synaptic vesicle clustering without directly participating in vesicle exocytosis. Proc. Natl. Acad. Sci. USA 107 (14), 6504–6509.
Neher, E., Marty, A., 1982. Discrete changes of cell membrane capacitance observed under conditions of enhanced secretion in bovine adrenal chromaffin cells. Proc. Natl. Acad. Sci. USA 79, 6712–6716.
Niemann, H., Blasi, J., Jahn, R., 1994. Clostridial neurotoxins: new tools for dissecting exocytosis. Trends Cell Biol. 4, 179–185.
Nishiki, T., Augustine, G.J., 2004. Synaptotagmin I synchronizes transmitter release in mouse hippocampal neurons. J. Neurosci. 24, 6127–6132.
Oldenziel, W.H., Westerink, B.H., 2005. Improving glutamate microsensors by optimizing the composition of the redox hydrogel. Anal. Chem. 77, 5520–5528.
Pantano, S., Montecucco, C., 2014. The blockade of the neurotransmitter release apparatus by botulinum neurotoxins. Cell. Mol. Life Sci. 71, 793–811.
Pfeffer, S.R., 1999. Transport-vesicle targeting: tethers before SNAREs. Nat. Cell. Biol. 1, E17–E22.
Plattner, H., Artalejo, A.R., Neher, E., 1997. Ultrastructural organization of bovine chromaffin cell cortex-analysis by cryofixation and morphometry of aspects pertinent to exocytosis. J. Cell Biol. 139, 1709–1717.
Poulain, B., Baux, G., Tauc, L., 1986. The quantal release at a neuro-neuronal synapse is regulated by the content of acetylcholine in the presynaptic cell. J. Physiol. (Paris) 81, 270–277.
Richmond, J.E., Davis, W.S., Jorgensen, E.M., 1999. UNC-13 is required for synaptic vesicle fusion in C. elegans. Nat. Neurosci. 2, 959–964.
Rothman, J.E., Orci, L., 1992. Molecular dissection of the secretory pathway. Nature 355, 409–415.
Ryu, J.K., Min, D., Rah, S.H., Kim, S.J., Park, Y., Kim, H., et al., 2015. Spring-loaded unraveling of a single SNARE complex by NSF in one round of ATP turnover. Science 347, 1485–1489.
Sarter, M., Parikh, V., Howe, W.M., 2009. Phasic acetylcholine release and the volume transmission hypothesis: time to move on. Nat. Rev. Neurosci. 10, 383–390.
Schekman, R., 1982. The secretory pathway in yeast. Trends Biochem. Sci. 7, 243–246.
Schekman, R., 1985. Protein localization and membrane traffic in yeast. Annu. Rev. Cell Biol. 1, 115–143.
Shen, J., Tareste, D.C., Paumet, F., Rothman, J.E., Melia, T.J., 2007. Selective activation of cognate SNAREpins by Sec1/Munc18 proteins. Cell 128, 183–195.
Shen, C., Rathore, S.S., Yu, H., Gulbranson, D.R., Hua, R., Zhang, C., et al., 2015. The trans-SNARE-regulating function of Munc18-1 is essential to synaptic exocytosis. Nat. Commun. 6, 8852.
Simpson, L.L., 1989. Botulinum Neurotoxin and Tetanus Toxin. Academic Press, San Diego, CA.
Sollner, T., Whiteheart, S.W., Brunner, M., Erdjument-Bromage, H., Geromanos, S., Tempst, P., et al., 1993. SNAP receptors implicated in vesicle targeting and fusion. Nature 362, 318–324.
Sudhof, T.C., 2002. Synaptotagmins: why so many? J. Biol. Chem. 277, 7629–7632.
Sudhof, T.C., Rothman, J.E., 2009. Membrane fusion: grappling with SNARE and SM proteins. Science 323, 474–477.

Sun, J.Y., Wu, X.S., Wu, L.G., 2002. Single and multiple vesicle fusion induce different rates of endocytosis at a central synapse. Nature 417, 555–559.

Sun, J., Pang, Z.P., Qin, D., Fahim, A.T., Adachi, R., Sudhof, T.C., 2007. A dual-Ca2+-sensor model for neurotransmitter release in a central synapse. Nature 450, 676–682.

Takamori, S., et al., 2006. Molecular anatomy of a trafficking organelle. Cell 127, 831–846.

Tauc, L., 1997. Quantal neurotransmitter release: vesicular or not vesicular? Neurophysiology 29, 219–226.

Ting, P.T., Freiman, A., 2004. The story of Clostridium botulinum: from food poisoning to Botox. Clin. Med. (Lond) 4, 258–261.

Toonen, R.F., Verhage, M., 2007. Munc18-1 in secretion: lonely Munc joins SNARE team and takes control. Trends Neurosci. 30, 564–572.

Torri-Tarelli, F., Grohovaz, F., Fesce, R., Ceccarelli, B., 1985. Temporal coincidence between synaptic vesicle fusion and quantal secretion of acetylcholine. J. Cell Biol. 101, 1386–1399.

Trimbuch, T., Rosenmund, C., 2016. Should I stop or should I go? The role of complexin in neurotransmitter release. Nat. Rev. Neurosci. 17, 118–125.

Varoqueaux, F., Sigler, A., Rhee, J.S., Brose, N., Enk, C., Reim, K., et al., 2002. Total arrest of spontaneous and evoked synaptic transmission but normal synaptogenesis in the absence of Munc13-mediated vesicle priming. Proc. Natl. Acad. Sci. USA 99, 9037–9042.

Watanabe, S., 2016. Flash-and-freeze: coordinating optogenetic stimulation with rapid freezing to visualize membrane dynamics at synapses with millisecond resolution. Front. Synaptic Neurosci. 8, 24.

Waters, M.G., Griff, I.C., Rothman, J.E., 1991. Proteins involved in vesicular transport and membrane fusion. Curr. Opin. Cell. Biol. 3, 615–620.

Weimer, R.M., Richmond, J.E., Davis, W.S., Hadwiger, G., Nonet, M.L., Jorgensen, E.M., 2003. Defects in synaptic vesicle docking in unc-18 mutants. Nat. Neurosci. 6, 1023–1030.

Wen, H., Linhoff, M.W., McGinley, M.J., Li, G.L., Corson, G.M., Mandel, G., et al., 2010. Distinct roles for two synaptotagmin isoforms in synchronous and asynchronous transmitter release at zebrafish neuromuscular junction. Proc. Natl. Acad. Sci. USA 107, 13906–13911.

Wu, D., Bacaj, T., Morishita, W., Goswami, D., Arendt, K.L., Xu, W., et al., 2017. Postsynaptic synaptotagmins mediate AMPA receptor exocytosis during LTP. Nature 544, 316–321.

Xu, J., Mashimo, T., Südhof, T.C., 2007. Synaptotagmin-1, -2, and -9: Ca(2+) sensors for fast release that specify distinct presynaptic properties in subsets of neurons. Neuron 54 (4), 567–581.

Yao, J., Gaffaney, J.D., Kwon, S.E., Chapman, E.R., 2011. Doc2 is a Ca2+ sensor required for asynchronous neurotransmitter release. Cell 147, 666–677.

Young, S.H., Chow, I., 1987. Quantal release of transmitter is not associated with channel opening on the neuronal membrane. Science 238, 1712–1713.

Zerial, M., McBride, H., 2001. Rab proteins as membrane organizers. Nat. Rev. Mol. Cell Biol. 2, 107–117.

Zhao, M., Wu, S., Zhou, Q., Vivona, S., Cipriano, D.J., Cheng, Y., et al., 2015. Mechanistic insights into the recycling machine of the SNARE complex. Nature 518, 61–67.

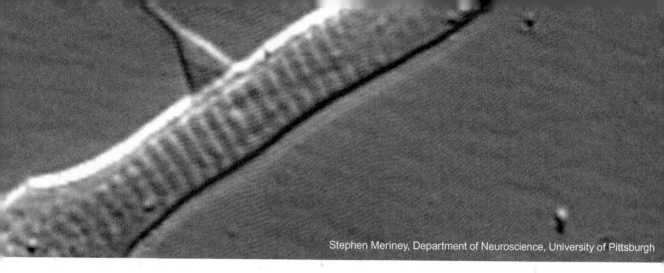

Stephen Meriney, Department of Neuroscience, University of Pittsburgh

CHAPTER 9

# Cellular and Molecular Mechanisms of Endocytosis and Synaptic Vesicle Trafficking

As we discussed in the previous chapter, chemical transmitter release occurs when synaptic vesicles fuse with the plasma membrane. Since synaptic vesicle membrane is incorporated into the plasma membrane of the nerve terminal, the plasma membrane increases in area with each vesicle fusion event. If the membrane at an axon terminal kept growing in size with the release of each vesicle, the axon terminal would eventually become physically distorted and would not be able to function properly. Instead, vesicular membrane is retrieved back into the cytoplasm by a process called endocytosis, ensuring that axon terminals continue to function properly even after sustained neurotransmitter release.

## RETRIEVAL AND REUSE OF SYNAPTIC VESICLE MEMBRANE

What is the evidence that endocytosis occurs? One way to indirectly observe endocytosis is to use electron microscopy to examine how repeated stimulation of a synapse changes the structure of the axon terminal. If a synapse is stimulated repeatedly at a low frequency (once every 5 seconds) for 30 minutes, there is no change in the appearance of the synapse, compared to synapses that were not stimulated. However, if the synapse is stimulated at a much faster rate (10 times per second; 10 Hz) for 30 minutes, the nerve terminal expands in area and the number of synaptic vesicles present in the cytoplasm is greatly reduced, often nearly eliminated. In these enlarged nerve terminals, there is an increase in the appearance of irregularly shaped cytoplasmic membrane inclusions called endosomes (see Fig. 9.1). If a nerve terminal is allowed to rest for 30 minutes following such stimulation, the changes reverse: the endosomes disappear, the nerve terminal returns to its normal size, and synaptic vesicles are present in their normal numbers (Heuser and Reese, 1973). These observations suggest that the membranes of synaptic vesicles join the plasma membrane during neurotransmitter release and that it takes time for the vesicular membrane to be retrieved back into the cytoplasm of the nerve terminal. These observations led to the speculation that after intense stimulation, vesicular membrane is retrieved in the form of large endosomes. Synaptic vesicles are then thought to reform by recycling vesicular membrane from the endosomes.

To provide evidence that endocytosed synaptic vesicles are actually reused, experimenters bathed the nerve terminal in an enzyme called horseradish peroxidase (HRP), which is electron-dense and therefore appears dark in electron microscope images after it has been processed. They then stimulated the axon, causing vesicle release. When vesicle membranes that fused with the plasma membrane were pulled back into the cytoplasm as endosomes, they pulled HRP molecules from the extracellular saline into the nerve terminal with them and reformed into HRP-containing synaptic vesicles (Heuser and Reese, 1973; see Fig. 9.2). In another set of experiments, this HRP loading procedure was repeated, but before the tissue was prepared for the electron microscope, the nerve terminal was bathed in a solution without HRP and stimulated again to cause a second round of exocytosis. After this second stimulus, the HRP label was no longer present in synaptic vesicles. These experiments provided evidence that recycled vesicles could be used again, dumping their HRP "cargo" after the second round of exocytosis.

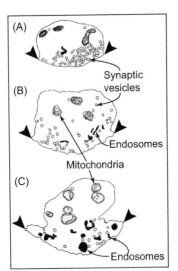

FIGURE 9.1 **Changes in the structure of frog motor nerve terminals after nerve stimulation.** In this figure, drawings have been made of electron micrographs. (A) A nerve terminal at rest has a defined contact area with the postsynaptic muscle cell (between large arrowheads), a large collection of synaptic vesicles, mitochondria, and only a few irregularly shaped membrane inclusions (endosomes; colored black). (B) After 1 min of stimulation at 10 Hz, the area where the nerve terminal contacts the postsynaptic membrane has increased (distance between arrowheads), the number of synaptic vesicles has decreased, and the number of endosomes has increased. (C) After 15 min of stimulation at 10 Hz, the area where the nerve terminal contacts the postsynaptic membrane has increased significantly (distance between arrowheads), there are very few synaptic vesicles in the nerve terminal, and the number of endosomes has increased dramatically. *Source: Adapted from Heuser, J.E., Reese, T.S., 1973. Evidence for recycling of synaptic vesicle membrane during transmitter release at the frog neuromuscular junction. J. Cell. Biol. 57, 315–344.*

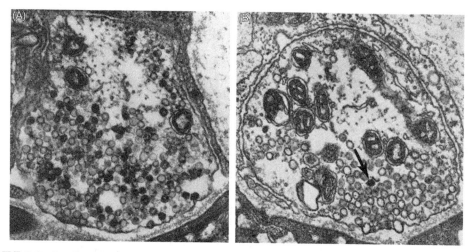

FIGURE 9.2 **Loading of recycled nerve terminal synaptic vesicles with horseradish peroxidase (HRP).** (A) When nerve terminals were bathed in HRP during 15 min of stimulation at 10 Hz and then allowed to rest for 1 h, the synaptic vesicles that reformed in the terminal contain the black reaction product of HRP (note the large fraction of synaptic vesicles that are black in this image). (B) After the initial HRP loading procedure, the stimulation protocol was repeated in the absence of HRP in the bathing medium, and the experimenters waited 1 h before preparing the tissue for the electron microscope so the vesicles would have time to reform. This image shows that the previously loaded HRP had been released from the recycled synaptic vesicles, indicating that the vesicle membrane that had been endocytosed was recycled to form new vesicles which were then released in subsequent rounds of exocytosis. *Source: Adapted from Heuser, J.E., Reese, T.S., 1973. Evidence for recycling of synaptic vesicle membrane during transmitter release at the frog neuromuscular junction. J. Cell. Biol. 57, 315–344.*

After a synaptic vesicle fuses with the plasma membrane, does the nerve terminal recycle that specific patch of membrane, or does it recycle pieces of the plasma membrane indiscriminately? One would expect it must recycle the patch that came from the fused vesicle, since vesicle membrane contains different proteins than plasma membrane. For example, vesicle membrane contains synaptobrevin/VAMP and synaptotagmin, while plasma membrane contains syntaxin and SNAP25. Therefore, if endocytosis retrieved a random patch of plasma membrane, recycled synaptic vesicles made from this membrane would not have the molecular components required to participate in subsequent rounds of exocytosis (see Chapter 8).

To demonstrate that endocytosis specifically retrieves vesicle membrane rather than random patches of plasma membrane from the axon terminal, experimenters used antibodies to identify vesicle membrane proteins. They showed that these antibodies only bound to the axon terminal membrane immediately after exocytosis, when vesicle membrane proteins were part of the plasma membrane. After endocytosis had occurred, the antibodies did not bind, indicating that vesicle membrane protein antigens were no longer present in the plasma membrane (Valtorta et al., 1988). These studies confirmed that endocytosis targets the selective retrieval of vesicle membrane from the nerve terminal plasma membrane after exocytosis.

The electron microscopic data discussed above (Figs. 9.1 and 9.2) are static images of synapses at particular time points in the experiment, relative to when the axon was stimulated. More recently, two new approaches have been developed that permit the experimenter to follow the process of exocytosis and endocytosis in living tissue using fluorescent molecules.

The first of these approaches is very similar to the HRP experiment, except that it uses a class of molecules called "FM dyes." These dyes are fluorescent molecules that integrate into lipid membranes. Upon membrane association, their fluorescence increases (Wu et al., 2009). If FM dyes are present in the extracellular saline during exocytosis, they bind to the newly exposed inside surfaces of the released vesicles and are pulled back into the nerve terminal when the vesicle membrane is endocytosed. When the dye is washed away from the extracellular space, the only remaining fluorescence is from the dye molecules bound to the lumen of recycled vesicles. If the nerve terminal is stimulated again, the dye trapped within synaptic vesicles is immediately released during exocytosis. This experimental approach shows that a population of synaptic vesicles can be labeled with a fluorescent indicator after transmitter release, and that these labeled vesicles can release this FM dye during subsequent nerve stimulation (Betz et al., 1992). These data provide evidence from a living synapse that recycled synaptic vesicles can be reused for action potential-triggered transmitter release right away.

The second approach takes advantage of the difference in pH between the inside of a synaptic vesicle (pH 5.5) and the extracellular space around cells (pH 7.2) (Miesenbock et al., 1998; Ashby et al., 2004). The inside of synaptic vesicles is acidic because synaptic vesicles express an ATP-dependent proton ($H^+$) transporter that pumps $H^+$ into the vesicle (see Chapter 16, for more details). To take advantage of this pH difference, investigators engineered pH-sensitive variants of a molecule called green fluorescent protein (GFP), which they connected to the intracellular portion of a synaptic vesicle transmembrane protein, creating a "fusion protein" that is both vesicular-membrane-bound and has a fluorescent component that is on the inside face of the vesicle membrane. The first of these fusion proteins, synaptopHluorin, is only weakly fluorescent at a pH of 5.5 but becomes strongly fluorescent at a pH of 7.2. When synaptopHluorin proteins are expressed inside vesicles at rest, and thus at a pH of 5.5, they are only weakly fluorescent; but after the synaptic vesicle fuses with the plasma membrane and the inside surface of the vesicle becomes pH 7.2, they become strongly fluorescent (see Fig. 9.3). The synaptopHluorin-expressing vesicle membrane remains fluorescent until the synaptic vesicle is endocytosed and reacidified by the proton pump.

SynaptopHluorin proteins expressed on the inside of vesicles allow investigators to visualize vesicle exocytosis and endocytosis in living synapses (see Fig. 9.4). An advantage of the synaptopHluorin approach is that the experimenter does not have to load the dye into the synaptic vesicle after each round of exocytosis and endocytosis, since this fluorescent marker is permanently attached to the inside of the synaptic vesicle (unlike FM dyes). The original synatopHluorin was a pH-sensitive GFP protein coupled to the vesicle membrane protein synaptobrevin (VAMP), but experimenters have also coupled it to other vesicle membrane proteins (synaptophysin, synaptotagmin, and the vesicular glutamate transporter; Balaji and Ryan, 2007; Diril et al., 2006; Granseth et al., 2006).

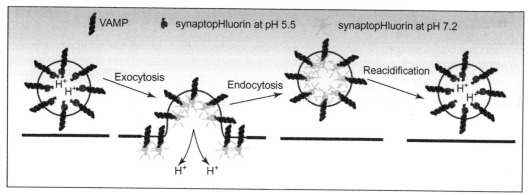

FIGURE 9.3  pH-sensitive synaptopHluorin proteins can be used to track exocytosis and endocytosis in living synapses. At rest, the inside of a synaptic vesicle is pH 5.5, due to a proton pump in the vesicular membrane that loads $H^+$ into synaptic vesicles. The synaptopHluorin inside these vesicles acts as a pH-sensitive marker that is weakly fluorescent at an acidic pH. When a synaptic vesicle fuses with the plasma membrane during exocytosis, the inside of the vesicle is exposed to the extracellular space (pH 7.2), causing the synaptopHluorin to fluoresce more strongly. These vesicles remain strongly fluorescent until the vesicle is endocytosed and reacidified.

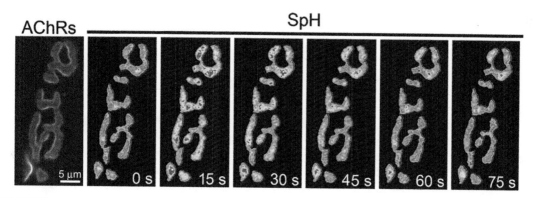

FIGURE 9.4  Use of synaptopHluorin to visualize exocytosis and endocytosis at a living mouse neuromuscular junction. The far left grayscale image shows the location of a single mouse neuromuscular junction under study, which was identified using fluorescently labeled α-bungarotoxin to indicate the location of acetylcholine receptors. The pseudocolor images show the amount of synaptopHluorin fluorescence (SpH) in the motor nerve terminal before (0 s) and at various times after stimulating the nerve at 50 Hz for 15 s (warmer colors indicate a higher level of fluorescence). Immediately after stimulation (15 s) there is a large increase in fluorescence as synaptic vesicles fuse with the plasma membrane. This fluorescence decreases gradually as the synaptic vesicles are internalized and reacidified over the next minute (30–75 s). *Source: Adapted from Wyatt, R.M., Balice-Gordon, R.J., 2008. Heterogeneity in synaptic vesicle release at neuromuscular synapses of mice expressing synaptopHluorin. J. Neurosci. 28, 325–335.*

# ENDOCYTOSIS OCCURS OUTSIDE THE ACTIVE ZONE

We know from Chapter 8, that synaptic vesicles fuse with the plasma membrane within the specialized region of the nerve terminal known as the active zone. However, it appears that endocytosis occurs primarily outside the active zone at regions that are adjacent to transmitter release sites (Miller and Heuser, 1984; Gad et al., 1998; Teng et al., 1999). Using the electron microscope to visualize vesicles undergoing endocytosis, experimenters noticed that vesicles that appeared to be pinching back into the nerve terminal had a coating on them. These coated, partially endocytosed patches of membrane were called "coated pits" when observed in electron micrographs (see Fig. 9.5). Coated pits appear in larger numbers after high-frequency presynaptic action potential activity, and the distribution of coated pits indicates that endocytosis occurs outside the active zone region of the nerve terminal (see Fig. 9.6).

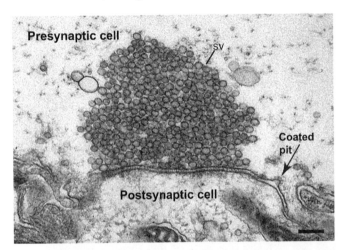

FIGURE 9.5 An electron micrograph of a transmitter release site from the lamprey giant reticulospinal synapse. At this synapse there is a prominent cloud of small clear synaptic vesicles (SV) surrounding the transmitter release sites in the presynaptic cell. The coated pits adjacent to these transmitter release sites are interpreted to represent synaptic vesicle membrane endocytosis. *Source: Adapted from Jakobsson, J., Gad, H., Andersson, F., Low, P., Shupliakov, O., Brodin, L., 2008. Role of epsin 1 in synaptic vesicle endocytosis. Proc. Natl. Acad. Sci. USA 105, 6445–6450. Copyright (2008) National Academy of Sciences, USA.*

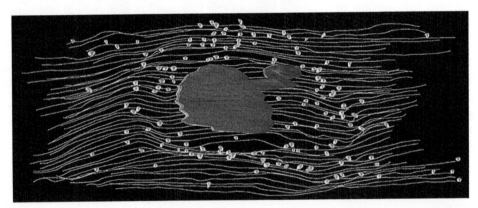

FIGURE 9.6 Endocytic zones are adjacent to the active zone. Serial reconstruction of electron microscopic images from the lamprey giant reticulospinal synapse after stimulation at high frequency for a prolonged period of time (each electron microscopic thin section is represented as a green line in this reconstructed image). The red region indicates the active zone of the nerve terminal, and the small blue dots represent locations where coated pits could be identified, indicating the location of endocytosis. This diagram shows that the sites of endocytosis (*blue dots*) are adjacent to the active zone (*red*). *Source: Adapted from Gad, H., Low, P., Zotova, E., Brodin, L., Shupliakov, O. 1998. Dissociation between Ca2+-triggered synaptic vesicle exocytosis and clathrin-mediated endocytosis at a central synapse. Neuron 21, 607–616.*

# MECHANISMS OF ENDOCYTOSIS

Nerve terminals seem to have several mechanisms they can use to recycle synaptic vesicles. The first is "**clathrin-mediated endocytosis**," which may be the predominant mechanism for endocytosis following high-frequency bursts of action potential activity. Following the experimental application of exceptionally high, nonphysiological stimulus activity (e.g., long bursts of action potentials or prolonged exposure to high extracellular potassium), synapses may employ a mechanism called "**bulk endocytosis**." Finally, following very infrequent exocytosis, perhaps as few as a single vesicle, nerve terminals may use a mechanism called "**kiss-and-run**."

# CLATHRIN-MEDIATED ENDOCYTOSIS

Once it was understood that vesicle membrane was specifically endocytosed, whereas nonvesicular plasma membrane was not, scientists asked how the vesicle membrane was identified during the process of endocytosis. It is hard to imagine how the nerve terminal would "know" which patch of membrane to retrieve without some sort of synaptic vesicle-specific marker (Wu et al., 2014a,b; Kononenko and Haucke, 2015).

There is now substantial evidence that at least one of these markers is synaptotagmin, which acts as a recognition site for an adapter protein that identifies vesicle membrane for endocytosis (Zhang et al., 1994; Li et al., 1995; Jorgensen et al., 1995; Chanaday and Kavalali, 2017). As described in Chapter 8, calcium ions that enter the nerve terminal during an action potential bind to synaptotagmin, which induces the synaptotagmin to bind phospholipids [in particular the specific lipid phosphatidylinositol (4,5)-bisphosphate, or "PIP2"]. The synaptotagmin–PIP2 lipid complex appears to recruit an adapter protein called **AP2** to join the complex (Takei and Haucke, 2001; see Fig. 9.7A), creating a binding site for a protein called **clathrin**.

Clathrin is a six-protein "triskelion" complex made up of three heavy chains and three light chains (see Fig. 9.7B). These triskelion complexes bind to the AP2 proteins collected at the vesicle membrane, and then the clathrin molecules link together to coat the membrane targeted for endocytosis. As the clathrin proteins link up, their structure naturally creates curvature and forms a rounded "cage" around the vesicle membrane, pinching it into the shape of a budding vesicle. The budding vesicle is called a "coated pit" because, in the electron microscope, these structures appear to have an electron-dense coating (see Fig. 9.5).

The clathrin-coated pit cannot complete the endocytosis process without another protein, **dynamin**, which creates the force required to pinch off the budding vesicle. However, dynamin can only aid in endocytosis when it has been dephosphorylated by the calcium-dependent phosphatase calcineurin (Ferguson and De Camilli, 2012), which is activated by the calcium that enters the nerve terminal during action potential activity. There is experimental evidence that as a clathrin-coated budding vesicle is forming, a protein called **endophilin** accumulates around the coated pit and may provide a template for dynamin to collect around the neck of the budding vesicle (Sundborger et al., 2011).

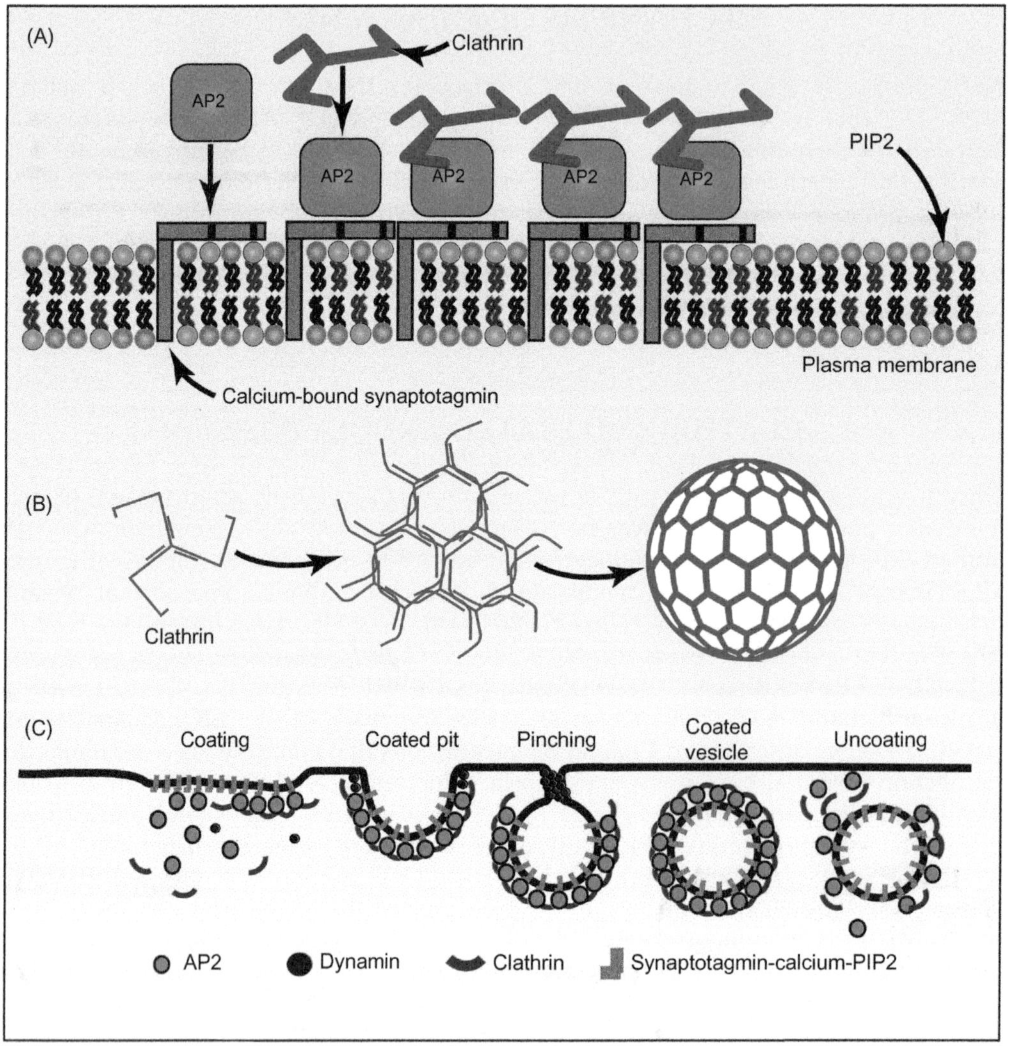


FIGURE 9.7 **Clathrin-mediated endocytosis.** (A) After exocytosis, the calcium-bound synaptotagmin is bound to PIP2 lipids, creating a binding site for the adapter protein AP2. AP2 attracts clathrin to coat the patch of membrane that has AP2 bound. (B) Multiple clathrin proteins form a matrix that creates membrane curvature and assembles into a lattice, which creates a "soccer ball"-like structure. (C) The sequence of events in clathrin-mediated endocytosis [as described in (A) and (B)], with the addition of the use of dynamin to form a ring around the base of the clathrin-coated pit and pinch off the clathrin-coated vesicle. After endocytosis, the clathrin-coated vesicle is uncoated.

Dynamin requires GTP to pinch off vesicle

Dynamin is a GTPase that forms a multimeric ring around the base of the budding vesicle, using the energy from cleaving GTP to pinch off the coated vesicle (see Fig. 9.7C). If an experimenter fills a neuron with an analog of GTP that cannot be cleaved by dynamin (GTP-γ-s), the dynamin will continue to form rings around the base of the coated pits. The pits will extend long dynamin-coated necks into the cytoplasm, but the dynamin rings never pinch the vesicle off. This result shows that dynamin requires GTP as a source of energy to pinch off clathrin-coated pits, completing the endocytosis process.

After endocytosis, the coated vesicle is uncoated with the aid of uncoating proteins (e.g., Hsc70 and synaptojanin; Xing et al., 2010; Yim et al., 2010; Verstreken et al., 2003). The newly recycled vesicle is then returned to the recycling pool of vesicles (see below) and refilled with neurotransmitter, allowing it to be used again in exocytosis. This clathrin-mediated endocytosis process appears to have a time constant of about 15–30 seconds and can be repeated many times, since the small clear synaptic vesicles can undergo many rounds of exocytosis and endocytosis (Ryan et al., 1996; Wu and Betz, 1996; Kim and Ryan, 2009).

## BULK ENDOCYTOSIS

Bulk endocytosis was first observed following nerve terminal stimulation that was intense enough to deplete the synaptic vesicles (Heuser and Reese, 1973; see Fig. 9.1). When experimenters used electron microscopy to image nerve terminals after intense stimulation, they observed an increased presence of endosomes (irregularly shaped cytoplasmic membrane inclusions; see Fig. 9.1). It is hypothesized that intense stimulation adds so much vesicle membrane to the plasma membrane that the nerve terminal needs to generate large membrane infoldings to recover this extra membrane quickly (Saheki and De Camilli, 2012).

The mechanism of this infolding is not completely understood, and while some earlier studies have proposed that dynamin is involved (Clayton and Cousin, 2009), subsequent studies have shown that infolding can proceed in the absence of clathrin and dynamin (Wu et al., 2014a,b). There is evidence that the actin cytoskeleton may work with a family of proteins (e.g., syndapin) to mediate bulk endocytosis (Andersson et al., 2008; Quan and Robinson, 2013). Syndapin proteins may link up to form scaffolds that bind lipids and mediate membrane curvature (see Fig. 9.8; Wang et al., 2009). This membrane curvature folds in to create large endosomes (membrane-bound structures that result from pinching in plasma membrane). After large endosomes pinch into the nerve terminal, synaptic vesicles are thought to bud from these endosomes and return to the recycling pool (see below). The budding of synaptic vesicles from endosomes may involve the clathrin-mediated mechanism described above (Kononenko et al., 2014), although this is debated (Wu et al., 2014a,b).

## KISS-AND-RUN

A third form of endocytosis was termed "kiss-and-run" by Ceccarelli et al. (1972, 1973) after they observed synaptic vesicles in the process of endocytosis using the electron microscope that did not have a coating on them. This form of endocytosis is characterized as the rapid opening of a fusion pore between the synaptic vesicle and the plasma

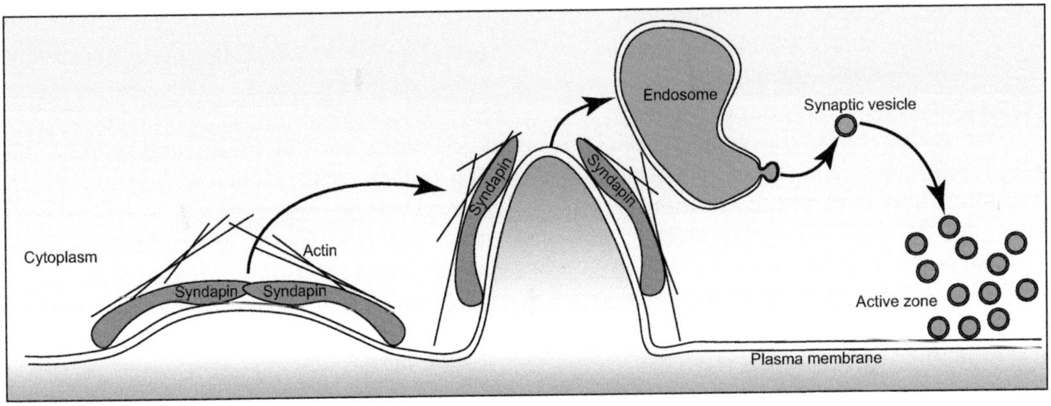

FIGURE 9.8 **The bulk endocytosis pathway.** Syndapin and actin may work together to create a large membrane curvature, resulting in large endosomes that are pinched into the nerve terminal. Synaptic vesicles bud off these endosomes and return to the vesicle pools in the nerve terminal.

membrane that allows neurotransmitter to diffuse out without causing the vesicle membrane to fully collapse into the plasma membrane (i.e., vesicles never became part of the plasma membrane; Alabi and Tsien, 2013). Instead, it is hypothesized that the synaptic vesicle remains in place within the active zone and that a fusion pore simply closes after chemical transmitter is released. In this scenario, the vesicle may not leave the docking site at all. After a kiss-and-run transmitter release event, the vesicle membrane is still intact and the vesicle only needs to be refilled with transmitter to be ready for subsequent release. Since the process of "endocytosis" would be very fast with this form of transmitter release, the rate-limiting step in recycling functional vesicles would be reacidification and refilling with transmitter.

Kiss-and-run transmitter release is controversial because it has been very difficult to study in detail (He and Wu, 2007; Alabi and Tsien, 2013; Kononenko and Haucke, 2015). It has been proposed that the numbers and types of protein complexes that form around a docked synaptic vesicle may determine whether a synaptic vesicle will fully collapse into the plasma membrane or simply open a fusion pore (Alabi and Tsien, 2013). The protein complexes involved may include SNARE protein complexes around the base of a vesicle, as well as other scaffold-like proteins found around vesicles. A larger number of SNARE complexes around the base of a vesicle may increase the likelihood of full collapse, since these complexes are hypothesized to create a force that pulls the vesicle membrane into the plasma membrane. In contrast, cytoskeletal elements and/or protein linkers between neighboring synaptic vesicles (as of yet undefined) may serve to prevent full collapse of the vesicle membrane into the plasma membrane (see Fig. 9.9; Alabi and Tsien, 2013). As such, the degree to which a synaptic vesicle is engaged with these other proteins prior to the arrival of an action potential may govern the occurrence of full collapse or "kiss-and-run" mechanisms for fusion and subsequent endocytosis.

Although the "kiss-and-run" process is difficult to study, some of the evidence for this form of endocytosis includes: (1) that the time course for recovery of membrane

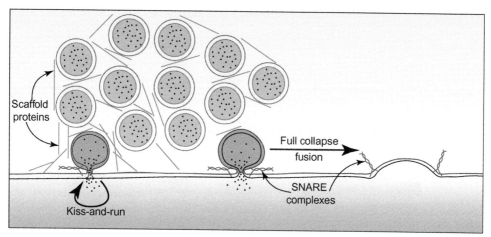

FIGURE 9.9  **Kiss-and-run versus full-collapse synaptic vesicle fusion.** The mode of fusion and subsequent endocytosis may be governed by the number of SNARE complexes and/or the number of scaffolding proteins around a docked synaptic vesicle. On the left side of this diagram, a synaptic vesicle is shown fusing with plasma membrane and releasing transmitter through a fusion pore. Relatively few SNARE complexes are hypothesized to exist around the fusion pore, and other proteins of unknown identity may form a scaffold network that supports the docked synaptic vesicle so that it does not fully collapse into the plasma membrane. One hypothesis proposes that vesicles which are strongly supported by scaffold proteins and have fewer SNARE protein forces pulling vesicle membrane into plasma membrane may not fully collapse, instead only undergoing kiss-and-run fusion (left). On the other hand, vesicles that are less supported by scaffold proteins and have a greater number of SNARE proteins creating stronger forces pulling vesicle membrane into plasma membrane may fully collapse into the plasma membrane (right).

capacitance values (which are reflective of the plasma membrane area) is sometimes too fast to be explained by clathrin-mediated mechanisms (He et al., 2006; Xu et al., 2008); and (2) that vesicles sometimes will not release experimentally loaded, very large fluorescent labels during neurotransmitter release, suggesting that these vesicles do not fully collapse into plasma membrane (Zhang et al., 2009). As experimenters design additional methods to study this interesting form of exocytosis and vesicle membrane recycling, "kiss-and-run" endocytosis can be studied more completely.

It is important to note that all of the endocytosis mechanisms described above are thought to apply to small clear vesicles that are recycled and reused in the nerve terminal. Nerve terminals also contain large dense-core vesicles that often contain peptide transmitters (see Chapter 21), but these vesicles are not recycled or reused.

## SYNAPTIC VESICLE POOLS

In Chapter 8, we discussed how synaptic vesicles dock in the active zone region of the nerve terminal and that other vesicles can be chaperoned along cytomatrix proteins to replace docked vesicles after they are used in exocytosis. Additionally, we discussed how vesicles could be recycled by endocytosis after fusion. We noted above that vesicles are thought to return to a "recycling pool" after endocytosis and could be reused after reloading with transmitter. Here, we will consider the different groups, or pools, within a nerve terminal in which synaptic vesicles can reside and some mechanisms that regulate where synaptic vesicles are stored and used within a nerve terminal.

Investigators have provided evidence that vesicles in a synaptic terminal can reside within pools that are differentiated according to whether the vesicles they contain are in storage, ready for fusion, or ready to take the place of a recently fused synaptic vesicle (Neher, 1998; Rizzoli and Betz, 2005; Harata et al., 2001; Richards et al., 2003). Furthermore, they showed that synaptic vesicles can move within and between these pools. For the purposes of this discussion, we will divide the vesicles within a presynaptic nerve terminal into three pools: "**reserve**," "**recycling**," and "**ready**" (often called the readily releasable pool; RRP; see Fig. 9.10; see Alabi and Tsien, 2012; Truckenbrodt and Rizzoli, 2015). The smallest pool of synaptic vesicles (consisting of only 1%–2% of all vesicles in the nerve terminal) is the **ready pool**, which contains all vesicles that are docked at the plasma membrane by the SNARE complex. Vesicles in the ready pool are the first to fuse with plasma membrane during action potential activity. The **recycling pool** includes all vesicles that are poised to participate in docking and fusion as soon as a docking site becomes available. This pool is intermediate in size (10%–20% of all vesicles), and is where newly recycled vesicles return to after endocytosis. Lastly, the **reserve pool** of synaptic vesicles is the largest (containing 80%–90% of all vesicles) and is sequestered from use because these vesicles are bound up in a cytoskeletal matrix. Some of these vesicles can be released from the cytoskeletal matrix during very high-frequency action

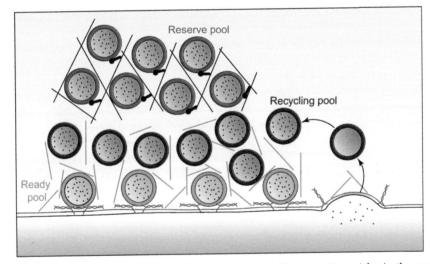

FIGURE 9.10 **Synaptic vesicle pools within the nerve terminal.** The synaptic vesicles in the nerve terminal can be divided into three pools. The "ready pool" (*light blue*) is docked and ready for rapid fusion during action potential depolarization. The "recycling pool" (*dark blue*) can replace vesicles that leave the ready pool after fusion, and is the pool that accepts vesicles after endocytosis. The "reserve pool" (*red*) is a storage pool of vesicles that are bound up in a cytoskeleton matrix and are only used during periods of very high demand. These pools represent a functional classification only, as the reserve and recycling pools are not as spatially segregated in the nerve terminal as diagrammed here.

potential activity, when there is an unusually high demand for synaptic vesicles to replace those in the ready and recycling pool that have recently been used and are still in the process of endocytosis and refilling with neurotransmitter.

## SYNAPTIC VESICLE TRAFFICKING IN THE NERVE TERMINAL

How do synaptic vesicles move between the pools described above? In order to answer this question, investigators have used an approach in which they illuminated a small group of vesicles containing a fluorescent marker. The illumination bleached the fluorescent molecules, making that area dark. The investigators then watched to see whether nearby labeled vesicles, which had not been bleached, would move in to fill this bleached area (Henkel et al., 1996; Jordan et al., 2005; Shtrahman et al., 2005). Using this technique, investigators studying frog neuromuscular synapses and hippocampal neurons found that the synaptic vesicles they observed did not drift around when the nerve terminal is at rest. However, when investigators treated synapses with a phosphatase inhibitor, shifting the balance in the nerve terminal to enhance phosphorylation of proteins, synaptic vesicles showed increased mobility (Betz and Henkel, 1994; Shtrahman et al., 2005). The vesicles observed in these early experiments were likely in the reserve pool, since this pool contains 80%–90% of synaptic vesicles in the nerve terminal.

A more recent study comparing the mobility of vesicles in the recycling and reserve pools found that while the reserve pool is relatively stable, vesicles in the recycling pool are more mobile at rest (Gaffield et al., 2006). Furthermore, after very high-frequency axon potential stimulation (several minutes of stimulation at 30 Hz), the reserve pool showed mobility similar to that of the recycling pool (Gaffield et al., 2006). These studies lead to the conclusion that synaptic vesicles in the reserve pool are normally fixed in position, as well as the conclusion that vesicle mobility can be influenced by the phosphorylation state of some nerve terminal proteins.

Vesicle mobility has been further explored with new super-resolution imaging techniques that allow tracking of very small fluorescent spots (Hell, 2007). Using super-resolution methods, several investigators have tracked the movement of synaptic vesicles within the neurites of cultured hippocampal neurons. In these cultured neurons, single vesicles could move along the axon, but often became trapped or immobile within structures that looked like presynaptic nerve terminals (Westphal et al., 2008; Shtrahman et al., 2005; Yeung et al., 2007). Kamin et al. (2010) found that very recently endocytosed vesicles are quite mobile, but that they can become more immobile after several hours as they associate with established pools of vesicles.

One protein that has been proposed to control synaptic vesicle mobility in the nerve terminal is **synapsin** (Navone et al., 1984; De Camilli and Greengard, 1986). The finding that synapsin is phosphorylated after strong depolarization of the nerve terminal, as well as the finding that it associates with synaptic vesicles, inspired interest in how synapsin affects synaptic transmission. Because synapsin was known to be phosphorylated in a calcium-dependent manner in response to neuronal activity, researchers investigated its role in transmission by injecting different forms of synapsin (phosphorylated or dephosphorylated) into the squid giant synapse nerve terminal (Llinas et al., 1991;

Greengard et al., 1993). After injection, the axon was stimulated to release transmitter at a low frequency, and the amount of released neurotransmitter was monitored over a 10-minute period. Results from these experiments showed that injecting the dephosphorylated form of synapsin caused a gradual decrease in the magnitude of transmitter released, but injecting the phosphorylated form of synapsin had no effect on transmitter release. Importantly, when these experimenters injected a kinase (calcium-calmodulin-dependent protein kinase II; CaMKII) into the nerve terminal, causing the phosphorylation of endogenous synapsin proteins, they observed a gradual increase in transmitter release.

Collectively, these results led to the hypothesis that the synapsin phosphorylation state regulates how many synaptic vesicles are available to participate in transmitter release (see Fig. 9.11), and that synapsin phosphorylation is controlled by calcium-regulated kinases in the nerve terminal cytoplasm (e.g., CaMKII) that are triggered by the large calcium influx induced by strong electrical activity. In other words, it has been proposed that the activity-dependent phosphorylation of synapsin regulates the number of synaptic vesicles in the reserve pool, relative to the number in the recycling pool. Dephosphorylated synapsin has a high affinity for both actin and synaptic vesicles, so dephosphorylated synapsin is proposed to cause vesicles to be bound to actin, making them part of the reserve pool. In contrast, activity-dependent phosphorylation is proposed to reduce synapsin binding to actin and synaptic vesicles, releasing vesicles from the reserve pool into the recycling pool (Hilfiker et al., 1999; Verstegen et al., 2014).

The idea that synaptic vesicles in the nerve terminal exist in different states has now been supported by experiments in many synaptic preparations. These include neuromuscular synapses in larval *Drosophila* (Delgato et al., 2000; Kuromi and Kidokoro, 2003), frog and mouse neuromuscular junctions (Richards et al., 2003; Reid et al., 1999;

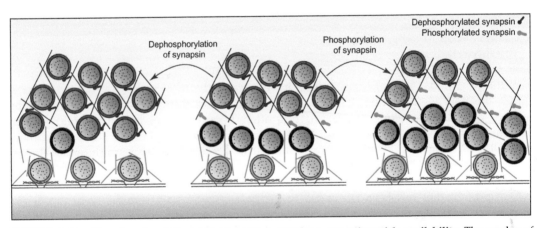

FIGURE 9.11 **The phosphorylation state of synapsin regulates synaptic vesicle availability.** The number of synaptic vesicles in the reserve pool (*red*) is regulated by the phosphorylation of synapsin [small black paddle-shaped structures linking the red vesicles to actin filaments (*black lines*)]. In the dephosphorylated state (left), synapsin binds actin to synaptic vesicles, making them a part of the reserve pool and reducing the size of the recycling pool (*dark blue vesicles*). In the phosphorylated state (right), synapsin releases some vesicles from the reserve pool, making them part of the recycling pool. Synapsin is phosphorylated in a calcium-dependent manner following high-frequency action potential activity.

Wang et al., 2016), goldfish retinal bipolar cells (von Gersdorff and Matthews, 1997), cultured CNS synapses (Rosenmund and Stevens, 1996; Mozhayeva et al., 2002; Ryan and Smith, 1995), and the mammalian calyx of Held synapse in the auditory brainstem (Schneggenburger et al., 1999, 2002). In all of these experiments, the synaptic vesicles can be divided into three major pools: the ready pool, the recycling pool, and the reserve pool (Rizzoli and Betz, 2005; Denker and Rizzoli, 2010). During low-to-moderate action potential activity that triggers the release of the ready pool of docked vesicles, vesicles in the recycling pool can easily dock at newly vacated docking sites and participate in vesicle fusion. The reserve pool contains vesicles that are bound up in an actin/synapsin network, and are therefore unavailable to participate in transmitter release until they are released from this reserve pool by intense, high-frequency activity that leads to the calcium-dependent phosphorylation of synapsin. Under conditions of high action potential frequency, vesicles in the recycling pool are quickly depleted, and it takes them some time to complete the process of endocytosis and refilling with transmitter. However, this high action potential frequency leads to significant calcium entry and the phosphorylation of synapsin, allowing reserve pool vesicles to move into the recycling pool where they can easily participate in transmitter release. This activity-dependent adaptation allows the synapse to maintain the ability to release transmitter during intense electrical activity.

# References

Alabi, A.A., Tsien, R.W., 2012. Synaptic vesicle pools and dynamics. Cold Spring Harb. Perspect. Biol. 4, a013680.
Alabi, A.A., Tsien, R.W., 2013. Perspectives on kiss-and-run: role in exocytosis, endocytosis, and neurotransmission. Annu. Rev. Physiol. 75, 393–422.
Andersson, F., Jakobsson, J., Löw, P., Shupliakov, O., Brodin, L., 2008. Perturbation of syndapin/PACSIN impairs synaptic vesicle recycling evoked by intense stimulation. J. Neurosci. 28 (15), 3925–3933.
Ashby, M.C., Ibaraki, K., Henley, J.M., 2004. It's green outside: tracking cell surface proteins with pH-sensitive GFP. Trends Neurosci. 27, 257–261.
Balaji, J., Ryan, T.A., 2007. Single-vesicle imaging reveals that synaptic vesicle exocytosis and endocytosis are coupled by a single stochastic mode. Proc. Natl. Acad. Sci. USA 104, 20576–20581.
Betz, W.J., Henkel, A.W., 1994. Okadaic acid disrupts clusters of synaptic vesicles in frog motor nerve terminals. J. Cell Biol. 124 (5), 843–854.
Betz, W.J., Mao, F., Bewick, G.S., 1992. Activity-dependent fluorescent staining and destaining of living vertebrate motor nerve terminals. J. Neurosci. 12, 363–375.
Ceccarelli, B., Hurlbut, W.P., Mauro, A., 1972. Depletion of vesicles from frog neuromuscular junctions by prolonged tetanic stimulation. J. Cell. Biol. 54, 30–38.
Ceccarelli, B., Hurlbut, W.P., Mauro, A., 1973. Turnover of transmitter and synaptic vesicles at the frog neuromuscular junction. J. Cell. Biol. 57, 499–524.
Chanaday, N.L., Kavalali, E.T., 2017. How do you recognize and reconstitute a synaptic vesicle after fusion? F1000Res. 6, 1734.
Clayton, E.L., Cousin, M.A., 2009. The molecular physiology of activity-dependent bulk endocytosis of synaptic vesicles. J. Neurochem. 111, 901–914.
De Camilli, P., Greengard, P., 1986. Synapsin I: a synaptic vesicle-associated neuronal phosphoprotein. Biochem. Pharmacol. 35, 4349–4357.
Delgado, R., Maureira, C., Oliva, C., Kidokoro, Y., Labarca, P., 2000. Size of vesicle pools, rates of mobilization, and recycling at neuromuscular synapses of a Drosophila mutant, shibire. Neuron 28, 941–953.
Denker, A., Rizzoli, S.O., 2010. Synaptic vesicle pools: an update. Front. Synaptic Neurosci. 2, 135.

Diril, M.K., Wienisch, M., Jung, N., Klingauf, J., Haucke, V., 2006. Stonin 2 is an AP-2-dependent endocytic sorting adaptor for synaptotagmin internalization and recycling. Dev. Cell. 10, 233–244.

Ferguson, S.M., De Camilli, P., 2012. Dynamin, a membrane-remodelling GTPase. Nat. Rev. Mol. Cell Biol. 13, 75–88.

Gad, H., Low, P., Zotova, E., Brodin, L., Shupliakov, O., 1998. Dissociation between $Ca^{2+}$-triggered synaptic vesicle exocytosis and clathrin-mediated endocytosis at a central synapse. Neuron 21, 607–616.

Gaffield, M.A., Rizzoli, S.O., Betz, W.J., 2006. Mobility of synaptic vesicles in different pools in resting and stimulated frog motor nerve terminals. Neuron 51 (3), 317–325.

Granseth, B., Odermatt, B., Royle, S.J., Lagnado, L., 2006. Clathrin-mediated endocytosis is the dominant mechanism of vesicle retrieval at hippocampal synapses. Neuron 51, 773–786.

Greengard, P., Valtorta, F., Czernik, A.J., Benfenati, F., 1993. Synaptic vesicle phosphoproteins and regulation of synaptic function. Science 259, 780–785.

Harata, N., Ryan, T.A., Smith, S.J., Buchanan, J., Tsien, R.W., 2001. Visualizing recycling synaptic vesicles in hippocampal neurons by FM 1-43 photoconversion. Proc. Natl. Acad. Sci. USA 98, 12748–12753.

He, L., Wu, L.G., 2007. The debate on the kiss-and-run fusion at synapses. Trends Neurosci. 30, 447–455.

He, L., Wu, X.S., Mohan, R., Wu, L.G., 2006. Two modes of fusion pore opening revealed by cell-attached recordings at a synapse. Nature 444, 102–105.

Hell, S.W., 2007. Far-field optical nanoscopy. Science 316 (5828), 1153–1158.

Henkel, A.W., Simpson, L.L., Ridge, R.M., Betz, W.J., 1996. Synaptic vesicle movements monitored by fluorescence recovery after photobleaching in nerve terminals stained with FM1-43. J. Neurosci. 16 (12), 3960–3967.

Heuser, J.E., Reese, T.S., 1973. Evidence for recycling of synaptic vesicle membrane during transmitter release at the frog neuromuscular junction. J. Cell. Biol. 57, 315–344.

Hilfiker, S., Pieribone, V.A., Czernik, A.J., Kao, H.T., Augustine, G.J., Greengard, P., 1999. Synapsins as regulators of neurotransmitter release. Philos. Trans. R. Soc. Lond. B. Biol. Sci. 354, 269–279.

Jakobsson, J., Gad, H., Andersson, F., Low, P., Shupliakov, O., Brodin, L., 2008. Role of epsin 1 in synaptic vesicle endocytosis. Proc. Natl. Acad. Sci. USA 105, 6445–6450. Copyright (2008) National Academy of Sciences, U.S.A.

Jordan, R., Lemke, E.A., Klingauf, J., 2005. Visualization of synaptic vesicle movement in intact synaptic boutons using fluorescence fluctuation spectroscopy. Biophys. J. 89 (3), 2091–2102.

Jorgensen, E.M., Hartwieg, E., Schuske, K., Nonet, M.L., Jin, Y., Horvitz, H.R., 1995. Defective recycling of synaptic vesicles in synaptotagmin mutants of Caenorhabditis elegans. Nature 378, 196–199.

Kamin, D., Lauterbach, M.A., Westphal, V., Keller, J., Schönle, A., Hell, S.W., Rizzoli, S.O., 2010. High- and low-mobility stages in the synaptic vesicle cycle. Biophys. J. 99 (2), 675–684.

Kim, S.H., Ryan, T.A., 2009. Synaptic vesicle recycling at CNS snapses without AP-2. J. Neurosci. 29, 3865–3874.

Kononenko, N.L., Haucke, V., 2015. Molecular mechanisms of presynaptic membrane retrieval and synaptic vesicle reformation. Neuron 85, 484–496.

Kononenko, N.L., Puchkov, D., Classen, G.A., Walter, A.M., Pechstein, A., Sawade, L., et al., 2014. Clathrin/AP-2 mediate synaptic vesicle reformation from endosome-like vacuoles but are not essential for membrane retrieval at central synapses. Neuron 82, 981–988.

Kuromi, H., Kidokoro, Y., 2003. Two synaptic vesicle pools, vesicle recruitment and replenishment of pools at the Drosophila neuromuscular junction. J. Neurocytol. 32, 551–565.

Li, C., Ullrich, B., Zhang, J.Z., Anderson, R.G., Brose, N., Sudhof, T.C., 1995. $Ca(2+)$-dependent and -independent activities of neural and non-neural synaptotagmins. Nature 375, 594–599.

Llinas, R., Gruner, J.A., Sugimori, M., McGuinness, T.L., Greengard, P., 1991. Regulation by synapsin I and Ca(2+)-calmodulin-dependent protein kinase II of the transmitter release in squid giant synapse. J. Physiol. 436, 257–282.

Miesenbock, G., De Angelis, D.A., Rothman, J.E., 1998. Visualizing secretion and synaptic transmission with pH-sensitive green fluorescent proteins. Nature 394, 192–195.

Miller, T.M., Heuser, J.E., 1984. Endocytosis of synaptic vesicle membrane at the frog neuromuscular junction. J. Cell. Biol. 98, 685–698.

Mozhayeva, M.G., Sara, Y., Liu, X., Kavalali, E.T., 2002. Development of vesicle pools during maturation of hippocampal synapses. J. Neurosci. 22, 654–665.

Navone, F., Greengard, P., De Camilli, P., 1984. Synapsin I in nerve terminals: selective association with small synaptic vesicles. Science 226, 1209–1211.

Neher, E., 1998. Vesicle pools and Ca2+ microdomains: new tools for understanding their roles in neurotransmitter release. Neuron 20, 389–399.

Quan, A., Robinson, P.J., 2013. Syndapin—a membrane remodelling and endocytic F-BAR protein. FEBS. J. 280, 5198–5212.

Reid, B., Slater, C.R., Bewick, G.S., 1999. Synaptic vesicle dynamics in rat fast and slow motor nerve terminals. J. Neurosci. 19, 2511–2521.

Richards, D.A., Guatimosim, C., Rizzoli, S.O., Betz, W.J., 2003. Synaptic vesicle pools at the frog neuromuscular junction. Neuron 39, 529–541.

Rizzoli, S.O., Betz, W.J., 2005. Synaptic vesicle pools. Nat. Rev. Neurosci. 6, 57–69.

Rosenmund, C., Stevens, C.F., 1996. Definition of the readily releasable pool of vesicles at hippocampal synapses. Neuron 16, 1197–1207.

Ryan, T.A., Smith, S.J., 1995. Vesicle pool mobilization during action potential firing at hippocampal synapses. Neuron 14, 983–989.

Ryan, T.A., Smith, S.J., Reuter, H., 1996. The timing of synaptic vesicle endocytosis. Proc. Natl. Acad. Sci. USA 93, 5567–5571.

Saheki, Y., De Camilli, P., 2012. Synaptic vesicle endocytosis. Cold Spring Harb. Perspect. Biol. 4, a005645.

Schneggenburger, R., Meyer, A.C., Neher, E., 1999. Released fraction and total size of a pool of immediately available transmitter quanta at a calyx synapse. Neuron 23, 399–409.

Schneggenburger, R., Sakaba, T., Neher, E., 2002. Vesicle pools and short-term synaptic depression: lessons from a large synapse. Trends Neurosci. 25, 206–212.

Shtrahman, M., Yeung, C., Nauen, D.W., Bi, G.Q., Wu, X.L., 2005. Probing vesicle dynamics in single hippocampal synapses. Biophys. J. 89 (5), 3615–3627.

Sundborger, A., Soderblom, C., Vorontsova, O., Evergren, E., Hinshaw, J.E., Shupliakov, O., 2011. An endophilin-dynamin complex promotes budding of clathrin-coated vesicles during synaptic vesicle recycling. J. Cell. Sci. 124, 133–143.

Takei, K., Haucke, V., 2001. Clathrin-mediated endocytosis: membrane factors pull the trigger. Trends. Cell Biol. 11, 385–391.

Teng, H., Cole, J.C., Roberts, R.L., Wilkinson, R.S., 1999. Endocytic active zones: hot spots for endocytosis in vertebrate neuromuscular terminals. J. Neurosci. 19, 4855–4866.

Truckenbrodt, S., Rizzoli, S.O., 2015. Synaptic vesicle pools: classical and emerging roles. In: Mochida, S. (Ed.), Presynaptic Terminals. Springer Japan, Tokyo, pp. 329–359.

Valtorta, F., Jahn, R., Fesce, R., Greengard, P., Ceccarelli, B., 1988. Synaptophysin (p38) at the frog neuromuscular junction: its incorporation into the axolemma and recycling after intense quantal secretion. J. Cell. Biol. 107, 2717–2727.

Verstegen, A.M., Tagliatti, E., Lignani, G., Marte, A., Stolero, T., Atias, M., Corradi, A., Valtorta, F., Gitler, D., Onofri, F., Fassio, A., Benfenati, F., 2014. Phosphorylation of synapsin I by cyclin-dependent kinase-5 sets the ratio between the resting and recycling pools of synaptic vesicles at hippocampal synapses. J. Neurosci. 34 (21), 7266–7280.

Verstreken, P., Koh, T.W., Schulze, K.L., Zhai, R.G., Hiesinger, P.R., Zhou, Y., et al., 2003. Synaptojanin is recruited by endophilin to promote synaptic vesicle uncoating. Neuron 40, 733–748.

von Gersdorff, H., Matthews, G., 1997. Depletion and replenishment of vesicle pools at a ribbon-type synaptic terminal. J. Neurosci. 17, 1919–1927.

Wang, Q., Navarro, M.V., Peng, G., Molinelli, E., Goh, S.L., Judson, B.L., et al., 2009. Molecular mechanism of membrane constriction and tubulation mediated by the F-BAR protein Pacsin/Syndapin. Proc. Natl. Acad. Sci. USA 106, 12700–12705.

Wang, X., Pinter, M.J., Rich, M.M., 2016. Reversible recruitment of a homeostatic reserve pool of synaptic vesicles underlies rapid homeostatic plasticity of quantal content. J. Neurosci. 36, 828–836.

Westphal, V., Rizzoli, S.O., Lauterbach, M.A., Kamin, D., Jahn, R., Hell, S.W., 2008. Video-rate far-field optical nanoscopy dissects synaptic vesicle movement. Science 320 (5873), 246–249.

Wu, L.G., Betz, W.J., 1996. Nerve activity but not intracellular calcium determines the time course of endocytosis at the frog neuromuscular junction. Neuron 17, 769–779.

Wu, L.G., Hamid, E., Shin, W., Chiang, H.C., 2014a. Exocytosis and endocytosis: modes, functions, and coupling mechanisms. Annu. Rev. Physiol. 76, 301–331.

Wu, Y., O'Toole, E.T., Girard, M., Ritter, B., Messa, M., Liu, X., et al., 2014b. A dynamin 1-, dynamin 3- and clathrin-independent pathway of synaptic vesicle recycling mediated by bulk endocytosis. eLife 3, e01621.

Wu, Y., Yeh, F.L., Mao, F., Chapman, E.R., 2009. Biophysical characterization of styryl dye-membrane interactions. Biophys. J. 97, 101–109.

Wyatt, R.M., Balice-Gordon, R.J., 2008. Heterogeneity in synaptic vesicle release at neuromuscular synapses of mice expressing synaptopHluorin. J. Neurosci. 28, 325–335.

Xing, Y., Bocking, T., Wolf, M., Grigorieff, N., Kirchhausen, T., Harrison, S.C., 2010. Structure of clathrin coat with bound Hsc70 and auxilin: mechanism of Hsc70-facilitated disassembly. EMBO J. 29, 655–665.

Xu, J., McNeil, B., Wu, W., Nees, D., Bai, L., Wu, L.G., 2008. GTP-independent rapid and slow endocytosis at a central synapse. Nat. Neurosci. 11, 45–53.

Yeung, C., Shtrahman, M., Wu, X.L., 2007. Stick-and-diffuse and caged diffusion: a comparison of two models of synaptic vesicle dynamics. Biophys. J. 92 (7), 2271–2280.

Yim, Y.I., Sun, T., Wu, L.G., Raimondi, A., De Camilli, P., Eisenberg, E., et al., 2010. Endocytosis and clathrin-uncoating defects at synapses of auxilin knockout mice. Proc. Natl. Acad. Sci. USA 107, 4412–4417.

Zhang, J.Z., Davletov, B.A., Sudhof, T.C., Anderson, R.G., 1994. Synaptotagmin I is a high affinity receptor for clathrin AP-2: implications for membrane recycling. Cell 78, 751–760.

Zhang, Q., Li, Y., Tsien, R.W., 2009. The dynamic control of kiss-and-run and vesicular reuse probed with single nanoparticles. Science 323, 1448–1453.

# PART III

# RECEPTORS AND SIGNALING

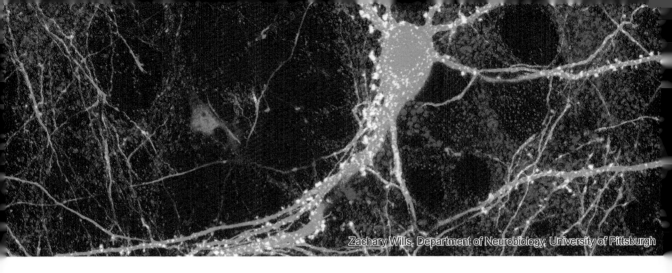

# CHAPTER 10

# Introduction to Receptors

Neurotransmitters exert their effects, or **actions**, by binding to specialized proteins on the surfaces of their target cells. These **neurotransmitter receptors** are capable of converting the chemical signal carried by a neurotransmitter molecule into either an electrical or a biochemical signal (depending on the type of receptor) that affects the function of the cell.

A given neurotransmitter chemical (e.g., glutamate) can bind to multiple types of receptors. It is common for discussions of neurotransmitter actions to include statements like "glutamate is an excitatory neurotransmitter" or "GABA is an inhibitory neurotransmitter." However, this phrasing should not be interpreted to mean that "excitatory" and "inhibitory" are properties of the neurotransmitters themselves. It is not the identity of a neurotransmitter that determines whether its effect is excitatory or inhibitory; instead, the action of a neurotransmitter is determined by the receptor to which it binds. For example, glutamate is excitatory at most synapses, but it can also be inhibitory depending on the type of glutamate receptor it binds to. With this in mind, Chapters 10–12 will discuss characteristics of neurotransmitter receptors, with an emphasis on how their activation affects cellular function at the electrical or biochemical level.

## NEUROTRANSMITTER RECEPTORS CAN BE DIVIDED INTO TWO GENERAL CLASSES: IONOTROPIC AND METABOTROPIC

**Ionotropic receptors** are ligand-gated ion channels, meaning that the receptor protein includes both a neurotransmitter binding site and an ion channel (see Fig. 10.1).

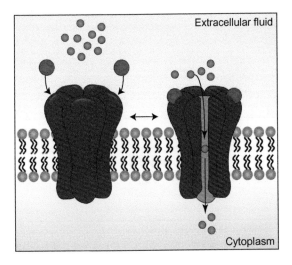

FIGURE 10.1  **Ionotropic receptors.** This class of receptor transduces neurotransmitter (red spheres) binding into the direct opening of an ion-conducting channel, which then can pass ions (green dots).

The binding of a neurotransmitter molecule (the ligand) to the binding site induces a conformational change in the receptor structure, which opens, or **gates**, the ion channel. The effect a neurotransmitter has on ionotropic receptors is often used to refer to the action of that neurotransmitter. For example, ionotropic glutamate receptors conduct a mix of ions (primarily sodium and potassium, often with some calcium) with an equilibrium potential of about 0 mV; so when glutamate binds to an ionotropic receptor while a cell is at rest (e.g., $V_m = -65$ mV), the membrane potential depolarizes as it moves closer to the equilibrium potential of the ionotropic glutamate receptor. Therefore, glutamate is commonly referred to as an excitatory neurotransmitter, despite the fact that it can also exert inhibitory action that is mediated by its binding to some metabotropic receptors.

**Metabotropic receptors** trigger second messenger-mediated effects within cells after neurotransmitter binding. When a neurotransmitter binds to a metabotropic receptor, this binding induces a conformational change in the receptor protein, which triggers a signaling cascade in the cytoplasm (sometimes described as a "biochemical event"). The term "metabotropic receptors" is typically used to refer to transmembrane G-protein-coupled receptors (see below), but it can also refer to transmembrane enzyme-linked receptors, in which the receptor complex itself is often a kinase, and cytoplasmic receptors, which often induce changes in gene expression. The different classes of metabotropic receptors are described in more detail below.

## GTP-Binding Protein (G-Protein)-Coupled Receptors

This is the largest neurotransmitter receptor class. It includes up to 2000 different specific receptors, making it the most diverse protein family in the mammalian genome (Kroeze et al., 2003). G-protein-coupled receptors all share a common structure and transduction mechanism, but are activated by a wide variety of ligands, including classical small-molecule neurotransmitters, neuropeptides, hormones, growth factors, nucleotides, chemokines, odorant molecules, and photons (light). When a ligand binds to a

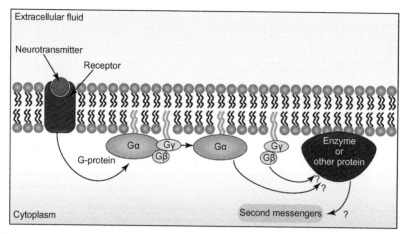

FIGURE 10.2  **G-protein-coupled receptors transform neurotransmitter binding into a conformational change in the receptor that activates G-proteins.** Once activated, G-protein subunits (α and β-γ) can bind to enzymes, ion channels, or other proteins to alter cellular function, often by affecting ion channel gating or generating second messengers. Question marks indicate possible interactions between proteins or possible signaling pathways.

G-protein-coupled receptor, it induces a conformational change that leads to the activation of a heterotrimeric GTP-binding protein (called a G-protein; Fig. 10.2). Depending on the type of G-protein, as well as the types of neighboring proteins that are affected by activated G-proteins, G-protein-coupled-receptor binding can initiate a wide variety of signaling cascades that alter neuronal function in different ways (Kroeze et al., 2003).

## Enzyme-Linked Receptors

This receptor class can be divided into three types: **receptor tyrosine kinases** (the most common type of growth factor receptor), **serine/threonine-specific protein kinases** (specific for some types of growth factors; ten Dijke et al., 1994), and **guanylate cyclase receptors** [e.g., the atrial natriuretic factor receptor that induces relaxation of smooth muscle (Hirose et al., 2001)]. Among the enzyme-linked receptors, the receptor tyrosine kinases are most common at synapses.

There are at least 20 classes of receptor tyrosine kinases (Hubbard and Till, 2000; Ségaliny et al., 2015). Receptor tyrosine kinases bind growth factors, extracellular adhesion molecules, and other trophic molecules that often induce cell differentiation and/or maturation (Fig. 10.3). Within the nervous system, the receptors in this group that are most relevant to synaptic transmission include neurotrophin receptors for neuronal growth factors [e.g., nerve growth factor (NGF), brain-derived neurotrophic factor (BDNF), neurotrophin-3 or -4 (NT3, NT4)]; MUSK receptors, which are required for the formation and maintenance of the neuromuscular junction (especially the organization of postsynaptic ionotropic acetylcholine receptors and the presynaptic active zone); and Ephrin receptors that bind to extracellular matrix proteins and mediate bidirectional signals between the pre- and postsynaptic cells at the synapse.

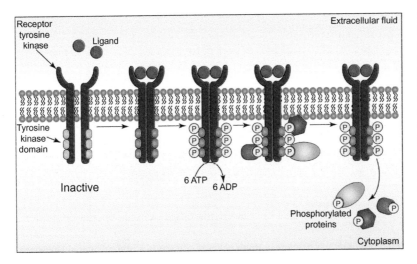

FIGURE 10.3 **After ligand binding, receptor tyrosine kinases form dimers that activate the tyrosine kinase activity.** Once activated, these kinases phosphorylate themselves, bind other cellular proteins, and then go on to phosphorylate other cellular proteins. The phosphorylated cellular proteins activate a variety of signaling cascades within the neuron.

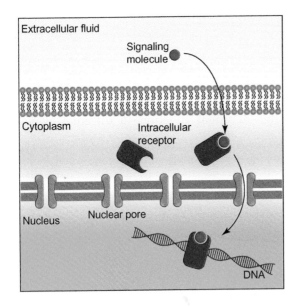

FIGURE 10.4 **Cytoplasmic receptors.** Signaling molecules that are hydrophobic and can diffuse across the plasma membrane bind to cytoplasmic receptors. After binding a ligand, most cytoplasmic receptors become active transcription factors and diffuse into the cell nucleus, where they bind to DNA to regulate gene expression.

## Cytoplasmic Receptors

As the name implies, these receptors are located within the cell cytoplasm. They bind ligands that can diffuse across the plasma membrane (e.g., steroid hormones). After these ligands bind to cytoplasmic receptors, the ligand–receptor complex often changes conformation to become an active transcription factor that moves into the cell nucleus and binds to DNA, altering the expression of genes (Fig. 10.4). As such, cytoplasmic receptors can induce changes in gene expression without having to use other signaling molecules in the cell.

Moving forward, our discussions of metabotropic receptors will focus primarily on G-protein-coupled receptors that mediate neurotransmitter action at synapses. We will often refer to these receptors simply as "metabotropic receptors," setting aside the fact that this category also includes enzyme-linked and cytoplasmic receptors.

# COMPARISON BETWEEN IONOTROPIC AND METABOTROPIC RECEPTORS

Table 10.1 contrasts several features of neurotransmitter action on ionotropic versus metabotropic receptors.

TABLE 10.1  A Comparison of Some Important Contrasting Features Between Ionotropic and Metabotropic Receptors

| Ionotropic Receptors | Metabotropic Receptors |
| --- | --- |
| Rapid onset of effects | Slow onset of effects |
| Rapid termination of effects | Slow termination of effects |
| 1:1 relationship between action and response | >1:1 amplification of response by G-proteins and second messengers |
| Effects limited by the type of ion channel that is part of the receptor protein* | Diverse possible effects from a single neurotransmitter due to a multitude of second messenger-mediated signaling pathways |
| Often bind neurotransmitter in the μM range | Usually bind neurotransmitter in the nM range |
| Often located near the site of neurotransmitter release | Often located at some distance from the site of neurotransmitter release |

*See Box 11.5 for possible exception to this generalization.

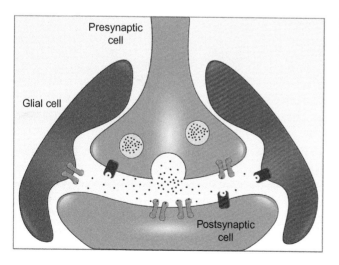

FIGURE 10.5  Both ionotropic (*orange*) and metabotropic (*purple*) neurotransmitter receptors can exist on presynaptic, postsynaptic, and glial cells.

As mentioned above, some neurotransmitters—including glutamate, acetylcholine, serotonin, and GABA—can activate both ionotropic and metabotropic receptors. Importantly, neurotransmitter receptors can be positioned both presynaptically and postsynaptically, as well as on glial cells that surround synapses (Fig. 10.5). In the following chapters, we will cover details of ionotropic (Chapter 11) and metabotropic receptors (Chapter 12) and their physiological signaling effects.

## References

Hirose, S., Hagiwara, H., Takei, Y., 2001. Comparative molecular biology of natriuretic peptide receptors. Can. J. Physiol. Pharmacol. 79, 665–672.
Hubbard, S.R., Till, J.H., 2000. Protein tyrosine kinase structure and function. Annu. Rev. Biochem. 69, 373–398.
Kroeze, W.K., Sheffler, D.J., Roth, B.L., 2003. G-protein-coupled receptors at a glance. J. Cell. Sci. 116, 4867–4869.
Ségaliny, A.I., Tellez-Gabriel, M., Heymann, M.F., Heymann, D., 2015. Receptor tyrosine kinases: characterisation, mechanism of action and therapeutic interests for bone cancers. J. Bone Oncol. 4, 1–12.
ten Dijke, P., Franzen, P., Yamashita, H., Ichijo, H., Heldin, C.H., Miyazono, K., 1994. Serine/threonine kinase receptors. Prog. Growth Factor Res. 5, 55–72.

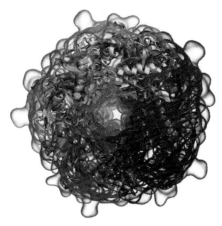

Rozita Laghaei, Pittsburgh Supercomputing Center, Carnegie Mellon University

# CHAPTER 11

# Ionotropic Receptors

In this chapter we cover the major classes of ionotropic receptors (also called ligand-gated ion channels; see Fig. 11.1) and discuss how they transduce neurotransmitter binding into functional changes in the neuron. We often characterize neurotransmitters by the effects of the ionotropic receptors they activate. For example, glutamate is often called an "excitatory" transmitter because ionotropic glutamate receptors depolarize neurons when activated. As discussed in Chapter 10, however, this is an oversimplification. It is the receptor type that determines neurotransmitter actions, and glutamate can have an inhibitory effect when acting on a subset of its metabotropic receptors (see Conn, 1997; Chapter 12).

All ionotropic receptors are ligand-gated ion channels that include a ligand-binding site for neurotransmitter molecules and an ion-conducting pore. When transmitter binds to the receptor, this causes a conformational change in the receptor protein, activating the receptor and opening the receptor channel. After the channel opens, ion flux is governed by multiple factors, including the driving force on the permeant ions. As with all ion channels, ligand-gated ion channels have an equilibrium potential that is determined by the electrochemical gradient for permeant ions (see Chapter 3). Neurotransmitter binding-induced channel opening occurs on a relatively rapid time scale (as compared with signaling induced by metabotropic receptors), so ligand-gated ion channels can mediate fast synaptic transmission. However, there are some types of ionotropic receptors that are not positioned immediately opposite transmitter release sites, but instead are positioned in an extrasynaptic location on the postsynaptic cell. Rather than directly mediating synaptic

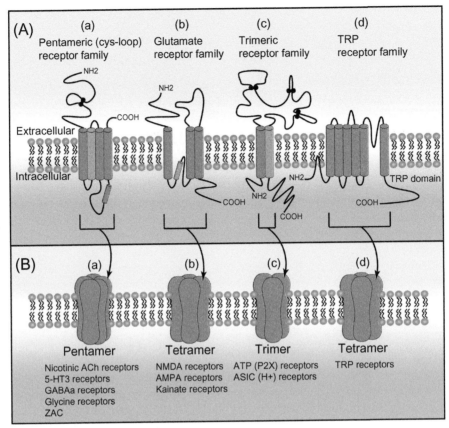

FIGURE 11.1 **Ionotropic receptor families.** (A) The structures of single subunit proteins that make up each ionotropic receptor family. (B) Diagrams of complete ionotropic receptors, indicating the number of subunits in each receptor. For both (A) and (B), lowercase letters provide more details for each type of receptor. (a) The pentameric ligand-gated ion channel (cys-loop) family of receptors (black dots indicate location of disulfide bonds between cysteines). These channels are constructed from five separate proteins, each containing four transmembrane segments. The second transmembrane segment in each of these subunit proteins lines the pore. This family includes receptors that bind acetylcholine, serotonin, GABA, glycine, and zinc. (b) The glutamate receptor family contains receptors that are constructed from four separate proteins, each of which contains three transmembrane segments. A reentrant loop, in combination with neighboring segments, forms the pore region. This family of receptors includes the NMDA, AMPA, and kainate receptors, all of which bind glutamate as a transmitter. (c) The trimeric receptors are constructed from three separate proteins, each containing only two transmembrane segments. The second transmembrane segment is thought to line the pore of these channels. This family of receptors includes the adenosine triphosphate (ATP) (P2X) and acid-sensing ion channels (ASIC; $H^+$ receptors). (d) The Transient Receptor Potential (TRP) channel family of receptors includes six subfamilies that make up a total of 29 different receptors, several of which may be ligand-gated ion channels for the cannabinoids. The TRP channel structure resembles the voltage-gated potassium channel structure, with four proteins forming the complete channel, and each of these four proteins contains six transmembrane segments. Like voltage-gated channels, the TRP channels have an S4 segment that can contain charged amino acids and a pore that is formed by segments 5 and 6 in combination with a P-loop.

transmission, these extrasynaptic receptors modulate, or alter, the fast synaptic transmission that is mediated by a different receptor.

The ionotropic receptors can be categorized into four major families (see Fig. 11.1): (1) the pentameric (cys-loop) family, (2) the glutamate receptor family, (3) the trimeric receptor family, and (4) the TRP receptor family.

# THE PENTAMERIC LIGAND-GATED ION CHANNEL FAMILY (CYS-LOOP RECEPTORS)

The first group of ionotropic receptors we will consider is the **pentameric** ligand-gated ion channel family, sometimes called "cys-loop receptors," which all share a common three-dimensional modular structure. These receptors all have a characteristic extracellular loop containing conserved amino acids that are held in a looped configuration by a disulfide bond between two cysteine (cys) amino acids (thus the name "cys-loop"; see Fig. 11.1; Miller and Smart, 2010; Thompson et al., 2010). As the name "pentameric" implies, these receptors are formed by five separate protein subunits, with each subunit containing four membrane-spanning segments (termed M1—M4). The pore of these receptors is lined by the second (M2) segments, one from each of the five subunits that make up the full receptor (see Fig. 11.2). The pentameric ligand-gated ion channel family of receptors includes acetylcholine, serotonin, zinc, GABA, and glycine-sensitive receptors.

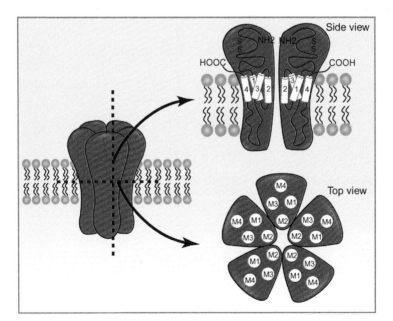

FIGURE 11.2 **Pentameric ligand-gated ion channel structure.** Each subunit that makes up the channel contains four transmembrane domains (M1—M4). The channel pore is lined by the M2 domains from each subunit.

## Nicotinic Acetylcholine Receptors

We will begin our discussion of the acetylcholine receptor family with the family member that we know the most about: the nicotinic acetylcholine receptor (Stroud and Finer-Moore, 1985). This receptor is named after the compound nicotine, a selective agonist of nicotinic receptors that is found in tobacco plants. Acetylcholine receptors can be divided into ionotropic receptors (selectively activated by nicotine) and metabotropic receptors (selectively activated by muscarine, see Chapter 12) and both receptor types are endogenously activated by acetylcholine. Nicotinic receptors are the most well-studied pentameric receptors for two reasons. First, there is a rich source of high-density nicotinic acetylcholine receptors in electric fishes and eels, and electric fishes have historically provided a good source material for structural studies (see Box 11.1). Second, there are highly selective toxins that bind strongly to nicotinic receptors, which allows experimenters to use affinity

---

### BOX 11.1

### ELECTRIC FISHES AND EELS AS AN EXPERIMENTAL SOURCE FOR ACETYLCHOLINE RECEPTORS

Electric fishes and eels provide a high-density source of acetylcholine receptors that aided in the initial study of nicotinic acetylcholine receptors. These species possess an electric organ that is believed to have evolved from muscle cells, and consists of stacks of "electrocytes" that are densely innervated by acetylcholine-releasing neurons (see Fig. 11.3). Because thousands of these electrocytes are stacked like plates in a battery, and because the density of acetylcholine receptors in these cells is so high, the total ion flux these organs generate creates up to 0.5 kW of electric power (hundreds of

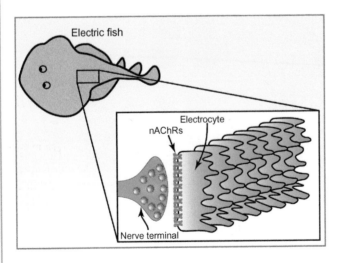

FIGURE 11.3 **The electric organ of the electric fish.** At the top, the side of the electric fish is boxed and the enlargement of the electric organ under the skin is drawn at the bottom. The electric organ contains stacks of electrocytes that have a very high density of nicotinic acetylcholine receptors (nAChRs) which are contacted by acetylcholine-releasing synapses.

## BOX 11.1 (cont'd)

volts and currents up to 1 A). The electric organ is used by these fishes and eels in cloudy and muddy water to communicate with other fish (perhaps for mating), to assist in navigation, to stun prey, and as a defense mechanism. For example, the muddy Amazon river is known to contain numerous electric fishes and eels.

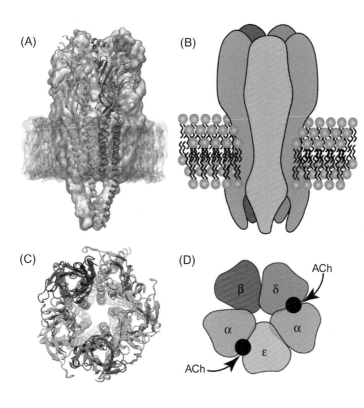

FIGURE 11.4 **Structure of the muscle nicotinic acetylcholine receptor.** (A) Side view of the structure of the channel protein with two of the subunits visualized (blue and orange). (B) Diagram of a side view of the channel protein. (C) Top view of the structure of the channel. (D) Diagram of a top view of the channel protein, with the subunits marked ($\alpha$, $\beta$, $\delta$, $\varepsilon$) and showing where the two binding sites for acetylcholine are located. *Source: (A) and (C) were generated by Rozita Laghaei using the software VMD and structures deposited in the RCSB protein data bank.*

purification methods to isolate these receptors for study (see Box 6.2 and Box 11.2). Historically, two receptor antagonist toxins have been used to study nicotinic acetylcholine receptors: $\alpha$-bungarotoxin, which is found in snake venom, and curare, which is a plant derivative (see Box 6.2). These compounds have been useful tools for labeling nicotinic receptors or pharmacologically blocking them during physiological experiments.

The first type of nicotinic receptor to be identified was the one found in electric fishes. Nicotinic acetylcholine receptors have five subunits (see Fig. 11.4), named α (two per receptor), β, ε, and δ (although in embryonic receptors and receptors in electric fishes and eels, the γ receptor protein takes the place of the ε protein, altering the conductance and mean open time of the channel; Schuetze, and Role, 1987). There are two acetylcholine-binding sites on each muscle nicotinic receptor (one for each α subunit), which bind acetylcholine with relatively low affinity (micromolar range). Both of these sites need to be occupied for the channel to open normally. Open nicotinic receptor channels predominately conduct sodium into the cell and potassium out of the cell (Fucile et al., 2003; Fucile, 2004), plus a small amount of calcium into the cell (representing only about 2% of the ion conductance in the muscle form of the nicotinic receptor). As such, these channels have an equilibrium potential of about −10 mV, which is halfway between the sodium (+60 mV) and potassium (−80 mV) equilibrium potentials. The small amount of calcium that enters through the channel is important for biochemical signaling at the synapse.

Neuronal membranes can also contain nicotinic acetylcholine receptors, but these receptors have a slightly different composition. The so-called "neuronal nicotinic receptors" include only α and β subunits. Some of these receptors are **homomeric** and contain only α subunits; for example, the "α-7" nicotinic receptor is so named because it is made up entirely of α-7 subunits. Others are **heteromeric** and include both α and β subunits (see Fig. 11.5). Neuronal nicotinic receptors usually have a higher affinity for acetylcholine than the muscle form, and a higher percentage of their total ionic flux consists of calcium (2%–5% in heteromeric neuronal nicotinic receptors, 6%–12% in homomeric α-7 neuronal nicotinic receptors), although sodium and potassium ions remain the primary conducting ions for these receptors (Fucile et al., 2003; Fucile, 2004; Dani, 2015).

Within the central nervous system, α-7 nicotinic receptors can be located both pre- and postsynaptically, but they are usually not positioned close to transmitter release sites. As such, α-7 nicotinic receptors respond primarily to **paracrine transmission**, a form of transmission in which molecules diffuse away from their site of release and bind to distant receptors (Lendvai and Vizi, 2008). Homomeric neuronal nicotinic receptors like

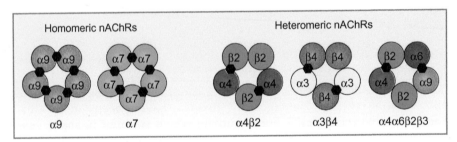

FIGURE 11.5 **Subunit configurations of neuronal nicotinic receptors.** Neuronal nicotinic receptors (nAChRs) can exist as homomeric channels, in which all five subunits are α, or as heteromeric channels, in which β subunits can also be included. The most common configurations in the nervous system are homomeric alpha-7 and heteromeric alpha-4, beta-2 receptors. Note that the examples shown here are a subset of the possible subunit configurations. Binding sites for acetylcholine are indicated with black hexagons.

α-7 have five identical acetylcholine-binding sites, one on each α subunit, though there is evidence that only one site needs to be occupied to trigger channel opening (Andersen et al., 2013). Because these acetylcholine-binding sites have a high affinity for acetylcholine, and only one needs to bind acetylcholine to open the channel, these receptors are very sensitive to the low concentration of acetylcholine that is present in **extrasynaptic regions** of the central nervous system.

Neuronal nicotinic receptors are potent modulators of signaling and neurotransmitter release in the nervous system. When neuronal nicotinic receptors are activated, the calcium ions that pass through them can influence the neuron's signaling properties in a number of ways, including enhancing neurotransmitter release (when neuronal nicotinic receptors are present in the presynaptic neuron), altering gene expression, altering synaptic plasticity mechanisms, and affecting receptor **desensitization** (Shen and Yakel, 2009).

The modulatory effects of neuronal nicotinic receptor activation are thought to underlie some of the behavioral and cognitive consequences of cigarette smoking. Smoking a single cigarette can cause blood levels of nicotine to reach between 20 and 60 ng/mL (equivalent to hundreds of nanomolar; Benowitz et al., 1982). This concentration is not high enough to reliably activate the low-affinity nicotinic receptors found on muscles, but it can potently activate the high-affinity neuronal nicotinic receptors. This means that even at blood levels that do not directly affect muscle contraction, nicotine from cigarette smoke can alter synaptic function in the central nervous system.

---

## BOX 11.2

## DANGERS IN NATURE: THE POISON HEMLOCK PLANT AND ACETYLCHOLINE RECEPTORS

### *Conium aculatum* (Poison Hemlock)

Though the dried flowers of *Conium maculatum* (see Fig. 11.6) are easily mistakable as anise fruit, this mix-up would result in serious poisoning. *Conium* contains the piperidine alkaloid coniine, which is structurally similar to nicotine and primarily acts as an antagonist to nicotinic acetylcholine receptors, with a minor influence on muscarinic acetylcholine receptors. Coniine exposure affects the periphery, blocking neuromuscular transmission onto skeletal and smooth muscles. As little as 0.5–1 g of coniine (about the same weight as a paperclip) is lethal for humans, and symptoms include drowsiness, burning of the mouth and throat, gradual asphyxia, ascending paralysis, and death. In 399 BCE, Socrates was forced to commit suicide by drinking a cup of *Conium* poison after he was convicted of corrupting the youth of ancient Greece.

## BOX 11.2 (cont'd)

FIGURE 11.6   *Conium maculatum.* Source: Javier Martin.

## The 5-HT$_3$ Serotonin Receptor

Serotonin (5-hydroxytryptamine; 5-HT; see Chapter 17) is a neurotransmitter that can influence a variety of behaviors. There are seven types of serotonin receptors (named 5-HT$_{1-7}$). Most are metabotropic receptors, except for the 5-HT$_3$ receptor, which is a ligand-gated ion channel.

Like nicotinic acetylcholine receptors, the 5-HT$_3$ receptor is a member of the pentameric ligand-gated ion channel family (Lummis, 2012). 5-HT$_3$ ionotropic receptors are nonselective monovalent cation channels that are permeable to sodium and potassium, and are thus excitatory when activated while a cell is at its resting membrane potential. Each 5-HT$_3$ receptor is made up of five subunits, and there are five different types of 5-HT$_3$ subunits (A–E). The "A" subunit is the one to which serotonin binds, so a functional receptor requires at least one of these, though they can have more than one or even be constructed entirely from "A" subunits. In some 5-HT$_3$ receptors the other four subunits are all "B" subunits, whereas others can contain a mixture of subunits (Jensen et al., 2008).

5-HT$_3$ receptors are located in several areas of the nervous system (including neurons that innervate the gastrointestinal tract). A high concentration of 5-HT$_3$ receptors is found in brainstem nuclei that are involved in control of the vomiting reflex (area postrema and the nucleus tractus solitarius; Tecott et al., 1993). Clinically, 5-HT$_3$ antagonists (e.g., Zofran) have been used to treat nausea and vomiting, especially in cancer patients (Aapro, 1991), and they can also be used to treat irritable bowel syndrome.

## GABA$_A$ Receptors → Primary inhb. transmitter in Mam. CNS

GABA is the primary inhibitory transmitter in the mammalian central nervous system. GABA can act on one of two types of receptors: GABA$_A$ (a ligand-gated ion channel) and GABA$_B$ (a metabotropic receptor).

GABA$_A$ receptors are pentameric ligand-gated chloride channels that are typically made up of α subunits along with β and/or γ subunits. GABA$_A$ receptors most often contain two α-, two β-, and one γ-subunit (Korpi et al., 2002; Olsen and Sieghart, 2008). There are multiple isoforms of each subunit (e.g. α1, α2, etc). Receptors that contain α1 and β2 are the most common, followed by α2- and β3-containing receptor isoforms. GABA binds to the α subunits at their interface with β subunits (see Fig. 11.7).

GABA$_A$ receptors inhibit adult neurons in two ways. The first is membrane potential hyperpolarization, which occurs if GABA$_A$ receptors open when the resting membrane potential is more positive (usually −60 or −65 mV) than the GABA$_A$ receptor equilibrium potential (usually about −70 mV). Remember that the membrane potential is always governed by the equilibrium potentials and relative proportions of the types of channels that are open. Thus, if a channel with an equilibrium potential more hyperpolarized than the membrane potential is opened, the membrane potential will be pulled in the hyperpolarizing direction. However, when the resting membrane potential is already at −70 mV, the opening of a GABA$_A$ receptor channel does not change the membrane potential because there is equal movement of chloride into and out of the cell at the chloride equilibrium potential. Under these conditions, a second form of inhibition can occur, called **shunting inhibition** (Borg-Graham et al., 1998; Kotak et al., 2017). Even though there is no net chloride flux while the membrane potential is at the GABA$_A$ receptor channel equilibrium potential, having more open channels in the cell membrane decreases the overall membrane resistance. This decrease in resistance creates pathways for charge redistribution across the membrane (in this context, a "shunt" is defined as a conductor in a circuit through which current can be diverted). Since the membrane

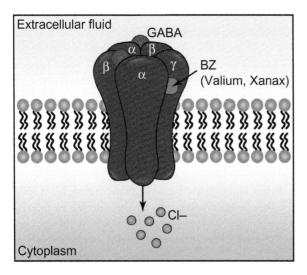

FIGURE 11.7 **Structure of the GABA$_A$ receptor.** GABA$_A$ receptors are ligand-gated ion channels that conduct chloride ions (orange dots) when GABA (red spheres) binds to the interface between the α and β subunits. Benzodiazepines (BZ; e.g., Valium and Xanax) bind to a different site on the channel (between the α and gamma subunits) and alter the gating such that channels open more easily (i.e., there is stronger receptor activation when an agonist binds).

voltage is governed by the equation $V = IR$, the reduced membrane resistance while $GABA_A$ receptors are open means that any simultaneous change in ionic current across the membrane (e.g., from the opening of nicotinic acetylcholine receptors) will result in a smaller change in the membrane voltage than it would if the $GABA_A$ receptors were not open. In other words, while $GABA_A$ receptor channels are open, excitatory neurotransmitters released onto the same cell are less effective at depolarizing the membrane. This reduction in excitation functions as an indirect form of inhibition.

Although $GABA_A$ receptor activation is inhibitory in adult neurons, in developing neurons it is initially excitatory. In adults, the $-70$ mV equilibrium potential of $GABA_A$ receptors is due to a strong chloride gradient established by chloride transporters that pump chloride out of the cell. However, during development, the chloride gradient is much smaller due to the expression of a sodium–potassium–chloride cotransporter (NKCC1), which creates a greater intracellular concentration of chloride compared to the extracellular fluid (Rivera et al., 1999). Under these conditions, chloride flows out of the cell when the $GABA_A$ receptor channel opens at the resting membrane potential, and this depolarizes the developing neuron. As neurons mature, they upregulate expression of the potassium–chloride extrusion cotransporter (KCC2), and can also downregulate NKCC1 expression in some cases, which drops intracellular chloride to low levels relative to the chloride concentration outside the cell. This change in the distribution of chloride across the cell membrane moves the $GABA_A$ receptor reversal potential to near $-70$ mV in the mature neuron, which means these receptors are hyperpolarizing at most resting membrane potentials. Because of their role in inhibitory transmission, $GABA_A$ receptors are a useful target for therapeutic drugs (see Box 11.3).

---

## BOX 11.3

## GABA-A RECEPTORS AS A THERAPEUTIC TARGET

The most commonly prescribed medications that target the $GABA_A$ receptor are the minor tranquilizers Valium (diazepam) and Xanax (alprazolam), which are classified as benzodiazepines. Benzodiazepines do not open $GABA_A$ receptors directly, and they do not bind to the receptors at the same location where GABA binds. Instead, they alter the gating of $GABA_A$ channels by binding to a separate site on the channel, which means they are referred to as **positive allosteric modulators**. Mechanistically, these drugs destabilize the closed state of the receptor, making it easier for the ligand-gated chloride channel to open when GABA is bound (see Fig. 11.7; Campo-Soria et al., 2006). As a result, benzodiazepines increase the number of times $GABA_A$ channels open when GABA is bound, though these drugs do not affect how long the channels remain open. The resulting increase in GABA-mediated inhibition of postsynaptic neurons reduces action potential firing. Xanax and Valium both act via this same mechanism, but Xanax is slower-acting than Valium and has longer-lasting effects. The differences in the time courses of action of these two benzodiazepines mean these drugs can be used

## BOX 11.3 (cont'd)

to treat different symptoms. In addition to their use in treating symptoms of anxiety, Xanax can be used to treat panic disorder, while Valium is indicated for muscle spasms, seizures, insomnia, restless leg syndrome, and withdrawal symptoms related to alcohol dependence.

An older class of antianxiety drugs called barbiturates (e.g., phenobarbital, pentobarbital) also bind to $GABA_A$ receptors and increases chloride currents. However, benzodiazepines and barbiturates bind to $GABA_A$ receptors at different sites and affect different characteristics of $GABA_A$ receptor chloride currents. Barbiturates can enhance the binding of GABA to the $GABA_A$ receptor, and at high concentrations they can also open $GABA_A$ channels in the absence of GABA. Single-channel recordings show that barbiturates increase the average amount of time a $GABA_A$ channel remains open (the open time) once it has been activated, but they do not affect the receptor conductance or the average number of times the channel opens (open frequency; Twyman and Macdonald, 1991). Thus, benzodiazepines enhance chloride flux by increasing $GABA_A$ channel open frequency, whereas barbiturates do so by affecting the channel open time.

Given that benzodiazepines and barbiturates both increase chloride flux through $GABA_A$ receptors, why do they not have identical therapeutic uses? Most barbiturate drugs were developed in the early 1900s, whereas the first benzodiazepine, chlordiazepoxide, became available in 1960. By 1970, benzodiazepines had largely replaced barbiturates for treatment of anxiety, though barbiturates are still used in some cases as anesthetics and antiseizure drugs. Barbiturates fell out of favor for treating anxiety both because they are addictive and because they have many side effects, some of which can be lethal at dosages near the therapeutic dose. One of the lethal side effects of barbiturates is respiratory depression, which occurs because brainstem neurons that help control respiration have a high concentration of $GABA_A$ receptors that contain barbiturate-binding sites but not benzodiazepine-binding sites. Additionally, barbiturates affect more than just $GABA_A$ receptors (they also affect other proteins in the nervous system), which contributes to their side effects. In contrast, benzodiazepines have fewer side effects, are usually not lethal when taken on their own, and may not be as addictive.

## Glycine Receptors

While GABA is the major inhibitory transmitter in higher centers of the CNS, glycine is the major inhibitory transmitter in the brainstem and spinal cord. Like $GABA_A$ receptors, inhibitory glycine receptors are pentameric ligand-gated chloride channels. The inhibitory glycine receptor can be homomeric (all five subunits are $\alpha$), or heteromeric, in which there are two or three $\alpha$-subunits (which all must be one of the four subtypes, $\alpha_{1-4}$) expressed in combination with two or three $\beta$-subunits (Grudzinska et al., 2005; Durisic et al., 2012). The kinetics of glycine and $GABA_A$ receptor channels differ, with glycine receptor channels gating (opening and closing) faster than $GABA_A$ receptor channels. When inhibitory

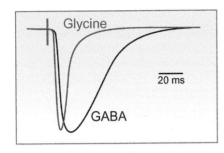

FIGURE 11.8 **GABA$_A$- and glycine-mediated synaptic currents.** Sample recording of synaptic currents following presynaptic nerve stimulation that triggers the release of GABA (*blue trace*) or glycine (*red trace*). These diagrammed recordings of synaptic current demonstrate that the glycine receptor channel gates faster than the GABA$_A$ receptor channel. *Source: Adapted from Grudt, T.J., Henderson, G., 1998. Glycine and GABAA receptor-mediated synaptic transmission in rat substantia gelatinosa: inhibition by mu-opioid and GABAB agonists. J. Physiol. 507 (Pt 2), 473–483.*

postsynaptic currents (IPSCs) are measured from each receptor type, glycine-mediated currents have rise times (time to reach peak current) with a time constant of less than 1 ms and decay time constants of about 4 ms (reflects rate of decay of the current), while GABA$_A$-mediated currents rise with a time constant of about 2 ms and decay with a time constant of 20–30 ms (see Fig. 11.8).

As mediators of inhibitory signaling in the spinal cord and brainstem, glycine receptors are important regulators of motor coordination, sensory processing, respiratory rhythms, and pain (Lynch, 2009; Callister and Graham, 2010; Zeilhofer et al., 2012). Mutations of the glycine receptor can cause a clinical syndrome called **hyperekplexia**, which is characterized by an exaggerated startle reflex in response to acoustic or tactile stimuli (Harvey et al., 2008). In addition, glycine receptor dysfunction has been linked to temporal lobe epilepsy, autism, breathing disorders, and chronic inflammatory pain (Lynch et al., 2017). Enhancement of glycine receptor current can alleviate chronic pain, and research is underway to develop glycine-receptor-targeted therapies for this condition (Lynch and Callister, 2006).

## Zinc-Activated Channel Receptors

The zinc ion ($Zn^{2+}$) is a trace element present in every part of the body. Zinc ions perform functions such as enhancing enzyme activity and regulating immune system function that are important in cell division, wound healing, and cell growth. Because zinc plays such a critical role in human health, many national and international organizations have set standards for dietary zinc intake (Roohani et al., 2013).

Zinc is also packaged into synaptic vesicles in some areas of the brain, and it has been shown to modulate glutamate, GABA$_A$, and glycine receptor activity. At some glutamate-releasing synapses, zinc is also released from synaptic vesicles and acts as an endogenous antagonist at NMDA receptors (Westbrook and Mayer, 1987), modulating NMDA receptor-mediated plasticity (Izumi et al., 2006). Zinc also acts as an endogenous inhibitor at GABA$_A$ receptors (Sharonova et al., 2000) and AMPA glutamate receptors (Kalappa et al., 2015). In addition, zinc can either enhance or inhibit glycine receptors, depending on the concentration of glycine present (Trombley et al., 2011; Burgos et al., 2016).

Interestingly, some synapses contain specific ionotropic receptors called "zinc-activated channels" (ZACs) that zinc binds to and activates directly, independent of its role in modulating the function of other neurotransmitter receptors. ZACs are the least well-understood members of the pentameric ligand-gated ion channel family (cys-loop receptors). Since they are nonselective monovalent cation channels that are permeable to sodium and potassium, they depolarize cells when activated at resting membrane potential. ZACs are present in opossums, dogs, cows, and primates (including humans), but not in mice or rats (Davies et al., 2003; Houtani et al., 2005).

More recent studies of the ZAC have demonstrated that it can be activated not only by zinc, but also by protons ($H^+$) and copper; in fact, protons are more potent and effective at activating ZAC receptors than either zinc or copper (Trattnig et al., 2016). Therefore, zinc-activated receptors may be more accurately described as zinc-, copper-, and proton-activated receptors, and they may be functionally activated by a variety of ligands present in their vicinity. More research is required to fully understand the function of these receptors in the nervous system, and the circumstances under which each ligand may endogenously activate ZACs.

## THE GLUTAMATE IONOTROPIC RECEPTOR FAMILY

Glutamate is the most common neurotransmitter in the central nervous system, used at over 50 trillion synapses in the human brain, and glutamate-mediated synaptic transmission regulates a large variety of nervous system functions. How can a single neurotransmitter molecule perform such a broad array of actions? The answer lies in the wide range of glutamate receptors in the brain and spinal cord, which include both ionotropic (discussed here) and metabotropic (discussed in Chapter 12) families of glutamate receptors.

The ionotropic glutamate receptor family can be broken into two groups that are structurally related to one another: NMDA and non-NMDA receptors. The non-NMDA receptors are broken down further into AMPA and kainate receptors, which are more similar to one another than either of them is to NMDA receptors.

Each of the three ionotropic glutamate receptor groups is named for a pharmacological agent that can be used as its selective artificial agonist: NMDA receptors are activated by N-methyl-D-aspartate (NMDA); AMPA receptors are activated by α-amino-3-hydroxy-5-methyl-4-isoxazoleproprionic acid (AMPA); and kainate receptors are activated by kainate, also called kainic acid (Dingledine et al., 1999; Lodge, 2009; Traynelis et al., 2010). Each of these receptor types consists of four subunits, and each subunit contains three transmembrane regions and a large extracellular loop region. The extracellular regions are organized into two domains that are each thought to form a clamshell-like structure (Fig. 11.9). The first of the clamshell-like domains, the amino-terminal domain, can sometimes bind ligands that inhibit receptor function. The second clamshell-like domain is called the ligand-binding domain and, as the name implies, it binds the agonists that activate the receptor and open the channel (Fig. 11.9).

One difference between the AMPA, kainate, and NMDA ionotropic glutamate receptor families is the speed at which the receptor channels open and close. AMPA receptors gate fastest, and kainate receptors are only slightly slower than AMPA receptors (Cossart et al., 2002), whereas NMDA receptors gate exceptionally slowly (Iacobucci and Popescu, 2017; see Fig. 11.10). Synapses that use glutamate as a neurotransmitter can express different

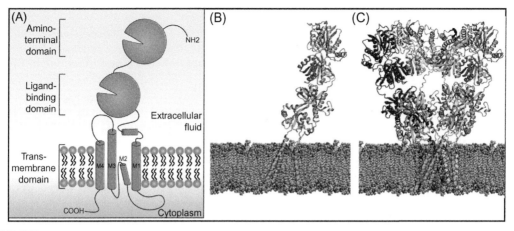

FIGURE 11.9  **Structure of ionotropic glutamate receptors.** (A) Diagram of a single glutamate receptor subunit showing the transmembrane domains (M1, M3, and M4), ligand-binding domain, and the amino-terminal domain. The ligand-binding domain and the amino-terminal domains are thought to function like clam shells. (B) Visualization of one protein subunit of the AMPA receptor (*yellow*) as it is thought to be positioned in the plasma membrane. (C) Visualization of the complete AMPA receptor positioned within the plasma membrane, and formed by four individual protein subunits (each in a different color). *Source: (B) and (C) were generated by Rozita Laghaei using the software VMD and structures deposited in the RCSB protein data bank.*

complements of ionotropic glutamate receptor family members to create synaptic responses with different time courses. We outline some of the details of each type of ionotropic glutamate receptor below.

## NMDA Receptors

The NMDA glutamate receptor is a both a critically important regulator of synaptic plasticity and a site of action for drugs used to treat a variety of neurological and psychiatric conditions. NMDA receptors are tetrameric, meaning they are made up of four protein subunits (see Fig. 11.11). There are two types of subunits that make up NMDA receptors, GluN1 subunits and GluN2 subunits, and most NMDA receptors are thought to be constructed from two of each type of subunit. NMDA receptors are permeable to sodium, potassium, and calcium, and the relatively large calcium flux through NMDA receptor channels (~12%–14% of total ionic flux) has been shown to be a trigger for various forms of synaptic plasticity (see Chapter 14).

Although they are categorized as glutamate receptors, NMDA receptors are activated by the simultaneous binding of two different agonists: glutamate binds to a GluN2 subunit while either glycine or D-serine binds to a GluN1 subunit (Johnson and Ascher, 1987; Kleckner and Dingledine, 1988). However, it is thought that under normal circumstances the extracellular concentrations of glycine or D-serine are high enough that one of these ligands is usually bound to GluN1 subunits even under resting conditions. Therefore, despite the need for glycine or D-serine to be present, activation of NMDA receptors

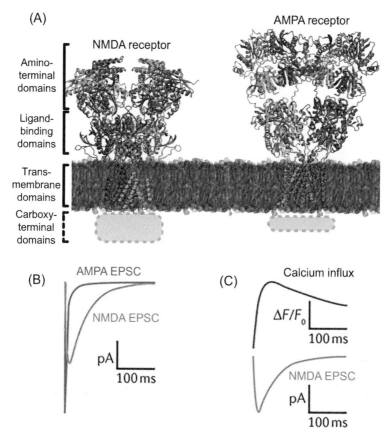

FIGURE 11.10  **A comparison of NMDA and AMPA receptors.** (A) Structural models reveal that NMDA and AMPA receptors have a very similar organization. Each receptor type is constructed from four protein subunits (color coded) with the same general regions: extracellular amino-terminal domains and ligand-binding domains, transmembrane domains, and carboxy-terminal domains [for which we do not have detailed structural information (*green dotted line boxes*)]. (B) Despite similarities in structure, these receptors have very different physiological characteristics. When activated by glutamate, NMDA receptors are slow to open and close (*light blue traces*), while AMPA receptors are very fast to open and close (*red trace*). (C) NMDA receptors conduct a relatively large amount of calcium when they open (*dark blue trace*). Most AMPA receptors do not conduct significant amounts of calcium ions, though there are some members of this receptor family that do conduct calcium, the so-called calcium-permeable AMPA receptors. *Source: Structural models of NMDA and AMPA receptors in (A) were generated by Rozita Laghaei using the software VMD and structures deposited in the RCSB protein data bank; (B) and (C) were adapted from Iacobucci, G.J., Popescu, G.K., 2017. NMDA receptors: linking physiological output to biophysical operation. Nat. Rev. Neurosci. 18, 236–249.*

primarily depends on the glutamate concentration and is closely tied to the release of glutamate from nerve terminals.

One of the most important characteristics of NMDA receptors is that the ion pore is predominately blocked by magnesium ions at rest, and this block is removed after sufficient depolarization of the presynaptic membrane (see Fig. 11.12). At the resting membrane

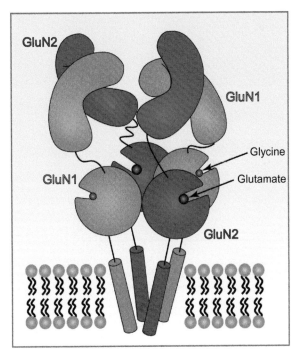

FIGURE 11.11 **NMDA receptor agonist binding sites.** NMDA receptors contain four protein subunits, including GluN1 subunits (which bind glycine or D-serine) and GluN2 subunits (which bind glutamate). This diagram focuses on the extracellular regions of the NMDA receptor, and, for simplicity, does not represent all of the transmembrane or cytoplasmic domains.

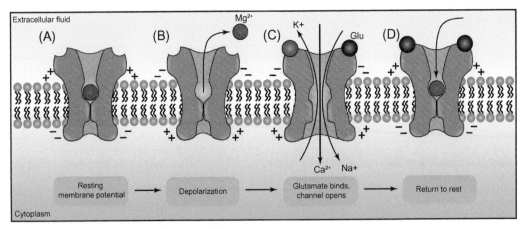

FIGURE 11.12 **Magnesium block of NMDA receptors.** NMDA receptor channels only conduct ions when glutamate binds to the receptor while the membrane is depolarized. If glutamate binds to the NMDA receptor at the resting membrane potential, ion flux is prevented because a magnesium ion blocks the pore (A). When the membrane depolarizes, the magnesium ion is displaced (B). If glutamate binds to the NMDA receptor at a depolarized membrane potential, sodium ($Na^+$) and calcium ($Ca^{2+}$) flow into the cell, and potassium ($K^+$) flows out (C). When the membrane potential returns to a resting value, the pore is once again blocked by a magnesium ion (D).

potential, the negative charge inside the membrane attracts a magnesium ion to a binding site inside the NMDA receptor channel pore. Once bound, the magnesium ion blocks the pore, inhibiting ion flux even when glutamate is bound to the receptor. When the membrane depolarizes, the magnesium ion is repelled out of the channel, clearing the way for ion flux. Therefore, the NMDA receptor channel conducts ions only when glutamate binding (to open the channel) and membrane depolarization (to remove the magnesium block) occur simultaneously. Since it only functions with the coincident release of glutamate from the presynaptic cell and depolarization of the postsynaptic cell, the NMDA receptor is often called a **coincidence detector** (see Fig. 11.12).

When the current flux through the NMDA receptor channel in the presence of glutamate, glycine, and magnesium is plotted as a function of voltage and compared to the current flux in the absence of magnesium, the effect of the magnesium block can be seen (see Fig. 11.13). For example, at a membrane potential of $-80$ mV, the magnesium block is so strong that almost all of the NMDA receptor channels are blocked and little current flows. When the membrane potential is depolarized to $-30$ mV, a significant portion of the magnesium block is relieved. When the membrane is depolarized to 0 mV, magnesium has been repelled from the pore in essentially all of the NMDA receptor channels.

NMDA receptors are modulated by a large number of molecules, including endogenous ligands, such as zinc, polyamines, and protons. They also have binding sites for a variety of drugs, such as ifenprodil, ketamine, memantine, phencyclidine, and MK-801, which have important psychiatric and neurological function (see Fig. 11.14). Some of these bind in the amino-terminal domain of the channel (zinc, ifenprodil, and protons). Most of these ligands inhibit the NMDA channel (Westbrook and Mayer, 1987; Lodge, 1997; Huggins and Grant, 2005; Hansen et al., 2010; Furukawa, 2012; Johnson et al., 2015; Zhu and Paoletti, 2015), but polyamines can act as either agonists or antagonists, depending on the subunits present (see Traynelis et al., 2010). An interesting clinical aside related to NMDA receptors is described in Box 11.4.

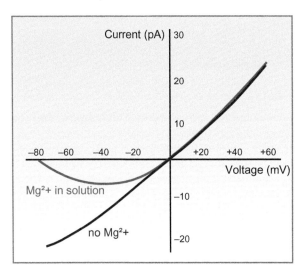

FIGURE 11.13 **Current−voltage relationship for NMDA receptor channels in the presence and absence of the magnesium block.** These curves compare the current−voltage relations for NMDA receptor channels in the presence (*red*) or absence (*blue*) of extracellular magnesium. The difference between any two points on the curves at each voltage indicates the percentage of channels that are blocked by magnesium at that voltage. For example, at $-80$ mV roughly 98% of the current is blocked, while at $-30$ mV, only about 30% of current is blocked. At $+20$ mV and above, there is no magnesium block. *Source: Adapted from Nowak, L., Bregestovski, P., Ascher, P., Herbet, A., Prochiantz, A., 1984. Magnesium gates glutamate-activated channels in mouse central neurones. Nature 307, 462−465.*

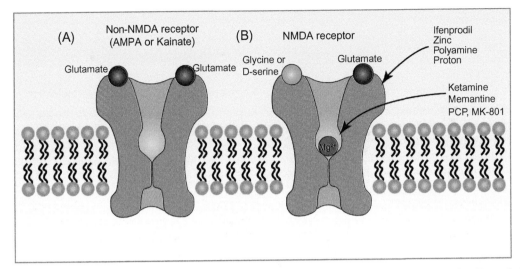

FIGURE 11.14  **NMDA and non-NMDA receptors shown in cross-section.** (A) Non-NMDA (AMPA or kainate) receptors are formed from four subunits, are activated by glutamate alone, and have fewer modulatory sites than NMDA receptors. AMPA and kainate receptors are formed from a variety of specific subunits. (B) NMDA receptors contain four subunits (two GluN1 subunits and two GluN2 subunits; GluN 2 can be made up using subtypes A–D) and have multiple binding sites that can modulate their gating properties. In addition to agonist sites for glutamate (on the GluN2 subunits) and glycine or D-serine (on the GluN1 subunits), there are also modulatory sites for magnesium ions ($Mg^{2+}$), ifenprodil, protons, polyamines, zinc ions, ketamine, memantine, phencyclidine (PCP), and MK-801.

---

BOX 11.4

## ANTI-NMDA RECEPTOR ENCEPHALITIS

Imagine you are a doctor, assessing a previously healthy patient in her 20s who has the following medical history. About 2 months ago she experienced flu-like symptoms, but soon began having extreme symptoms of mood disorders such as depression and mania, as well as psychotic symptoms such as delusions and hallucinations. These were followed by seizures, and then progressed to problems with movement, and eventually catatonia. She is now experiencing an inability to regulate body temperature, blood pressure, and heart rate. Her symptoms began fairly suddenly and progressed to this point over the course of just a couple months. Does your patient have a psychiatric disorder? Epilepsy? A stroke? Malfunction of the brainstem or hypothalamus?

In fact, these symptoms are characteristic of a disorder called anti-NMDA receptor encephalitis. For many patients with this disorder, the symptoms described above, along with other related symptoms, including amnesia and cognitive deterioration, commonly follow a progression from psychiatric symptoms to nonpsychiatric neurological symptoms to problems with autonomic function, and this progression can occur over as little as 1–2 months. Without treatment, death can result. Treatments include corticosteroids and immunotherapy, including removal of antibodies via intravenous immunoglobulin or plasma exchange. Many patients experience significant recovery, but it is not always complete and relapses do occur.

## BOX 11.4 (cont'd)

What could cause such a devastating disorder? In 2007, Dalmau and colleagues first characterized anti-NMDA receptor encephalitis, an autoimmune disorder in which antibodies attack NMDA receptors in the brain. It is hypothesized that when antibodies attack NMDA receptors, the receptors are internalized and neurons are no longer able to maintain sufficient amounts of NMDA receptors in their membranes. In a study of 100 patients with anti-NMDA receptor encephalitis, all patients had antibodies that targeted the NR1 subunit of NMDA receptors, and when the cerebrospinal fluid from these patients was incubated with hippocampal neurons, which normally express large numbers of NMDA receptors, the numbers of NMDA receptors were well below normal levels (see Fig. 11.15; Dalmau et al., 2008). Anti-NMDA receptor encephalitis can be diagnosed by testing cerebrospinal fluid for antibodies against NMDA receptors.

One of the mysterious aspects of anti-NMDA receptor encephalitis is that it leads to such a wide range of neurological symptoms, many of which are not usually directly related. Another mystery is that in many patients the symptoms described above occur in a stereotyped sequence: psychiatric, neurological, autonomic. Even more intriguingly, the symptoms patients experience during the long recovery period (usually many months) often recur in roughly the reverse order of their onset. What this tells us about neurological and psychiatric symptoms, of any cause, is not yet known, but anti-NMDA receptor encephalitis is a reminder of how finely tuned synaptic communication in the brain is and how easily its function can be compromised.

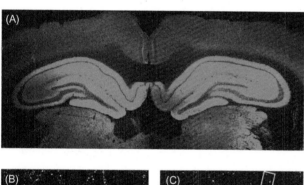

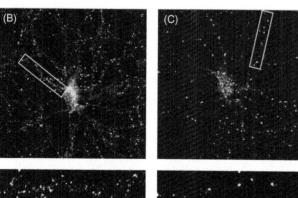

FIGURE 11.15 **Anti-NMDA receptor antibodies target NMDA receptors that contain the NR1 subunit.** (A) In a coronal section of a mouse brain, green staining indicates areas where antibodies from patient cerebrospinal fluid and serum targeted NR1-containing NMDA receptors. Staining is especially high in the dentate gyrus of the hippocampus. Cultured hippocampal neurons were exposed to cerebrospinal fluid either from controls (B) or from patients with anti-NMDA receptor encephalitis (C), and then labeled for clusters of receptors containing the NR1 NMDA receptor subunit [*white spots* in (B) and (C)]. After 7–14 days of exposure to patients' cerebrospinal fluid, there was a >50% reduction in NR1-labeled clusters. Boxed areas in (B) and (C) indicate areas enlarged below. *Source: Adapted from Dalmau, J., Gleichman, A.J., Hughes, E.G., Rossi, J.E., Peng, X., Lai, M., et al., 2008. Anti-NMDA-receptor encephalitis: case series and analysis of the effects of antibodies. Lancet. Neurol. 7, 1091–1098.*

## AMPA and Kainate Glutamate Receptors

The two types of non-NMDA glutamate receptors, AMPA and kainate receptors, are responsible for most of the fast excitatory neurotransmission in the brain, and it is their excitatory action that leads glutamate to often be characterized as an excitatory transmitter. In particular, the AMPA receptor is the most common glutamate receptor that mediates excitatory postsynaptic potentials at synapses. AMPA receptors are tetrameric and are made up of various combinations of the subunits GluA1, GluA2, GluA3, and GluA4 (Dingledine et al., 1999). Most AMPA receptors conduct only sodium and potassium, but types that do not contain the GluA2 subunit are also calcium-permeable (Hollmann et al., 1991). Unlike NMDA receptors, AMPA receptors are not blocked by magnesium, which means they are available to open and pass current whenever glutamate binds to them. Furthermore, these receptors are activated by glutamate alone and do not have binding sites for glycine (or D-serine; see Fig. 11.14). AMPA receptor-mediated ionic current has very fast kinetics (see Fig. 11.10), which leads to a rapid rise and decay of the excitatory postsynaptic currents mediated by this receptor.

A postsynaptic terminal membrane can contain both AMPA and NMDA receptors in different ratios from synapse to synapse. Because magnesium ions block NMDA receptors at the resting membrane potential, if both NMDA and AMPA receptors are present at a synapse, release of glutamate in response to a single action potential will lead predominately to AMPA receptor activation since the postsynaptic cell will be at or near its resting membrane potential when the glutamate binds. However, if the same synapse is activated by a high-frequency burst of presynaptic action potentials, the resulting postsynaptic AMPA receptor currents might depolarize the postsynaptic neuron sufficiently to relieve the magnesium block from NMDA receptors, allowing NMDA receptor channels to contribute to the postsynaptic current.

What would happen if the postsynaptic membrane only expressed NMDA receptors? In this scenario, the presynaptic release of glutamate would barely affect the postsynaptic membrane potential, as long as the postsynaptic cell was at its resting membrane potential when the glutamate bound. Therefore, synapses at which the postsynaptic neuron has only NMDA receptors are often called **silent synapses**. These silent synapses can only contribute to postsynaptic activity if the postsynaptic cell is in a depolarized state, which might only happen during particular patterns of neuronal activity and has therefore been proposed to be a mechanism for synaptic plasticity (see Chapter 14).

Most excitatory synapses in the central nervous system contain AMPA and NMDA receptors. However, some synapses in the hippocampus, cerebellum, amygdala, retina, and spinal cord use kainate receptors (Straub et al., 2016; Contractor et al., 2011; Yan et al., 2013; DeVries and Schwartz, 1999; Li and Rogawski, 1998; Li et al., 1999). Kainate receptors are made up of GluK1–3 or 4/5 subunits. They mediate smaller and slower excitatory postsynaptic potentials than AMPA receptors, and therefore create a longer-lasting synaptic current. Overall, both AMPA and kainate receptors are less susceptible to modulation than NMDA receptors. These non-NMDA receptors can be modulated by zinc ions (although the relevance of zinc modulation to physiological conditions is debated; see Mott et al., 2008; Kalappa et al., 2015), but not by the large number of modulators that have been described for NMDA receptors (see Fig. 11.14). Unconventional signaling mediated by ionotropic glutamate receptors is described in Box 11.5.

# BOX 11.5

## UNCONVENTIONAL IONOTROPIC GLUTAMATE RECEPTOR SIGNALING

Interestingly, some studies have provided evidence that in addition to functioning as traditional ligand-gated ion channels, ionotropic glutamate receptors might also link to the activation of biochemical cascades within the cytoplasm (see Fig. 11.16).

### Intracellular Signaling Cascades Associated With Non-NMDA Receptors

The specific mechanism by which non-NMDA receptors trigger cytoplasmic messengers after agonist binding is not well understood, but may involve the activation of a cascade that is similar to the traditional metabotropic GTP-binding protein receptors discussed in detail in Chapter 12. This unusual additional signaling mechanism for a ligand-gated ion channel might mean that AMPA and kainate receptors can function as both ionotropic and metabotropic receptors (Rodriguez-Moreno and Lerma, 1998; Wang et al., 1997; Kawai and Sterling, 1999; Takago et al., 2005). Further details of this novel type of signaling will require additional studies.

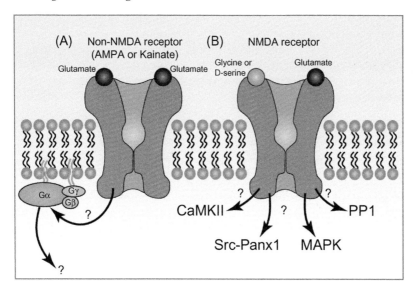

FIGURE 11.16  **Potential dual function of NMDA, AMPA, and kainate receptors.** (A) AMPA and kainate ionotropic receptors are known to pass sodium and potassium ions after binding glutamate, but some subtypes of these receptors appear to also be metabotropic, activating G-proteins within the cytoplasm that trigger biochemical cascades in the cell. (B) In some cases, ionotropic NMDA receptors may couple to intracellular signaling cascades. In addition to conducting sodium, potassium, and calcium, NMDA receptors may couple through their intracellular domains to protein phosphatase 1 (PP1), calcium-calmodulin-dependent protein kinase II (CaMKII), p38 Mitogen-activated protein kinase (MAPK), sarcoma kinase (Src), and pannexin-1 (Panx1).

## BOX 11.5 (cont'd)

### Intracellular Signaling Cascades Associated With NMDA Receptors

Although still debated, recent experiments have shown that glutamate can activate signals through the NMDA receptor independent of ionic flux (Dore et al., 2017). The experimental test of this is often based on the observation that pharmacologically blocking NMDA receptor pores using MK-801 can prevent ionic flux-related functions of these receptors but not block all downstream effects of receptor activation by agonists. For example, NMDA receptor activation during high-frequency stimulation has been shown to induce long-term potentiation (LTP) that is dependent on calcium flux through the NMDA receptor channel (see Chapter 14). However, when the pore is blocked by MK-801, glutamate binding to the NMDA receptor during low-frequency stimulation was reported to induce long-term depression, independent of ionic flux (Nabavi et al., 2013). This "metabotropic" signaling is thought to be mediated by protein phosphatase 1 (PP1) and calcium-calmodulin-dependent protein kinase II (CaMKII). There is also evidence that this form of "metabotropic" NMDA receptor signaling can activate p38 mitogen-activated protein kinase (MAPK) to induce structural changes in dendritic spines (Stein et al., 2015). Lastly, sarcoma (Src) kinase appears to be coupled to NMDA receptors and can activate **pannexin channels** (large transmembrane channels that allow the passage of ions and small molecules) in the membrane during excitotoxic stimuli (Weilinger et al., 2016). All of these signaling molecules are thought to be activated when the receptor undergoes a conformational change in response to glutamate binding.

## THE TRIMERIC RECEPTOR FAMILY

It might be surprising to imagine that adenosine triphosphate (ATP) and protons ($H^+$) could be neurotransmitters, since we usually think of these molecules in the context of other roles they play in cells (energy for ATP, and pH for $H^+$). As will be discussed in more detail in Chapter 21, all synaptic vesicles contain ATP and $H^+$. Therefore, when synaptic vesicles fuse with the plasma membrane, these ATP and $H^+$ molecules are released along with the more traditional neurotransmitter molecules.

Of course, just because a neuron releases a molecule does not necessarily mean that molecule is a neurotransmitter. So what does determine whether a molecule is a neurotransmitter? The set of criteria that define neurotransmitters will be covered in Chapter 15, but one of the most important criteria is the presence of a receptor for the molecule at the synapse. Because $H^+$ and ATP are released from neurons and there are receptors for these molecules in the vicinity of release sites, these molecules are considered neurotransmitters.

## P2X Receptors for ATP

ATP is classified as a purine, and some breakdown products of ATP are also purine neurotransmitters (e.g., adenosine; see Chapter 21, for more details). The purine neurotransmitter family includes both metabotropic and ionotropic receptors (Abbracchio et al., 2009). We will focus here on the P2X receptors, ionotropic receptors that are selective for ATP. There are seven isoforms of P2X receptors ($P2X_{1-7}$), which are made up of three protein subunits, each of which has only two transmembrane domains (see Fig. 11.17). P2X receptors are ligand-gated nonselective cation channels that pass sodium, potassium, and calcium. In fact, the calcium permeability for these receptor channels is relatively high. Therefore, these receptors depolarize neurons and increase the cytoplasmic calcium concentration. They can be located either pre- or postsynaptically.

In the central nervous system, P2X receptors are usually expressed at relatively low levels at synapses, and they are often not positioned immediately opposite the sites of neurotransmitter release. Further, many P2X receptors have a low affinity for ATP, which, combined with their extrasynaptic location, means P2X receptors are activated only during bursts of high-frequency presynaptic action potentials (Richler et al., 2008; Khakh and North, 2012). These properties suggest that ATP activation of P2X receptors often only results in modulation of synaptic transmission during high-frequency activity.

In contrast, in the peripheral nervous system, ATP can act as a fast neurotransmitter that has much stronger effects. For example, in the sympathetic innervation of the vas deferens, ATP is released with norepinephrine to mediate strong postsynaptic effects (Navarrete et al., 2014; Sneddon et al., 1982). In addition, in the gastrointestinal tract, ATP and acetylcholine are co-released and both mediate fast excitatory postsynaptic potentials

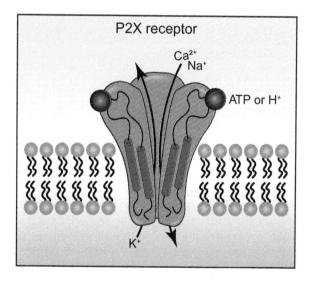

FIGURE 11.17 Diagram of the structure of trimeric ionotropic receptors that bind ATP or $H^+$ as ligands. The P2X receptor is shown as an example.

in the myenteric plexus, which controls gut motility, and in the submucous plexus, which controls secretion and absorption (Galligan and Bertrand, 1994; Monro et al., 2004). ATP-activated P2X receptors carry a major component of excitation at these peripheral nervous system synapses.

## Acid-Sensing Ion Channels

Acid-sensing ion channels (ASIC) are members of the epithelial sodium channel/degenerin (ENaC/DEG) family. Most members of the ENaC/DEG family are expressed primarily in mammalian non-neuronal tissue (kidney and lungs) or in the nervous systems of some invertebrates (insects, worms, and mollusks) (Kellenberger and Schild, 2002), but the ASIC is expressed in the mammalian nervous system. The ASIC is a ligand-gated ion channel that is activated by binding protons ($H^+$), which means it is activated more at low pH, when the proton concentration is higher. Activated ASICs are more permeable to sodium than to potassium, but some types of ASIC can also conduct protons and calcium ions.

Four ASIC genes (*ASIC 1—4*) have been identified and have a widespread expression pattern in the nervous system. These genes code for six different protein subunits. Functional ASIC receptors contain three subunits, and can either be homomeric (containing three of the same protein subunit type) or heteromeric (containing a mix of different protein subunits). High levels of ASIC expression have been reported in the hippocampus, cerebellum, olfactory bulb, amygdala, pituitary gland, and sensory neurons (Kellenberger and Schild, 2002).

An important question related to the activation of these ASICs is: When would the pH in the nervous system drop sufficiently to activate them? We think of pH as being tightly regulated in the body. The average pH of human blood serum is about 7.4, and the range is 7.35—7.45, making it clear that $H^+$ ion concentration is tightly regulated. This tight regulation is needed because global changes outside this range can have devastating effects on the body. Our environment and the food we eat can affect blood pH, so the body normally regulates pH by using a bicarbonate buffer in the blood, altering blood $CO_2$ concentration using the lungs (since breathing affects $CO_2$ blood concentration), and regulating $H^+$ ion absorption or excretion in the kidneys. However, the above factors only control the pH in the blood. In the context of ASIC function in the brain, we are more concerned with local regulation of proton concentration within the small compartment of the synaptic cleft.

In the nervous system, pH is controlled at a synaptic level via local release of $H^+$ ions by synaptic vesicles, proton pumps, $Na^+/H^+$ exchangers, and $Na^+$/bicarbonate exchangers (Schlue and Dörner, 1992; Obara et al., 2008). Changes to pH in the nervous system can be pathological, such as when **acidosis** produces pain in peripheral tissues, when cerebral **ischemia** causes cell damage, or when **epileptic activity** occurs (Chesler and Kaila, 1992; Lipton, 1999; Reeh and Steen, 1996; Sherwood et al., 2012). However, at the subcellular level in the brain, there can be significant normal physiological alterations in synaptic pH (but not blood pH) following high-frequency neuronal activity and synaptic vesicle exocytosis (Krishtal et al., 1987). These local changes in pH are important in the physiological activation of ASICs in the healthy nervous system following neurotransmitter release, and have been shown to affect synaptic communication (Gonzalez-Inchauspe et al., 2017).

# THE TRANSIENT RECEPTOR POTENTIAL CHANNEL FAMILY

The transient receptor potential (TRP) group of channels includes six families of about 29 different channels that are part of a signaling system used in sensing light, sound, chemicals, temperature, and touch (Venkatachalam and Montell, 2007). All TRP channels are tetrameric, and each of their four protein subunits has a characteristic structure of six transmembrane segments. In fact, the structure of TRP subunits resembles that of voltage-gated ion channel subunits, since TRP family members can possess a voltage-sensitive S4 segment and include a "P" loop between segments 5 and 6 (see Fig. 11.1). Nevertheless, TRP channels are unusual among channel families because they have a very large array of activation mechanisms and ion selectivity, and in some cases a single TRP channel can be activated by a variety of ligands and cellular signaling mechanisms that often appear unrelated. Some neuroscientists refer to these channels as "multiple signal integrators" because a TRP channel's response to one molecule can be modified by other molecular interactions.

Neurons use TRP channels for various cellular signaling functions in the brain. Most often, TRP channels are activated downstream of second-messenger pathways that are activated either by growth factors (e.g., BDNF and epidermal growth factor; EGF) or by transmitters acting on metabotropic receptors. Related to this function is the fact that most TRP channels require phospholipase C (PLC) activation in order to open. The PLC enzyme is usually activated by growth factors or by other metabotropic receptors, and it triggers the generation of diacylglycerol (DAG) and inositol triphosphate ($IP_3$) by cleaving lipids in the plasma membrane (see Chapter 12, for more details). The generation of DAG plus the $IP_3$-triggered calcium release from intracellular stores often activates TRP channels, and the ion flux mediated by TRP channels then participates in the signal transduction pathway for metabotropic signals.

Given that the mechanisms discussed above are usually associated strictly with metabotropic receptors, why are we including TRP channels in our section on ligand-gated ion channels? In fact, this family of channels is so diverse that there is at least one example of TRP channels apparently being used as an extracellular ligand-gated ion channel: the response of neurons to the endocannabinoids anandamide and 2-arachidonoylglycerol (2-AG). Classically, endocannabinoid receptors are known as metabotropic G-protein-coupled receptors (CB1 and CB2). There is now experimental evidence that several TRP channel family members (e.g., TRPV1 and TRPA1) might also be sensitive to extracellular endocannabinoids (Di Marzo and De Petrocellis, 2010; Storozhuk and Zholos, 2018). As a result, these TRP channels are now called ionotropic cannabinoid receptors, and they add to the complexity of endocannabinoid signaling.

## References

Aapro, M.S., 1991. 5-HT3 receptor antagonists. An overview of their present status and future potential in cancer therapy-induced emesis. Drugs 42, 551–568.

Abbracchio, M.P., Burnstock, G., Verkhratsky, A., Zimmermann, H., 2009. Purinergic signalling in the nervous system: an overview. Trends Neurosci. 32, 19–29.

Andersen, N., Corradi, J., Sine, S.M., Bouzat, C., 2013. Stoichiometry for activation of neuronal α7 nicotinic receptors. Proc. Natl. Acad. Sci. USA 110, 20819–20824.

Benowitz, N.L., Kuyt, F., Jacob 3rd, P., 1982. Circadian blood nicotine concentrations during cigarette smoking. Clin. Pharmacol. Ther. 32, 758–764.

Borg-Graham, L.J., Monier, C., Fregnac, Y., 1998. Visual input evokes transient and strong shunting inhibition in visual cortical neurons. Nature 393, 369–373.

Burgos, C.F., Yevenes, G.E., Aguayo, L.G., 2016. Structure and pharmacologic modulation of inhibitory glycine receptors. Mol. Pharmacol. 90, 318–325.

Callister, R.J., Graham, B.A., 2010. Early history of glycine receptor biology in mammalian spinal cord circuits. Front. Mol. Neurosci. 3, 13.

Campo-Soria, C., Chang, Y., Weiss, D.S., 2006. Mechanism of action of benzodiazepines on GABAA receptors. Br. J. Pharmacol. 148, 984–990.

Chesler, M., Kaila, K., 1992. Modulation of pH by neuronal activity. Trends Neurosci. 15, 396–402.

Conn, P.J., 1997. Pharmacology and functions of metabotropic glutamate receptors. Annu. Rev. Pharmacol. Toxicol. 37, 205–237.

Contractor, A., Mulle, C., Swanson, G.T., 2011. Kainate receptors coming of age: milestones of two decades of research. Trends Neurosci. 34, 154–163.

Cossart, R., Epsztein, J., Tyzio, R., Becq, H., Hirsch, J., Ben-Ari, Y., et al., 2002. Quantal release of glutamate generates pure kainate and mixed AMPA/kainate EPSCs in hippocampal neurons. Neuron 35, 147–159.

Dalmau, J., Gleichman, A.J., Hughes, E.G., Rossi, J.E., Peng, X., Lai, M., et al., 2008. Anti-NMDA-receptor encephalitis: case series and analysis of the effects of antibodies. Lancet. Neurol. 7, 1091–1098.

Dani, J.A., 2015. Neuronal nicotinic acetylcholine receptor structure and function and response to nicotine. Int. Rev. Neurobiol. 124, 3–19.

Davies, P.A., Wang, W., Hales, T.G., Kirkness, E.F., 2003. A novel class of ligand-gated ion channel is activated by $Zn2+$. J. Biol. Chem. 278, 712–717.

DeVries, S.H., Schwartz, E.A., 1999. Kainate receptors mediate synaptic transmission between cones and 'Off' bipolar cells in a mammalian retina. Nature 397, 157–160.

Di Marzo, V., De Petrocellis, L., 2010. Endocannabinoids as regulators of transient receptor potential (TRP) channels: a further opportunity to develop new endocannabinoid-based therapeutic drugs. Curr. Med. Chem. 17, 1430–1449.

Dingledine, R., Borges, K., Bowie, D., Traynelis, S.F., 1999. The glutamate receptor ion channels. Pharmacol. Rev. 51, 7–61.

Dore, K., Stein, I.S., Brock, J.A., Castillo, P.E., Zito, K., Sjostrom, P.J., 2017. Unconventional NMDA receptor signaling. J. Neurosci. 37, 10800–10807.

Durisic, N., Godin, A.G., Wever, C.M., Heyes, C.D., Lakadamyali, M., Dent, J.A., 2012. Stoichiometry of the human glycine receptor revealed by direct subunit counting. J. Neurosci. 32, 12915–12920.

Fucile, S., 2004. $Ca2+$ permeability of nicotinic acetylcholine receptors. Cell Calcium 35, 1–8.

Fucile, S., Renzi, M., Lax, P., Eusebi, F., 2003. Fractional $Ca(2+)$ current through human neuronal α7 nicotinic acetylcholine receptors. Cell Calcium 34, 205–209.

Furukawa, H., 2012. Structure and function of glutamate receptor amino terminal domains. J. Physiol. 590, 63–72.

Galligan, J.J., Bertrand, P.P., 1994. ATP mediates fast synaptic potentials in enteric neurons. J. Neurosci. 14, 7563–7571.

Gonzalez-Inchauspe, C., Urbano, F.J., Di Guilmi, M.N., Uchitel, O.D., 2017. Acid-sensing ion channels activated by evoked released protons modulate synaptic transmission at the mouse calyx of held synapse. J. Neurosci. 37, 2589–2599.

Grudt, T.J., Henderson, G., 1998. Glycine and GABAA receptor-mediated synaptic transmission in rat substantia gelatinosa: inhibition by mu-opioid and GABAB agonists. J. Physiol. 507 (Pt 2), 473–483.

Grudzinska, J., Schemm, R., Haeger, S., Nicke, A., Schmalzing, G., Betz, H., et al., 2005. The β subunit determines the ligand binding properties of synaptic glycine receptors. Neuron 45, 727–739.

Hansen, K.B., Furukawa, H., Traynelis, S.F., 2010. Control of assembly and function of glutamate receptors by the amino-terminal domain. Mol. Pharmacol. 78, 535–549.

Harvey, R.J., Topf, M., Harvey, K., Rees, M.I., 2008. The genetics of hyperekplexia: more than startle!. Trends Genet. 24, 439–447.

Hollmann, M., Hartley, M., Heinemann, S., 1991. Ca2+ permeability of KA-AMPA—gated glutamate receptor channels depends on subunit composition. Science 252, 851–853.

Houtani, T., Munemoto, Y., Kase, M., Sakuma, S., Tsutsumi, T., Sugimoto, T., 2005. Cloning and expression of ligand-gated ion-channel receptor L2 in central nervous system. Biochem. Biophys. Res. Commun. 335, 277–285.

Huggins, D.J., Grant, G.H., 2005. The function of the amino terminal domain in NMDA receptor modulation. J. Mol. Graph. Model. 23, 381–388.

Iacobucci, G.J., Popescu, G.K., 2017. NMDA receptors: linking physiological output to biophysical operation. Nat. Rev. Neurosci. 18, 236–249.

Izumi, Y., Auberson, Y.P., Zorumski, C.F., 2006. Zinc modulates bidirectional hippocampal plasticity by effects on NMDA receptors. J. Neurosci. 26, 7181–7188.

Jensen, A.A., Davies, P.A., Brauner-Osborne, H., Krzywkowski, K., 2008. 3B but which 3B and that's just one of the questions: the heterogeneity of human 5-HT3 receptors. Trends Pharmacol. Sci. 29, 437–444.

Johnson, J.W., Ascher, P., 1987. Glycine potentiates the NMDA response in cultured mouse brain neurons. Nature 325, 529–531.

Johnson, J.W., Glasgow, N.G., Povysheva, N.V., 2015. Recent insights into the mode of action of memantine and ketamine. Curr. Opin. Pharmacol. 20, 54–63.

Kalappa, B.I., Anderson, C.T., Goldberg, J.M., Lippard, S.J., Tzounopoulos, T., 2015. AMPA receptor inhibition by synaptically released zinc. Proc. Natl. Acad. Sci. USA 112, 15749–15754.

Kawai, F., Sterling, P., 1999. AMPA receptor activates a G-protein that suppresses a cGMP-gated current. J. Neurosci. 19, 2954–2959.

Kellenberger, S., Schild, L., 2002. Epithelial sodium channel/degenerin family of ion channels: a variety of functions for a shared structure. Physiol. Rev. 82, 735–767.

Khakh, B.S., North, R.A., 2012. Neuromodulation by extracellular ATP and P2X receptors in the CNS. Neuron 76, 51–69.

Kleckner, N.W., Dingledine, R., 1988. Requirement for glycine in activation of NMDA-receptors expressed in Xenopus oocytes. Science 241, 835–837.

Korpi, E.R., Grunder, G., Luddens, H., 2002. Drug interactions at GABA(A) receptors. Prog. Neurobiol. 67, 113–159.

Kotak, V.C., Mirallave, A., Mowery, T.M., Sanes, D.H., 2017. GABAergic inhibition gates excitatory LTP in perirhinal cortex. Hippocampus 27, 1217–1223.

Krishtal, O.A., Osipchuk, Y.V., Shelest, T.N., Smirnoff, S.V., 1987. Rapid extracellular pH transients related to synaptic transmission in rat hippocampal slices. Brain Res. 436, 352–356.

Lendvai, B., Vizi, E.S., 2008. Nonsynaptic chemical transmission through nicotinic acetylcholine receptors. Physiol. Rev. 88, 333–349.

Li, P., Wilding, T.J., Kim, S.J., Calejesan, A.A., Huettner, J.E., Zhuo, M., 1999. Kainate-receptor-mediated sensory synaptic transmission in mammalian spinal cord. Nature 397, 161–164.

Lipton, P., 1999. Ischemic cell death in brain neurons. Physiol. Rev. 79, 1431–1568.

Lodge, D., 1997. Subtypes of glutamate receptors. In: Monaghan, D.T., Wenthold, R.J. (Eds.), The Ionotropic Glutamate Receptors. Humana Press, Totowa, NJ, pp. 1–38.

Lodge, D., 2009. The history of the pharmacology and cloning of ionotropic glutamate receptors and the development of idiosyncratic nomenclature. Neuropharmacology 56, 6–21.

Lummis, S.C., 2012. 5-HT(3) receptors. J. Biol. Chem. 287, 40239–40245.

Lynch, J.W., 2009. Native glycine receptor subtypes and their physiological roles. Neuropharmacology 56, 303–309.

Lynch, J.W., Callister, R.J., 2006. Glycine receptors: a new therapeutic target in pain pathways. Curr. Opin. Investig. Drugs 7, 48–53.

Lynch, J.W., Zhang, Y., Talwar, S., Estrada-Mondragon, A., 2017. Glycine receptor drug discovery. Adv. Pharmacol. 79, 225–253.

Miller, P.S., Smart, T.G., 2010. Binding, activation and modulation of Cys-loop receptors. Trends Pharmacol. Sci. 31, 161–174.

Monro, R.L., Bertrand, P.P., Bornstein, J.C., 2004. ATP participates in three excitatory postsynaptic potentials in the submucous plexus of the guinea pig ileum. J. Physiol. 556, 571–584.

Mott, D.D., Benveniste, M., Dingledine, R.J., 2008. pH-dependent inhibition of kainate receptors by zinc. J. Neurosci. 28, 1659–1671.

Nabavi, S., Kessels, H.W., Alfonso, S., Aow, J., Fox, R., Malinow, R., 2013. Metabotropic NMDA receptor function is required for NMDA receptor-dependent long-term depression. Proc. Natl. Acad. Sci. USA 110, 4027–4032.

Navarrete, L.C., Barrera, N.P., Huidobro-Toro, J.P., 2014. Vas deferens neuro-effector junction: from kymographic tracings to structural biology principles. Auton. Neurosci. 185, 8–28.

Nowak, L., Bregestovski, P., Ascher, P., Herbet, A., Prochiantz, A., 1984. Magnesium gates glutamate-activated channels in mouse central neurones. Nature 307, 462–465.

Obara, M., Szeliga, M., Albrecht, J., 2008. Regulation of pH in the mammalian central nervous system under normal and pathological conditions: facts and hypotheses. Neurochem. Int. 52, 905–919.

Olsen, R.W., Sieghart, W., 2008. International Union of Pharmacology. LXX. Subtypes of gamma-aminobutyric acid(A) receptors: classification on the basis of subunit composition, pharmacology, and function. Update. Pharmacol. Rev. 60, 243–260.

Reeh, P.W., Steen, K.H., 1996. Tissue acidosis in nociception and pain. Prog. Brain. Res. 113, 143–151.

Richler, E., Chaumont, S., Shigetomi, E., Sagasti, A., Khakh, B.S., 2008. Tracking transmitter-gated P2X cation channel activation in vitro and in vivo. Nat. Methods 5, 87–93.

Rivera, C., Voipio, J., Payne, J.A., Ruusuvuori, E., Lahtinen, H., Lamsa, K., et al., 1999. The K + /Cl- co-transporter KCC2 renders GABA hyperpolarizing during neuronal maturation. Nature 397, 251–255.

Rodriguez-Moreno, A., Lerma, J., 1998. Kainate receptor modulation of GABA release involves a metabotropic function. Neuron 20, 1211–1218.

Roohani, N., Hurrell, R., Kelishadi, R., Schulin, R., 2013. Zinc and its importance for human health: an integrative review. J. Res. Med. Sci. 18, 144–157.

Schlue, W.R., Dörner, R., 1992. The regulation of pH in the central nervous system. Can J Physiol Pharmacol 70 (Suppl), S278–S285.

Schuetze, S.M., Role, L.W., 1987. Developmental regulation of nicotinic acetylcholine receptors. Annu. Rev. Neurosci. 10, 403–457.

Sharonova, I.N., Vorobjev, V.S., Haas, H.L., 2000. Interaction between copper and zinc at GABA(A) receptors in acutely isolated cerebellar Purkinje cells of the rat. Br. J. Pharmacol. 130, 851–856.

Shen, J.X., Yakel, J.L., 2009. Nicotinic acetylcholine receptor-mediated calcium signaling in the nervous system. Acta Pharmacol. Sin. 30, 673–680.

Sherwood, T.W., Frey, E.N., Askwith, C.C., 2012. Structure and activity of the acid-sensing ion channels. Am. J. Physiol. Cell. Physiol. 303, C699–C710.

Sneddon, P., Westfall, D.P., Fedan, J.S., 1982. Cotransmitters in the motor nerves of the guinea pig vas deferens: electrophysiological evidence. Science 218, 693–695.

Stein, I.S., Gray, J.A., Zito, K., 2015. Non-ionotropic NMDA receptor signaling drives activity-induced dendritic spine shrinkage. J. Neurosci. 35, 12303–12308.

Storozhuk, M.V., Zholos, A.V., 2018. TRP channels as novel targets for endogenous ligands: focus on endocannabinoids and nociceptive signalling. Curr. Neuropharmacol. 16, 137–150.

Straub, C., Noam, Y., Nomura, T., Yamasaki, M., Yan, D., Fernandes, H.B., et al., 2016. Distinct subunit domains govern synaptic stability and specificity of the kainate receptor. Cell Rep. 16, 531–544.

Stroud, R.M., Finer-Moore, J., 1985. Acetylcholine receptor structure, function, and evolution. Annu. Rev. Cell. Biol. 1, 317–351.

Takago, H., Nakamura, Y., Takahashi, T., 2005. G protein-dependent presynaptic inhibition mediated by AMPA receptors at the calyx of Held. Proc. Natl. Acad. Sci. USA 102, 7368–7373.

Tecott, L.H., Maricq, A.V., Julius, D., 1993. Nervous system distribution of the serotonin 5-HT3 receptor mRNA. Proc. Natl. Acad. Sci. USA 90, 1430–1434.

Thompson, A.J., Lester, H.A., Lummis, S.C., 2010. The structural basis of function in Cys-loop receptors. Q. Rev. Biophys. 43, 449–499.

Trattnig, S.M., Gasiorek, A., Deeb, T.Z., Ortiz, E.J., Moss, S.J., Jensen, A.A., et al., 2016. Copper and protons directly activate the zinc-activated channel. Biochem. Pharmacol. 103, 109–117.

Traynelis, S.F., Wollmuth, L.P., McBain, C.J., Menniti, F.S., Vance, K.M., Ogden, K.K., et al., 2010. Glutamate receptor ion channels: structure, regulation, and function. Pharmacol. Rev. 62, 405–496.

Trombley, P.Q., Blakemore, L.J., Hill, B.J., 2011. Zinc modulation of glycine receptors. Neuroscience 186, 32–38.

Twyman, R.E., Macdonald, R.L., 1991. Kinetic properties of the glycine receptor main- and sub-conductance states of mouse spinal cord neurones in culture. J. Physiol. 435, 303–331.

Venkatachalam, K., Montell, C., 2007. TRP channels. Annu. Rev. Biochem. 76, 387–417.

Wang, Y., Small, D.L., Stanimirovic, D.B., Morley, P., Durkin, J.P., 1997. AMPA receptor-mediated regulation of a Gi-protein in cortical neurons. Nature 389, 502–504.

Weilinger, N.L., Lohman, A.W., Rakai, B.D., Ma, E.M., Bialecki, J., Maslieieva, V., et al., 2016. Metabotropic NMDA receptor signaling couples Src family kinases to pannexin-1 during excitotoxicity. Nat. Neurosci. 19, 432–442.

Westbrook, G.L., Mayer, M.L., 1987. Micromolar concentrations of $Zn^{2+}$ antagonize NMDA and GABA responses of hippocampal neurons. Nature 328, 640–643.

Yan, D., Yamasaki, M., Straub, C., Watanabe, M., Tomita, S., 2013. Homeostatic control of synaptic transmission by distinct glutamate receptors. Neuron 78, 687–699.

Zeilhofer, H.U., Wildner, H., Yevenes, G.E., 2012. Fast synaptic inhibition in spinal sensory processing and pain control. Physiol. Rev. 92, 193–235.

Zhu, S., Paoletti, P., 2015. Allosteric modulators of NMDA receptors: multiple sites and mechanisms. Curr. Opin. Pharmacol. 20, 14–23.

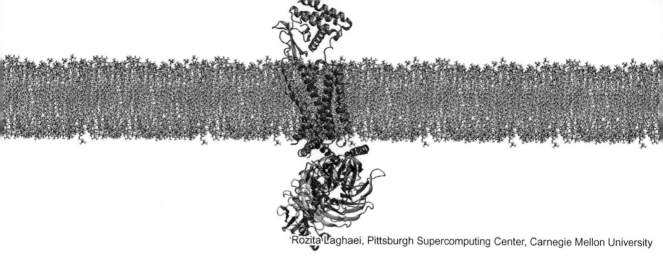

Rozita Laghaei, Pittsburgh Supercomputing Center, Carnegie Mellon University

CHAPTER 12

# Metabotropic G-Protein-Coupled Receptors and Their Cytoplasmic Signaling Pathways

As was briefly discussed in Chapter 10, metabotropic G-protein-coupled receptors link the binding of an extracellular ligand with the generation of cellular signaling cascades in the cell cytoplasm. This form of signaling requires a receptor protein on the plasma membrane to transduce the binding of a neurotransmitter to its extracellular surface (the "first signal") into a "second signal" inside the cell. The concept that the activation of receptors on the plasma membrane triggers intracellular "**second messengers**" was first pioneered by Earl Sutherland and colleagues in the 1950s. In experiments using liver tissue, Sutherland discovered that a soluble cytoplasmic molecule called **cyclic AMP** (cAMP) was generated by the binding of a ligand to a plasma membrane receptor, and that this binding triggered downstream biochemical changes in the cell (Sutherland and Rall, 1957, 1958). In recognition of his identifying a

*Liver Tissue: Cyclic AMP Experiment*

molecule that acts as a second messenger and introducing the scientific community to the concept of metabotropic signaling, Sutherland received the Nobel Prize in Physiology or Medicine in 1971 (see Box 12.1).

The cellular enzyme **adenylyl cyclase** was shown to produce cAMP, but the details of how the cAMP second messenger was generated following the stimulation of a plasma membrane receptor was only later elucidated by other researchers. Alfred Gilman and Martin Rodbell discovered that the link between metabotropic receptors and adenylyl cyclase was a GTP-binding protein made up of three different subunits (termed a "heterotrimeric G-protein"). This heterotrimeric G-protein contains one α (α) subunit, one β (β) subunit, and one γ (γ) subunit. The G-protein subunits are not transmembrane proteins, but are associated with the plasma membrane by post-translational modifications to the α and γ subunits (e.g., palmitoylation, myristoylation, isoprenylation, or farnesylation, which are modifications made to amino acids in the protein to add a hydrophobic chain which then buries in the membrane; Milligan and Kostenis, 2006).

---

## BOX 12.1

### EARL SUTHERLAND NOBEL PRIZE IN PHYSIOLOGY OR MEDICINE, 1971

The defining experiment that led to the discovery of cAMP as a soluble "second messenger" was performed using homogenized liver tissue (physically disrupted cells to break them up into small pieces). Sutherland (see Fig. 12.1) and colleagues were trying to determine how the hormone epinephrine could activate the cytoplasmic enzyme liver phosphorylase, which mediates the breakdown of glycogen into glucose during activation of the sympathetic division of the autonomic nervous system (i.e., the "fight or flight response"). Sutherland and colleagues showed that the interaction of epinephrine with the membrane fraction of the liver homogenate induced the formation of the soluble messenger cAMP. When cAMP was added to the whole liver homogenate, it could induce phosphorylation of enzymes such as liver phosphorylase, even in the absence of epinephrine (Fig. 12.2). In 1971, Earl Sutherland received the Nobel Prize in Physiology or Medicine "for his discoveries concerning the mechanisms of the action of hormones."

https://www.nobelprize.org/nobel_prizes/medicine/laureates/1971/.

FIGURE 12.1 Earl J. Sutherland, c. 1970. Source: http://resource.nlm.nih.gov/101441404.

## BOX 12.1 (cont'd)

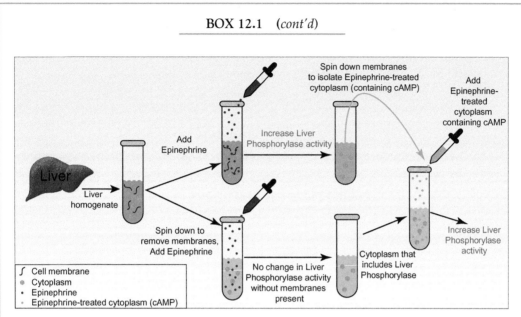

FIGURE 12.2  Diagram of the sequence of experiments that were used to show that a "second messenger" was generated by the interaction of epinephrine with cell membranes, and could then go on to activate cytoplasmic enzymes. (Moving from left to right) First, liver tissue was homogenized and put into a test tube. Then, this test tube could either be spun down to remove the cell membranes or left intact with the cell membranes. Next, epinephrine (*red dots*) was added to each sample. In the sample containing the cell membranes, adding epinephrine caused the generation of a second messenger (cAMP; *green*), which in turn caused increased activity of liver phosphorylase. In contrast, in the sample with no membranes, adding epinephrine did not generate a second messenger, and there was no change in liver phosphorylase activity. Lastly, when the epinephrine-treated cytoplasm containing the second messenger cAMP was transferred to the liver homogenate sample lacking cell membranes, liver phosphorylase activity increased.

Since the β subunit does not exist in isolation (it is coupled to γ), the γ membrane anchor holds the β-γ two-protein complex in association with the membrane.

The α subunit is the only one that binds to GTP, and once GTP is bound, the heterotrimer is activated and breaks apart to form two dimers, α-GTP and β-γ (the β-γ pair remains together; see Fig. 12.5). Rodbell and colleagues showed that when ligand binds to a metabotropic receptor, the receptor undergoes a conformational change, which causes heterotrimeric G-proteins to bind GTP and dissociate from the receptor, becoming active signaling molecules (Rodbell, 1995; Gilman, 1984). They termed these G-proteins "transducers," because activated heterotrimeric G-proteins *transduce* ligand binding to the receptor into activation of the cAMP signaling cascade. Although there are different types of G-proteins, heterotrimeric G-protein transducer activation turned out to be a conserved mechanism by which all metabotropic receptors trigger signaling cascades, and for this work, Gilman and Rodman shared the Nobel Prize in Physiology or Medicine in 1994 (see Box 12.2).

Heterotrimeric G-proteins are used as molecular switches that can be turned "on" and "off" by ligand binding to metabotropic receptors. The process of flipping this molecular switch on and off is often called the GTP cycle (see Fig. 12.5).

---

BOX 12.2

## ALFRED GILMAN AND MARTIN RODBELL NOBEL PRIZE IN PHYSIOLOGY OR MEDICINE, 1994

In the 1970s, Martin Rodbell (see Fig. 12.3) demonstrated that there were three components to metabotropic receptor signaling: a receptor, a transducer, and a generator of second messengers. In further studies of the "transducer," he determined that guanosine 5'-triphosphate (GTP) was required to bind to the transducer.

Starting in the later 1970s, Alfred Gilman (see Fig. 12.4) set out to determine the chemical nature of the "transducer" described by Rodbell. By mutating various proteins related to cell signaling, he discovered a set of GTP-binding proteins that were critical for metabotropic receptor function. These were shown to be a set of three different protein families that assembled to create heterotrimeric GTP-binding proteins (termed "G-proteins").

G-proteins were found to act as transducers of metabotropic receptor activation, using a cycle of GTP binding to the α subunits to activate the G-protein and subsequently being hydrolyzed to GDP to return the G-protein to its inactive state (see Fig. 12.5). This discovery was critical to our understanding of metabotropic receptor signaling.

https://www.nobelprize.org/nobel_prizes/medicine/laureates/1994/

FIGURE 12.3  Martin Rodbell. *Source: NIH historic image courtesy of Andrew M. Rodbell.*

FIGURE 12.4  Alfred Gilman. *Source: NIH historic image - http://profiles.nlm.nih.gov/ps/retrieve/Series/4729.*

BOX 12.2 (cont'd)

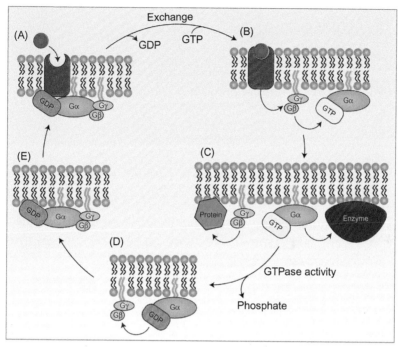

FIGURE 12.5 **The GTP cycle for G-protein-coupled metabotropic receptors.** (A) The resting metabotropic receptor (purple) with the heterotrimeric G-protein associated with the receptor, and the α subunit bound to GDP. (B) Upon ligand binding, the α-bound GDP is exchanged for GTP, the heterotrimer dissociates, and becomes unbound from the receptor. α-GTP and β-γ are now active signaling transducers. (C) α-GTP and β-γ remain membrane-bound and can associate with other proteins to initiate a signaling cascade. (D) When α-GTP is hydrolyzed to α-GDP, the active state of the G-proteins is terminated as the α-GDP binds β-γ to reform the heterotrimer (E). This heterotrimer with GDP bound to α (E) then reassociates with a receptor that is in the resting state (A). The cycle can be repeated when ligand again binds to the receptor.

When there is no ligand bound to a metabotropic receptor (the resting state), the α, β, and γ subunits of the heterotrimeric G-protein are bound together; the heterotrimer is associated with the receptor and the α subunit is bound to GDP. GDP is less abundant than GTP, but when the receptor is in the resting state, GDP has a higher affinity than GTP for the α subunit of the heterotrimeric G-protein.

When ligand binds to a metabotropic receptor, it induces a conformational change in the receptor that alters the heterotrimeric G-protein and its association with the receptor. First, the α subunit of the heterotrimeric G-protein increases its affinity for GTP, causing it to unbind GDP in exchange for GTP. The heterotrimeric G-protein then breaks apart into α-GTP and β-γ dimers, which are now both active signaling transducers (i.e., the molecular switch is now "on" and these G-proteins are able to participate in downstream signaling events). The active G-protein subunits remain associated with the membrane, but they dissociate from the receptor. This allows them to interact with other proteins (e.g., adenylyl cyclase) and initiate a biochemical signaling cascade of events (e.g., by generating the second messenger cAMP).

The G-proteins remain in the active state until the α subunit hydrolyzes the bound GTP back to GDP. Once the α subunit is bound to GDP, it is rendered inactive and now has a high affinity for β-γ subunits. When α-GDP binds to an active β-γ dimer, the heterotrimer reforms, rendering the β-γ subunits inactive. At this point, both G-protein signaling complexes (α and β-γ) have been turned "off." This heterotrimer then reassociates with a resting receptor (not bound to ligand), and the whole process can begin again.

Once it had been determined that metabotropic receptors could induce the activation of heterotrimeric G-proteins, which could in turn activate adenylyl cyclase to generate cAMP, the only part of this first historical pathway left to investigate was how cAMP changes the function of downstream proteins in the cell. A group of laboratories went on to show that cAMP binds to a regulatory subunit on a protein kinase that leads to the activation of the catalytic subunit of this kinase and the phosphorylation of proteins (Fischer, 1993; Krebs, 1993; Soderling et al., 1970). This kinase specifically phosphorylates serine or threonine amino acids within proteins. Since it is activated by cAMP, it was named the **cAMP-dependent protein kinase**, although it is more commonly abbreviated as PKA for **"protein kinase A."**

Through the historical experiments described above, researchers outlined the steps of the cAMP signaling pathway from beginning to end: ligand binding to a plasma membrane receptor, leading to the exchange of GDP for GTP on the α subunit of the heterotrimeric G-protein, causing the separation and activation of α-GTP and β-γ. Then, α-GTP activates adenylyl cyclase, leading to the generation of the second messenger cAMP, which activates cAMP-dependent protein kinase (PKA). PKA can then phosphorylate proteins in the cell.

Importantly, this signaling cascade can amplify the intracellular response to receptor activation. For each molecule of ligand that binds to a receptor, α-GTP interacting with adenylyl cyclase can lead to the generation of many copies of cAMP, cAMP molecules can activate multiple copies of PKA, and each active PKA enzyme can phosphorylate many proteins. We now know that similar mechanisms of amplification are a hallmark of all metabotropic G-protein-coupled receptor signaling systems. As such, in contrast to ionotropic signaling where a single ligand binds to a single receptor and only affects ion flux through that one receptor channel, for G-protein-coupled receptors, a single ligand–receptor binding event can lead to a large and significant change in signaling. This allows metabotropic receptors, which are not usually present in very large numbers and are not often positioned close to the sites of neurotransmitter release, to have a significant modulatory impact on synaptic transmission.

# COMMON THEMES IN RECEPTOR COUPLING TO HETEROTRIMERIC G-PROTEINS

The family of G-protein-coupled receptors includes a staggering up to 2000 different members, which are highly diverse and include receptors that can selectively be activated by small chemical molecules, peptides, large proteins, odor molecules, photons (light), lipids, and ions (Bockaert and Pin, 1999). In fact, this group includes over 1000 receptors that are sensitive to different odors and pheromones alone.

How can all of the thousands of G-protein-coupled receptors tap into a common signal transduction system—the heterotrimeric G-proteins—to activate downstream biochemical changes inside cells? This is possible because metabotropic G-protein-coupled receptors combine incredible diversity in their extracellular (ligand-binding) regions with a conservation of structure and function (G-protein binding) in their transmembrane and intracellular domains (Strader et al., 1994). Despite the fact that different G-protein-coupled receptors may have very little similarity in the amino acid sequence that makes up the protein, all of these receptors have a common **secondary structure**, which includes seven transmembrane domains connected by three extracellular and three intracellular loops (Katritch et al., 2012). Most of these receptors also have a conserved set of two cysteine residues that form a disulfide bond to assist in restricting the conformation (or shape) that these receptors can form. The conserved transmembrane structure is thought to be important in the conformational changes that are associated with activation of the receptor when a ligand binds to the extracellular portion of the protein, and the conserved intracellular cytoplasmic loop structures interact with heterotrimeric G-proteins. Thus, these receptor proteins have highly variable extracellular regions that allow for a large range of ligand specificities, coupled with a highly conserved transmembrane and intracellular domain structure that allows all of them to couple to heterotrimeric G-proteins (Katritch et al., 2012). This receptor class is one of the best examples of how evolution has created amazing molecular diversity while maintaining a conservation of function (Jacob, 1977; see Fig. 12.6).

## Families of Heterotrimeric G-Proteins

If all G-protein-coupled metabotropic receptors use heterotrimeric G-proteins to transduce ligand binding to the receptor into biochemical signaling within the cell, how does that cell receive an independent, unique signal from each receptor? In other words, if a presynaptic nerve terminal at a synapse has two different metabotropic receptors present (e.g., one that binds glutamate and another that binds acetylcholine), how do those two receptors create a signal within that nerve terminal that does not overlap, or get mixed together? One reason is that although all heterotrimeric G-protein complexes consist of one $\alpha$, one $\beta$, and one $\gamma$ subunit, they can be constructed using different family members of each subunit. There are 16 genes that code for different $\alpha$ subunit types, 5 genes that code for different $\beta$ subunits, and 14 genes that code for different $\gamma$ subunits (Milligan and Kostenis, 2006).

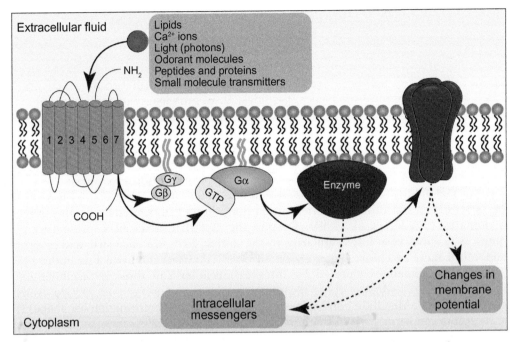

FIGURE 12.6  **Metabotropic G-protein-coupled receptors.** All G-protein-coupled receptors have a common structure that includes seven transmembrane domains, a variable extracellular region that includes a specific binding site for one or more of a wide variety of ligands, and a common intracellular region that couples to heterotrimeric G-proteins. These G-proteins, once activated by ligand binding, go on to activate downstream enzymes or ion channels.

The α subunit genes can be broken down into four major α protein families (termed αi/o, αs, αq, α12; see Fig. 12.22), with subtypes within these families being labeled, for example, αi1, αi2, etc. α, β, and γ variants can mix and match in different combinations to create a large number of possible **heterotrimers** (although not every possible combination is thought to exist in nature). Bacterial toxins that target heterotrimeric G-proteins are highlighted in Box 12.3.

How can we identify experimentally the specific G-protein heterotrimer that is coupled to each receptor? Kleuss et al. (1991, 1992, 1993) used a technique called antisense oligonucleotide knockdown to block the translation of specific α, β, or γ family member proteins, one at a time, to explore which ones were coupled to particular metabotropic receptors. What this series of experiments showed was that individual metabotropic receptors were each coupled to a different heterotrimeric combination of G-proteins (see Fig. 12.22). This result highlighted an important mechanism to prevent interactions between signaling pathways of two distinct receptors that might both be functioning in the same synapse: the use of different G-protein family members by distinct receptors.

After a ligand binds to a metabotropic receptor, both the α-GTP and βγ dimers are "active" and able to bind to other proteins and activate them. Proteins that are

> BOX 12.3
>
> ## BACTERIAL TOXINS THAT TARGET HETEROTRIMERIC G-PROTEINS
>
> Some bacterial toxins modify heterotrimeric G-proteins as part of their host–pathogen interaction. These toxins have been harnessed by scientists as selective experimental tools that can be used to dissect signaling mechanisms within neurons. The toxins listed here, which are selective for subsets of heterotrimeric G-proteins, are but a few of the many bacterial toxins that have been used as experimental tools to study a variety of biological processes (Schiavo and van der Goot, 2001).
>
> The *Pasteurella multocida* toxin causes persistent activation of $G\alpha_q$, $G\alpha 13$, and $G\alpha_i$ by deamidation. Clinically, this bacterium is the most common cause of infection from wounds caused by dog or cat bites.
>
> Cholera toxin is secreted by the bacterium *Vibrio cholera*. This toxin selectively adds a sugar that is naturally found in the body (ribose) to $G\alpha_s$, causing a persistent activation. Clinically, persistent activation of $G\alpha_s$ by cholera toxin leads to a continuous stimulation of adenylate cyclase in intestinal epithelial cells that persistently activates the cystic fibrosis transmembrane conductance regulator (CFTR), causing a dramatic efflux of ions and water that leads to diarrhea.
>
> Pertussis toxin, which is the clinical cause of whooping cough, is produced by the bacterium *Bordetella pertussis*. Pertussis toxin selectively adds a ribose sugar to the $G\alpha_{i/o}$ subfamily of heterotrimeric G-proteins. This covalent modification persistently inactivates these G-proteins.

activated after binding to either α or βγ are called **"effector proteins."** When G-protein signaling was first discovered, investigators hypothesized that only α-GTP initiated signaling, with βγ only serving to bind up the α subunits after they hydrolyzed GTP to GDP. A signaling role for βγ was first discovered through exploration of the signaling mechanism responsible for the acetylcholine-mediated vagus nerve control over the heart rate. (This was the system Otto Löwi had used to provide the earliest evidence that neurons use chemical transmitters to communicate with postsynaptic partners; see Chapter 16.) As experimenters were trying to elucidate the signal cascade used by muscarinic receptors on cardiac muscle cells that leads to the slowing of the heart rate, they came to the conclusion that βγ subunits directly bind to potassium channels to open them and hyperpolarize the muscle cell (Logothetis et al., 1987). This type of direct action of an activated G-protein onto an ion channel came to be known as the "direct pathway" because no second messenger was required and the "primary effector" was, in this case, an ion channel, which was the end of the signaling pathway (see below).

Since these early experiments, βγ subunits have been shown to activate many other primary effectors, including adenylyl cyclase (Tang and Gilman, 1991), phospholipase C (Camps et al., 1992), phosphatidylinositol 3 kinase (Hawes et al., 1996), Src tyrosine kinase

(Luttrell et al., 1997), and ion channels (e.g., voltage-gated calcium channels; Ikeda, 1996). Therefore, βγ subunits are just as likely as α-GTP subunits to activate signaling cascades after ligand binds to a metabotropic receptor (see Khan et al., 2013 for review).

## How Is Metabotropic Signaling Terminated?

Earlier we discussed how the effects of a ligand-gated ion channel receptor are terminated when the ligand unbinds from the receptor. This occurs because the conformational change that opens the receptor's ion channel pore depends on the ligand being bound. In contrast, the conformational change induced by ligand binding to a G-protein-coupled metabotropic receptor activates the G-proteins rather than opening a pore, and since activated G-proteins dissociate from the receptor and move on to interact with other proteins, ligand unbinding from the receptor has no effect on the signaling system that was initiated by ligand binding. The receptor-mediated activation of the G-proteins triggers a "**domino effect**," a signaling cascade that proceeds without interaction with or subsequent influence by the receptor. Therefore, the G-proteins that are activated by ligand binding to the receptor (α-GTP and βγ) continue to independently activate primary effectors until the α-GTP subunit hydrolyzes the GTP back to GDP.

α-GTP hydrolysis can happen in one of three ways (see Fig. 12.7). First, the α subunit is itself a slow **GTPase** and will eventually cleave its bound GTP to GDP. Second, the binding of α-GTP subunits to their primary effectors often speeds up the GTPase activity. Lastly, there are distinct cellular proteins called "regulators of G-protein signaling" (**RGS proteins**), whose sole function is to regulate the time course of G-protein signaling. RGS proteins are not primary effectors because they do not mediate any downstream effects on the cell. Instead, they speed up the intrinsic GTPase activity of the α subunit when they bind to α-GTP.

There are more than 20 members of the RGS protein family (De Vries et al., 2000). Because RGS proteins bind α-GTP but do not mediate signaling, they can compete with primary effectors for binding α-GTP, and this competitive binding limits the activation of these primary effectors. The expression of individual RGS protein family members can be limited to specific cell types and they can be restricted to targeting particular signaling complexes in those cells. RGS proteins are often post-translationally modified with membrane anchors (e.g., palmitoylation, myristoylation) and are frequently closely associated with specific signaling cascades. For example, within sensory neurons, norepinephrine inhibits one particular type of voltage-gated calcium channel (N-type; Cav2.2) using two different signaling cascades (Diversé-Pierluissi et al., 1999). Interestingly, the termination mechanisms of these two cascades work at different rates. This difference occurs because the termination mechanism for each cascade is facilitated by a different RGS protein that is selectively tethered within each signaling complex and hydrolyzes GTP at different speeds (see Fig. 12.8). Thus, by coupling particular RGS proteins with signaling complexes, the time course of each neurotransmitter-mediated effect can be regulated.

Activated G-proteins are "turned off" by two related processes. First, when α-GTP is hydrolyzed to α-GDP, this renders the α subunit inactive. Second, the newly formed α-GDP dimer quickly binds βγ, which inactivates this dimer as well, and reforms the heterotrimer (see Fig. 12.5).

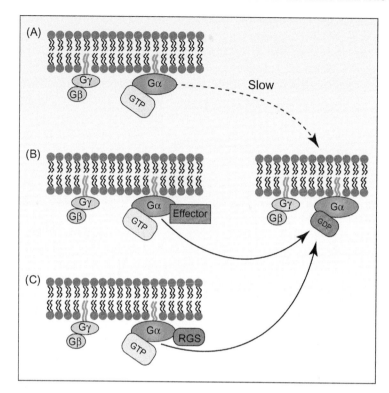

FIGURE 12.7 Three mechanisms of GTP hydrolysis that "turn off" activated α subunits. (A) α-GTP has intrinsic GTPase activity that is relatively slow, compared to the mechanisms presented in (B) and (C). (B) When α-GTP binds to effector proteins, this speeds up the intrinsic GTPase activity of the α subunit. (C) RGS proteins can bind to α-GTP with the sole purpose of speeding up GTPase activity.

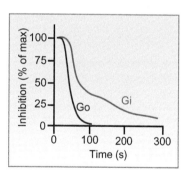

FIGURE 12.8 Different time courses of signaling determined by particular RGS family protein members coupled to distinct pathways. In this example, norepinephrine mediates inhibition of voltage-gated calcium channels within sensory neurons via two different G-proteins that initiate separate signaling cascades. The inhibition mediated by norepinephrine receptors coupled to Gαo subunits (*blue*; Go) does not last as long as the inhibition mediated by norepinephrine receptors coupled to Gαi subunits (*red*; Gi). These differences in the time course of inhibition are mediated by distinct RGS family members that are selectively coupled to each signaling cascade. (adapted from Diversé-Pierluissi et al., 1999).

# THE FOUR MOST COMMON G-PROTEIN-COUPLED SIGNALING PATHWAYS IN THE NERVOUS SYSTEM

As discussed above, G-proteins are the transducers (or molecular switches) that link ligand binding to metabotropic receptors via a signaling cascade within the cell. An advantage of the G-protein system is that it allows a small collection of biochemical signaling cascades to mediate the effects of hundreds or thousands of distinct metabotropic G-protein-coupled receptors. Activated G-proteins initiate intracellular signaling cascades by binding primary effector proteins, with different primary effectors yielding different cascades. Although there are a variety of signaling cascades that can be initiated by G-protein-coupled receptors, we will focus primarily on the four pathways that are most often used in synaptic transmission: the direct ion channel pathway, the cAMP pathway, the phosphoinositol pathway, and the arachidonic acid pathway. Metabotropic receptors can be coupled to any one of the signaling systems described in this chapter, and the identity of the signaling system attached to a particular receptor can only be determined based on the identity of the primary effector, second messenger, or secondary effector.

### Direct Ion Channel Pathway

The "direct" G-protein pathway is so named because its primary effector is an ion channel. As a result, activation of the G-protein can *directly* alter the function of an ion channel, and no further downstream signaling occurs (e.g., no second messenger is generated; see Fig. 12.9).

The direct pathway was first identified as the signaling mechanism that mediates the effects of acetylcholine on the heart. How can experimenters provide evidence that the direct pathway is used, and how can they be sure that there are no proteins or signaling molecules between the G-protein and the ion channel? The two most convincing experiments to provide evidence for this pathway are **reconstitution studies**, which show that a G-protein in isolation can alter the function of an ion channel, and **in vitro binding studies**, which identify specific binding sites for the G-protein on the ion channel.

Reconstitution experiments have often been performed using two techniques (see Fig. 12.10). First, using inside-out patches of cardiac muscle membrane that contain the

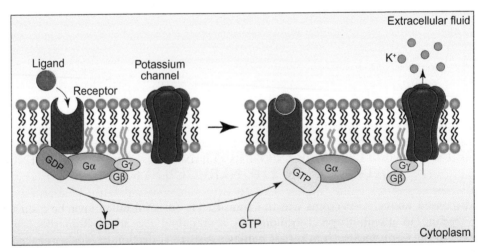

FIGURE 12.9  **The direct pathway for G-protein signaling.** In this pathway, the primary effector is an ion channel (for example a potassium channel), and there is no second messenger generated.

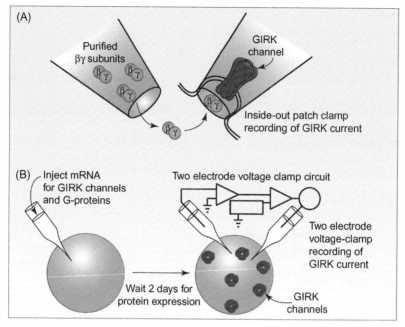

FIGURE 12.10 **Experimental methods used to provide evidence for the direct pathway of G-protein-coupled receptor signaling.** Both panels represent experiments that were performed with the diagramed elements submerged in saline (not pictured). (A) Purified activated Gβγ subunits, when applied to the inner surface of an isolated GIRK channel, can directly activate the channel. (B) Expression of activated Gβγ subunits and GIRK channels in frog oocytes results in the activation of GIRK channels.

cardiac muscarinic receptor-activated potassium channel (sometimes called the **GIRK channel**), investigators were able to perfuse purified G-protein subunits onto the inner surface of the membrane (Logothetis et al., 1987). They showed that purified βγ subunits activate the GIRK channel, while α-GTP subunits do not. Second, a **heterologous expression system** was used to experimentally induce the expression of ion channels in a type of cell that does not naturally express these proteins (in this case, a frog oocyte). Experimenters injected mRNA that codes for the GIRK channel plus mRNA for one of the candidate G-proteins (e.g., βγ subunits), waited several days for the proteins to express, and then recorded the activity of the GIRK ion channel using voltage clamp techniques (Kofuji et al., 1995). Using this method, they showed that the GIRK channel has very low activity when expressed alone, higher activity when coexpressed with βγ subunits, but very low activity when coexpressed with α subunits. The same approach can be used with other candidate ion channels and G-proteins.

To conduct in vitro binding studies, scientists incubate G-protein subunits with pieces of the ion channel of interest to determine whether they can bind together. To confirm these proteins are in fact binding together, pieces of the channel are bound to beads, mixed with the G-protein of interest, and centrifuged into a pellet at the bottom of a test tube. Then, to determine whether the G-protein subunit is bound to the ion channel component used in the incubation, a labeled antibody to the G-protein subunit is used to determine whether that G-protein subunit was precipitated (spun to the bottom of the test tube) with the piece of the channel. This experimental approach is often called a **pull-down assay**. When a pull-down assay was used to investigate whether the α-GTP or βγ subunit was capable of binding to the GIRK channel, binding sites for βγ were identified on both the C- and N-termini of the GIRK channel (Huang et al., 1995; Krapivinsky et al., 1995; Cohen et al., 1996; He et al., 2002; Ivanina et al., 2003). Interestingly, a binding site for the complete heterotrimer (Gαβγ) was also found on the N-terminus (Huang et al., 1995; Cohen et al., 1996). It is possible that this Gαβγ binding site serves to colocalize the receptor-bound heterotrimer with the GIRK channel for effective signaling after activation of the βγ subunits. The presence of binding sites for activated βγ subunits on GIRK channels is further evidence for the direct pathway of metabotropic receptor signaling (see Fig. 12.11).

Since these early studies, this pathway for direct βγ modulation of ion channels has been proposed for many signaling systems, including the inhibitory effects of norepinephrine on voltage-gated calcium channels in chicken sensory neurons (Dunlap and Fischbach, 1981; reviewed by Currie, 2010; Jewell and Currie, 2013).

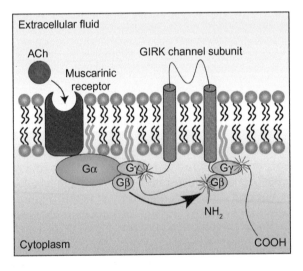

FIGURE 12.11 **Direct binding sites between G-proteins and the GIRK potassium channel.** The GIRK channel has binding sites (*red starbursts*) for the activated βγ subunit on both the C- and N-termini, which activate the channel when the βγ subunit binds. In addition, there is a binding site for the Gαβγ heterotrimer that may serve to colocalize the receptor–G-protein complex with the GIRK channel. In this diagram, only one of the four domains (two transmembrane segmentsin orange) of the GIRK channel is shown.

## cAMP Pathway

The first second messenger to be discovered, cAMP, is part of a very common signaling system in cells (see Fig. 12.12). In the cAMP pathway, the primary effector is **adenylyl cyclase**. Adenylyl cyclase is stimulated by activated Gαs subunits, but adenylyl cyclase activity can also be affected by other activated G-protein subunits (e.g., Gαi/o or βγ). PKA contains both regulatory subunits (which bind cAMP and regulate when catalytic subunits

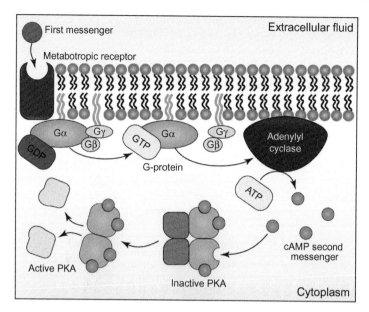

FIGURE 12.12 **The cAMP signaling pathway.** The primary effector is adenylyl cyclase (*brown*), which, when activated by a G-protein, converts ATP to cAMP (dark green spheres), which is the second messenger in this pathway. cAMP then activates cAMP-dependent protein kinase (PKA), the secondary effector for this pathway, by binding to regulatory subunits (*light green squares*) and activating catalytic subunits (*gray*).

can be active), and catalytic subunits (which actually perform the phosphorylation). cAMP binding to the regulatory subunits of PKA activates the catalytic subunits of this protein, allowing PKA to phosphorylate a large variety of proteins. The particular proteins that are phosphorylated are usually those that are nearby in the small subcompartmental space within a neuron where PKA is activated (see discussion of macromolecular complexes below). When proteins become phosphorylated, the phosphate added to a serine or threonine amino acid changes their structure and results in a change in protein function. This cAMP-mediated activation of PKA is terminated by **phosphodiesterases** within cells that cleave cAMP to AMP.

In order to determine whether a particular neurotransmitter receptor is coupled to the cAMP system, scientists can either block the primary effector (adenylyl cyclase) or the secondary effector (PKA) and determine whether the neurotransmitter action is blocked. Alternatively, drugs such as forskolin stimulate adenylyl cyclase independently of G-proteins, leading to the generation of cAMP, which can mimic the action of the neurotransmitter if the cAMP system is coupled to that neurotransmitter receptor. If a particular neurotransmitter receptor is coupled to the cAMP system, the action of forskolin on adenylyl cyclase should not only mimic the action of the neurotransmitter, but also prevent the application of neurotransmitter from mediating further activation, since forskolin maximally activates the pathway that the neurotransmitter is trying to use. These experiments can provide evidence of whether a particular neurotransmitter receptor is using the cAMP system.

## Phosphoinositol Pathway

Another common pathway that is coupled to many metabotropic receptors is the phosphoinositol (PI) pathway (see Fig. 12.13). In this pathway, activated G-proteins bind to the

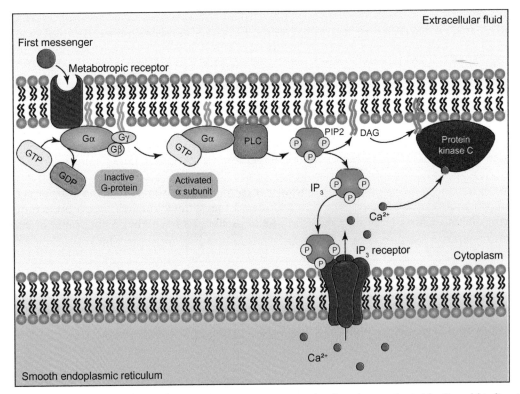

FIGURE 12.13  **The phosphoinositol (PI) pathway.** G-proteins that have been activated by ligand binding to a receptor activate the primary effector for this pathway, phospholipase C (PLC). This allows PLC to cleave the membrane lipid phosphatidylinositol 4,5-bisphosphate (PIP2) into inositol trisphosphate ($IP_3$) and diacylglycerol (DAG), which both act as second messengers in the PI pathway. $IP_3$ activates $IP_3$ receptors on the endoplasmic reticulum, which release calcium into the cytoplasm. DAG binds to and activates the kinase Protein Kinase C.

primary effector, an enzyme called **phospholipase C** (PLC). PLC is anchored in the membrane, and when activated by G-proteins it cleaves a specific type of membrane lipid called **phosphatidylinositol 4,5-bisphosphate** (PIP2). When PLC cleaves PIP2, the two molecules that result, **inositol trisphosphate** ($IP_3$) and **diacylglycerol (DAG)**, are both considered second messengers.

$IP_3$ is a ligand for the $IP_3$ receptor, which is located on the endoplasmic reticulum (see Chapter 7). [We usually think of the endoplasmic reticulum as being localized in the cell body, but it is now clear that the endoplasmic reticulum actually exists in both presynaptic nerve terminals and postsynaptic dendritic spines in the nervous system where it is involved in synaptic function and plasticity (de Juan-Sanz et al., 2017; Segal and Korkotian, 2014; Ramirez and Couve, 2011).] The $IP_3$ receptor is an intracellular ligand-gated ion channel that is selective for calcium ions, and activating it leads to the release of calcium ions that are stored in the endoplasmic reticulum. The calcium ions released into the cytoplasm can then serve as another messenger to initiate cellular events.

The other second messenger, DAG, binds to and activates a kinase called **protein kinase C** (PKC). The calcium that is released from the endoplasmic reticulum following $IP_3$ receptor activation also promotes the activation of PKC. PKC is considered the secondary effector for this PI pathway because it is activated by the second messengers.

In order to determine experimentally whether a particular neurotransmitter receptor is coupled to the PI signaling pathway a researcher can either block the primary effector (in this case, PLC) or the secondary effector (PKC) and measure whether neurotransmitter action is blocked. Alternatively, it is possible to target one of the second messenger actions by administering 1-oleoyl-2-acetyl-sn-glycerol (OAG), a membrane-permeable analog of DAG that activates PKC. If the PI pathway is coupled to a neurotransmitter receptor, application of OAG should both mimic the action of the neurotransmitter that activates that receptor and prevent the application of neurotransmitter from mediating further activation, since OAG has already maximally activated the pathway that the neurotransmitter is trying to use. These experiments can provide evidence of whether a neurotransmitter receptor uses the PI pathway.

## Arachidonic Acid Pathway

In the PI pathway described above, PLC cleaves PIP2 lipids to create two messengers ($IP_3$ and DAG). In the arachidonic acid pathway, that same lipid (PIP2) is cleaved by a different lipase called **phospholipase A2 (PLA2)**. When PIP2 is cleaved by PLA2 rather than PLC, the cleavage occurs at a different location and thus creates a different second messenger called **arachidonic acid** (see Fig. 12.14).

Historically, there has been some debate as to which signaling pathways can activate phospholipase A2 (PLA2). Some reports provide evidence that the G-proteins can directly activate

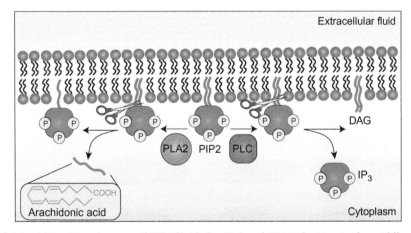

FIGURE 12.14 **Alternative cleavage of PIP2 lipids by PLC and PLA2.** Starting in the middle of this figure, PIP2 can be cleaved (diagramed by scissors) by two different enzymes to generate different second messengers. PLA2 cleaves PIP2 in one location to generate arachidonic acid, while PLC cleaves PIP2 in a different location to generate $IP_3$ and DAG.

PLA2 after metabotropic receptor activation (Axelrod, 1995; Burch, 1989). Axelrod and colleagues first used a thyroid cell line to show that norepinephrine could induce both PLC and PLA2 activity, but that each was activated by a different G-protein-coupled pathway. In particular, they showed that norepinephrine-induced arachidonic acid generation by PLA2, but not norepinephrine-induced $IP_3$ generation by PLC, was blocked by pertussis toxin (a selective inhibitor of Gαi/o family G-proteins). These results were evidence that PLA2 is a primary effector for norepinephrine receptors in thyroid cells.

In addition to arachidonic acid acting as a second messenger, it may also be acted upon by any of several enzymes (lipoxygenases or oxygenases). These enzymes convert arachidonic acid into active metabolites, which can act as soluble messengers that affect protein function. For example, the 12-HPETE metabolite of arachidonic acid has been shown to mediate the action of a peptide neurotransmitter and increase the probability that potassium channels open in sensory neurons of the sea slug (Buttner et al., 1989; see Fig. 12.15). An example of a drug that affects the synthesis of one of these metabolites is described in Box 12.4.

In addition to the potential for direct activation of membrane-bound PLA2 by G-proteins, a cytoplasmic form of PLA2 can be activated by phosphorylation and induced by elevated cytosolic calcium concentrations to translocate to membranes where it can cleave PIP2 (see Fig. 12.16). Activation of PKC can be the result of another signaling pathway interacting with the arachidonic acid pathway. For example, the PI pathway described in the previous section leads to the activation of PKC, which can activate another kinase called **mitogen-activated protein kinase (MAP kinase or MAPK)**. MAP kinase can then phosphorylate cytoplasmic PLA2, which, when combined with elevated

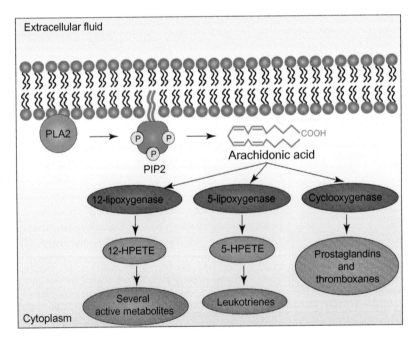

FIGURE 12.15 **The G-protein-coupled arachidonic acid pathway.** After PLA2 cleaves PIP2 to generate arachidonic acid, this second messenger can be metabolized (acted upon by other enzymes; lipoxygenases or cyclooxygenases) to create a variety of downstream metabolites that are themselves active messengers.

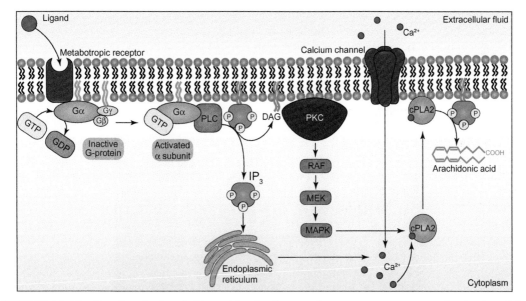

FIGURE 12.16  **Cytoplasmic phospholipase A2 activation as a downstream consequence of activation of the PI pathway.** The PI pathway outlined in Fig. 12.13 leads to activation of PKC. PKC can phosphorylate RAF, triggering a sequence of events that can activate MAPK. MAPK can phosphorylate cytoplasmic PLA2 which leads to the generation of arachidonic acid. This sequence of events illustrates how arachidonic acid generation can be a downstream consequence of the PI pathway.

---

### BOX 12.4

### DRUGS FROM NATURE: WILLOW TREES AND PROSTAGLANDIN HORMONES

#### *Salix* (Willow Tree)

The benefits of extracts from *Salix* (see Fig. 12.17) have been used medicinally in Assyrian, Babylonian, and Ancient Egyptian cultures, as long ago as 3500–2000 BCE, but it was nearly 4000 years later that the beneficial constituent of *Salix* was discovered. Willow trees contain the molecule salicin (named after the *Salix* willow tree), which is broken down in the body into salicylic acid. Scientists in the 1890s decided to acetylate the salicylic acid to develop a pain-relieving chemical with stronger effects (creating acetylsalicylic acid), and in 1899 marketed the drug under the name Aspirin. However, it wasn't until the 1970s

## BOX 12.4 *(cont'd)*

FIGURE 12.17  *Salix, the willow tree. Source: By Alvesgaspar (CC BY 2.5).*

that scientists discovered that salicylic and acetylsalicylic acids are inhibitors of prostaglandin synthesis (with acetylsalicylic acid causing a stronger depression of prostaglandin expression compared to salicylic acid). Prostaglandins are hormones synthesized from arachidonic acid and use paracrine signaling to transmit pain information to the brain, among other diverse roles in the body. Aspirin is one of the most widely used medicines in the world, and is on the World Health Organization's List of Essential Medicines for use in management of pain, fever, and inflammation, and as a treatment for heart dysfunction and cancer prevention. Despite these benefits, sick children (especially those suffering from influenza or chicken pox viruses) who are given aspirin are susceptible to developing Reye's syndrome, a rare and serious disorder featuring symptoms of encephalitis and fatty liver.

cytosolic calcium, initiates the arachidonic acid pathway (Leslie, 1997). There is also evidence that PLA2 can be activated by downstream pathways other than the PI pathway (e.g., Rho-mediated p38 activation; Kurrasch-Orbaugh et al., 2003). Therefore, investigations of distinct signaling pathways that can be activated by G-protein-coupled metabotropic receptors have come together to reveal that (1) these pathways can interact, and (2) there are multiple kinases in neurons that can also affect signaling (see discussion of other pathways below).

## OTHER G-PROTEIN-COUPLED SIGNALING PATHWAYS IN THE NERVOUS SYSTEM

In addition to the four most common metabotropic G-protein-coupled receptor signaling pathways (the direct ion pathway, the cAMP pathway, the PI pathway, and the arachidonic acid pathway), there are less common pathways that are also important in neuronal signaling. Some of these pathways are outlined below.

### Phosphoinositide 3 Kinase (PI3K) Pathway

PI3K is named for its ability to phosphorylate the third hydroxyl group of the inositol ring within the PIP2 lipid of the plasma membrane (the same lipid that PLC and PLA2 cleave; see above). When PIP2 is phosphorylated, it becomes PIP3, which recruits a molecule called **Akt** (also called protein kinase B) to the plasma membrane to be activated by phosphorylation. Activated Akt dissociates from the plasma membrane and then phosphorylates a number of cytoplasmic and nuclear proteins, which can alter neuronal function (Manning and Cantley, 2007; see Fig. 12.18).

The PI3K pathway is commonly used during the development of the nervous system (e.g., in mechanisms that support cell survival, growth, and proliferation). For example, many growth factor receptors and some metabotropic receptors have been shown to mediate actions through PI3K (Cantley, 2002; Brazil et al., 2004). Independent of its functions in nervous system development, PI3K is also important in synaptic plasticity (see Chapter 14). In particular, PI3K has been implicated in long-term potentiation (LTP) in the hippocampus (Opazo et al., 2003; Horwood et al., 2006) and fear conditioning in the amygdala (Sui et al., 2008). These functions are mediated by Akt effects on CREB, mTOR, and GSK-3, which are downstream effectors of the PI3K/Akt pathway that can alter protein synthesis by regulating gene expression (Tang et al., 2002; Cammalleri et al., 2003).

### MAP Kinase and Rho Pathways

The MAP kinase (MAPK) and Rho pathways are each initiated by a different small GTPase: "Ras" in the MAP kinase pathway, and "Rho" in the Rho pathway. Ras and Rho are inactive when bound to GDP and active when bound to GTP. The active GTP-bound state is regulated by guanine nucleotide exchange factor (GEF) proteins, which activate Ras and Rho signaling proteins, and by GTPase-activating proteins (GAPs), which induce the inactivation of Ras and Rho by accelerating their intrinsic GTPase activity.

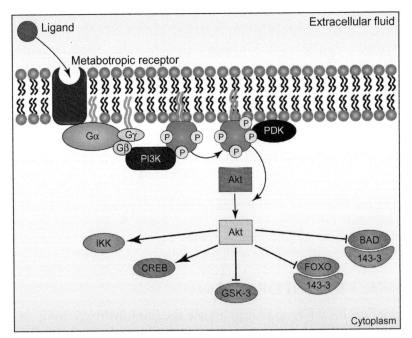

FIGURE 12.18  **The PI3K pathway.** Some metabotropic receptors can activate PI3 kinase (PI3K) which can convert PIP2 to PIP3. PIP3 and phosphoinositide-dependent kinase (PDK), can activate a protein called Akt which in turn can activate (arrows) or inhibit (T-shaped endings) a variety of downstream regulators of cell growth, protein synthesis, survival, and metabolism.

The MAP kinase pathway is named after MAPK's ability to activate a nuclear transcription factor called mitogen-activated protein (MAP). The MAPKs themselves are downstream secondary effectors (e.g. MEK and ELK) in the pathway that is initiated by receptor activation of Ras GTPases (see Fig. 12.19). It is well known that Ras is activated by Ras GEFs recruited by growth factor receptors (a subcategory of tyrosine kinase receptors), but metabotropic G-protein-coupled receptors also appear to be able to directly activate Ras (perhaps by binding of activated $\beta\gamma$ subunits; Gutkind, 1998). Ras can also be activated as a downstream consequence of the PI pathway (Blaukat et al., 2000) or the cAMP pathway (Bourne et al., 1990).

The Rho GEFs that activate Rho are known to be activated by $\alpha$ or $\beta\gamma$ subunits of a variety of G-protein families (Seasholtz et al., 1999). Once activated, Rho can go on to activate a variety of effector proteins, including Rho kinase, which can phosphorylate cytoskeletal proteins affecting actin polymerization, cell migration, and adhesion (see Fig. 12.20). Other effectors can modulate membrane trafficking, microtubule stability, and cell division (Schwartz, 2004).

## The Src Pathway

Sarcoma (Src) family kinases are nonreceptor kinases that phosphorylate tyrosine residues. Src kinases are activated when they **autophosphorylate** themselves. This activation

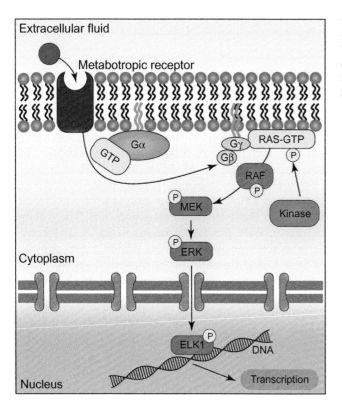

FIGURE 12.19 **The MAP kinase pathway.** Receptor activation can trigger β-γ binding to Ras, which can trigger a cascade of signaling molecules that enter the nucleus and result in the regulation of gene transcription.

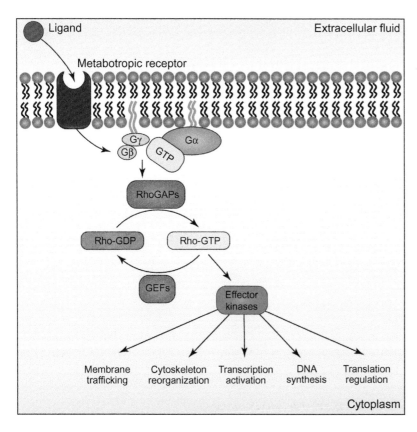

FIGURE 12.20 **The Rho pathway.** Like Ras, the Rho GTPase protein is activated by GEFs that induce Rho to bind GTP. When GTP-bound, Rho activates effector kinases (such as Rho kinase) that go on to phosphorylate a variety of proteins, including transcription factors.

can be induced by growth factor tyrosine kinase receptors, but there is also evidence that they can be activated by metabotropic G-protein-coupled receptors (see Fig. 12.21). Early evidence for a direct role of G-protein-coupled receptors in the activation of Src kinases came when scientists discovered that G-protein-coupled receptors could mediate the tyrosine phosphorylation of effector proteins (Thomas and Brugge, 1997). Chen et al. (1994) were perhaps the first to show that G-protein-coupled metabotropic receptors could directly activate Src, when they showed that this pathway linked thrombin receptors in platelets with Src. To carefully examine their hypothesis of direct activation of Src by G-proteins, Ma et al. (2000) carried out an in vitro assay using purified G-proteins and Src. They first expressed G-proteins by transforming DNA for each specific G-protein subunit into bacteria and purifying these proteins from the bacteria after expression. Then, they purified Src from cells that naturally express it. With purified proteins in hand, they combined each purified G-protein in a test tube with purified Src, and added a Src peptide substrate (to measure kinase activity of Src). Using this approach, they showed that Gαs and Gαi proteins, but not Gαq, Gα12, or Gβγ proteins, could directly stimulate the kinase activity of Src. These results showed that Src kinases could be primary effectors of Gαs and Gαi proteins (Ma et al., 2000; Luttrell and Luttrell, 2004).

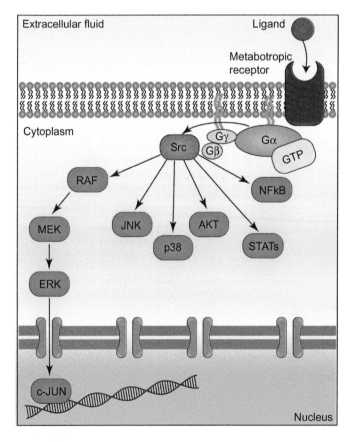

FIGURE 12.21 **The Src kinase pathway.** Some metabotropic receptors are thought to be able to activate Src, which in turn triggers a downstream cascade of events that can modulate gene expression.

# SPECIFICITY OF COUPLING BETWEEN RECEPTORS AND G-PROTEIN-COUPLED SIGNALING CASCADES

In our discussion of the major G-protein-coupled receptor signaling pathways above, we have presented examples of particular neurotransmitters acting on their metabotropic receptors to activate particular signaling cascades. However, it is important to note that metabotropic receptors selective for particular neurotransmitter ligands can potentially be coupled to a wide variety of signaling cascades (see Fig. 12.22). For example, epinephrine receptors are not always coupled to the cAMP pathway, and acetylcholine receptors are not always coupled to the direct pathway. Along these lines, as we have already discussed, particular metabotropic receptors can be coupled to a variety of G-protein family members (see Fig. 12.22). Therefore, one cannot use the type of metabotropic receptor (e.g., muscarinic receptors) to predict the specific G-protein subunits that make up the heterotrimeric G-protein coupled to that receptor, or to predict which intracellular signaling pathway is initiated by activation of that receptor.

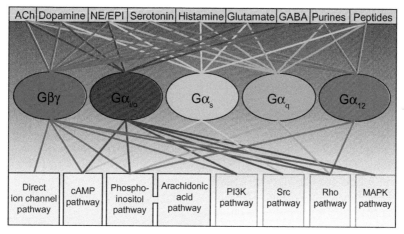

FIGURE 12.22 **Diversity in the coupling of neurotransmitter receptors with subtypes of activated G-proteins and with intracellular signaling cascades.** This diagram represents some of the known connections between neurotransmitter receptors (top), specific G-proteins (middle), and intracellular signaling cascades (bottom; bright yellow pathways are the most common ones coupled to G-protein-coupled receptors). Lines are colored to match each major family of G-proteins and connect them to receptors and cascades. This figure is intended to make the point that each of the three generic elements in G-protein-coupled receptor signaling pathways (receptor, G-protein, and signaling cascade) can be mixed and matched within particular cells to achieve a great diversity of signaling systems within neurons.

## Mechanisms That Prevent Unintended Crosstalk Between Receptor-Mediated Signaling Systems

It should be evident from this chapter that there is a large variety of receptors, G-proteins, primary effectors, second messengers, and signaling kinases that can be mixed and matched to create very specific effects within neurons. It is important to remember that metabotropic G-protein-coupled receptors have a highly conserved receptor structure that allows all of them to use heterotrimeric G-proteins as signaling switches to trigger biochemical cascades within cells. Furthermore, because these G-proteins can activate a large variety of signaling cascades within cells, and some of these cascades can interact with one another, G-protein-coupled receptor activation can mediate a broad array of potential biochemical signals.

One potential complication to using a conserved signaling switch, like the heterotrimeric G-proteins, is that receptors must avoid mixing their signals in an uncontrolled way. How does a neuron that expresses multiple types of metabotropic receptors distinguish the signals received from each type of receptor? For example, how might it distinguish metabotropic acetylcholine signals from metabotropic dopamine signals? It turns out that there are at least three major mechanisms by which neurons distinguish between different metabotropic receptor-triggered signaling cascades.

First, each type of metabotropic receptor in a neuron is coupled to one or a small subset of heterotrimeric G-protein types ($\alpha$, $\beta$, and $\gamma$ family members). As discussed above, there are many family members for each heterotrimeric G-protein, and coupling each metabotropic receptor type to a different subset can prevent unintended crosstalk between receptor signaling systems. For example, if a metabotropic acetylcholine receptor uses G$\alpha$i and a dopamine receptor in the same cell uses G$\alpha$q, these receptors can avoid activating one another's signal transducers.

Second, as was also discussed above, each metabotropic receptor can be coupled to a different signaling cascade. For example, if a metabotropic acetylcholine receptor uses the PI pathway, and a dopamine receptor in the same cell couples to the cAMP pathway, this will avoid confusion between their signals.

Third, it is common for signaling cascade effectors and messengers triggered by a particular receptor to be confined within a small subcellular compartment so that their signaling is restricted in space within the cytoplasm. Furthermore, the proteins within a signaling cascade that are coupled to a particular receptor are often tethered together in a macromolecular complex (see Fig. 12.23). These spatial restrictions prevent interactions between different receptors in a single neuron.

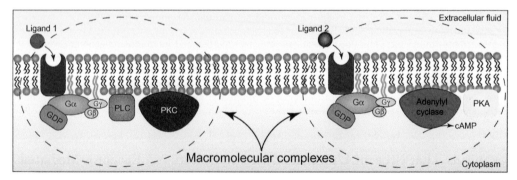

FIGURE 12.23 **The proteins that constitute particular G-protein-coupled receptor signaling cascades are often coupled together within different compartments of the neuron.** These coupled proteins form a macromolecular complex that signals within these restricted regions of the cell, which can prevent interactions between the downstream pathways of particular G-protein-coupled receptors.

These three mechanisms for segregation of signals from each receptor on one neuron allow neurons to use a large number of receptors and signaling systems so they can be selectively sensitive to a wide diversity of communicating events.

# References

Axelrod, J., 1995. Phospholipase A2 and G proteins. Trends Neurosci. 18, 64–65.

Blaukat, A., Barac, A., Cross, M.J., Offermanns, S., Dikic, I., 2000. G protein-coupled receptor-mediated mitogen-activated protein kinase activation through cooperation of Gα(q) and Gα(i) signals. Mol. Cell. Biol. 20, 6837–6848.

Bockaert, J., Pin, J.P., 1999. Molecular tinkering of G protein-coupled receptors: an evolutionary success. EMBO J. 18, 1723–1729.

Bourne, H.R., Sanders, D.A., McCormick, F., 1990. The GTPase superfamily: a conserved switch for diverse cell functions. Nature 348, 125–132.

Brazil, D.P., Yang, Z.Z., Hemmings, B.A., 2004. Advances in protein kinase B signalling: AKTion on multiple fronts. Trends. Biochem. Sci. 29, 233–242.

Burch, R.M., 1989. G protein regulation of phospholipase A2. Mol. Neurobiol. 3, 155–171.

Buttner, N., Siegelbaum, S.A., Volterra, A., 1989. Direct modulation of Aplysia S-K+ channels by a 12-lipoxygenase metabolite of arachidonic acid. Nature 342, 553–555.

Cammalleri, M., Lütjens, R., Berton, F., King, A.R., Simpson, C., Francesconi, W., et al., 2003. Time-restricted role for dendritic activation of the mTOR-p70S6K pathway in the induction of late-phase long-term potentiation in the CA1. Proc. Natl. Acad. Sci. USA 100, 14368–14373.

Camps, M., Carozzi, A., Schnabel, P., Scheer, A., Parker, P.J., Gierschik, P., 1992. Isozyme-selective stimulation of phospholipase C-β 2 by G protein β γ-subunits. Nature 360, 684–686.

Cantley, L.C., 2002. The phosphoinositide 3-kinase pathway. Science 296, 1655–1657.

Chen, Y.H., Pouysségur, J., Courtneidge, S.A., Van Obberghen-Schilling, E., 1994. Activation of Src family kinase activity by the G protein-coupled thrombin receptor in growth-responsive fibroblasts. J. Biol. Chem. 269, 27372–27377.

Cohen, N.A., Sha, Q., Makhina, E.N., Lopatin, A.N., Linder, M.E., Snyder, S.H., et al., 1996. Inhibition of an inward rectifier potassium channel (Kir2.3) by G-protein βγ subunits. J. Biol. Chem. 271, 32301–32305.

Currie, K.P., 2010. G protein modulation of CaV2 voltage-gated calcium channels. Channels (Austin) 4, 497–509.

de Juan-Sanz, J., Holt, G.T., Schreiter, E.R., de Juan, F., Kim, D.S., Ryan, T.A., 2017. Axonal endoplasmic reticulum Ca2+ content controls release probability in CNS nerve terminals. Neuron 93, 867–881. e866.

De Vries, L., Zheng, B., Fischer, T., Elenko, E., Farquhar, M.G., 2000. The regulator of G protein signaling family. Annu. Rev. Pharmacol. Toxicol. 40, 235–271.

Diversé-Pierluissi, M.A., Fischer, T., Jordan, J.D., Schiff, M., Ortiz, D.F., Farquhar, M.G., et al., 1999. Regulators of G protein signaling proteins as determinants of the rate of desensitization of presynaptic calcium channels. J. Biol. Chem. 274, 14490–14494.

Dunlap, K., Fischbach, G.D., 1981. Neurotransmitters decrease the calcium conductance activated by depolarization of embryonic chick sensory neurones. J. Physiol. 317, 519–535.

Fischer, E.H., 1993. Protein phosphorylation and cellular regulation II (Nobel lecture). Angew. Chem. Int. Ed. Engl. 32, 1130–1137.

Gilman, A.G., 1984. Guanine nucleotide-binding regulatory proteins and dual control of adenylate cyclase. J. Clin. Invest. 73, 1–4.

Gutkind, J.S., 1998. The pathways connecting G protein-coupled receptors to the nucleus through divergent mitogen-activated protein kinase cascades. J. Biol. Chem. 273, 1839–1842.

Hawes, B.E., Luttrell, L.M., van Biesen, T., Lefkowitz, R.J., 1996. Phosphatidylinositol 3-kinase is an early intermediate in the G β γ-mediated mitogen-activated protein kinase signaling pathway. J. Biol. Chem. 271, 12133–12136.

He, C., Yan, X., Zhang, H., Mirshahi, T., Jin, T., Huang, A., et al., 2002. Identification of critical residues controlling G protein-gated inwardly rectifying K(+) channel activity through interactions with the β γ subunits of G proteins. J. Biol. Chem. 277, 6088–6096.

Horwood, J.M., Dufour, F., Laroche, S., Davis, S., 2006. Signalling mechanisms mediated by the phosphoinositide 3-kinase/Akt cascade in synaptic plasticity and memory in the rat. Eur. J. Neurosci. 23, 3375–3384.

Huang, C.L., Slesinger, P.A., Casey, P.J., Jan, Y.N., Jan, L.Y., 1995. Evidence that direct binding of G β γ to the GIRK1 G protein-gated inwardly rectifying K+ channel is important for channel activation. Neuron 15, 1133–1143.

Ikeda, S.R., 1996. Voltage-dependent modulation of N-type calcium channels by G-protein beta gamma subunits. Nature 380 (6571), 255–258.

Ivanina, T., Rishal, I., Varon, D., Mullner, C., Frohnwieser-Steinecke, B., Schreibmayer, W., et al., 2003. Mapping the Gβγ-binding sites in GIRK1 and GIRK2 subunits of the G protein-activated K+ channel. J. Biol. Chem. 278, 29174–29183.

Jacob, F., 1977. Evolution and tinkering. Science 196, 1161–1166.

Jewell, M.L., Currie, K.P.M., 2013. Control of CaV2 calcium channels and neurosecretion by heterotrimeric G protein coupled receptors. In: Stephens, G., Mochida, S. (Eds.), Modulation of Presynaptic Calcium Channels. Springer Netherlands, Dordrecht, pp. 101–130.

Katritch, V., Cherezov, V., Stevens, R.C., 2012. Diversity and modularity of G protein-coupled receptor structures. Trends Pharmacol. Sci. 33, 17–27.

Khan, S.M., Sleno, R., Gora, S., Zylbergold, P., Laverdure, J.P., Labbe, J.C., et al., 2013. The expanding roles of Gβγ subunits in G protein-coupled receptor signaling and drug action. Pharmacol. Rev. 65, 545–577.

Kleuss, C., Hescheler, J., Ewel, C., Rosenthal, W., Schultz, G., Wittig, B., 1991. Assignment of G-protein subtypes to specific receptors inducing inhibition of calcium currents. Nature 353, 43–48.

Kleuss, C., Scherubl, H., Hescheler, J., Schultz, G., Wittig, B., 1992. Different β-subunits determine G-protein interaction with transmembrane receptors. Nature 358, 424–426.

Kleuss, C., Scherubl, H., Hescheler, J., Schultz, G., Wittig, B., 1993. Selectivity in signal transduction determined by γ subunits of heterotrimeric G proteins. Science 259, 832–834.

Kofuji, P., Davidson, N., Lester, H.A., 1995. Evidence that neuronal G-protein-gated inwardly rectifying K+ channels are activated by G β γ subunits and function as heteromultimers. Proc. Natl. Acad. Sci. USA 92, 6542–6546.

Krapivinsky, G., Krapivinsky, L., Wickman, K., Clapham, D.E., 1995. G β γ binds directly to the G protein-gated K+ channel, IKACh. J. Biol. Chem. 270, 29059–29062.

Krebs, E.G., 1993. Nobel Lecture. Protein phosphorylation and cellular regulation I. Biosci. Rep. 13, 127–142.

Kurrasch-Orbaugh, D.M., Parrish, J.C., Watts, V.J., Nichols, D.E., 2003. A complex signaling cascade links the serotonin2A receptor to phospholipase A2 activation: the involvement of MAP kinases. J. Neurochem. 86, 980–991.

Leslie, C.C., 1997. Properties and regulation of cytosolic phospholipase A2. J. Biol. Chem. 272 (27), 16709–16712.

Logothetis, D.E., Kurachi, Y., Galper, J., Neer, E.J., Clapham, D.E., 1987. The β γ subunits of GTP-binding proteins activate the muscarinic K+ channel in heart. Nature 325, 321–326.

Luttrell, D.K., Luttrell, L.M., 2004. Not so strange bedfellows: G-protein-coupled receptors and Src family kinases. Oncogene 23, 7969–7978.

Luttrell, L.M., Della Rocca, G.J., van Biesen, T., Luttrell, D.K., Lefkowitz, R.J., 1997. Gβγ subunits mediate Src-dependent phosphorylation of the epidermal growth factor receptor. A scaffold for G protein-coupled receptor-mediated Ras activation. J. Biol. Chem. 272, 4637–4644.

Ma, Y.C., Huang, J., Ali, S., Lowry, W., Huang, X.Y., 2000. Src tyrosine kinase is a novel direct effector of G proteins. Cell 102, 635–646.

Manning, B.D., Cantley, L.C., 2007. AKT/PKB signaling: navigating downstream. Cell 129, 1261–1274.

Milligan, G., Kostenis, E., 2006. Heterotrimeric G-proteins: a short history. Br. J. Pharmacol. 147 (Suppl 1), S46–S55.

Opazo, P., Watabe, A.M., Grant, S.G., O'Dell, T.J., 2003. Phosphatidylinositol 3-kinase regulates the induction of long-term potentiation through extracellular signal-related kinase-independent mechanisms. J. Neurosci. 23, 3679–3688.

Ramirez, O.A., Couve, A., 2011. The endoplasmic reticulum and protein trafficking in dendrites and axons. Trends. Cell Biol. 21, 219–227.

Rodbell, M., 1995. Nobel Lecture. Signal transduction: evolution of an idea. Biosci. Rep. 15, 117–133.

Schiavo, G., van der Goot, F.G., 2001. The bacterial toxin toolkit. Nat. Rev. Mol. Cell Biol. 2, 530–537.

Schwartz, M., 2004. Rho signalling at a glance. J. Cell. Sci. 117, 5457–5458.

Seasholtz, T.M., Majumdar, M., Brown, J.H., 1999. Rho as a mediator of G protein-coupled receptor signaling. Mol. Pharmacol. 55, 949–956.

Segal, M., Korkotian, E., 2014. Endoplasmic reticulum calcium stores in dendritic spines. Front. Neuroanat. 8, 64.

Soderling, T.R., Hickenbottom, J.P., Reimann, E.M., Hunkeler, F.L., Walsh, D.A., Krebs, E.G., 1970. Inactivation of glycogen synthetase and activation of phosphorylase kinase by muscle adenosine 3′,5′-monophosphate-dependent protein kinases. J. Biol. Chem. 245, 6317–6328.

Strader, C.D., Fong, T.M., Tota, M.R., Underwood, D., Dixon, R.A., 1994. Structure and function of G protein-coupled receptors. Annu. Rev. Biochem. 63, 101–132.

Sui, L., Wang, J., Li, B.M., 2008. Role of the phosphoinositide 3-kinase-Akt-mammalian target of the rapamycin signaling pathway in long-term potentiation and trace fear conditioning memory in rat medial prefrontal cortex. Learn. Mem. 15, 762–776.

Sutherland, E.W., Rall, T.W., 1957. The properties of an adenine ribonucleotide produced with cellular particles, ATP, $Mg++$, and epinephrine or glucagon. J. Am. Chem. Soc. 79, 3608.

Sutherland, E.W., Rall, T.W., 1958. Fractionation and characterization of a cyclic adenine ribonucleotide formed by tissue particles. J. Biol. Chem. 232, 1077–1091.

Tang, W.J., Gilman, A.G., 1991. Type-specific regulation of adenylyl cyclase by G protein $\beta\gamma$ subunits. Science 254, 1500–1503.

Tang, S.J., Reis, G., Kang, H., Gingras, A.C., Sonenberg, N., Schuman, E.M., 2002. A rapamycin-sensitive signaling pathway contributes to long-term synaptic plasticity in the hippocampus. Proc. Natl. Acad. Sci. USA 99, 467–472.

Thomas, S.M., Brugge, J.S., 1997. Cellular functions regulated by Src family kinases. Annu. Rev. Cell Dev. Biol. 13, 513–609.

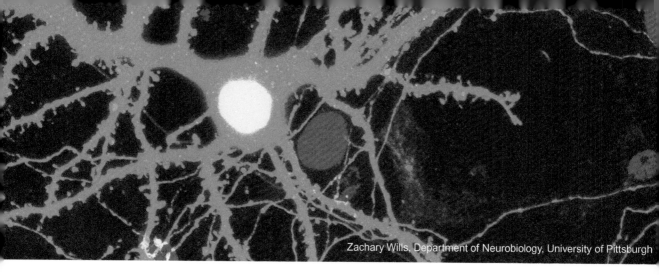

Zachary Wills, Department of Neurobiology, University of Pittsburgh

# CHAPTER 13

# Synaptic Integration Within Postsynaptic Neurons

When considering the issues that govern synaptic integration, it is important to keep in mind the details of neuronal anatomy. Neurons are structurally segregated cells with specialized subcellular compartments, including the soma, dendrites, axon, and nerve terminal that are each specialized for particular functions. For example, neurons receive most synaptic input at the dendrites and soma (Fig. 13.1).

Most neurons receive hundreds or thousands of synaptic inputs and thus need to possess a relatively large amount of membrane surface area to receive these inputs. Therefore, dendrites have extensive, branched neuronal processes (often called a **dendritic tree**) that provide surface area for thousands of nerve terminal contact sites. These elaborate processes of the neuron were first recognized by Santiago Ramón y Cajal in sections of central nervous system tissue stained using the Golgi method (Fig. 13.2). The role of the dendrites was not explored until the 1950s, when experimentalists began to hypothesize how inputs to specific locations within dendrites would affect the membrane potential in the cell body.

Neurons sum all of the inputs they receive at any one point in time to determine if the membrane potential reaches the threshold to fire an action potential. Because the axon hillock (the base of the axon, near the cell body) contains the neuron's highest concentration

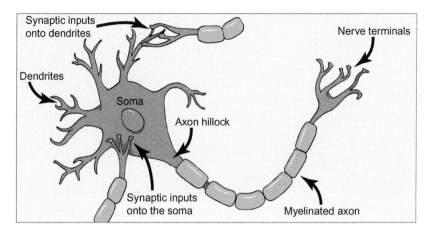

FIGURE 13.1  **Diagram of a neuron emphasizing the specialization within different parts of the cell.** This specialization includes the soma, myelinated axon, nerve terminals, axon hillock, and dendrites. Neurons receive input via synapses onto dendrites.

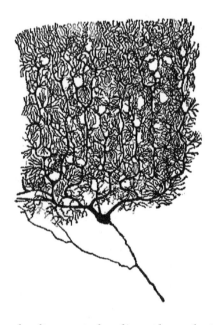

FIGURE 13.2  **Golgi stain of a Purkinje neuron with extensive dendritic branches.** *Source: Illustration based on Santiago Ramón y Cajal, the Spanish neuroscientist, from the book "The Beautiful Brain: The Drawings of Santiago Ramón y Cajal"; Public Domain, Wikipedia Commons.*

of voltage-gated sodium channels, the axon potential threshold is lowest there, making the axon hillock the site where the summation or integration of inputs is most likely to generate an action potential. If the sum of all inputs that reach the axon hillock depolarizes the membrane potential above the action potential threshold, the neuron fires an action potential.

The relatively simple concept of summing all of the depolarizing and hyperpolarizing inputs to a cell is made more complex by the properties of neuronal dendrites. There are a number of dendritic characteristics that influence how an EPSP caused by a postsynaptic current changes on its way to the site of integration at the axon hillock. We will cover these characteristics in turn, first explaining how a given characteristic would be predicted to regulate synaptic integration in isolation, and then introducing subsequent discoveries that complicate these predictions (see Stuart and Spruston, 2015 for review).

## PASSIVE MEMBRANE PROPERTIES

The summation of hyperpolarizing and depolarizing inputs onto a cell is complicated by the fact that the many synapses onto a neuron (which can number in the hundreds or even thousands) occur at different distances from the axon hillock (Fig. 13.1). Depolarizations or hyperpolarizations that are initiated at different locations in the dendritic tree need to propagate to the axon hillock in order to be included in the electrical summation of excitatory and inhibitory inputs onto that neuron. For the purposes of illustrating the main issues related to synaptic integration we will focus the discussion below exclusively on excitatory postsynaptic potentials (EPSPs).

Most presynaptic inputs onto postsynaptic partners in the central nervous system arise from relatively small presynaptic nerve terminals that may release as few as one synaptic vesicle following each presynaptic action potential. This limited synaptic vesicle release results in a relatively small change in postsynaptic membrane potential. The EPSPs that derive from individual presynaptic nerve terminals might depolarize the local dendritic branch on which they directly synapse by 0.5–5 mV (Magee and Cook, 2000; Berger et al., 2001; Williams and Stuart, 2002; Nevian et al., 2007), but these local EPSPs must propagate to the cell body and axon hillock in order to trigger an action potential. Due to several passive electrical properties of the cell membrane (see below), and without considering any other characteristics of dendrites that affect EPSP propagation, an EPSP would be expected to decay in amplitude as it passively propagates away from the site of initiation in the dendrite. This process is called "**dendritic filtering**" (Fig. 13.3).

The initial size of the EPSP is determined by the current that crosses the membrane (e.g., through ligand-gated ion channels) multiplied by the membrane resistance. Remember from Chapter 3, that membrane resistance at resting potential is determined by the number of available pathways, such as open ion channels, through which charge can distribute across the membrane. After the synaptic potential is generated, it spreads along the dendritic membrane while decaying at a rate determined in part by three passive properties of the membrane: **membrane conductance, membrane capacitance**, and **axial resistance** (Fig. 13.3). Membrane conductance is the sum of all pathways for charge movement across the membrane; membrane capacitance is the membrane's ability to separate and store charge; and axial resistance is the resistance to ionic current flow along the length of the dendrite. The last property, axial resistance, is higher in smaller-diameter dendrites and increases with the distance from the site of initiation to the soma.

The effect of these membrane properties on the passive decay of synaptic potentials can be estimated mathematically (see Magee, 2000; Rall and Rinzel, 1973). The

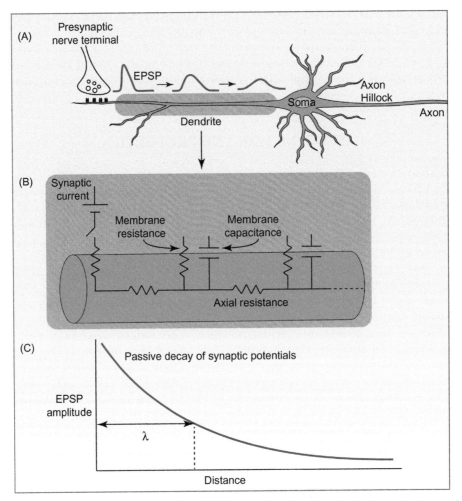

FIGURE 13.3  **Predicted passive spread of an EPSP away from the site of initiation and along a dendrite.** (A) Diagram of a neuron receiving a depolarizing synaptic input along the dendrite at a distance from the soma and axon hillock. The resulting excitatory postsynaptic potential (EPSP; *red trace*) is shown propagating passively toward the cell body as it decays at a predicted rate based on passive membrane properties. (B) An illustration of the electrical properties of the dendrite predicted to give rise to the changes in the EPSP as it passively travels down the dendrite. The synaptic current that initially generated the EPSP at the site of the synaptic contact is represented as a battery, with the amplitude of the EPSP governed by the size of the current injection and the local resistance of the membrane. As the EPSP passively travels down the dendrite, the axial resistance, membrane conductance, and membrane capacitance change the size and shape of the EPSP. (C) A plot of the predicted decay in amplitude of the EPSP as it travels passively down the dendrite. The distance traveled along the dendrite at which the EPSP has decayed to 37% of its original amplitude is defined as the length constant of the dendrite ($\lambda$).

properties that govern the predicted decay in passively propagating synaptic potentials are sometimes called the "**cable properties**" of the membrane, and the distance over which a synaptic potential decays to about 37% of its initial amplitude is called the **length constant** ($\lambda$; Fig. 13.3). Based on calculations accounting only for the cable properties, small synaptic potentials that are generated at very distant locations in the dendritic tree would be predicted to have very little influence on the membrane potential in the soma and at the axon hillock.

Although the "filtered" synaptic events that propagate to the soma from distant sites on the dendritic tree are predicted to be smaller in amplitude, they are also predicted to be slower in time course, which would increase their **temporal summation** with other events. Temporal summation could be important during bursts of synaptic activity, when the predicted slowing caused by dendritic filtering would cause each individual EPSP in the burst to last longer and thus overlap to a greater extent with the other EPSPs in the burst, increasing the amplitude of the summed depolarization (Fig. 13.4). These effects could have a strong impact on the membrane potential at the axon hillock where action potentials are generated. Synaptic potentials from distant synaptic inputs show greater temporal summation, which may partially offset the decay in amplitude of individual events. This temporal summation could assist in bringing the cell closer to threshold in the case of excitatory events, and in holding the cell further from threshold for longer periods of time in the case of inhibitory events.

Taking the issues discussed above into consideration, the mechanisms that govern effective signaling (events that lead to action potential generation at the axon hillock) might vary depending on the specific site of synaptic input to a neuron (Magee, 2000). Subthreshold events that originate near the axon hillock may only trigger an action potential if there are several synaptic events that arrive at nearly the same time—essentially requiring the coincident occurrence of inputs from multiple synapses. On the other hand, subthreshold synaptic events that elicit EPSPs at a distance from the soma can summate even if they do not all arrive at exactly the same time (due to dendritic filtering that prolongs their time course). Below we will discuss why assumptions about

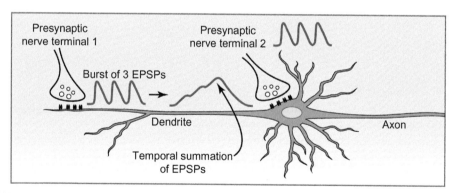

FIGURE 13.4 **Temporal summation of synaptic potentials and the predicted impact of dendritic filtering.** As EPSPs travel passively down a dendrite, the predicted changes in the time course of the EPSP spread the depolarization out in time. If there are multiple EPSPs generated in rapid succession, they are predicted to overlap (temporal summation), and the extent to which they do so is enhanced when EPSPs have broadened in time course due to dendritic filtering. These factors may govern the size and shape of the summed depolarization that reaches the axon hillock. Distant inputs (presynaptic nerve terminal #1) would be predicted to have enhanced temporal summation compared to local inputs (presynaptic nerve terminal #2).

EPSP propagation based solely on the passive properties of cell membranes are oversimplified and therefore, may not accurately predict the observed integration of synaptic events in neurons (see **Active membrane properties** section).

## SPINES ARE SPECIALIZED POSTSYNAPTIC COMPARTMENTS ON DENDRITES

One feature of many dendrites that is not considered in our discussion above is the presence of specialized subcompartments within dendrites called spines (Fig. 13.5). These protrusions from the dendrite were first identified by Ramón y Cajal in Golgi-stained sections of central nervous system tissue and represent postsynaptic specializations onto which nerve terminals form synapses. Dendritic spines serve to increase the surface area of dendrites, allowing for a higher density of presynaptic nerve terminals per unit length of dendrite. Dendritic spines also create a partially isolated postsynaptic compartment that allows for local biochemical events to occur within spines that can regulate neurotransmitter release within that subcompartment (often leading to synaptic plasticity; see Chapter 14). Over years of observation, scientists have identified a variety of spine morphologies, including various dimensions of the spine shaft and head (Fig. 13.5). Furthermore, it appears that spines can be morphologically plastic, changing size and shape under different conditions (including development and plasticity; Holtmaat et al., 2005; Hofer et al., 2009; Holtmaat and Svoboda, 2009; see Chapter 14).

The dendritic spines are membrane protrusions that alter the electrical properties of the dendrite and create subcompartments within it. How does the morphology of a dendritic spine affect the voltage change that occurs after postsynaptic ionotropic receptors are activated? Furthermore, how does the spine neck (the thin connection between the spine head and the dendritic shaft) alter the propagation of synaptic events out of the spine and into the dendrite? Spine necks can be thin and thus have a relatively high axial resistance. This means that the dendritic spine is an electrically isolated subcompartment (Yuste, 2013, 2015). Under these conditions, the change in membrane potential caused by a postsynaptic current within a spine head might be predicted to be relatively large.

There has been considerable debate over the degree to which the spine neck influences the propagation of synaptic potentials into the dendrite, primarily because this has been difficult to measure directly (Tonnesen and Nagerl, 2016; Jayant et al., 2017). Recently, however, fine-tipped quartz nanopipette electrodes have been used to record directly from these very small spine head compartments in hippocampal neuron cultures. Consistent with the prediction above, results from these experiments confirmed that EPSPs within the spine heads were relatively large (25–30 mV; Jayant et al., 2017). These large local dendritic spine EPSPs were attenuated greatly after they propagated to the soma, where they were usually only 0.5–1 mV in amplitude. While this study focused on recording from spines with long necks (as these were more easily identified), which might select for spine necks with especially high resistance, the large difference between the measured spine head EPSP (25 mV) and the soma EPSP (0.5–1 mV) implies that there is a high spine neck

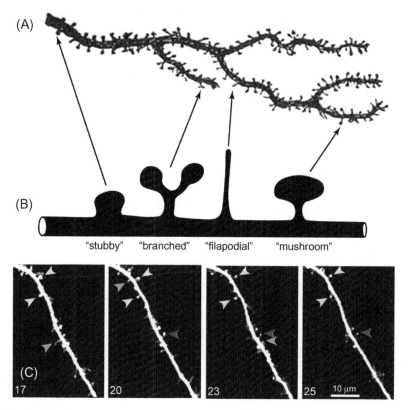

FIGURE 13.5  **Dendritic spines.** (A) Original drawing by Ramon y Cajal of dendritic spines from a Golgi-stained cerebellar Purkinje cell. (B) Drawing of some of the different types of dendritic spines, named for their shape. (C) Plasticity in dendritic spines as documented by repeated in vivo imaging over time in mouse cortex. In this example, some spines are stable during the imaging period (*yellow arrowheads*), others transiently appear and disappear (*blue arrowheads*), and still others are newly developed and persist throughout the imaging time period (*red arrowheads*). The age in postnatal days is shown in the bottom left corner of each image. *Source: (A) From Cajal, 1899; (B) and (C) Adapted with permission from Holtmaat, A.J., Trachtenberg, J.T., Wilbrecht, L., Shepherd, G.M., Zhang, X., Knott, G.W., et al., 2005. Transient and persistent dendritic spines in the neocortex in vivo. Neuron 45, 279–291.*

resistance which significantly attenuates the propagation of synaptic potentials out of the spine.

Differently shaped spines are predicted to alter the propagation of synaptic events into the dendrite in different ways. Spine head volumes can range from 0.01 to 1 $\mu m^3$, and spine necks can be 50–500 nm in diameter and from 0 to 3 $\mu m$ in length (Tonnesen and Nagerl, 2016). Therefore, it is likely that amount of spine neck filtering of synaptic events will vary (Fig. 13.6). Additional studies will be required to confirm these results and further our understanding of the impact of dendritic spines on the propagation of synaptic events to the dendritic shaft, en route to the soma.

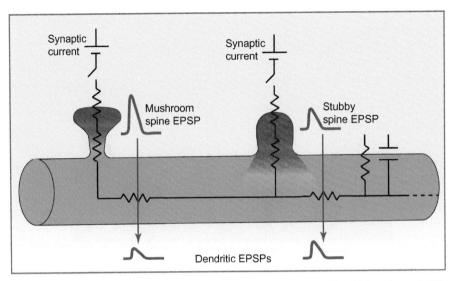

FIGURE 13.6  **The effects of spine neck resistances on synaptic events.** Dendritic spines of different morphology likely result in varied postsynaptic EPSPs after synaptic input. For example, mushroom spines (left) have a thin spine neck with relatively high resistance that may electrically isolate the spine head environment (depicted as a restriction in *red* color spread to the *blue* dendrite). In this scenario, synaptic input is predicted to result in a relatively large EPSP within the spine, but this event likely decays significantly as it propagates to the dendritic shaft. On the other hand, stubby spines have very low-resistance necks, which does not electrically isolate the spine head from the dendrite (depicted as a spread of color to the dentrite), and may result in smaller spine EPSPs (right). These stubby spine EPSPs are predicted to propagate to the dendritic shaft with little attenuation through the low-resistance stubby spine neck. This diagram utilizes the electrical circuit elements defined in Fig. 13.3B.

The discussion above, centered around the passive properties of cell membranes and the impact of spine morphology on the passive propagation of synaptic events within dendrites, leads to the expectation that synaptic events that originate at different distances along the dendritic tree from the neuron cell body will result in vastly different synaptic events measured at the axon hillock. For example, one might predict that synaptic events that originate at very distant dendritic sites onto mushroom spines with high-resistance necks would not have a significant impact on the cell body membrane potential. As scientists began to test these hypotheses experimentally, sometimes the effects of dendritic filtering did reveal EPSPs that were very small when originating from very distal dendrites as compared with EPSPs that originated from close to the soma (e.g., when recorded from neocortical pyramidal cells; Williams and Stuart, 2002). On the other hand (e.g., in hippocampal CA1 pyramidal cells), experimenters were surprised to find that the size and shape of synaptic events measured at the soma (originating from different distances along the dendritic tree) were more similar to one another than expected (Magee, 2000). For example, Andersen et al. (1980) reported that EPSPs that originated from proximal versus distal Schaffer collaterals onto hippocampal pyramidal cells had very similar amplitudes and time courses. Similar findings were also reported for synaptic input to spinal motoneurons (Jack et al., 1981; Iansek and Redman, 1973) and hippocampal CA1 pyramidal neurons (Magee and Cook, 2000). To further evaluate these observations, experimenters have applied glutamate (using glutamate uncaging techniques) to dendrites at different distances from the soma in several preparations and confirmed that the amplitude of the current recorded at the soma is relatively independent of where on the dendritic tree the glutamate was applied (see, e.g., Pettit and Augustine, 2000). Given the known passive properties of membranes, how can this happen?

Using measurements near the site of synaptic input within dendrites, experimenters found evidence that the size of the synaptic input is scaled to be larger at more distant sites and smaller at closer sites, so that by the time each propagates to the soma, EPSPs do not vary much in size (Magee and Cook, 2000). Furthermore, their decay time course also seems to be altered appropriately in the distal dendrites to counterbalance the effects of dendritic filtering (Magee, 1999; Williams and Stuart, 2000). What occurs within these distal dendrites to create these alterations in EPSP amplitude and time course?

First, dendrites tend to taper such that they have larger diameters near the soma, and smaller diameters at sites more distant from the soma. This difference in morphology effectively creates a small cellular compartment at distant sites (with a higher membrane resistance and lower membrane capacitance) that promotes signal propagation with less filtering from more distant sites (Rall and Rinzel, 1973; Rinzel and Rall, 1974; Jaffe and Carnevale, 1999). Second, the synapses themselves can vary in their properties (e.g., number of presynaptic release sites, probability of release, postsynaptic receptor density and properties) at different synaptic sites along the dendrite. These properties may scale with distance from the soma to make more distant synapses stronger and produce synaptic events with faster kinetics (Magee and Cook, 2000; Korn et al., 1993; Takumi et al., 1999).

## ACTIVE MEMBRANE PROPERTIES

Perhaps the most exciting development in the study of dendrites came when investigators discovered that dendrites are not constructed of just passive membranes, but express a large variety of ion channels, or active conductances. If synaptic events trigger the opening of local dendritic ion channels, then instead of only the passive properties of dendritic membrane governing EPSP propagation, the ion channels could actively increase or decrease the size of the local dendritic depolarization as synaptic events move along the dendrite.

Dendritic ion channels also allow action potentials that were generated at the axon hillock to not only propagate along the axon, but also "back-propagate" into the dendrites, using the dendritic ion channels to maintain the action potential depolarization as it moves. A back-propagated action potential often decreases in amplitude as it moves into the dendrites because the density of dendritic ion channels is not high enough to fully regenerate the action potential. Back-propagating action potentials will be discussed further in the context of synaptic plasticity in Chapter 14.

Patch clamp recording is the most direct experimental approach to measuring voltage-gated ion channel flux within dendrites (Stuart et al., 1993). A large variety of voltage-gated ion channels have now been mapped to dendrites, including voltage-gated sodium, calcium, and potassium channels and a hyperpolarization-activated cation channel (Migliore and Shepherd, 2002). The subtypes of these voltage-gated channels that are present in dendrites are often those that are activated near resting membrane potential or after mild depolarization (e.g., "T-type" calcium channels). In general, voltage-gated sodium and calcium channels open during EPSPs in the dendrite and serve to bring in more cations, "boosting" (or enlarging) the EPSP amplitude. Active boosting in dendrites following glutamate synaptic activation can also be due to the presence of NMDA receptors (large conductance ligand-gated ion channels with slow kinetics that depend on the coincident presence of glutamate and depolarization; see Chapter 11).

In contrast, the hyperpolarization-activated cation channels are open at resting membrane potentials and close during EPSPs. Since these channels are relatively slow to activate, the delayed closure of a cation channel in the dendrite during an EPSP tends to repolarize the EPSP faster. Importantly, in the dendrites of cortical pyramidal cells, the density of the hyperpolarization-activated cation channels increases with distance from the soma (Magee, 1999; Williams and Stuart, 2000). Therefore, these channels speedup the repolarization of EPSPs from very distant synapses to a greater extent than EPSPs from synapses closer to the soma. Functionally, this helps to counteract the effects of dendritic

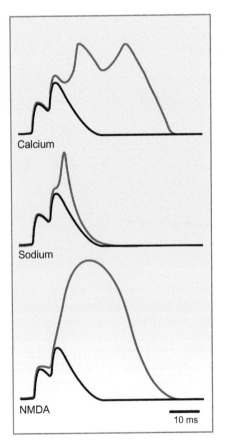

FIGURE 13.7  **Dendritic spikes.** Drawings of sample dendritic spikes (*red*) mediated by voltage-gated calcium channels (top), voltage-gated sodium channels (middle), and NMDA receptors (bottom). Black traces represent dendritic EPSPs when each channel type is pharmacologically blocked.

filtering on the time course of synaptic events. The presence of each of these voltage-gated ion channel subtypes, and their distribution along the dendrites of neurons, is variable and depends on the specific neuron type.

As a result of dendritic ion channel activity, postsynaptic EPSPs can be enlarged as they propagate down the dendrite toward the soma. In some cases, if the EPSP is large enough, or when a burst of synaptic inputs occurs, a "dendritic spike" can be generated. Dendritic spikes are similar to action potentials but are triggered in the dendritic membrane before propagating to the axon hillock. They have been shown to be mediated by voltage-gated sodium or calcium channels and NMDA receptors (Fig. 13.7). These dendritic spikes are often not maintained as they propagate to the soma, but the enhanced depolarization they provide further aids in propagating the synaptic event that caused them to the soma and axon hillock, sometimes resulting in a prolonged somal depolarization, which can have a strong influence on synaptic integration at the axon hillock.

## References

Andersen, P., Silfvenius, H., Sundberg, S.H., Sveen, O., 1980. A comparison of distal and proximal dendritic synapses on CAi pyramids in guinea-pig hippocampal slices in vitro. J. Physiol. 307, 273–299.

Berger, T., Larkum, M.E., Luscher, H.R., 2001. High I(h) channel density in the distal apical dendrite of layer V pyramidal cells increases bidirectional attenuation of EPSPs. J. Neurophysiol. 85, 855–868.

Hofer, S.B., Mrsic-Flogel, T.D., Bonhoeffer, T., Hubener, M., 2009. Experience leaves a lasting structural trace in cortical circuits. Nature 457, 313–317.

Holtmaat, A., Svoboda, K., 2009. Experience-dependent structural synaptic plasticity in the mammalian brain. Nat. Rev. Neurosci. 10, 647–658.

Holtmaat, A.J., Trachtenberg, J.T., Wilbrecht, L., Shepherd, G.M., Zhang, X., Knott, G.W., et al., 2005. Transient and persistent dendritic spines in the neocortex in vivo. Neuron 45, 279–291.

Iansek, R., Redman, S.J., 1973. The amplitude, time course and charge of unitary excitatory post-synaptic potentials evoked in spinal motoneurone dendrites. J. Physiol. 234, 665–688.

Jack, J.J., Redman, S.J., Wong, K., 1981. The components of synaptic potentials evoked in cat spinal motoneurones by impulses in single group Ia afferents. J. Physiol. 321, 65–96.

Jaffe, D.B., Carnevale, N.T., 1999. Passive normalization of synaptic integration influenced by dendritic architecture. J. Neurophysiol. 82, 3268–3285.

Jayant, K., Hirtz, J.J., Plante, I.J., Tsai, D.M., De Boer, W.D., Semonche, A., et al., 2017. Targeted intracellular voltage recordings from dendritic spines using quantum-dot-coated nanopipettes. Nat. Nanotechnol. 12, 335–342.

Korn, H., Bausela, F., Charpier, S., Faber, D.S., 1993. Synaptic noise and multiquantal release at dendritic synapses. J. Neurophysiol. 70, 1249–1254.

Magee, J.C., 1999. Dendritic Ih normalizes temporal summation in hippocampal CA1 neurons. Nat. Neurosci. 2, 848.

Magee, J.C., 2000. Dendritic integration of excitatory synaptic input. Nat. Rev. Neurosci. 1, 181–190.

Magee, J.C., Cook, E.P., 2000. Somatic EPSP amplitude is independent of synapse location in hippocampal pyramidal neurons. Nat. Neurosci. 3, 895–903.

Migliore, M., Shepherd, G.M., 2002. Emerging rules for the distributions of active dendritic conductances. Nat. Rev. Neurosci. 3, 362–370.

Nevian, T., Larkum, M.E., Polsky, A., Schiller, J., 2007. Properties of basal dendrites of layer 5 pyramidal neurons: a direct patch-clamp recording study. Nat. Neurosci. 10, 206–214.

Pettit, D.L., Augustine, G.J., 2000. Distribution of functional glutamate and GABA receptors on hippocampal pyramidal cells and interneurons. J. Neurophysiol. 84, 28–38.

Rall, W., Rinzel, J., 1973. Branch input resistance and steady attenuation for input to one branch of a dendritic neuron model. Biophys. J. 13, 648–687.

Ramón y Cajal, S., 1899. La Textura del Sistema Nerviosa del Hombre y losVertebrados. Moya (Primera Edicion), Madrid.
Rinzel, J., Rall, W., 1974. Transient response in a dendritic neuron model for current injected at one branch. Biophys. J. 14, 759–790.
Stuart, G.J., Spruston, N., 2015. Dendritic integration: 60 years of progress. Nat. Neurosci. 18, 1713–1721.
Stuart, G.J., Dodt, H.U., Sakmann, B., 1993. Patch-clamp recordings from the soma and dendrites of neurons in brain slices using infrared video microscopy. Pflugers Arch. 423, 511–518.
Takumi, Y., Ramirez-Leon, V., Laake, P., Rinvik, E., Ottersen, O.P., 1999. Different modes of expression of AMPA and NMDA receptors in hippocampal synapses. Nat. Neurosci. 2, 618–624.
Tonnesen, J., Nagerl, U.V., 2016. Dendritic spines as tunable regulators of synaptic signals. Front. Psychiatry 7, 101.
Williams, S.R., Stuart, G.J., 2000. Site independence of EPSP time course is mediated by dendritic I(h) in neocortical pyramidal neurons. J. Neurophysiol. 83, 3177–3182.
Williams, S.R., Stuart, G.J., 2002. Dependence of EPSP efficacy on synapse location in neocortical pyramidal neurons. Science 295, 1907–1910.
Yuste, R., 2013. Electrical compartmentalization in dendritic spines. Annu. Rev. Neurosci. 36, 429–449.
Yuste, R., 2015. The discovery of dendritic spines by Cajal. Front. Neuroanat. 9, 18.

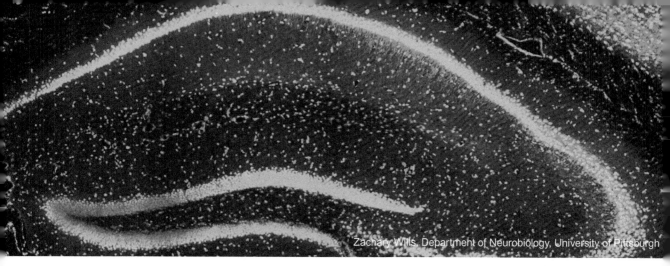

# CHAPTER 14

# Synaptic Plasticity

Everything we do is the result of electrical and chemical communication between neurons, which triggers action potential activity in subsets of neurons. When our behavior changes (e.g., if we learn something, or alter our response to a stimulus as a result of experience) there are alterations in synaptic strength and/or connections that influence the communication between neurons and, in so doing, influence their activity. In fact, the nervous system's plasticity (its ability to change based on experience) underlies most of the complex behaviors that it controls, including learning, memory, adaptation, sensitization, sensory processing, addiction, and cognition. The mechanisms that underlie synaptic plasticity are critical to our understanding of the complex functions of the nervous system. These changes can be a result of structural alterations in presynaptic nerve terminals and/or postsynaptic specializations (e.g., dendritic spines), or in the function of those synaptic compartments.

Many forms of synaptic plasticity have been discovered since Cajal first proposed its existence. These range from small transient changes in the probability of chemical neurotransmitter release at existing synapses, to large and long-lasting structural changes in the physical connections between neurons. As mentioned above, synaptic plasticity often results from experience, but *how* does our experience affect synapses? Below we will discuss the different "experiences" that a neuron and a synapse can be exposed to and cover some of the major forms of synaptic plasticity and the mechanisms that underlie these changes.

## SHORT-TERM SYNAPTIC PLASTICITY

Put simply, the "experience" of a synapse is a series of events, initiated by action potential invasion of the nerve terminal, that includes opening of voltage-gated ion channels, influx of calcium ions, binding of these calcium ions to presynaptic proteins, fusion of synaptic vesicles, release of chemical neurotransmitter, activation of postsynaptic receptors for this neurotransmitter, and postsynaptic consequences of receptor activation. Therefore, every time that a synapse experiences an action potential invasion, it triggers a cascade of events that may lead to changes in the way the synapse will respond to subsequent action potentials (Zucker and Regehr, 2002).

Some forms of synaptic plasticity can be initiated by only two action potentials that arrive at the synapse in rapid succession (less than 100 ms between them). The plasticity that results from just a pair of action potentials is called **paired-pulse plasticity**. The magnitude of neurotransmitter released changes after the second action potential in a pair, because after the first action potential invades the nerve terminal, it triggers events that have not completely reversed back to normal by the time the second action potential arrives. The most important of these events are (1) the calcium ions binding to proteins in the nerve terminal, and (2) the fusion of docked and primed synaptic vesicles. These two events create changes in the nerve terminal that require some time to revert back to the resting state.

The first event, calcium ion binding, can *increase* the amount of neurotransmitter released following the second action potential. When a second action potential arrives before calcium ions that entered during the first action potential have had time to dissociate from proteins in the nerve terminal, the calcium ions that enter the nerve terminal during that second action potential add to the calcium ions already present from the first action potential. Because chemical neurotransmitter release is triggered by calcium ions, the so-called residual bound calcium from the first action potential increases the probability of chemical neurotransmitter release during the second action potential in the pair.

The second event, vesicle fusion, can *reduce* the amount of neurotransmitter released following the second action potential. After the fusion of docked and primed synaptic vesicles during the first action potential, it takes time for the fused vesicles to either be recycled or be replaced by another synaptic vesicle. Therefore, when a second action potential invades the nerve terminal before the recently fused vesicles have been replaced, there are fewer synaptic vesicles available for release during that second action potential depolarization. This phenomenom is often called vesicle depletion.

Since these two residual effects from the first action potential have opposite effects, changes in the magnitude of neurotransmitter release after the second action potential in a pair are governed by the balance between the magnitude of residual bound calcium and the magnitude of vesicle depletion (Fig. 14.1). The relative weight of each factor depends heavily on the strength of the synapse (the probability of vesicle fusion after an action potential).

At synapses with a low probability of vesicle fusion after an action potential, vesicle depletion is minimal, so the effect of residual bound calcium predominates. Since residual bound calcium ions within the nerve terminal increase the probability that the second action potential will trigger vesicle fusion, there tends to be more neurotransmitter release

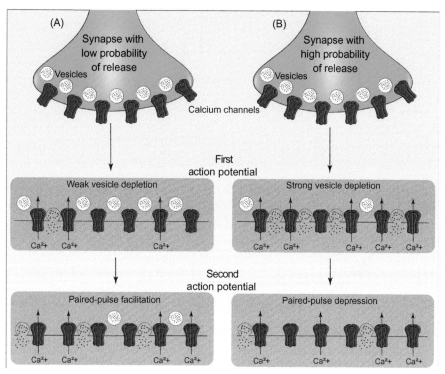

FIGURE 14.1  Residual bound calcium and vesicle depletion are opposing influences that govern short-term synaptic plasticity during a pair of action potential stimuli. Two nerve terminals are drawn at the top with many single vesicle release sites. One nerve terminal has a low probability of vesicle release following action potential stimulation (A), and the other has a high probability of vesicle release following action potential stimulation (B). (A) At weak synapses, the first action potential in a pair (middle) only opens a few calcium channels, so only a small number of synaptic vesicles fuse to release neurotransmitter. In addition, calcium entry during the first action potential leaves some of the vesicles that did not fuse during the first action potential with residual bound calcium. These synaptic vesicles with residual bound calcium are easier to trigger to fuse during the second action potential. Since there is very little synaptic vesicle depletion following the first action potential, the residual bound calcium leads to paired-pulse facilitation of release during the second action potential. (B) At a strong synapse, the first action potential in a pair (middle) opens a large percentage of voltage-gated calcium channels, leading to a large number of calcium-bound vesicles, many of which fuse and release neurotransmitter. During the second action potential, the number of depleted synaptic vesicle release sites is larger than the number of vesicles that fuse after calcium entry, leading to paired-pulse depression.

at these weak synapses after the second action potential in a pair than after the first action potential. This form of short-term plasticity is often called "paired-pulse facilitation" (Figs. 14.1B and 14.2).

At synapses with a high probability of vesicle fusion after an action potential, vesicle depletion predominates, so there tends to be less neurotransmitter released at these strong synapses after the second action potential in a pair. In this scenario, since fewer vesicles are available for release (having already been released in response to the first action potential), is a lower probability of vesicle fusion. In addition, because there are significantly

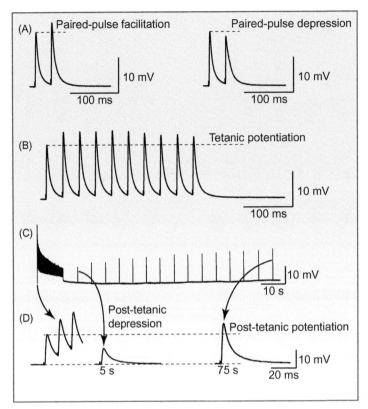

FIGURE 14.2  A comparison of paired-pulse, tetanic, and post-tetanic plasticity at the frog NMJ. (A) An example of paired-pulse facilitation (left). In this example, the second endplate potential in a pair is larger than the first. On the right is an example of paired-pulse depression in which the second endplate potential in a pair is smaller than the first. (B) An example of tetanic potentiation. During a short train of high-frequency activity (30 action potentials per second; 30 Hz) the endplate potentials (EPP) increase in amplitude compared to the first EPP. (C) An example of post-tetanic changes in neurotransmitter release. In this case, a long high-frequency burst of activity (200 Hz for 10 s) leads to an immediate depression of neurotransmitter release (post-tetanic depression), followed by an increase in neurotransmitter release (post-tetanic potentiation). Post-tetanic depression occurs due to synaptic vesicle depletion from the nerve terminal. Post-tetanic potentiation occurs when residual bound calcium remains in the terminal after synaptic vesicle depletion recovers. Panel D shows an enlargement of endplate potentials in C at time points defined by the arrows.

fewer docked synaptic vesicles available to respond to calcium, residual bound calcium ions don't have the opportunity to increase the probability of vesicle fusion (because there are fewer docked synaptic vesicles present). As a result, the magnitude of neurotransmitter release after the second action potential in a pair is smaller than after the first action potential. This form of short-term synaptic plasticity is often called "paired-pulse depression" (see Figs. 14.1 and 14.2).

The same balance of influences discussed above for paired-pulse plasticity are also the major factors governing short-term synaptic plasticity when more than two action potentials invade a synapse with short intervals between them (less than 100 ms). For example, many neurons communicate using a short burst of action potentials that fire at short time intervals (e.g., 5–10 action potentials with 10–50 ms between each). When such a burst of action potentials reaches the nerve terminal, residual bound calcium can build up between each action potential, gradually increasing the probability of neurotransmitter release. Opposing the influence of residual bound calcium is the gradual depletion of docked synaptic vesicles, decreasing the probability of neurotransmitter release. As in paired-pulse facilitation, either residual bound calcium or vesicle depletion can be the dominant factor in how the magnitude of neurotransmitter release changes over the course of the burst, depending on the initial probability of neurotransmitter release at the synapse. This form of short-term plasticity is called "tetanic potentiation" if the magnitude of neurotransmitter release increases during the burst of action potentials (Fig. 14.2), or "tetanic depression" if the magnitude of neurotransmitter release decreases during the burst of action potentials. Measurements of neurotransmitter release during bursts of electrical stimulation are relevant to clinical tests of synaptic strength and plasticity in peripheral neuromuscular junction synapses (Box 14.1).

---

### BOX 14.1

## THE REPETITIVE NERVE STIMULATION TEST (RNS) FOR PERIPHERAL NEUROMUSCULAR DISEASES

Neuromuscular disorders are diagnosed using a combination of clinical observations and laboratory tests. One tool that is used by clinicians to aid in this diagnosis is the **repetitive nerve stimulation test** (RNS). The RNS test is performed by attaching a pair of adhesive skin electrodes to the inner wrist to stimulate the median nerve (which runs just under the skin in that location, and between the two prominent tendons that appear when someone makes a fist), and another pair of adhesive skin electrodes to the belly of the thumb muscle (see Fig. 14.3). A single electric shock to the median nerve will elicit a contraction of the thumb muscle and the electrodes over the belly of that muscle can record a **compound muscle action potential** (CMAP), whose amplitude is proportional to the number of muscle fibers under the electrode that fired an action potential.

In healthy individuals, the CMAP is very large and reflects the fact that all of the muscle cells under the recording electrodes are normally brought to threshold by the motor nerve action potential's release of acetylcholine. This is because the neuromuscular junction normally releases more than enough acetylcholine to sufficiently excite the postsynaptic muscle cell. This extra acetylcholine release acts as a so-called "safety factor" that ensures that every time a motor neuron fires, the membrane of the postsynaptic muscle cell depolarizes to above threshold to fire an action potential. Even after a burst of action potentials, the safety factor ensures that the resulting balance of residual bound calcium and vesicle depletion at the neuromuscular junction does not allow the amount of neurotransmitter released by the *next* action potential in the burst to drop below the threshold amount required to fire a postsynaptic action potential (mammalian neuromuscular synapses usually only experience mild tetanic depression). As a result, when healthy individuals receive a burst of high-frequency stimulation at the median nerve, the CMAP is of the same large size following each action potential in the burst.

## BOX 14.1 (cont'd)

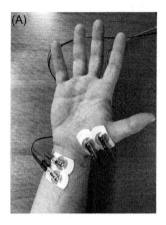

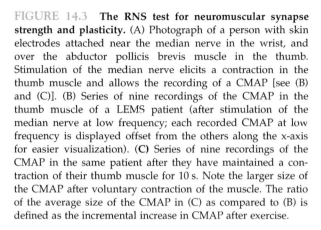

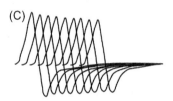

FIGURE 14.3 **The RNS test for neuromuscular synapse strength and plasticity.** (A) Photograph of a person with skin electrodes attached near the median nerve in the wrist, and over the abductor pollicis brevis muscle in the thumb. Stimulation of the median nerve elicits a contraction in the thumb muscle and allows the recording of a CMAP [see (B) and (C)]. (B) Series of nine recordings of the CMAP in the thumb muscle of a LEMS patient (after stimulation of the median nerve at low frequency; each recorded CMAP at low frequency is displayed offset from the others along the x-axis for easier visualization). (C) Series of nine recordings of the CMAP in the same patient after they have maintained a contraction of their thumb muscle for 10 s. Note the larger size of the CMAP after voluntary contraction of the muscle. The ratio of the average size of the CMAP in (C) as compared to (B) is defined as the incremental increase in CMAP after exercise.

However, in some neuromuscular disorders (particularly **Lambert–Eaton myasthenic syndrome**; LEMS; see Box 7.3), the neuromuscular junction has been weakened and releases very little neurotransmitter following a motor nerve action potential. Under these conditions, the CMAP is small even at the beginning of a burst, since not all muscle fibers under the electrodes are brought to threshold. However, the CMAP increases in amplitude during repeated high-frequency median nerve stimulation. This is because more and more postsynaptic muscle cells are brought to threshold by the tetanic potentiation of neurotransmitter release that dominates at weakened presynaptic motor nerve terminals, since they do not experience significant vesicle depletion. Another way to test this, that is more comfortable for patients, is to record the CMAP repeated at low frequency, ask the patient to maintain a strong and sustained contraction of the hand muscles for 10–20 seconds, and then record the CMAP again at low frequency. Using this approach, LEMS patients will show a small CMAP before exercise of the hand muscles (Fig. 14.3B), and a much larger CMAP after exercise (Fig. 14.3C). In both cases, residual calcium facilitates neurotransmitter release, whether during the prolonged high-frequency burst of action potentials (tetanic potentiation) or for a short time after a maintained contraction (post-tetanic potentiation).

> **BOX 14.1** *(cont'd)*
>
> If, however, the patient is experiencing neuromuscular weakness due to a problem on the postsynaptic muscle cell, and there is no presynaptic nerve terminal weakness, the CMAP will not increase in amplitude with repeated stimulation of the median nerve at high frequency or voluntary contraction of the thumb muscle for 10–20 seconds. Therefore, using this approach, the RNS test can be used to determine if the source of neuromuscular weakness is presynaptic, and this information can be used in combination with other clinical tests for the diagnosis of neuromuscular diseases and selection of appropriate treatments.

Longer bursts of action potentials can also induce synaptic plasticity. When a long action potential burst ends, the excessive calcium entry that occurred during the burst heavily loads intracellular organelles (e.g., mitochondria) with calcium ions. Mitochondria serve as a temporary buffer for calcium loading within the nerve terminal and will slowly release that calcium back into the cytoplasm (see Chapter 7). Thus, even after the calcium entry through voltage-gated calcium channels ends with the conclusion of the stimulated train of action potentials, there is a prolonged elevation of calcium, in part due to mitochondria dumping calcium back into the cytoplasm (Tang and Zucker, 1997; David and Barrett, 2000; David et al., 1998). At some synapses, calcium stores in the endoplasmic reticulum can also contribute to elevated calcium levels after a burst of action potentials (Narita et al., 2000). Due to the heightened levels of calcium in the terminal, a single action potential that invades the nerve terminal at some time after a burst of action potentials has ended results in elevated neurotransmitter release. This is called "post-tetanic potentiation" (PTP) and can persist for many seconds after a train of action potential stimuli (Fig. 14.2).

In experimental contexts, scientists often deliver action potential bursts that last longer and are more regularly spaced than is typical physiologically, in order to provide insight into the balance of influences on neurotransmitter release at the nerve terminal under study. This often means delivering 20–100 action potentials in a steady "stimulus train" with a constant interstimulus time interval. If this stimulus train is long enough, the influences that impact short-term plasticity eventually appear to come to an equilibrium and the magnitude of neurotransmitter release levels off to a relatively constant rate during the train. Under these conditions, vesicle depletion is balanced by vesicle replenishment (due to recycling and replacement of fused vesicles) at each vesicle docking site in the synapse, allowing scientists to obtain an estimate of the size of the recycling pool of vesicles (Schneggenburger et al., 1999).

Within the central nervous system, many synapses are small and have a low probability of neurotransmitter release during a single action potential. As such, vesicle depletion during repeated stimulation is minimal at these synapses. Small central nervous system synapses therefore often undergo short-term synaptic facilitation during pairs and short bursts of action potentials. This raises the question: What roles can short-term synaptic facilitation play in the central nervous system?

First, because residual bound calcium builds up more when the time interval between action potentials is shorter, short-term facilitation increases with action potential frequency. This can offset the partial depletion of docked synaptic vesicles, even though vesicle depletion is also more pronounced at higher action potential frequencies. The fact that short-term synaptic facilitation can offset vesicle depletion allows synapses to use their firing rate as a communication signal without loss in their strength of communication (Jackman and Regehr, 2017). For example, circuits in the avian auditory system and the rodent cerebellum have been shown to encode information by changing their action potential firing rate, while maintaining their synaptic transmission response amplitudes (Sullivan and Konishi, 1984; Takahashi et al., 1984; Heck et al., 2013; Turecek et al., 2016).

Second, synapses that normally have a very low probability of neurotransmitter release (in response to a single action potential) don't communicate effectively with their postsynaptic partners until there is a burst of action potentials to facilitate or increase their strength of communication. Under these conditions, short-term synaptic facilitation filters the effects of action potential stimulation such that it is ineffective at low frequencies, but effective at high frequencies (Jackman and Regehr, 2017; Fig. 14.4B). This effect is sometimes called high-pass filtering (Atluri and Regehr, 1996) because these synapses only communicate effectively during high-frequency activity. Of course, the opposite can also occur if synapses display short-term synaptic depression. Under this scenario, effective communication only occurs during low-frequency action potential activity because vesicle depletion dominates during high-frequency action potential activity (Fig. 14.4A). When this occurs, short-term synaptic depression functions as a low-pass filter for effective synaptic communication (Abbott et al., 1997; Rose and Fortune, 1999). These high- or low-pass filtering effects on synaptic communication represent a mechanism that neurons can use to integrate their synaptic inputs and decide when this information is significant enough to be passed on in the form of an action potential.

Lastly, short-term synaptic facilitation is hypothesized to be important in the circuits that control working memory (Jackman and Regehr, 2017). Working memory has been hypothesized to be stored by groups of synapses that are active in a repeating pattern (perhaps with repeated short bursts of activity), either within a particular brain area (such as the prefrontal cortex; Roux and Uhlhaas, 2014) or between two brain areas (e.g., between the thalamus and prefrontal cortex; Bolkan et al., 2017). This repeated bursting activity is interspersed with time periods of inactivity (Sreenivasan et al., 2014; Stokes, 2015; Stokes et al., 2013). This activity pattern has been used as a basis for building models of working memory in which short-term synaptic facilitation, combined with post-tetanic potentiation to maintain facilitation during the periods of inactivity, are hypothesized to be important components (Itskov et al., 2011; Mongillo et al., 2008).

## Mechanisms of Residual Calcium Effects on Transmitter Release

Due to fast-acting calcium buffering and handling mechanisms in the nerve terminal, it is unlikely that free ionic calcium could remain in the cytoplasm for long enough to explain short-term synaptic facilitation that lasts for ~100 ms. Therefore, the residual calcium ions responsible for synaptic facilitation must be bound to proteins, hence the term "residual bound calcium." Intense experimental investigation has been focused on

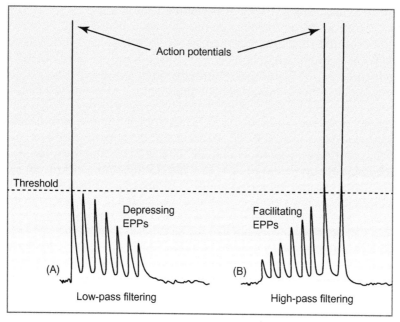

FIGURE 14.4  **Short-term synaptic plasticity can allow synapses to function as either high-pass or low-pass filters.** Both (A) and (B) are examples of recordings of membrane potential in a postsynaptic neuron during a short burst of action potential activity in the presynaptic nerve terminal. In one case, the synapse displays strong tetanic depression (A), while in the other the synapse displays strong tetanic potentiation (B). In the example of tetanic depression (A), only the first stimulus produces an excitatory postsynaptic potential that is large enough to trigger a postsynaptic action potential. In this case, the depressing synapse is acting as a low-pass filter by preventing high-frequency activity from triggering action potentials in the postsynaptic cell. However, in the example of tetanic potentiation (B), this synapse only triggers postsynaptic action potentials after a train of presynaptic action potential activity, thus acting like a high-pass filter (only allowing effective communication during high-frequency activity).

identifying proteins that bind calcium ions in the nerve terminal to mediate the ability of residual bound calcium to increase the probability of neurotransmitter released during short-term synaptic plasticity. What are these proteins and how do they increase the probability of neurotransmitter release?

As described in Chapter 7 and Chapter 8, after an action potential invades a nerve terminal, voltage-gated calcium channels open in the active zone and this leads to a local rise in calcium ion concentration within the nerve terminal cytoplasm. These calcium ions bind to a variety of proteins within the active zone region of nerve terminals before they are removed from the cytoplasm by the calcium buffering and handling mechanisms. Synaptotagmin 1 is one of the calcium-binding proteins that is positioned on docked synaptic vesicles and has been shown to increase the probability of vesicle fusion (see Chapter 8). This protein has a low affinity for calcium ions, and very rapid kinetics of binding [calcium ions bind quickly ($\sim 2$ ions/ms) and unbind quickly ($\sim 1$ ion/ms); Hui et al, 2005; Jackman and Regher, 2017]. The short-lived presynaptic calcium transient after an action potential, coupled with these properties of synaptotagmin 1, leads to the fast but brief increase in the probability of vesicle fusion that is responsible for fast synchronous action potential-triggered neurotransmitter release. However, synaptotagmin 1's calcium affinity is hypothesized to be too low for it to participate in residual bound calcium-triggered facilitation of neurotransmitter release. Therefore, it is unlikely that calcium ions could remain bound to synaptotagmin 1's low-affinity binding sites for very long.

It was proposed that there must be other proteins in the nerve terminal that can bind calcium ions, increase the probability of neurotransmitter release, and mediate short-term synaptic facilitation. These proteins would be expected to bind calcium with a slower **on-rate** (so as not to participate in fast synchronous release) and a slower **off-rate** (to allow calcium ions to remain bound between stimuli that cause short-term synaptic facilitation). There are a variety of proteins present in the nerve terminal that might fit these criteria (Corbalan-Garcia and Gomez-Fernandez, 2014). A subset of these proteins that have been shown to play a role in short-term synaptic facilitation are outlined below.

First, could the simple presence of calcium buffer proteins mediate short-term synaptic facilitation? This hypothesis was proposed based on the concept that local calcium buffers normally bind the calcium ions that enter during a single action potential, effectively competing with synaptotagmin 1. Therefore, if these calcium buffer proteins become saturated with calcium ions following the first action potential in a burst, subsequent calcium ion entry would not be bound as effectively by these calcium-bound buffer proteins and, as a result, there might be more calcium ions available for binding to synaptotagmin 1 (Klingauf and Neher, 1997; Neher, 1998; Rozov et al., 2001). This excess calcium might create a higher probability of vesicle fusion during subsequent action potentials in a burst (Fig. 14.5A). This mechanism may contribute to short-term synaptic facilitation at synapses where calcium buffers are present that have fast calcium-binding kinetics (e.g., **calbindin**), and in which docked synaptic vesicles may not be as tightly coupled to presynaptic calcium channels. This mechanism has been proposed to occur at hippocampal mossy fiber synapses between dentate gyrus neurons and CA3 pyramidal cells (Blatow et al., 2003). At this synapse, the calcium buffer calbindin may compete significantly with synaptotagmin 1 for calcium binding after an action potential. Partial saturation of calbindin calcium ion binding during the first action potential in a burst may increase the calcium binding to synaptotagmin 1 during subsequent action potential stimuli and thus increase the probability of vesicle fusion.

A second mechanism for short-term plasticity is that calcium entry during the first action potential in a burst might modulate presynaptic voltage-gated calcium channel function, such that more calcium ions are conducted through these channels during subsequent action potential stimuli (Fig. 14.5B). This hypothesis has been tested at a variety of synapses and found to occur at a limited few (the rodent calyx of Held synapse, cultured superior ganglion neuron synapses, some hippocampal synapses, and cultured synapses between Purkinje neurons; Cuttle et al., 1998; Muller et al., 2008; Yan et al., 2014; Nanou et al., 2016), but not in the majority of synapses (in other words, the calcium entry during each action potential in a burst is usually unchanged; Charlton et al., 1982; Atluri and Regehr, 1996; Jackman et al., 2016). When calcium flux through channels does increase during a burst of activity, it appears that calcium-binding proteins interact selectively with

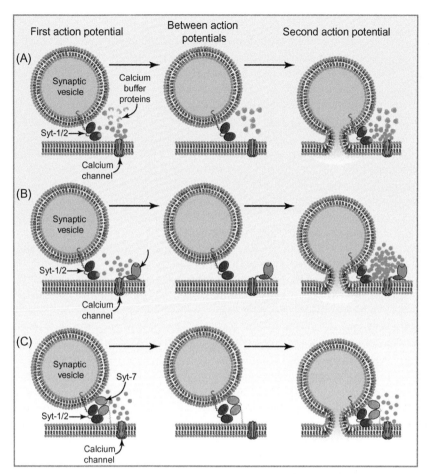

FIGURE 14.5  **Hypothesized mechanisms by which residual bound calcium can affect short-term synaptic plasticity.** Three columns are diagramed to represent proposed mechanisms of paired-pulse facilitation in the active zone during a pair of action potentials delivered with a short interstimulus interval (<100 ms). The first column depicts events that occur within the active zone during the first action potential stimulus. In these diagrams, not enough calcium ions bind to synaptotagmin (Syt-1/2) to trigger vesicle fusion. The second column depicts events that occur during the interval between stimuli, and the third column depicts events that occur during the second action potential in a pair. In each diagram, synaptic vesicles are drawn with synaptotagmin 1 or 2 (*blue* structure with two calcium-binding domains, Syt-1/2; the fast calcium sensors for neurotransmitter release), and a calcium channel is drawn positioned near the synaptic vesicle. During each action potential, calcium ions (*green dots*) enter through the calcium channel near the synaptic vesicle. (A) In the "buffer saturation" hypothesis for short-term plasticity, during the first action potential, calcium ions enter the nerve terminal and are bound by calcium buffer proteins (*orange*). Calcium remains bound to these buffer proteins between stimuli such that during the second action potential, the new calcium ions that enter cannot bind buffer proteins and therefore are more available to bind Syt-1/2 and trigger vesicle fusion. (B) In the "calcium channel modulation" hypothesis, calcium-binding proteins that bound calcium during the first action potential directly alter voltage-gated calcium channel function. After modulation by calcium-bound calmodulin and/or neuronal calcium sensor-1, the calcium channel conducts more calcium ions than normal during the second action potential, leading to an increased probability that these calcium ions will bind Syt1/2 and trigger vesicle fusion. (C) In the "synaptotagmin 7 (Syt-7)" hypothesis, Syt-7, which is hypothesized to be expressed in the plasma membrane and has a higher affinity for calcium than Syt1/2, acts as a second calcium sensor that increases the probability of vesicle fusion. After binding calcium during the first action potential, Syt-7 remains calcium-bound and increases the probability of vesicle fusion during the second action potential.

the **carboxy tail** of Cav2.1 (P/Q-type) voltage-gated calcium channels to increase the calcium flux. Proteins that have been hypothesized to mediate this increase in P/Q-type calcium channel current during short-term synaptic facilitation include **calmodulin** and **neuronal calcium sensor-1** (Nanou et al., 2016; DeMaria et al., 2001; Lee et al., 2003; Dason et al., 2012). Therefore, it appears that facilitation of current flux through voltage-gated calcium channels can underlie at least some forms of short-term synaptic facilitation, but perhaps only for synapses that express the specific type of calcium channel (P/Q-type; Cav2.1) which has been shown to undergo this type of modulation.

Despite evidence for a role of buffer saturation and voltage-gated calcium current facilitation at some synapses (discussed above), another protein has received considerable attention as a prominent trigger for short-term synaptic facilitation at many synapses. This protein is called synaptotagmin 7 (Fig. 14.5C). While we have previously discussed the role of synaptotagmin 1 and 2 in triggering fast, synchronous vesicle fusion (see Chapter 8; Sudhof, 2013), the synaptotagmin proteins are a large family of proteins with different calcium-binding properties and expression patterns in the nervous system. Synaptotagmin 7 is the only member of this family whose properties make it attractive as a candidate mediator of short-term synaptic facilitation (Zucker and Regehr, 2002). Synaptotagmin 7 has the highest affinity for calcium of all the synaptotagmins (Sugita et al., 2002), has the slowest on- and off-rates for calcium binding (binds calcium at $\sim 0.2$ per ms, and unbinds very slowly at $\sim 1.5$ ions over a 100 ms time period; Brandt et al., 2012; Hui et al., 2005; Jackman and Regher, 2017), and may be expressed in the active zone as a plasma membrane protein (as opposed to synaptotagmin 1 and 2, which are vesicle membrane proteins). Perhaps the best evidence that synaptotagmin 7 is a prominent calcium-binding protein that mediates short-term synaptic facilitation comes from studies of mice in which the synaptotagmin 7 protein is knocked out (absent). When Jackman et al. (2016) examined four different synapses chosen for their prominent short-term synaptic facilitation (hippocampal Schaffer collateral synapses, hippocampal mossy fiber synapses, lateral perforant path synapses onto dentate gyrus neurons, and cortico-thalamic synapses), they reported that paired-pulse facilitation was eliminated in synaptotagmin 7 knockout mice. Tetanic potentiation was also eliminated in three of these synapses, and greatly reduced in the fourth (Jackman et al., 2016). However, when synapses that normally show strong tetanic depression were examined, no effects of synaptotagmin 7 knockout on short-term synaptic plasticity were observed, and instead it appeared that synaptotagmin 7's contribution was to slow asynchronous vesicle fusion for some time after an action potential stimulus (Bacaj et al., 2013, 2015; Luo and Sudhof, 2017). In summary, synaptotagmin 7 seems to be a prominent mediator of short-term synaptic facilitation, at least at some synapses. Further experiments will be required to completely understand the mechanisms by which residual bound calcium mediates short-term synaptic facilitation.

Although we have focused on how residual bound calcium ions can mediate short-term synaptic plasticity, there are two other mechanisms that are independent of residual bound calcium that may contribute to short-term synaptic facilitation under special circumstances (e.g., especially during prolonged trains of action potential activity).

The first mechanism is broadening of the action potential. As described in Chapter 7, action potential shape is an important modulator of voltage-gated calcium channel function. When action potential shape changes at the nerve terminal, the amount of calcium influx can change. Action potential broadening at the synapse can occur when the voltage-gated potassium channels that normally speed repolarization of the action potential **inactivate** during a long burst of action potentials. Since the terminal remains depolarized for a longer time during these broadened action potentials, voltage-gated calcium channels are more likely to open, leading to more calcium influx. Synaptic facilitation via action potential broadening has been demonstrated only in specialized presynaptic terminals of the neurohypophysis (Dreifuss et al., 1971; Gainer et al., 1986; Jackson et al, 1991) and at hippocampal mossy fiber synapses (Geiger and Jonas, 2000).

In the second mechanism, metabotropic receptor-mediated effects can contribute to short-term plasticity either when the stimulus train that triggers tetanic potentiation extends longer than several hundred milliseconds (the minimum time required for metabotropic signaling) or in the case of post-tetanic potentiation (where the time between the train and subsequent single action potentials that trigger neurotransmitter release is sufficiently long). As will be discussed in Chapter 21, synapses can release more than one neurotransmitter, and the chemical released to act on ionotropic receptors is usually accompanied by other chemicals released to act on metabotropic receptors in the synapse. Sometimes the same chemical (e.g., glutamate or acetylcholine) can act on both ionotropic and metabotrobic receptors at the synapse. While ionotropic receptors mediate the fast responses at the synapse, metabotropic receptors initiate slower biochemical changes that can modify the ionotropic signaling (see Chapter 12). When metabotropic signaling cascades are inhibitory, they can contribute to short-term synaptic depression during a long burst of action potentials. When they are excitatory, metabotropic signaling can lead to short-term facilitation.

## METABOTROPIC RECEPTOR-MEDIATED PLASTICITY OF IONOTROPIC SIGNALING

One of the most common forms of metabotropic short-term synaptic plasticity is sometimes called **autocrine inhibitory feedback**. As a point of definition, "autocrine" refers to chemical communication between the releasing neuron and itself. For example, at synapses that release acetylcholine or glutamate, the presynaptic nerve terminal may contain presynaptic inhibitory metabotropic receptors. Neurotransmitter release from the terminal activates these receptors, which inhibit further release as a negative feedback mechanism.

A related term, "paracrine," refers to chemical communication in which a neuron releases a chemical neurotransmitter that diffuses in the extracellular space around cells. Sometimes, **paracrine transmission** is referred to as "volume transmission" because it is not confined to a synaptic cleft, but rather diffuses in the space around cells (Fig. 14.6). The terms "autocrine" and "paracrine" are both related to the term "endocrine," which is commonly used to describe chemical communication through the bloodstream.

Furthermore, not all chemicals involved in communication at the synapse are released from the nerve terminal. The term "retrograde messenger" refers to chemicals that are released by the postsynaptic cell and diffuse "backwards" to the presynaptic cell.

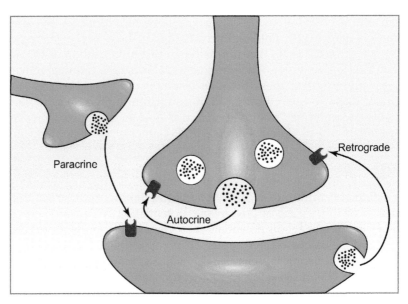

FIGURE 14.6  **Autocrine, paracrine, and retrograde signaling at synapses.** In this diagram, neurotransmitter is shown being released from both pre- (top) and postsynaptic (bottom) neurons, and from a neighboring axon (left). When neurotransmitter released from the presynaptic neuron binds onto receptors on that same presynaptic neuron, this is termed autocrine transmission. When neurotransmitter is released from a postsynaptic neuron and binds to a presynaptic neuron, this is termed retrograde transmission. When neurotransmitter is released from a neighboring axon and diffuses to a neighboring synapse, this is termed paracrine transmission.

Because metabotropic receptors generally have a very high affinity for their ligands (usually in the nM range), the chemical transmitters that activate these receptors often employ autocrine or paracrine means of communication. This allows these transmitters to diffuse some distance away from their site of release and still have functional effects on their receptors.

A classic example of metabotropic receptor-mediated signaling that modifies the strength of ionotropic signaling comes from studies of the sympathetic chain ganglion adjacent to the spinal cord (see Fig. 14.7 for a graphical representation of the details discussed below). These ganglia each receive synaptic input from neighboring ganglia in the chain, as well as from the spinal cord. The input from the superior ganglion in the chain is cholinergic, while the input from the spinal cord releases a peptide called "luteinizing hormone-releasing hormone" (LHRH).

The acetylcholine released from the superior ganglion in the chain acts on "B cells" within the ganglion by binding to two different acetylcholine receptor types. Acetylcholine acts within the synaptic cleft on nicotinic ionotropic receptors to generate a large "fast EPSP," but a burst of action potentials can trigger the release of enough acetylcholine to diffuse out of the synaptic cleft and act on extrasynaptic muscarinic metabotropic receptors. Activation of the muscarinic receptors triggers the closing of some of the B cells' resting potassium channels, leading to a slowly developing and long-lasting mild

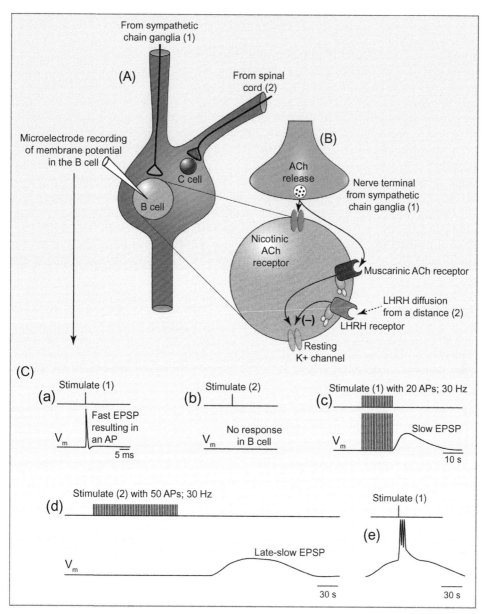

FIGURE 14.7 **Metabotropic modulation of ionotropic signaling within the sympathetic chain ganglion.** (A) The sympathetic ganglion shown contains larger B cells and smaller C cells. The B cells receive a synaptic input that comes through the connection to other neighboring sympathetic ganglia, and the C cells receive a synaptic input that comes from the spinal cord. To study synaptic transmission and modulation in this system, experimenters have used intracellular microelectrodes to record from the larger B cells and stimulate either the synaptic input from the neighboring sympathetic ganglia (1) or from the spinal cord (2). (B) Enlargement of the diagram detailing the synaptic input and receptors on the B cell. The direct synaptic input onto B cells releases the neurotransmitter acetylcholine (ACh), which binds to nicotinic ACh receptors on the postsynaptic B cell. In addition, extrasynaptic muscarinic ACh receptors on B cells bind ACh following bursts of action potential activity that lead to the diffusion of ACh to these more distant receptors. B cells also contain luteinizing hormone-releasing hormone (LHRH) receptors, which bind LHRH when it diffuses from nerve terminals that synapse directly onto the more distant C cells. Both muscarinic and LHRH receptors are metabotropic and their signaling

depolarization of the membrane potential (the "slow EPSP"). The slow EPSP is **subthreshold**, meaning it does not depolarize the membrane enough to generate an action potential.

Meanwhile, the input to the sympathetic ganglion from the spinal cord releases LHRH directly onto smaller neurons in this ganglion ("C cells"). A single stimulus of this spinal input does not affect the B cells. However, when the spinal input is stimulated with a burst of action potentials, LHRH diffuses away from the C cells and acts in a paracrine fashion on the more distant B cells, binding to metabotropic LHRH receptors that are coupled to resting potassium channels (the same type of potassium channels that the B cells' muscarinic receptors act on). LHRH binding to its receptors closes the coupled potassium channels, causing a "late-slow EPSP." This late-slow EPSP is similar to the muscarinic slow EPSP, but is delayed due to the time it takes for LHRH to diffuse from a distance within the ganglion to act on the LHRH receptors on B cells (Fig. 14.7).

Interestingly, although these slow and late-slow EPSPs are subthreshold, they modify the resting state of the B cells, altering the ionotropic nicotinic signaling when it occurs at the same time as these subthreshold slow EPSPs. With some of the B cell's resting potassium channels blocked, the ionotropic fast EPSP is larger (because the membrane resistance of the B cell is higher) and lasts longer (because there are fewer open resting potassium channels to aid in pulling the B cell's membrane potential back to rest after a depolarization). Unlike a fast EPSP that occurs in the absence of a slow or late-slow EPSP, which triggers a single action potential, this larger, longer-lasting modified fast EPSP can remain above threshold long enough to trigger a short burst of action potentials. Therefore, the muscarinic and LHRH metabotropic receptors, despite not generating an action potential themselves, modify the B cell membrane in a way that alters the normal ionotropic signaling. This is a classic example of metabotropic receptor-mediated plasticity in ionotropic signaling (Weight and Votava, 1970; Schulman and Weight, 1976; Jan et al., 1979; Jan and Jan, 1982).

The above example in sympathetic chain ganglia is only one of a growing list of ways in which metabotropic signaling can modify ionotropic signaling. Because metabotropic

cascade inhibits resting potassium channels on the B cells. (C) Diagrams that represent recordings of membrane potential from B cells after stimulating either the synaptic input from the sympathetic chain (1) or from the spinal cord (2) with either a single action potential or a train of action potentials. (Ca) When a single stimulus is delivered to the sympathetic input (1) to the B cells, investigators record a fast excitatory postsynaptic potential (EPSP) that is so large that it is above threshold and elicits an action potential in the B cell. (Cb) A single stimulus to the spinal cord input does not lead to any synaptic potential in B cells (because the spinal cord input does not synapse directly onto B cells). (Cc) When a train of action potentials is delivered to the sympathetic input (1) to the B cell, the fast EPSP-mediated train of action potentials is followed by a slowly rising and falling EPSP (slow EPSP). This slow EPSP is caused by activation of muscarinic receptors, which close resting potassium channels, leading to a depolarization of B cell membrane potential (this is a subthreshold response). (Cd) When a train of action potentials is delivered to the spinal cord input (2), there are no fast events in the B cell (because this input does not synapse directly onto the B cell), but with some delay, there is a slowly rising and falling EPSP (late-slow EPSP). This late-slow EPSP is caused by activation of LHRH receptors (due to LHRH release onto neighboring C cells that diffuses to B cells in a paracrine fashion), which close resting potassium channels, leading to a depolarization of B cell membrane potential (this is a subthreshold response). (Ce) When a single stimulus is delivered to the sympathetic chain input during a slow or late-slow EPSP (generated by prior stimulus activity), the fast EPSP that is generated is larger and lasts longer, leading to the generation of a short burst of action potentials [instead of the one action potential that is normally generated during resting conditions—see (Ca)].

receptors can couple to so many pathways within neurons (see Chapter 12), and these pathways can impact many pre- and postsynaptic targets, the range of synaptic plasticity that can be accomplished is extraordinary. The targets for metabotropic modulation of synapses include voltage-gated ion channels (especially calcium and potassium), SNARE proteins, calcium-binding proteins (e.g., synaptotagmins), ionotropic receptors, kinases, and other enzymes. Metabotropic modulation of so many proteins at the synapse can mediate a wide array of plastic changes.

## HABITUATION AND SENSITIZATION

There are also longer-term plastic events in the nervous system that underlie learned behaviors. Perhaps the most well studied of these are the relatively simple learning paradigms termed "habituation" and "sensitization." The underlying cellular and synaptic mechanisms that lead to habituation and sensitization were first studied in the sea slug *Aplysia* (Carew and Kandel, 1973; Castellucci et al., 1978). This invertebrate has a relatively simple nervous system, but it can still exhibit learned behaviors. Investigators used this model system in the hope that it would permit a cellular and synaptic dissection of behavioral responses (although the early hypotheses developed to explain *Aplysia* behaviors may be more complicated than anticipated; see Burrell and Sahley, 2001; Roberts and Glanzman, 2003).

**Habituation** is one of the simplest behaviors used to study memory (Thompson and Spencer, 1966). Habituation is defined as a reduction in responsiveness to an environmental stimulus after repeated or prolonged exposure. For example, you don't sense that your socks are on after walking around in them for a while. Touch receptors in your skin detect the presence of the socks when you first put them on, but gradually, the sensory response habituates. Depression of touch sensory synapses during habituation is adaptive because it allows us to ignore irrelevant sensory stimuli and focus on more important sensations.

Experiments in *Aplysia* have provided valuable historical data on habituation and sensitization. *Aplysia* is a soft, shell-less mollusk, and thus protects itself by withdrawing its delicate siphon and gill when touched. However, this withdrawal reflex is subject to habituation. Kandel and colleagues showed that a single tactile stimulus to the siphon of the *Aplysia* sea slug led to a strong gill-withdrawal behavior (Fig. 14.8); however, after repeated low-frequency tactile stimulation to the siphon ($\sim$ once per second), the gill-withdrawal behavior became less pronounced and would gradually vanish.

Both short-term and long-term forms of habituation have been demonstrated in *Aplysia*. Here, we will focus only on the short-term form of habituation, which occurs during low-frequency repeated stimulation and has been shown to last for several minutes. Short-term habituation has been hypothesized to be mediated by presynaptic changes in the terminal of the sensory neuron that excites motor neurons responsible for the withdrawal response. Early work identified that the mechanism for short-term habituation involved a decrease in the probability of neurotransmitter release at sensory nerve terminals (Castellucci and Kandel, 1974; Castellucci et al., 1978). A presynaptic locus for change is supported by experiments showing that blocking postsynaptic glutamate receptors fails to prevent

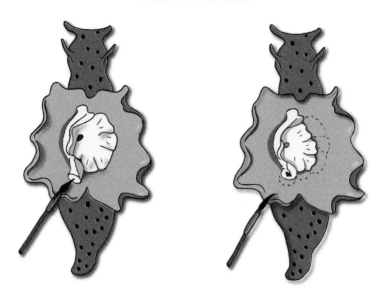

FIGURE 14.8 **Gill and siphon withdrawal in *Aplysia*.** The left panel shows an *Aplysia* with its gill (*white*) fully extended as a small point brush approaches the siphon. (Right panel) After the paint brush gently touches the siphon, the siphon and gill withdraws.

reductions in presynaptic neurotransmitter release during repeated stimulation (Armitage and Siegelbaum, 1998).

What causes a reduction in the magnitude of neurotransmitter released from the sensory nerve terminal during repeated stimulation? Early theories centered on the possibility that presynaptic calcium channels might experience use-dependent inactivation (Klein et al., 1980), or that there might be vesicle depletion at neurotransmitter release sites (Bailey and Chen, 1988). However, neither of these mechanisms appears responsible for short-term habituation at this synapse (Armitage and Siegelbaum, 1998; Jiang and Abrams, 1998; Byrne, 1982). Recall that the magnitude of quantal neurotransmitter release can be defined by the equation $m = n \times p \times q$ (see Chapter 6). With this in mind, Royer et al. (2000) used a binomial analysis to show that the probability of release ($p$) does not change during habituation; however, the number of functional release sites ($n$) does change. Because habituation occurs rapidly (within several seconds), it was hypothesized that the nerve terminal was not physically changing size (or number of release sites), but rather that existing release sites were being silenced completely. How do individual release sites become silenced or turned off? While it is known that presynaptic calcium is required to initiate habituation, the details of how this occurs are still under investigation. One hypothesis is that calcium entry during low-frequency action potential activity triggers the activation of small GTP-binding proteins that shut down neurotransmitter release sites (Gover and Abrams, 2009; Fig. 14.9). Furthermore, since high-frequency action potential activity does not induce habituation, habituation is proposed to be prevented by protein kinase C (PKC) activation that only occurs during high-frequency activity (Fig. 14.9). In summary, this hypothesis proposes a synaptic switch that silences release sites to

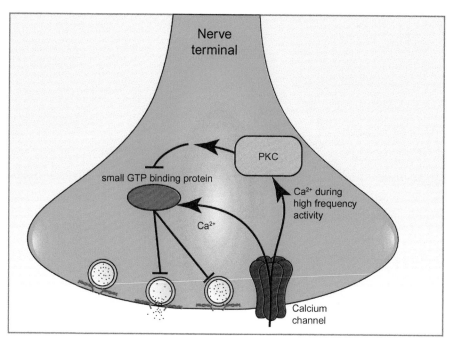

FIGURE 14.9 **Proposed habituation mechanisms in *Aplysia*.** Low-frequency activity is proposed to cause calcium influx that activates small GTP-binding proteins in the nerve terminal. These GTP-binding proteins are thought to silence neurotransmitter release sites. This mechanism may underlie short-term habituation in *Aplysia* as sensitivity to infrequent stimuli leads to a reduction in presynaptic neurotransmitter release. During high-frequency (intense) activity, calcium entry is much larger and hypothesized to activate protein kinase C (PKC), which leads to the inhibition of the GTP-binding proteins, preventing habituation. During intense activity, the synapse does not habituate because PKC prevents GTP-binding proteins from silencing neurotransmitter release sites.

habituate neurotransmitter release during low-frequency activity, but prevents this silencing of release sites when sensory neuron activity is very high, thus maintaining sensitivity to presumably important stimuli that intensely activate the sensory neuron (Gover and Abrams, 2009; Gover et al., 2002).

**Sensitization** is the process by which repeated exposure to a stimulus can lead to an increase in responsiveness. Sensitization can also occur when an irritating or alarming stimulus precedes an innocuous environmental stimulus, leading to an enhanced response to the innocuous environmental stimulus. For example, your startle response to a loud noise can be enhanced by a sense of fear. This is another "simple" behavior that has been studied using the gill withdrawal response in the sea slug *Aplysia*. Investigators discovered that the magnitude of the gill withdrawal response to light touch of the siphon could be enhanced (or sensitized) by a preceding strong stimulus (e.g., electric shock) to another part of the *Aplysia* body (e.g., the tail). The relatively simple nervous system in this animal aided in the discovery that electric shock to the tail activated sensory neurons that could "sensitize" the normal gill withdrawal synaptic circuit triggered by light touch to the

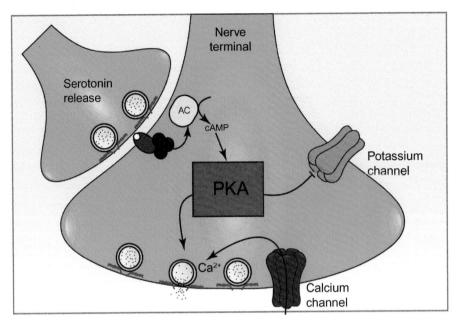

FIGURE 14.10 **Sensitization mechanisms enhancing neurotransmitter release at the presynaptic nerve terminal that activates the motor response in the *Aplysia* gill-withdrawal behavior.** Serotonin release from a facilitating interneuron activates the cAMP signaling system, which enhances neurotransmitter release to excite motor neurons for gill withdrawal in *Aplysia*. Neurotransmitter release is enhanced by cAMP-dependent protein kinase activity (PKA) that blocks potassium channels (broadening the presynaptic action potential) and phosphorylates active zone proteins to increase the probability of neurotransmitter release.

siphon (Fig. 14.10). The pathway that underlies this sensitization appears to involve a sensory neuron in the tail of the animal that excites a "facilitating" interneuron, which in turn synapses onto the presynaptic nerve terminal that normally activates the motor response to trigger the gill withdrawal. The facilitating neuron synapse releases the neurotransmitter serotonin onto the presynaptic nerve terminal activating gill withdrawal and increases the magnitude of neurotransmitter released to excite the gill withdrawal (Fig. 14.10).

How does serotonin increase neurotransmitter release in the motor circuit that triggers gill withdrawal? Serotonin acts on metabotropic receptors on the presynaptic nerve terminals that normally excite the motor neurons in this system. These metabotropic receptors are coupled to the cAMP signaling system (see Chapter 12) within the nerve terminal and the elevated presynaptic cAMP leads to the phosphorylation of proteins in the nerve terminal that increase the probability of neurotransmitter release by several mechanisms (e.g., blocking potassium channels to broaden the presynaptic action potential, and an alteration of unknown active zone proteins that increase neurotransmitter release; Fig. 14.10; Glanzman et al., 1989; Mackey et al., 1989; Abrams et al, 1984).

These short-term forms of habituation and sensitization in *Aplysia* can change synaptic function for several minutes. What regulates the time course of these plastic events? In the case of habituation, the small GTP-binding proteins that are thought to inhibit release sites are active in their GTP bound form. They remain active as long as GTP is bound and they have very slow intrinsic GTPase activity, which governs the time course of their effects. In the case of sensitization, serotonin activates metabotropic receptors, and the downstream signaling cascade in this case is regulated by the lifetime of cAMP (governed by the phosphodiesterase activity in the cell) and the balance between PKA activity to phosphorylate proteins and the opposing influences of cytoplasmic phosphatases that will dephosphorylate those proteins. Because these studies in *Aplysia* led to a molecular and cellular understanding of behavior, in 2000, Eric Kandel was awarded the Nobel Prize for this work (Box 14.2).

---

### BOX 14.2

### NOBEL PRIZE TO ERIC KANDEL FOR STUDIES ELUCIDATING THE CELLULAR AND SIGNALING SYSTEMS THAT REGULATE BEHAVIOR

Eric Kandel (Fig. 14.11) shared the Nobel Prize in Physiology or Medicine in 2000 with Arvid Carlsson and Paul Greengard "for their discoveries concerning signal transduction in the nervous system" (Nobel.org). Eric Kandel's work focused on using the relatively simple nervous system of the sea slug Aplysia to elucidate the molecular and cellular mechanisms for the synaptic plasticity that was responsible for learning and memory in this animal. In particular, Kandel showed that protein phosphorylation played an important role in changing synaptic communication in the Aplysia nervous system within simple neuronal circuits that mediated these behaviors.

FIGURE 14.11 Eric Kandel. *Source: http://bigthink.com/think-again-podcast/think-again-podcast-62-eric-kandel-the-eye-ofthe-beholder*

## LONG-TERM SYNAPTIC PLASTICITY

Because we know that experience can lead to long-lasting changes in behavior (e.g., memory of an event in your life), there must be mechanisms for synaptic plasticity whose effects are equally long-lasting. In fact, even in the "simple" nervous system of the *Aplysia*, investigators discovered synaptic mechanisms for long-term plasticity. In *Aplysia*, if the siphon is stimulated repeatedly at low frequency for a prolonged period of time, there is a long-lasting form of habituation of the gill-withdrawal response (lasting many hours or days) that is dependent on calcium flux through postsynaptic glutamate receptors that triggers a long-term depression (LTD) of the sensory–motor synapse. This LTD is dependent on activating calcium-sensitive pathways in the postsynaptic cell and on new protein synthesis (Ezzeddine and Glanzman, 2003; Esdin et al., 2010).

Plasticity in synaptic circuits underlying simple behaviors in the sea slug *Aplysia* provided the foundation for our understanding of the neural basis of behavior (see Box 14.2), but how are complex behaviors in mammals regulated by synaptic plasticity? If synaptic plasticity did underlie complex behavior in the mammalian brain, where would one look for these changes?

## CLINICAL CASES THAT FOCUSED THE INVESTIGATION OF LONG-TERM SYNAPTIC PLASTICITY

There are several clinical cases that identified brain regions critical for long-term memory, which focused attention on the temporal lobe of the brain and the **hippocampus** in particular. The most famous of these was the case of Henry Molaison (often referred to as "H.M."; see Box 14.3). This patient had surgery to bilaterally removed large parts of his temporal lobes including large sections of the hippocampi, in order to cure intractable seizures. As a result, H.M. suffered from an inability to form new **declarative memories**. He could still remember much of his past and could learn new non-declarative memory tasks. This clinical case led to a focus on neural circuits in the hippocampus as being critical for this type of memory formation.

With the hippocampus as a focus for investigation, investigators began looking for synaptic plasticity that could help explain the formation of memories. The hippocampus is organized in a **trisynaptic loop** that aids in the study of the synaptic circuit (Fig. 14.12). In this circuit, neurons in the **entorhinal cortex** extend a fiber pathway (the **perforant path**) that synapses with neurons in the **dentate gyrus**. These dentate gyrus neurons then send axons along the mossy fiber pathway to synapse with **pyramidal cells** in the **CA3 region of the hippocampus**, which in turn send axons along the Schaffer collateral pathway to synapse with pyramidal neurons in the **CA1 region of the hippocampus**.

## LONG-TERM POTENTIATION

Timothy Bliss and Terje Lomo were the first to identify a form of long-term synaptic plasticity in the hippocampus (Bliss and Lomo, 1973). They used recordings of synaptic

## BOX 14.3

## THE CLINICAL CASE OF HENRY MOLAISON (H.M.)

At the age of 9, Henry Molaison sustained a head injury that caused a severe case of epilepsy. His epilepsy became so difficult to manage that the decision was made to surgically remove the region of the brain that was the focus for this epileptic activity (the hippocampi). This bilateral removal of most of his hippocampi and a small amount of the surrounding cortical areas of the brain was successful in that it reduced his epilepsy. Furthermore, this surgery did not affect his cognitive abilities, working memory, and skill learning, but it left him with the inability to form new long-term memories. This unexpected sideeffect became critical to our understanding of the role of hippocampal function in memory formation because the details of the surgery were well documented (we knew exactly which brain areas were removed), and the surgeon (Dr. William Scoville) had a PhD student (Brenda Milner) who carefully studied the behavioral deficits and capabilities of this patient (Scoville and Milner, 1957; Milner, 1970). As a result of these studies, it was determined that this patient suffered from severe **anterograde amnesia** (the loss of the ability to create new memories after the surgery). It became clear that the hippocampus was critical to forming new memories, and this significantly influenced basic research into the neural basis of memory.

In later years, H.M.'s brain was imaged using **magnetic resonance imaging** (MRI) and this revealed that the extent of the damage from the surgery was more widespread than had been expected (Corkin et al., 1997). After his death in 2008, H.M.'s brain was studied anatomically (Worth and Annese, 2012), and a detailed digital reconstruction of H.M.'s brain was performed. Based on these studies, it was realized that the surgery was not a "pure hippocampal lesion" as had been suspected. In fact, about half of his hippocampus had survived the surgery. While these more recent findings will lead investigators to re-evaluate some of the behavioral consequences of the surgery on this patient, the basic finding that the hippocampus is a critical site for the formation of new memories continues to drive basic research.

---

activity in the intact rabbit brain to look for synaptic plasticity in this circuit. When they stimulated the perforant pathway, they recorded synaptic activity in the dentate gyrus (see Fig. 14.12). When they stimulated at low frequency, they recorded a stable synaptic response amplitude in the dentate gyrus. However, when they delivered a high-frequency stimulation to the perforant pathway, they discovered that the strength of communication at synapses in the dentate gyrus gradually increased, and that this increase lasted for up to 10 hours! They concluded that at least one group of synapses in the hippocampus can be affected by neuronal activity that occurred several hours earlier, and that this time scale of synaptic change might be useful for information storage (Bliss and Lomo, 1973). This long-term change in synaptic strength was termed "long-term potentiation" or LTP (Fig. 14.13). Because the same synapses that were stimulated were the ones that were strengthened, this

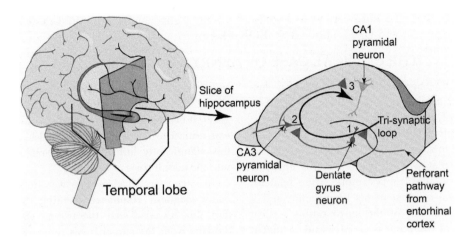

FIGURE 14.12  **Hippocampal trisynaptic circuit.** (Left panel) The hippocampus (*blue*) exists within the temporal lobe of the brain. If a slice of this area is removed, the synapses in the hippocampus can be studied more easily in vitro. (Right panel) Hippocampal slice showing the trisynaptic loop. Each of the three synapses in the loop are labeled. In particular, a popular synapse for study in this loop occurs between the axon branches of CA3 neurons (Schaffer collateral) and CA1 neurons (labeled "3"). For the purposes of illustration, the human brain is depicted on the left, and a hippocampal slice from a rodent brain is shown on the right (since most studies of the hippocampal trisynaptic circuit are performed in slices from rodent brain).

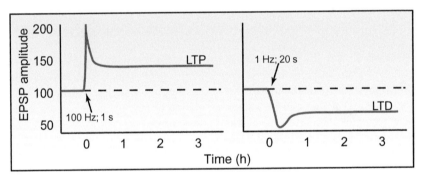

FIGURE 14.13  **Long-term potentiation (LTP) and long-term depression (LTD).** Plots of excitatory postsynaptic potential amplitude at a hippocampal synapse over time during two different stimulus patterns. (Left panel) Following a long burst of high-frequency stimulation (100 Hz for 1 s), synapses strengthen, leading to a larger EPSP amplitude, and this is maintained for hours (LTP). The transient spike in strengthening that occurs immediately after the 100 Hz stimulus train results from post-tetanic potentiation. (Right panel) Following a low-frequency train of activity (1 Hz for 20 s), synapses weaken persistently, leading to a smaller EPSP amplitude (LTD).

form of plasticity is termed "homosynaptic plasticity." These experiments confirmed a theory that had been postulated by D.O. Hebb in 1949 to explain how the nervous system might change in response to experience, often called "Hebbian plasticity" (see Box 5.2 in Chapter 5; Hebb, 1949). Essentially, Hebb proposed that when pre- and postsynaptic neurons were active simultaneously, they would strengthen their connections with one another ("neurons that fire together, wire together"). This theory had been difficult to validate until Bliss and Lomo (1973) showed that this phenomenon was true in the hippocampus.

Since these early recordings of LTP by Bliss and Lomo (1973), others have used hippocampal slices (see Fig. 14.12; which can be kept alive in a dish and present the opportunity for easier recording of synaptic activity) and identified LTP at other synapses in the trisynaptic loop. Subsequently, LTP has been discovered that conforms to Hebbian rules in a variety of brain areas (e.g., cortex, amygdala, and midbrain; Artola and Singer, 1987; Iriki et al., 1989; Hirsch et al., 1992; Chapman et al., 1990; Clugnet and LeDoux, 1990; Liu et al., 2005; Pu et al., 2006). Furthermore, variations in the cells studied and stimulus patterns used have led to the identification of a variety of types of LTP.

## PHYSIOLOGICAL STIMULUS PATTERNS THAT CAN INDUCE LONG-TERM POTENTIATION

After the discovery of LTP, it was tempting to speculate that this form of plasticity could underlie learning and memory. However, one concern was the nonphysiological patterns of stimulation that were traditionally used to induce LTP (e.g., a 100-Hz burst of action potentials for 1 second). As investigators began to explore other conditions that could induce LTP, it was discovered that LTP could in fact be induced by physiological patterns of stimulation. One such pattern was called "theta burst stimulation." Theta burst stimulation is a short burst of high-frequency activity (four action potentials at 100 Hz) that repeats about every 100 ms. This pattern of activity is thought to underlie the theta rhythm of brain activity characterized in electroencephalography (EEG) recordings as a burst of activity at 3–10 Hz (Fig. 14.14). This rhythm was of interest to investigators because it had been shown that theta rhythms are associated with cognitive processing in the brain, including learning and memory (Colgin, 2013). Interestingly, if 10–30 such theta bursts are delivered to synapses, LTP can be induced (Larson et al., 1986; Larson and Munkacsy, 2015).

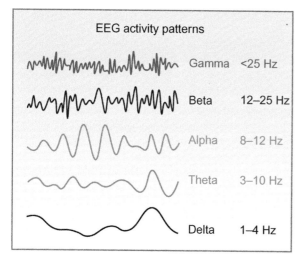

FIGURE 14.14 **EEG activity patterns recorded from the scalp.** Using a technique called EEG, rapid changes in the patterns of electrical activity of many neurons can be detected from electrodes on the scalp. These patterns are referred to as "brain waves," and the patterns are named using Greek letters, according to the frequencies of the voltage changes.

## ASSOCIATIVE LONG-TERM POTENTIATION

A type of long-term synaptic plasticity called "associative LTP" was identified at the thousands of synapses formed between the Schaffer collateral pathway and the CA1 pyramidal cells. These synapses are studied experimentally by stimulating the fiber tract (Schaffer collaterals) that runs between CA3 and CA1 regions of the hippocampus and recording from pyramidal neurons in the CA1 region (see Fig. 14.12). Depending on the intensity of the stimulus delivered to these Schaffer collaterals, a different number of axons are recruited to fire an action potential and this varies the strength of the postsynaptic response measured in the CA1 pyramidal cell. Barrionuevo and Brown (1983) found that they could place two different stimulating electrodes in the Schaffer collateral pathway in different locations and stimulate one very weakly (generating a small postsynaptic response that was always far below threshold) and the other very strongly (generating a very large postsynaptic response that was often above threshold and elicited an action potential in the CA1 pyramidal cell). If they stimulated the weak connection repeatedly at high frequency (100-Hz train for 1 second), they could not elicit **homosynaptic LTP**. Furthermore, if they stimulated the strong input pathway at high frequency (100-Hz train for 1 second) there was also no change in the strength of the weak input [indicating that there was no **heterosynaptic LTP**—defined as LTP induced in neighboring synapses that were not stimulated (see below)]. Interestingly, if they stimulated both the weak and strong Schaffer collateral inputs to a CA1 pyramidal neuron at the same time using the 100-Hz train for one second, then the weak synapses exhibited a robust increase in their synaptic strength that was stable for hours. Since this long-term plasticity in the weak synapses required the combined stimulation of both sets of inputs, it was termed **associative LTP** (Barrionuevo and Brown, 1983). This type of LTP is still considered to be governed by Hebbian rules because it is proposed that the simultaneous stimulation of the strong input with the weak input provides the necessary depolarization of the postsynaptic neuron to allow weak synapses to fire an action potential at the same time as the postsynaptic neuron to which they are synapsing, thus inducing a strengthening of the weak input.

Importantly, associative LTP does not require "cooperation" between the weak and strong inputs to depolarize the postsynaptic membrane and induce LTP. The "weak" input can actually only contain NMDA receptors postsynaptically (which require strong postsynaptic depolarization to open due to the $Mg^{2+}$ block; see Chapter 11). Synapses that only contain NMDA receptors are often called "silent synapses" because they lack the AMPA receptors that would allow them to respond during glutamate release while the postsynaptic membrane was near resting membrane potential. In this case, the stimulation of weak "silent" synapses at the same time as strong synapses onto one pyramidal cell can induce the weak "silent" synapses to display associative LTP because the strong synapses provide the postsynaptic depolarization that leads to Hebbian plasticity (Liao et al., 1995; Isaac et al., 1995).

Associative LTP has been proposed to underlie the behavioral learning paradigm called "classical conditioning" (Kelso and Brown, 1986). During classical conditioning, an animal is trained to associate a neutral stimulus (e.g., the ringing of a bell) with a strong motivational stimulus (e.g., food) when both are presented together. This effect was most famously demonstrated by Ivan Pavlov in his studies of salivation in dogs

(Pavlov et al., 1928). When comparing associative LTP with classical conditioning, the weak stimulus in associative LTP is likened to the neutral stimulus (bell ringing) in the behavioral paradigm, and the strong stimulus in associative LTP is likened to the strong motivational stimulus (food) in the behavioral paradigm. However, there is some debate regarding whether associative LTP is a substrate for classical conditioning (Diamond and Rose, 1994).

## SPIKE TIMING-DEPENDENT PLASTICITY

Because **LTP induction** (mechanisms that trigger LTP) had been shown to be dependent on simultaneous activity in both pre- and postsynaptic cells (as postulated by Hebb, see Box 5.1 in Chapter 5), investigators began examining various scenarios in which this would be the case. For example, with the discovery of active conductances in dendrites (see Chapter 13), the concept of a "back-propagating action potential" added another possible mechanism for postsynaptic depolarization. What would happen if a back-propagating action potential arrived at a dendritic spine at nearly the same time as an action potential arrived at the nerve terminal synapsing onto that spine? Surprisingly, several investigators showed that the timing of the two action potentials was critical (Levy and Steward, 1983; Bi and Poo, 1998; Debanne et al., 1998; Magee and Johnston, 1997; Markram et al., 1997). Considering how a synapse normally triggers an action potential in a postsynaptic neuron, we know that the presynaptic action potential occurs first, triggers neurotransmitter release, and then if that release is large enough to bring the postsynaptic membrane to threshold, the postsynaptic neuron fires an action potential about 1−2 ms later. In this normal sequence, the presynaptic action potential precedes the postsynaptic action potential in time. This sequence is consistent with Hebb's postulate because the presynaptic neuron is causing the postsynaptic neuron to be active, thus predicting a strengthening of this synapse (Caporale and Dan, 2008). In fact, experimentalists can vary the timing of these two action potentials, and have shown that if a presynaptic action potential precedes a postsynaptic action potential, and this paired stimulus is delivered many times at low frequency, that synapse is potentiated (LTP). Furthermore, the smaller the time difference, the larger the potentiation (Fig. 14.15). In contrast, if the postsynaptic action potential precedes the presynaptic action potential (e.g., if there is a back-propagating action potential that was initiated by some prior synaptic event), then the synapse under study becomes weaker, or depresses (LTD), and the magnitude of this depression is dependent on the time interval between each action potential (Fig. 14.15). With very short intervals between the postsynaptic action potential and the presynaptic action potential, the depression is stronger. These opposite effects on synaptic plasticity (potentiation vs depression) that are dependent on the timing of the pre- and postsynaptic action potentials have been termed "spike timing-dependent plasticity," or STDP. Interestingly, even with these pairs of action potentials pre- and postsynaptically, the plasticity has been shown to be very long-lasting.

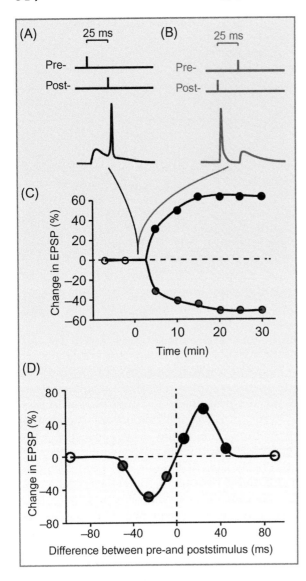

FIGURE 14.15 **Spike timing-dependent plasticity.** (A) The top two lines indicate the timing of a stimulus (vertical line) to the presynaptic (Pre-) and postsynaptic (Post-) neuron. Below these stimulus indicators is a drawing of the expected voltage recording in the postsynaptic neuron. This recording shows an EPSP triggered by the Pre- stimulus followed 25 ms later by an action potential triggered by the Post- stimulus. (B) The top two lines indicate the timing of a stimulus (vertical line) to the presynaptic (Pre-) and postsynaptic (Post-) neuron. Below these stimulus indicators is a drawing of the expected voltage recording in the postsynaptic neuron. This recording shows an action potential triggered by the Post- stimulus followed 25 ms later by an EPSP triggered by the Pre- stimulus. (C) When the stimulus pairs shown in (A) and (B) are delivered 50 times, once every 200 ms, the size of the EPSP changes persistently. If the Pre-stimulus is delivered before the Post-stimulus (A), the EPSP increases in amplitude (*black points on the curve*), however, if the Post- stimulus is delivered before the Pre- stimulus (B), the EPSP decreases in amplitude (*red points on the curve*). (D) Summary plot of the persistent changes in EPSP amplitude that occur depending on the time interval between Pre- and Post- stimuli. *Source: Adapted from Sgritta, M., Locatelli, F., Soda, T., Prestori, F., D'Angelo, E.U., 2017. Hebbian spike-timing dependent plasticity at the cerebellar input stage. J. Neurosci. 37, 2809–2823.*

## LONG-TERM DEPRESSION

The STDP described above that can lead to depression when the postsynaptic action potential precedes the presynaptic action potential highlights the fact that long-term synaptic plasticity can result in either potentiation or depression, depending on the conditions. In fact, *homosynaptic* **long-term depression** (LTD) also may result from low-frequency (1 Hz) stimulation to a synapse (in contrast to the high-frequency bursts that triggered LTP; Wagner and Alger, 1996; Lee at al., 1998, 2000; Fig. 14.13). LTP of Schaffer collaterals in the hippocampus is hypothesized to be important in memory formation (discussed above), but in order to undergo repeated episodes of learning, these potentiated synapses need a way to reset. One mechanism for this "resetting" is "homosynaptic depotentiation" (a form of LTD) in which the same inputs that undergo LTP can return to baseline levels of neurotransmitter release (reset) during low-frequency activity (Barrionuevo et al., 1980; Dudek and Bear, 1992, 1993).

We now know that there are various types of LTD (as there are for LTP), and that this general form of synaptic plasticity is present broadly across the central nervous system. LTD appears to represent a family of long-term mechanisms responsible for reducing neurotransmitter release, and is implicated in nervous system development (Bienenstock et al., 1982), addiction (Thomas et al., 2001), learning and memory, and a variety of neurological disorders, among others.

## HETEROSYNAPTIC PLASTICITY

The homosynaptic plasticity discussed above, while having powerful mechanisms for the refinement of connections, is not sufficient to explain the full extent of experience-dependent plasticity that underlies complex behavior. In addition, homosynaptic Hebbian plasticity by itself is predicted to create runaway changes in synaptic strength that could ultimately lead to either maximum potentiation or depression, depending on the stimulus patterns (Chistiakova et al., 2015). For example, Hebbian LTP in isolation will strengthen synapses and this will make them more likely to be strengthened further (leading to overexcitation), while homosynaptic LTD will weaken synapses such that it will make them less likely to experience Hebbian LTP (leading to near silencing of synapses). Therefore, additional plasticity mechanisms are required to balance homosynaptic LTP and LTD.

**Heterosynaptic plasticity** is defined as changes in the strength of a synapse induced by activity in neighboring synapses. This form of plasticity was first recognized during homosynaptic LTP induced by Schaffer collateral stimulation in CA1 of the hippocampus (see above). In addition to the homosynaptic LTP induced by Schaffer collateral stimulation, investigators noticed that there was heterosynaptic LTD at nearby commissural synapses that were not stimulated (Lynch et al., 1977). In fact, in some brain areas (hippocampus and amygdala), there was a type of center-surround pattern of LTP and LTD (Fig. 14.16). For example, if homosynaptic LTP was generated at a group of inputs, immediately adjacent there would be a weak form of heterosynaptic LTP, but further away there would be heterosynaptic LTD. The converse was also observed: homosynaptic LTD is surrounded by heterosynaptic LTD very close by, and heterosynaptic LTP further away (White et al., 1990; Royer and Pare, 2003; see Fig. 14.16). These examples demonstrate that activity in the presynaptic neuron is not required for synaptic plasticity, and provide a mechanism to balance synaptic strength of many inputs onto one postsynaptic neuron. Mechanistically, experimenters have shown that long-term plasticity can be induced by simply elevating the calcium concentration in the postsynaptic neuron independent of presynaptic activity (Neveu and Zucker, 1996a,b; Yang et al., 1999).

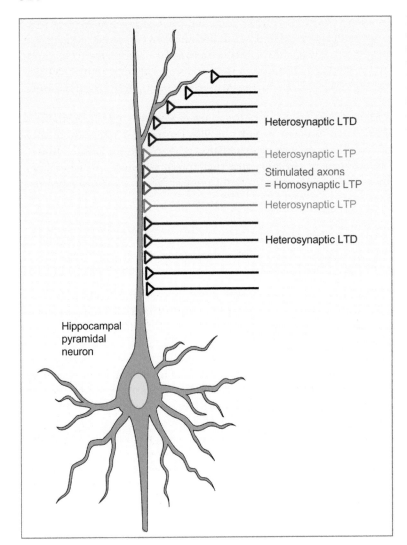

FIGURE 14.16 Patterns of homosynaptic and heterosynaptic plasticity following activation of a select group of synapses in the hippocampus. When an set of inputs to the apical dendrites are stimulated with a pattern of activity that leads to homosynaptic LTP (*red*), there are effects that surround these stimulated inputs that lead to heterosynaptic LTP (*yellow*) close by, and heterosynaptic LTD further away (*blue*). At greater distances, there are no changes in synaptic strength (*black*).

# SYNAPTIC SIGNALING MECHANISMS OF LONG-TERM POTENTIATION AND LONG-TERM DEPRESSION

How can different patterns of electrical activity lead to opposite synaptic plasticity effects? For example, a 100-Hz burst of action potentials can lead to homosynaptic LTP, while a series of low-frequency action potentials (1 Hz for several seconds) can lead to homosynaptic LTD. Interestingly, both of these stimulus-triggered plastic events can be triggered by calcium influx through the glutamate NMDA receptor (which requires the coincident presence of presynaptically released glutamate and postsynaptic depolarization to remove the magnesium block; see Chapter 11). In fact, postsynaptic calcium is a major signaling ion involved in plasticity. In one case (LTP) the calcium influx is large and occurs quickly, while in the other (LTD) the calcium influx is small and spread out over time. Therefore, for these types of plasticity, the magnitude and time course of calcium entry in the postsynaptic spine of a dendrite through the NMDA receptor is critical for determining whether the synapse experiences LTP or LTD (Bliss and Cooke, 2011).

How can the synapse use these different patterns of calcium entry to guide plasticity? The hypothesis is that large brief calcium entry activates calcium-calmodulin-dependent protein kinase II (CaMKII), while small prolonged calcium entry activates protein phosphatases (PP) in dendritic spines (see Fig. 14.17A). These two different enzymes can regulate the phosphorylation state of glutamate AMPA receptors (CaMKII can increase AMPA receptor function, while PP can decrease it).

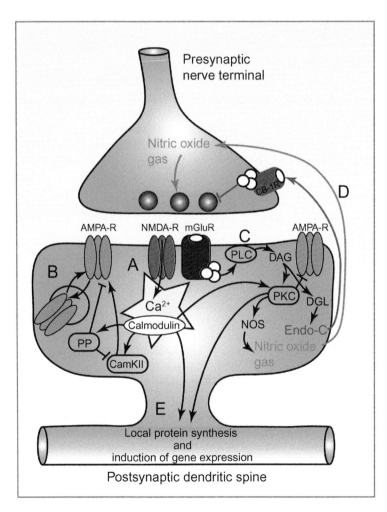

FIGURE 14.17 **Synaptic signaling mechanisms that underlie LTP and LTD.** Details of these pathways (A–E) are discussed in the text. Overall, pathway lines that end with an arrow represent a positive (stimulatory) influence, while lines that end with a bar represent a negative (inhibitory) influence. (A) Depending on the pattern of activation, NMDA receptors can either stimulate or inhibit AMPA receptor function via calcium-dependent pathways. (B) AMPA receptor number can be regulated. (C) Metabotropic glutamate receptors can modulate the function of AMPA receptors. (D) Retrograde messengers generated in postsynaptically can act back on the presynaptic nerve terminal to alter the probability of transmitter release. (E) The maintenance of long-term changes in synaptic function can depend on local protein synthesis and changes in gene expression of the neuron.

In addition to changing the function of AMPA receptors, LTP and LTD signaling systems can change the number of AMPA receptors on the cell membrane. There is an intraspine population of AMPA receptors that exists in transport vesicles that can be inserted into the spine plasma membrane during LTP. In contrast, during LTD, AMPA receptors can be removed from the spine plasma membrane and internalized into storage vesicles (Fig. 14.17B). These plasticity-induced changes in the number of AMPA receptors on the spine plasma membrane can change the strength of synaptic transmission as AMPA receptors often carry the majority of the synaptic current.

Metabotropic glutamate receptors can also be important regulators of long-term synaptic plasticity (Fig. 14.17C). In dendritic spines these metabotropic glutamate receptors are often coupled to the phosphoinositide system (see Chapter 12). This signaling system activates protein kinase C (PKC), which can alter the phosphorylation state of many proteins, including AMPA receptors—altering their function in a different manner than phosphorylation by CaMKII (because each kinase targets different amino acids in the AMPA receptor sequence). As a result, metabotropic glutamate receptor-mediated long-term plasticity can affect the strength of communication at synapses.

So far, we have focused on postsynaptic mechanisms of long-term synaptic plasticity, but there is evidence of presynaptic changes as well. Because long-term synaptic plasticity appears to be triggered by postsynaptic receptor function, how can the presynaptic nerve terminal be affected? The hypothesis is that there are diffusible, **retrograde messengers** (meaning they signal in the backward direction; postsynaptic to presynaptic) that are activated in the postsynaptic spine and travel through the extracellular space back to the presynaptic nerve terminal to exert their action (Fig. 14.17D). Two such candidates that have been studied include the gas **nitric oxide** and **endocannabinoids**.

Nitric oxide is generated by activation of its synthetic enzyme **nitric oxide synthase** (see Chapter 19). Once generated, nitric oxide is a freely diffusible reactive gas (even across cell membranes). This gas can enter the presynaptic nerve terminal and either activate **guanylyl cyclase** (increasing cGMP concentration) or **s-nitrosylate proteins** on cysteine amino acids (altering protein function). Either or both of these actions can alter the function of the surrounding cells in a paracrine manner, including the presynaptic nerve terminal, where these changes could contribute to long-term synaptic plasticity.

Endocannabinoids (e.g., 2-arachidonoylglycerol; 2-AG) are metabolites of the lipid phosphatidylinositol. After PLC generates diacylglycerol (DAG), diacylglycerol lipase (DGL) can convert DAG to 2-AG (an active endocannabinoid) in the postsynaptic spine (Katona et al., 2006; Yoshida et al., 2006). 2-AG then diffuses to the presynaptic nerve terminal where it binds to a selective metabotropic receptor (CB1R) to induce a signaling cascade within the nerve terminal that reduces the probability of neurotransmitter release (see Fig. 14.17C). Therefore, endocannabinoids are known primarily as potent mediators of synaptic depression.

Lastly, the *maintenance* of long-term synaptic plasticity (e.g., mechanisms that allow plastic changes to last especially long) is thought to be dependent on local dendritic protein synthesis (Sutton and Schuman, 2006) and gene expression (Silva et al., 1998). Changes in the genetic expression of so-called "immediate early genes" (a family of about

40 genes that can be activated transiently and rapidly following cellular activity) are often turned on by the cellular activity associated with long-term plasticity (Dragunow, 1996; Lanahan and Worley, 1998; Miyashita et al., 2008). Many of these immediate early genes code for the expression of transcription factors. There is now strong evidence that synaptic activity triggers signals in the postsynaptic spine that can be transferred to the cell nucleus and induce the expression of a variety of transcription factors (Richter and Klann, 2009). The hypothesis is that these transcription factors induce the expression of a variety of synaptic proteins and enzymes that are critical to maintaining long-term changes in synaptic function. With regard to local protein expression in the dendrites, historical dogma had led to the general view that all protein synthesis took place in the cell body. This view changed when polyribosomes were first reported at the base of dendritic spines (Steward and Levy, 1982). It now appears that there are many proteins that can be translated locally at the dendritic spine in response to very local activity signals (Fig. 14.17E), allowing for rapid input-specific changes to local dendritic spines that contribute significantly to long-term synaptic plasticity. Local proteins for which mRNA has been shown to be located at the spine include kinases (CamKII and an atypical isoform of PKC critical to long-term plasticity; PKM$\zeta$), brain-derived neurotrophic factor (BDNF) and its receptor (TrkB), subunits of the NMDA and glycine receptors, cytoskeletal associated proteins (MAP2 and Arc), and the intracellular IP3 receptor (Kleiman et al., 1990; Miyashiro et al., 1994; Steward and Schuman, 2003; Lyford et al., 1995). The details regarding how these mRNAs are translated into proteins are still being investigated.

It is important to note that the same stimulus patterns that might induce LTP or LTD in one region of the brain may not be effective in another. Furthermore, even within one brain region, the effectiveness of a particular stimulus paradigm is influenced by the recent history of signaling in that system (Abraham and Bear, 1996; Ngezahayo et al., 2000). Therefore, the ability of synapses to experience LTP or LTD can be regulated by either prior electrical activity (**metaplasticity**) or metabotropic signaling (**plasticity modulation**).

## METAPLASTICITY

Metaplasticity (sometimes called the "plasticity of synaptic plasticity") can determine the conditions under which long-term plasticity can take place. This gating of synaptic plasticity is important because we don't want to learn everything we experience. In other words, the context within which we experience something might be important in determining if we want a long-term memory of that event. In metaplasticity, prior activity at the synapse can influence whether LTP or LTD is generated by subsequent activity (Abraham and Bear, 1996). For example, a brief burst of activity that generates short-term plasticity (e.g., tetanic potentiation) at a hippocampal synapse doesn't change baseline synaptic transmission for very long (hundreds of milliseconds), but this activity can have prolonged effects that inhibit the ability to generate LTP and enhance the ability to generate LTD during subsequent activity (reviewed in Abraham and Bear, 1996). This metaplasticity can last for up to an hour before conditions are reverse back to the control state.

How can prior activity of a synapse influence whether long-term synaptic plasticity can occur? There are likely several mechanisms that can underlie metaplasticity, and these can vary at different synapses. In some cases, calcium flux through NMDA receptors can be the trigger than initiates metaplasticity, and could involve the activity of calcium-dependent kinases (e.g., calcium-calmodulin-dependent protein kinase), and other kinases that can be activated by the short burst of activity that induced metaplasticity (Deisseroth et al., 1995; Abraham, 2008).

## PLASTICITY MODULATION

Above we discussed how metabotropic receptors could contribute to, or modify, short-term synaptic plasticity. Therefore it is not a stretch to imagine that metabotropic receptors might influence long-term synaptic plasticity as well. In fact, since metabotropic receptors can essentially change the resting biochemical state of a cell, they can influence the synaptic plasticity that occurs.

One example of such plasticity modulation in the hippocampus is demonstrated by the actions of the neurotransmitter norepinephrine. Norepinephrine acts on metabotropic **β-adrenergic receptors** in the hippocampus to increase the likelihood of synaptic plasticity (Katsuki et al., 1997; Gelinas and Nguyen, 2007). These adrenergic receptors can induce the phosphorylation of dendritic potassium channels, reducing the likelihood that they open near resting membrane potential, and thus increasing membrane excitability of the dendrite. This increase in membrane excitability increases back-propagating action potentials in these dendrites, making it more likely that NMDA receptors will be depolarized (relieving the magnesium block). These events allow postsynaptic NMDA receptors to flux calcium ions into the dendritic spine after presynaptic glutamate release, which essentially makes it more likely that spike timing-dependent LTP can occur (Hoffman et al., 1997; Yuan et al., 2002; Maity et al., 2015; Sweatt, 2016).

## HOMEOSTATIC SYNAPTIC PLASTICITY

In the nervous system, network activity regulates the output and behavioral consequences of brain function. Individual neurons participate in network activity based on the reliability of their communication with one another. When neurons either become too active or too silent, there is another form of synaptic plasticity that serves to balance this activity, returning it to a set point at which neurons and circuits function well. This form of plasticity is called **homeostatic synaptic plasticity**. **Homeostasis** is defined as the ability to adjust in order to maintain a stable state. With regard to neuronal activity, if neurons had a way to sense their average level of activity, they could adjust the strength of their communication with synaptic partners to reach a set point at which network activity would be stable (Miller, 1996; Sullivan and de Sa, 2006). This concept essentially requires a negative feedback system in which too much overall activity could be reduced, and too little overall activity could be boosted. Importantly, these activity levels need to be sensed over the time course of minutes to hours (despite the fact that neurons and synapses communicate on a time scale of milliseconds) and some aspect of synaptic communication needs to be adjusted to get back to the set point (Davis, 2006).

The first experimental measure of homeostatic plasticity was performed in cultured cortical neurons in which synaptic connections were formed between excitatory pyramidal neurons and inhibitory interneurons. After a long-term block (hours or days) of action potential activity in these cultures using a sodium channel toxin (tetrodotoxin, TTX; see Box 6.2), Turrigiano et al. (1998) found that neurons increased the number of postsynaptic receptors at excitatory synapses in an attempt to compensate for the reduced network activity. This increase in the number of postsynaptic receptors served to increase the strength of excitatory communication at synapses and was termed "synaptic scaling" (Fig. 14.18). Synaptic scaling can be detected as an increase in the size of miniature synaptic events (an increase in "q" due to a larger postsynaptic response to a single quantum; see Chapter 6, for a discussion of quantal analysis). This synaptic scaling effect was also shown to be bidirectional, such that manipulations that increased network activity led to a decrease in the size of miniature synaptic events. The advantage of synaptic scaling is that based on the level of network activity, the excitatory synapses in the network can all be scaled up or down without changing the relative level of connectivity between neurons in the network, preserving information processing in the network. Since this first report, other mechanisms have been identified in different neuronal populations and following different activity manipulations (Turrigiano, 2007, 2008).

In a variety of studies, synaptic activity block by one method or another has not only been shown to lead to an increase in postsynaptic glutamate receptors as described above, but has also been shown to increase presynaptic function (Turrigiano, 2007). This phenomenon has been most extensively studied at the *Drosophila* neuromuscular junction (Davis, 2006). First shown by genetic reductions in postsynaptic glutamate receptor function at this synapse (*Drosophila* neuromuscular synapses use glutamate as a transmitter), researchers found that this reduction led to an increase in presynaptic quantal content (Paradis et al., 2001; Fig. 14.19). A change in presynaptic function that results from a manipulation of postsynaptic function implies that there must be a retrograde message conveyed between pre- and postsynaptic compartments. More recently, this presynaptic form of homeostatic plasticity was shown to occur quickly (within 10 minutes) after only a mild (partial) pharmacological block of postsynaptic receptors (Frank et al., 2006). The advantage of studying this form of plasticity in *Drosophila* is that investigators could use genetic screens to relatively quickly identify candidate proteins in this process. As a result, many proteins have been shown to be critical for homeostatic plasticity in *Drosophila* (Davis and Muller, 2015). One presynaptic target that has been identified is the voltage-gated calcium channel that is expressed within neurotransmitter release sites. Because calcium influx is such a critical trigger for neurotransmitter release, this is a logical target for plasticity. The precise manner by which a retrograde signal manipulates presynaptic function to mediate this form of plasticity is still under study.

At mammalian central synapses, presynaptic changes resulting from homeostatic plasticity have been shown to include changes in presynaptic release probability (Murthy et al., 2001), spontaneous release frequency (Bacci et al., 2001), and the size of the synapse (Murthy et al., 2001). Overall, it is clear that homeostatic synaptic plasticity is an

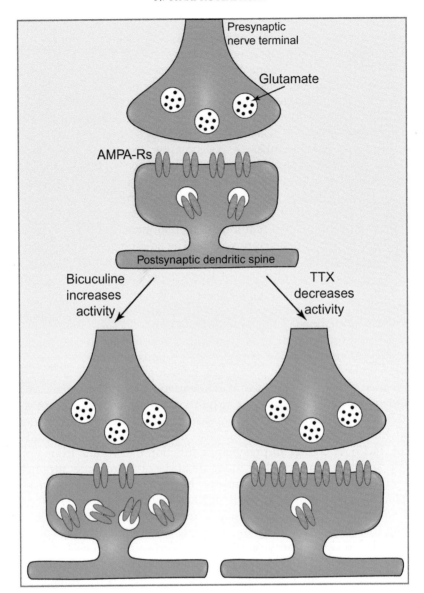

FIGURE 14.18 **Synaptic scaling as a homeostatic mechanism to adjust excitatory synaptic strength in response to changes in action potential activity in the system.** When system activity is increased by blocking GABA receptors with bicuculline, there is a compensatory decrease in the postsynaptic glutamate receptor density (to decrease excitation caused by glutamate release). In contrast, when activity is decreased by blocking voltage-gated sodium channels (using tetrodotoxin which blocks action potential activity), there is a compensatory increase in the postsynaptic glutamate receptor density (to increase excitation caused by glutamate release).

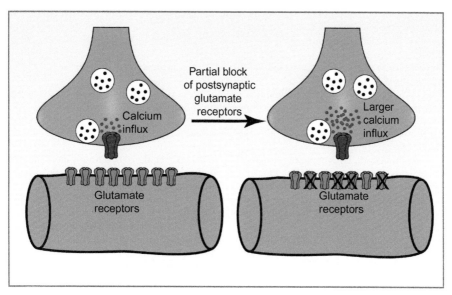

FIGURE 14.19  **Presynaptic homeostatic plasticity at the *Drosophila* neuromuscular junction.** When postsynaptic glutamate receptors are partially blocked, the presynaptic nerve terminal compensates, in part, by increasing calcium influx into the presynaptic nerve terminal.

important form of compensation by neurons to adapt to changes in the set point of activity in the system.

Homeostatic plasticity is also an important aspect of diseases that attack synapses. For example, the autoimmune disease myasthenia gravis is known to be caused by autoantibodies that patients make which attack their own postsynaptic acetylcholine receptors at the neuromuscular junction (Berrih-Aknin and Le Panse, 2014). This autoimmune attack leads to neuromuscular weakness caused by a reduction in the number of postsynaptic acetylcholine receptors. In response to this attack, the presynaptic nerve terminal of the neuromuscular junction increases the magnitude of neurotransmitter released, including a mechanism that shifts the calcium-sensitivity of the release apparatus (Cull-Candy et al., 1980; Wang and Rich, 2018). It is now clear that homeostatic synaptic plasticity is associated with a large number of neurological and psychiatric diseases (Wondolowski and Dickman, 2013).

# References

Abbott, L.F., Varela, J.A., Sen, K., Nelson, S.B., 1997. Synaptic depression and cortical gain control. Science 275, 220–224.

Abraham, W.C., Bear, M.F., 1996. Metaplasticity: the plasticity of synaptic plasticity. Trends Neurosci. 19, 126–130.

Abraham, W.C., 2008. Metaplasticity: tuning synapses and networks for plasticity. Nat. Rev. Neurosci. 9 (5), 387.

Abrams, T.W., Castellucci, V.F., Camardo, J.S., Kandel, E.R., Lloyd, P.E., 1984. Two endogenous neuropeptides modulate the gill and siphon withdrawal reflex in Aplysia by presynaptic facilitation involving cAMP-dependent closure of a serotonin-sensitive potassium channel. Proc. Natl. Acad. Sci. USA 81, 7956–7960.

Armitage, B.A., Siegelbaum, S.A., 1998. Presynaptic induction and expression of homosynaptic depression at Aplysia sensorimotor neuron synapses. J. Neurosci. 18, 8770–8779.

Artola, A., Singer, W., 1987. Long-term potentiation and NMDA receptors in rat visual cortex. Nature 330, 649–652.

Atluri, P.P., Regehr, W.G., 1996. Determinants of the time course of facilitation at the granule cell to Purkinje cell synapse. J. Neurosci. 16, 5661–5671.

Bacaj, T., Wu, D., Yang, X., Morishita, W., Zhou, P., Xu, W., et al., 2013. Synaptotagmin-1 and synaptotagmin-7 trigger synchronous and asynchronous phases of neurotransmitter release. Neuron 80, 947–959.

Bacaj, T., Wu, D., Burre, J., Malenka, R.C., Liu, X., Sudhof, T.C., 2015. Synaptotagmin-1 and -7 are redundantly essential for maintaining the capacity of the readily-releasable pool of synaptic vesicles. PLoS Biol. 13, e1002267.

Bacci, A., Coco, S., Pravettoni, E., Schenk, U., Armano, S., Frassoni, C., et al., 2001. Chronic blockade of glutamate receptors enhances presynaptic release and downregulates the interaction between synaptophysin-synaptobrevin-vesicle-associated membrane protein 2. J. Neurosci. 21, 6588–6596.

Bailey, C.H., Chen, M., 1988. Morphological basis of short-term habituation in Aplysia. J. Neurosci. 8, 2452–2459.

Barrionuevo, G., Brown, T.H., 1983. Associative long-term potentiation in hippocampal slices. Proc. Natl. Acad. Sci. USA 80, 7347–7351.

Barrionuevo, G., Schottler, F., Lynch, G., 1980. The effects of repetitive low frequency stimulation on control and "potentiated" synaptic responses in the hippocampus. Life. Sci. 27, 2385–2391.

Berrih-Aknin, S., Le Panse, R., 2014. Myasthenia gravis: a comprehensive review of immune dysregulation and etiological mechanisms. J. Autoimmun. 52, 90–100.

Bi, G.Q., Poo, M.M., 1998. Synaptic modifications in cultured hippocampal neurons: dependence on spike timing, synaptic strength, and postsynaptic cell type. J. Neurosci. 18, 10464–10472.

Bienenstock, E.L., Cooper, L.N., Munro, P.W., 1982. Theory for the development of neuron selectivity: orientation specificity and binocular interaction in visual cortex. J. Neurosci. 2, 32–48.

Blatow, M., Caputi, A., Burnashev, N., Monyer, H., Rozov, A., 2003. Ca2+ buffer saturation underlies paired pulse facilitation in calbindin-D28k-containing terminals. Neuron 38, 79–88.

Bliss, T.V., Lomo, T., 1973. Long-lasting potentiation of synaptic transmission in the dentate area of the anaesthetized rabbit following stimulation of the perforant path. J. Physiol. 232, 331–356.

Bliss, T.V., Cooke, S.F., 2011. Long-term potentiation and long-term depression: a clinical perspective. Clinics (Sao Paulo) 66 (Suppl 1), 3–17.

Bolkan, S.S., Stujenske, J.M., Parnaudeau, S., Spellman, T.J., Rauffenbart, C., Abbas, A.I., Harris, A.Z., Gordon, J.A., Kellendonk, C., 2017. Thalamic projections sustain prefrontal activity during working memory maintenance. Nat. Neurosci. 20 (7), 987–996.

Brandt, D.S., Coffman, M.D., Falke, J.J., Knight, J.D., 2012. Hydrophobic contributions to the membrane docking of synaptotagmin 7 C2A domain: mechanistic contrast between isoforms 1 and 7. Biochemistry 51, 7654–7664.

Burrell, B.D., Sahley, C.L., 2001. Learning in simple systems. Curr. Opin. Neurobiol. 11, 757–764.

Byrne, J.H., 1982. Analysis of synaptic depression contributing to habituation of gill-withdrawal reflex in Aplysia californica. J. Neurophysiol. 48, 431–438.

Caporale, N., Dan, Y., 2008. Spike timing-dependent plasticity: a Hebbian learning rule. Annu. Rev. Neurosci. 31, 25–46.

Carew, T.J., Kandel, E.R., 1973. Acquisition and retention of long-term habituation in Aplysia: correlation of behavioral and cellular processes. Science 182, 1158–1160.

Castellucci, V.F., Kandel, E.R., 1974. A quantal analysis of the synaptic depression underlying habituation of the gill-withdrawal reflex in Aplysia. Proc. Natl. Acad. Sci. USA 71, 5004–5008.

Castellucci, V.F., Carew, T.J., Kandel, E.R., 1978. Cellular analysis of long-term habituation of the gill-withdrawal reflex of Aplysia californica. Science 202, 1306–1308.

Chapman, P.F., Kairiss, E.W., Keenan, C.L., Brown, T.H., 1990. Long-term synaptic potentiation in the amygdala. Synapse 6, 271–278.

Charlton, M.P., Smith, S.J., Zucker, R.S., 1982. Role of presynaptic calcium ions and channels in synaptic facilitation and depression at the squid giant synapse. J. Physiol. 323, 173–193.

Chistiakova, M., Bannon, N.M., Chen, J.Y., Bazhenov, M., Volgushev, M., 2015. Homeostatic role of heterosynaptic plasticity: models and experiments. Front. Comput. Neurosci. 9, 89.

Clugnet, M.C., LeDoux, J.E., 1990. Synaptic plasticity in fear conditioning circuits: induction of LTP in the lateral nucleus of the amygdala by stimulation of the medial geniculate body. J. Neurosci. 10, 2818–2824.

Colgin, L.L., 2013. Mechanisms and functions of theta rhythms. Annu. Rev. Neurosci. 36, 295–312.

Corbalan-Garcia, S., Gomez-Fernandez, J.C., 2014. Signaling through C2 domains: more than one lipid target. Biochim. Biophys. Acta 1838, 1536–1547.

Corkin, S., Amaral, D.G., Gonzalez, R.G., Johnson, K.A., Hyman, B.T., 1997. H. M.'s medial temporal lobe lesion: findings from magnetic resonance imaging. J. Neurosci. 17, 3964–3979.

Cull-Candy, S.G., Miledi, R., Trautmann, A., Uchitel, O.D., 1980. On the release of transmitter at normal, myasthenia gravis and myasthenic syndrome affected human end-plates. J. Physiol. 299, 621–638.

Cuttle, M.F., Tsujimoto, T., Forsythe, I.D., Takahashi, T., 1998. Facilitation of the presynaptic calcium current at an auditory synapse in rat brainstem. J. Physiol. 512 (Pt 3), 723–729.

Dason, J.S., Romero-Pozuelo, J., Atwood, H.L., Ferrus, A., 2012. Multiple roles for frequenin/NCS-1 in synaptic function and development. Mol. Neurobiol. 45, 388–402.

David, G., Barrett, E.F., 2000. Stimulation-evoked increases in cytosolic [Ca(2 + )] in mouse motor nerve terminals are limited by mitochondrial uptake and are temperature-dependent. J. Neurosci. 20, 7290–7296.

David, G., Barrett, J.N., Barrett, E.F., 1998. Evidence that mitochondria buffer physiological Ca2 + loads in lizard motor nerve terminals. J. Physiol. 509 (Pt 1), 59–65.

Davis, G.W., 2006. Homeostatic control of neural activity: from phenomenology to molecular design. Annu. Rev. Neurosci. 29, 307–323.

Davis, G.W., Muller, M., 2015. Homeostatic control of presynaptic neurotransmitter release. Annu. Rev. Physiol. 77, 251–270.

Debanne, D., Gahwiler, B.H., Thompson, S.M., 1998. Long-term synaptic plasticity between pairs of individual CA3 pyramidal cells in rat hippocampal slice cultures. J. Physiol. 507 (Pt 1), 237–247.

Deisseroth, K., Bito, H., Schulman, H., Tsien, R.W., 1995. Synaptic plasticity: A molecular mechanism for metaplasticity. Curr. Biol. 5 (12), 1334–1338.

DeMaria, C.D., Soong, T.W., Alseikhan, B.A., Alvania, R.S., Yue, D.T., 2001. Calmodulin bifurcates the local Ca2 + signal that modulates P/Q-type Ca2 + channels. Nature 411, 484–489.

Diamond, D.M., Rose, G.M., 1994. Does associative LTP underlie classical conditioning? Psychobiology 22, 263–269.

Dragunow, M., 1996. A role for immediate-early transcription factors in learning and memory. Behav. Genet. 26, 293–299.

Dreifuss, J.J., Kalnins, I., Kelly, J.S., Ruf, K.B., 1971. Action potentials and release of neurohypophysial hormones in vitro. J. Physiol. 215, 805–817.

Dudek, S.M., Bear, M.F., 1992. Homosynaptic long-term depression in area CA1 of hippocampus and effects of N-methyl-D-aspartate receptor blockade. Proc. Natl. Acad. Sci. USA 89, 4363–4367.

Dudek, S.M., Bear, M.F., 1993. Bidirectional long-term modification of synaptic effectiveness in the adult and immature hippocampus. J. Neurosci. 13, 2910–2918.

Esdin, J., Pearce, K., Glanzman, D.L., 2010. Long-term habituation of the gill-withdrawal reflex in aplysia requires gene transcription, calcineurin and L-type voltage-gated calcium channels. Front. Behav. Neurosci. 4, 181.

Ezzeddine, Y., Glanzman, D.L., 2003. Prolonged habituation of the gill-withdrawal reflex in Aplysia depends on protein synthesis, protein phosphatase activity, and postsynaptic glutamate receptors. J. Neurosci. 23, 9585–9594.

Frank, C.A., Kennedy, M.J., Goold, C.P., Marek, K.W., Davis, G.W., 2006. Mechanisms underlying the rapid induction and sustained expression of synaptic homeostasis. Neuron 52, 663–677.

Gainer, H., Wolfe Jr., S.A., Obaid, A.L., Salzberg, B.M., 1986. Action potentials and frequency-dependent secretion in the mouse neurohypophysis. Neuroendocrinology 43, 557–563.

Geiger, J.R., Jonas, P., 2000. Dynamic control of presynaptic Ca(2 + ) inflow by fast-inactivating K(+) channels in hippocampal mossy fiber boutons. Neuron 28, 927–939.

Gelinas, J.N., Nguyen, P.V., 2007. Neuromodulation of hippocampal synaptic plasticity, learning, and memory by noradrenaline. Cent. Nerv. Syst. Agents Med. Chem. 7, 17–33.

Glanzman, D.L., Mackey, S.L., Hawkins, R.D., Dyke, A.M., Lloyd, P.E., Kandel, E.R., 1989. Depletion of serotonin in the nervous system of Aplysia reduces the behavioral enhancement of gill withdrawal as well as the heterosynaptic facilitation produced by tail shock. J. Neurosci. 9, 4200–4213.

Gover, T.D., Abrams, T.W., 2009. Insights into a molecular switch that gates sensory neuron synapses during habituation in Aplysia. Neurobiol. Learn. Mem. 92, 155–165.

Gover, T.D., Jiang, X.Y., Abrams, T.W., 2002. Persistent, exocytosis-independent silencing of release sites underlies homosynaptic depression at sensory synapses in Aplysia. J. Neurosci. 22, 1942–1955.

Hebb, D.O., 1949. The Organization of Behavior; A Neuropsychological Theory. Wiley, New York.

Heck, D.H., De Zeeuw, C.I., Jaeger, D., Khodakhah, K., Person, A.L., 2013. The neuronal code(s) of the cerebellum. J. Neurosci. 33, 17603–17609.

Hirsch, J.C., Barrionuevo, G., Crepel, F., 1992. Homo- and heterosynaptic changes in efficacy are expressed in prefrontal neurons: an in vitro study in the rat. Synapse 12, 82–85.

Hoffman, D.A., Magee, J.C., Colbert, C.M., Johnston, D., 1997. K+ channel regulation of signal propagation in dendrites of hippocampal pyramidal neurons. Nature 387, 869–875.

Hui, E., Bai, J., Wang, P., Sugimori, M., Llinas, R.R., Chapman, E.R., 2005. Three distinct kinetic groupings of the synaptotagmin family: candidate sensors for rapid and delayed exocytosis. Proc. Natl. Acad. Sci. USA 102, 5210–5214.

Iriki, A., Pavlides, C., Keller, A., Asanuma, H., 1989. Long-term potentiation in the motor cortex. Science 245, 1385–1387.

Isaac, J.T., Nicoll, R.A., Malenka, R.C., 1995. Evidence for silent synapses: implications for the expression of LTP. Neuron 15, 427–434.

Itskov, V., Hansel, D., Tsodyks, M., 2011. Short-term facilitation may stabilize parametric working memory trace. Front. Comput. Neurosci. 5, 40.

Jackson, M.B., Konnerth, A., Augustine, G.J., 1991. Action potential broadening and frequency-dependent facilitation of calcium signals in pituitary nerve terminals. Proc. Natl. Acad. Sci. USA 88, 380–384.

Jackman, S.L., Turecek, J., Belinsky, J.E., Regehr, W.G., 2016. The calcium sensor synaptotagmin 7 is required for synaptic facilitation. Nature 529 (7584), 88–91.

Jackman, S.L., Regehr, W.G., 2017. The mechanisms and functions of synaptic facilitation. Neuron. 94 (3), 447–464.

Jan, L.Y., Jan, Y.N., 1982. Peptidergic transmission in sympathetic ganglia of the frog. J. Physiol. 327, 219–246.

Jan, Y.N., Jan, L.Y., Kuffler, S.W., 1979. A peptide as a possible transmitter in sympathetic ganglia of the frog. Proc. Natl. Acad. Sci. USA 76, 1501–1505.

Jiang, X.Y., Abrams, T.W., 1998. Use-dependent decline of paired-pulse facilitation at Aplysia sensory neuron synapses suggests a distinct vesicle pool or release mechanism. J. Neurosci. 18, 10310–10319.

Katona, I., Urban, G.M., Wallace, M., Ledent, C., Jung, K.M., Piomelli, D., et al., 2006. Molecular composition of the endocannabinoid system at glutamatergic synapses. J. Neurosci. 26, 5628–5637.

Katsuki, H., Izumi, Y., Zorumski, C.F., 1997. Noradrenergic regulation of synaptic plasticity in the hippocampal CA1 region. J. Neurophysiol. 77, 3013–3020.

Kelso, S.R., Brown, T.H., 1986. Differential conditioning of associative synaptic enhancement in hippocampal brain slices. Science 232, 85–87.

Kleiman, R., Banker, G., Steward, O., 1990. Differential subcellular localization of particular mRNAs in hippocampal neurons in culture. Neuron 5, 821–830.

Klein, M., Shapiro, E., Kandel, E.R., 1980. Synaptic plasticity and the modulation of the $Ca^{2+}$ current. J. Exp. Biol. 89, 117–157.

Klingauf, J., Neher, E., 1997. Modeling buffered $Ca^{2+}$ diffusion near the membrane: implications for secretion in neuroendocrine cells. Biophys. J. 72, 674–690.

Lanahan, A., Worley, P., 1998. Immediate-early genes and synaptic function. Neurobiol. Learn. Mem. 70, 37–43.

Larson, J., Munkacsy, E., 2015. Theta-burst LTP. Brain Res. 1621, 38–50.

Larson, J., Wong, D., Lynch, G., 1986. Patterned stimulation at the theta frequency is optimal for the induction of hippocampal long-term potentiation. Brain Res. 368, 347–350.

Lee, H.K., Kameyama, K., Huganir, R.L., Bear, M.F., 1998. NMDA induces long-term synaptic depression and dephosphorylation of the GluR1 subunit of AMPA receptors in hippocampus. Neuron 21, 1151–1162.

Lee, H.K., Barbarosie, M., Kameyama, K., Bear, M.F., Huganir, R.L., 2000. Regulation of distinct AMPA receptor phosphorylation sites during bidirectional synaptic plasticity. Nature 405, 955–959.

Lee, A., Zhou, H., Scheuer, T., Catterall, W.A., 2003. Molecular determinants of Ca(2+)/calmodulin-dependent regulation of Ca(v)2.1 channels. Proc. Natl. Acad. Sci. USA 100, 16059–16064.

Levy, W.B., Steward, O., 1983. Temporal contiguity requirements for long-term associative potentiation/depression in the hippocampus. Neuroscience 8, 791–797.

Liao, D., Hessler, N.A., Malinow, R., 1995. Activation of postsynaptically silent synapses during pairing-induced LTP in CA1 region of hippocampal slice. Nature 375, 400–404.

Liu, Q.S., Pu, L., Poo, M.M., 2005. Repeated cocaine exposure in vivo facilitates LTP induction in midbrain dopamine neurons. Nature 437, 1027–1031.

Luo, F., Sudhof, T.C., 2017. Synaptotagmin-7-mediated asynchronous release boosts high-fidelity synchronous transmission at a central synapse. Neuron 94, 826–839. e823.

Lyford, G.L., Yamagata, K., Kaufmann, W.E., Barnes, C.A., Sanders, L.K., Copeland, N.G., et al., 1995. Arc, a growth factor and activity-regulated gene, encodes a novel cytoskeleton-associated protein that is enriched in neuronal dendrites. Neuron 14, 433–445.

Lynch, G.S., Dunwiddie, T., Gribkoff, V., 1977. Heterosynaptic depression: a postsynaptic correlate of long-term potentiation. Nature 266, 737–739.

Mackey, S.L., Kandel, E.R., Hawkins, R.D., 1989. Identified serotonergic neurons LCB1 and RCB1 in the cerebral ganglia of Aplysia produce presynaptic facilitation of siphon sensory neurons. J. Neurosci. 9, 4227–4235.

Magee, J.C., Johnston, D., 1997. A synaptically controlled, associative signal for Hebbian plasticity in hippocampal neurons. Science 275, 209–213.

Maity, S., Rah, S., Sonenberg, N., Gkogkas, C.G., Nguyen, P.V., 2015. Norepinephrine triggers metaplasticity of LTP by increasing translation of specific mRNAs. Learn. Mem. 22, 499–508.

Markram, H., Lubke, J., Frotscher, M., Sakmann, B., 1997. Regulation of synaptic efficacy by coincidence of postsynaptic APs and EPSPs. Science 275, 213–215.

Miller, K.D., 1996. Synaptic economics: competition and cooperation in synaptic plasticity. Neuron 17, 371–374.

Milner, B., 1970. Memory and the medial temporal regions of the brain. In: Pribram, K.H., Broadbent, D.E. (Eds.), Biology of Memory. Academic Press, pp. 29–50.

Miyashita, T., Kubik, S., Lewandowski, G., Guzowski, J.F., 2008. Networks of neurons, networks of genes: an integrated view of memory consolidation. Neurobiol. Learn. Mem. 89, 269–284.

Miyashiro, K., Dichter, M., Eberwine, J., 1994. On the nature and differential distribution of mRNAs in hippocampal neurites: implications for neuronal functioning. Proc. Natl. Acad. Sci. USA 91, 10800–10804.

Mongillo, G., Barak, O., Tsodyks, M., 2008. Synaptic theory of working memory. Science 319, 1543–1546.

Muller, M., Felmy, F., Schneggenburger, R., 2008. A limited contribution of Ca2+ current facilitation to paired-pulse facilitation of transmitter release at the rat calyx of Held. J. Physiol. 586, 5503–5520.

Murthy, V.N., Schikorski, T., Stevens, C.F., Zhu, Y., 2001. Inactivity produces increases in neurotransmitter release and synapse size. Neuron 32, 673–682.

Nanou, E., Sullivan, J.M., Scheuer, T., Catterall, W.A., 2016. Calcium sensor regulation of the CaV2.1 Ca2+ channel contributes to short-term synaptic plasticity in hippocampal neurons. Proc. Natl. Acad. Sci. USA 113, 1062–1067.

Narita, K., Akita, T., Hachisuka, J., Huang, S., Ochi, K., Kuba, K., 2000. Functional coupling of Ca(2+) channels to ryanodine receptors at presynaptic terminals. Amplification of exocytosis and plasticity. J. Gen. Physiol. 115, 519–532.

Neher, E., 1998. Usefulness and limitations of linear approximations to the understanding of Ca++ signals. Cell Calcium 24, 345–357.

Neveu, D., Zucker, R.S., 1996a. Long-lasting potentiation and depression without presynaptic activity. J. Neurophysiol. 75, 2157–2160.

Neveu, D., Zucker, R.S., 1996b. Postsynaptic levels of [Ca2+]i needed to trigger LTD and LTP. Neuron 16, 619–629.

Ngezahayo, A., Schachner, M., Artola, A., 2000. Synaptic activity modulates the induction of bidirectional synaptic changes in adult mouse hippocampus. J. Neurosci. 20, 2451–2458.

Paradis, S., Sweeney, S.T., Davis, G.W., 2001. Homeostatic control of presynaptic release is triggered by postsynaptic membrane depolarization. Neuron 30, 737–749.

Pavlov, I.P., Gantt, W.H., Volborth, G., Cannon, W.B., 1928. Lectures on Conditioned Reflexes Twenty-five Years of Objective Study of the Higher Nervous Activity (Behaviour) of Animals. Historical Medical Books. Book 35. http://digitalcommons.hsc.unt.edu/hmedbks/35.

Pu, L., Liu, Q.S., Poo, M.M., 2006. BDNF-dependent synaptic sensitization in midbrain dopamine neurons after cocaine withdrawal. Nat. Neurosci. 9, 605–607.

Richter, J.D., Klann, E., 2009. Making synaptic plasticity and memory last: mechanisms of translational regulation. Genes Dev. 23, 1–11.

Roberts, A.C., Glanzman, D.L., 2003. Learning in Aplysia: looking at synaptic plasticity from both sides. Trends Neurosci. 26, 662–670.

Rose, G.J., Fortune, E.S., 1999. Frequency-dependent PSP depression contributes to low-pass temporal filtering in Eigenmannia. J. Neurosci. 19, 7629–7639.

Roux, F., Uhlhaas, P.J., 2014. Working memory and neural oscillations: α-γ versus θ-γ codes for distinct WM information? Trends Cogn. Sci. 18 (1), 16–25.

Royer, S., Pare, D., 2003. Conservation of total synaptic weight through balanced synaptic depression and potentiation. Nature 422, 518–522.

Royer, S., Coulson, R.L., Klein, M., 2000. Switching off and on of synaptic sites at aplysia sensorimotor synapses. J. Neurosci. 20, 626–638.

Rozov, A., Burnashev, N., Sakmann, B., Neher, E., 2001. Transmitter release modulation by intracellular Ca2+ buffers in facilitating and depressing nerve terminals of pyramidal cells in layer 2/3 of the rat neocortex indicates a target cell-specific difference in presynaptic calcium dynamics. J. Physiol. 531, 807–826.

Schulman, J.A., Weight, F.F., 1976. Synaptic transmission: long-lasting potentiation by a postsynaptic mechanism. Science 194, 1437–1439.

Schneggenburger, R., Meyer, A.C., Neher, E., 1999. Released fraction and total size of a pool of immediately available transmitter quanta at a calyx synapse. Neuron 23 (2), 399–409.

Scoville, W.B., Milner, B., 1957. Loss of recent memory after bilateral hippocampal lesions. J. Neurol. Neurosurg. Psychiatry 20, 11–21.

Sgritta, M., Locatelli, F., Soda, T., Prestori, F., D'Angelo, E.U., 2017. Hebbian spike-timing dependent plasticity at the cerebellar input stage. J. Neurosci. 37, 2809–2823.

Silva, A.J., Kogan, J.H., Frankland, P.W., Kida, S., 1998. CREB and memory. Annu. Rev. Neurosci. 21, 127–148.

Sreenivasan, K.K., Curtis, C.E., D'Esposito, M., 2014. Revisiting the role of persistent neural activity during working memory. Trends. Cogn. Sci. 18, 82–89.

Steward, O., Levy, W.B., 1982. Preferential localization of polyribosomes under the base of dendritic spines in granule cells of the dentate gyrus. J. Neurosci. 2, 284–291.

Steward, O., Schuman, E.M., 2003. Compartmentalized synthesis and degradation of proteins in neurons. Neuron 40, 347–359.

Stokes, M.G., 2015. 'Activity-silent' working memory in prefrontal cortex: a dynamic coding framework. Trends. Cogn. Sci. 19, 394–405.

Stokes, M.G., Kusunoki, M., Sigala, N., Nili, H., Gaffan, D., Duncan, J., 2013. Dynamic coding for cognitive control in prefrontal cortex. Neuron 78, 364–375.

Südhof, T.C., 2013. A molecular machine for neurotransmitter release: synaptotagmin and beyond. Nat. Med. 19 (10), 1227–1231.

Sugita, S., Shin, O.H., Han, W., Lao, Y., Sudhof, T.C., 2002. Synaptotagmins form a hierarchy of exocytotic Ca (2+) sensors with distinct Ca(2+) affinities. EMBO J. 21, 270–280.

Sullivan, T.J., de Sa, V.R., 2006. Homeostatic synaptic scaling in self-organizing maps. Neural Netw. 19, 734–743.

Sullivan, W.E., Konishi, M., 1984. Segregation of stimulus phase and intensity coding in the cochlear nucleus of the barn owl. J. Neurosci. 4, 1787–1799.

Sutton, M.A., Schuman, E.M., 2006. Dendritic protein synthesis, synaptic plasticity, and memory. Cell 127, 49–58.

Sweatt, J.D., 2016. Neural plasticity and behavior—sixty years of conceptual advances. J. Neurochem. 139 (Suppl 2), 179–199.

Takahashi, T., Moiseff, A., Konishi, M., 1984. Time and intensity cues are processed independently in the auditory system of the owl. J. Neurosci. 4, 1781–1786.

Tang, Y., Zucker, R.S., 1997. Mitochondrial involvement in post-tetanic potentiation of synaptic transmission. Neuron 18, 483–491.

Thomas, M.J., Beurrier, C., Bonci, A., Malenka, R.C., 2001. Long-term depression in the nucleus accumbens: a neural correlate of behavioral sensitization to cocaine. Nat. Neurosci. 4, 1217–1223.

Thompson, R.F., Spencer, W.A., 1966. Habituation: a model phenomenon for the study of neuronal substrates of behavior. Psychol. Rev. 73, 16–43.

Turecek, J., Jackman, S.L., Regehr, W.G., 2016. Synaptic specializations support frequency-independent Purkinje cell output from the cerebellar cortex. Cell Rep. 17, 3256–3268.

Turrigiano, G., 2007. Homeostatic signaling: the positive side of negative feedback. Curr. Opin. Neurobiol. 17, 318–324.

Turrigiano, G.G., 2008. The self-tuning neuron: synaptic scaling of excitatory synapses. Cell 135, 422–435.

Turrigiano, G.G., Leslie, K.R., Desai, N.S., Rutherford, L.C., Nelson, S.B., 1998. Activity-dependent scaling of quantal amplitude in neocortical neurons. Nature 391, 892–896.

Wagner, J.J., Alger, B.E., 1996. Homosynaptic LTD and depotentiation: do they differ in name only? Hippocampus 6, 24–29.

Wang, X., Rich, M.M., 2018. Homeostatic synaptic plasticity at the neuromuscular junction in myasthenia gravis. Ann. NY Acad. Sci. 1412, 170–177.

Weight, F.F., Votava, J., 1970. Slow synaptic excitation in sympathetic ganglion cells: evidence for synaptic inactivation of potassium conductance. Science 170, 755–758.

White, G., Levy, W.B., Steward, O., 1990. Spatial overlap between populations of synapses determines the extent of their associative interaction during the induction of long-term potentiation and depression. J. Neurophysiol. 64, 1186–1198.

Wondolowski, J., Dickman, D., 2013. Emerging links between homeostatic synaptic plasticity and neurological disease. Front. Cell. Neurosci. 7, 223.

Worth R., Annese J. (2012) Brain observatory and the continuing study of H.M.: Interview with Jacopo Annese. Europe's J. Psychology 8 (2), 222–230.

Yan, J., Leal, K., Magupalli, V.G., Nanou, E., Martinez, G.Q., Scheuer, T., et al., 2014. Modulation of CaV2.1 channels by neuronal calcium sensor-1 induces short-term synaptic facilitation. Mol. Cell. Neurosci. 63, 124–131.

Yang, S.N., Tang, Y.G., Zucker, R.S., 1999. Selective induction of LTP and LTD by postsynaptic [Ca2 + ]i elevation. J. Neurophysiol. 81, 781–787.

Yoshida, T., Fukaya, M., Uchigashima, M., Miura, E., Kamiya, H., Kano, M., et al., 2006. Localization of diacylglycerol lipase-alpha around postsynaptic spine suggests close proximity between production site of an endocannabinoid, 2-arachidonoyl-glycerol, and presynaptic cannabinoid CB1 receptor. J. Neurosci. 26, 4740–4751.

Yuan, L.L., Adams, J.P., Swank, M., Sweatt, J.D., Johnston, D., 2002. Protein kinase modulation of dendritic K + channels in hippocampus involves a mitogen-activated protein kinase pathway. J. Neurosci. 22, 4860–4868.

Zucker, R.S., Regehr, W.G., 2002. Short-term synaptic plasticity. Annu. Rev. Physiol. 64, 355–405.

# PART IV

# CHEMICAL TRANSMITTERS

# CHAPTER 15

# Introduction to Chemical Transmitter Systems

In Part III, we discussed neurotransmitter receptors that exist on the pre- and postsynaptic compartments of synapses. In Part IV, we will discuss ligands for these receptors. **Endogenous ligands**, which are molecules in the body that act as agonists for neurotransmitter receptors, are referred to as neurotransmitters and/or neuromodulators. We will also discuss compounds that are agonists or antagonists for neurotransmitter receptors but are not endogenous. These compounds may be naturally occurring (e.g., made by another organism), or they may be synthetic ligands developed to have a specific action on the nervous system. Natural and synthetic receptor ligands that interact with neurotransmitter receptors and have therapeutic or recreational nervous system effects are generally referred to as drugs.

## NEUROTRANSMITTER VERSUS NEUROMODULATOR

The terms "neurotransmitter" and "neuromodulator" are similar, but they are not used in quite the same way. A **neurotransmitter** is a molecule released from a neuron that transmits a signal to another cell. Neurotransmitters are often released locally into the synaptic cleft, where they act on nearby receptors on the postsynaptic cell, and they directly mediate most "fast" synaptic transmission. Additionally, they usually do not diffuse long distances from the synapses at which they are released. Of course, there are exceptions to these generalizations that will be discussed below.

In contrast, the term **neuromodulator** is typically used to refer to a molecule that modifies the characteristics of fast synaptic transmission mediated by classical neurotransmitters. That is, rather than being involved directly in what we usually consider fast neuronal communication, neuromodulators modify the state of a neuron to alter its response to a neurotransmitter. Neuromodulators can be released into the synaptic cleft from a presynaptic neuron (like neurotransmitters), or they can be released more broadly into a region where they can diffuse to many synapses. This type of diffuse release is referred to as **volume transmission or paracrine transmission** and allows neuromodulators to exert widespread effects that last longer than the effects of fast synaptic transmission (see Fig. 15.1).

However, the neurotransmitter and neuromodulator categories of signaling molecules are not as distinct or mutually exclusive as the definitions above might suggest. This is because these definitions are functional, and a given chemical neurotransmitter may have different functions depending on the context in which it is released. If a signaling molecule is acting directly within a synapse and its effect is rapid, it is considered to be acting as a neurotransmitter. In contrast, if that same signaling molecule is modulating fast synaptic transmission, it is considered to be acting as a neuromodulator.

For example, the effect of acetylcholine on muscle cells at the neuromuscular junction is fast because acetylcholine receptors are located directly across the synapse from the sites of vesicle release. In this context, acetylcholine acts as a neurotransmitter. However, at a different type of synapse, such as in the ventral tegmental area (see Chapter 11, acetylcholine acts on presynaptic nicotinic acetylcholine receptors to increase the amount of glutamate released in response to an action potential. In *this* context, acetylcholine acts

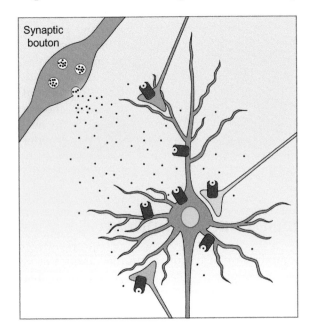

FIGURE 15.1 **Volume or paracrine transmission by neuromodulators.** Neuromodulators (*black dots*) can diffuse from their site of release (swelling on *blue axon*; "synaptic bouton") to multiple nearby synapses and cell bodies by a process called volume or paracrine transmission. Neurotransmitter receptors are shown in purple.

as a neuromodulator. You can see from these two examples that acetylcholine can act as a neurotransmitter at one type of synapse and as a neuromodulator at another (in fact, at some synapses it can act as both, if that synapse has both pre- and postsynaptic receptors).

The fact that one signaling molecule can act as a neurotransmitter, a neuromodulator, or both, depending on context, can lead to some ambiguity in the nomenclature. For this reason, as we discuss signaling molecules, both here and in other chapters, we will use the word "neurotransmitter" to describe any chemical that could be thought of a neurotransmitter or neuromodulator. This simplification is not only a linguistic shortcut, but also highlights the fact that it is often difficult to precisely define the role(s) of a given signaling molecule within the nervous system.

# CRITERIA USED TO CLASSIFY A SIGNALING MOLECULE AS A NEUROTRANSMITTER

Not all molecules that affect neurons are neurotransmitters. In fact, before specific endogenous neurotransmitters had been discovered, chemists had synthesized substances that could affect the nervous system by acting at neurotransmitter receptors (even though many of these receptors had not yet been discovered). An example is the synthetic molecule LSD, whose hallucinogenic effects are largely attributable to its action as an agonist at a particular type of serotonin receptor (see Chapter 17. There are also molecules produced naturally by nonhuman organisms that affect the nervous system and can, to some degree, mimic the effects of endogenous neurotransmitters. For example, nicotine is made by the plant *Nicotiana tabacum*, the leaves of which are processed to make tobacco products. Nicotine activates nicotinic acetylcholine receptors, mimicking the effect of acetylcholine on these receptors. However, because nicotine is not made by the body and is thus not an endogenous ligand for acetylcholine receptors, we do not refer to it as a neurotransmitter.

In order for scientists to agree on how to classify neuroactive substances, several criteria have traditionally been applied to determine whether a candidate neuroactive molecule can be considered to be a neurotransmitter, although experimental evidence for meeting these criteria can be indirect or difficult to obtain using current experimental techniques. The neurotransmitter criteria, as well as methods that have been used to assess them, are described here for two reasons. First, it is useful to outline important and/or common characteristics of neurotransmitter molecules, since these are used to inform ongoing searches for as-yet-unidentified neurotransmitters. Second, by presenting recent discoveries of chemicals in the nervous system that clearly mediate endogenous communication between neurons, yet do not satisfy the traditional neurotransmitter criteria, we will call attention to the fact that there may in fact be no list of specific criteria that can unambiguously classify *all* molecules that can act as neurotransmitters. As more molecules are identified as candidate neurotransmitters, researchers will undoubtedly continue to debate how best to classify them.

For these reasons, the list below is presented not as dogma, but as a historical framework for debate and consideration of the question: What defines a neurotransmitter?

1. *Presence of the molecule in synaptic vesicles of nerve terminals:* A candidate neurotransmitter molecule might be expected to be present inside the presynaptic neuron, usually packaged in synaptic vesicles. The presence of a neurotransmitter in a presynaptic neuron can be assessed using an antibody against the candidate neurotransmitter or a biochemical reaction that allows its presence and location within neurons to be visualized (see, e.g., Burger et al., 1991).
2. *"Identity of action":* Does the candidate chemical mimic the actions of nerve stimulation? If a candidate neurotransmitter is chemically isolated and applied to a synapse at which it is thought to normally be released, it might be expected to have the same effect on the postsynaptic cell as stimulating the presynaptic neuron that releases it (see Fig. 15.2;

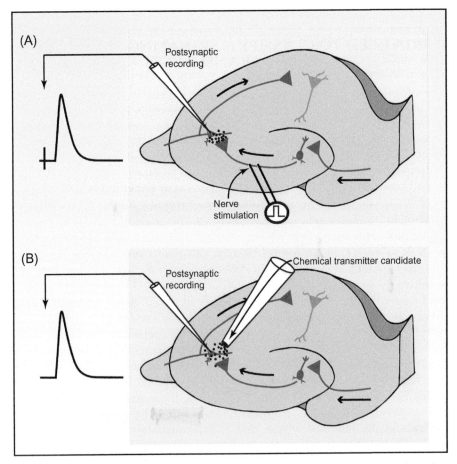

FIGURE 15.2 **Identity of action.** (A) When the presynaptic axon is stimulated with an electrode, it releases neurotransmitter, and the response can be recorded in the postsynaptic neuron. (B) The candidate neurotransmitter can then be applied to the same synapse through a pipette, and the response can be recorded in the postsynaptic cell. If this response is similar to the one evoked by electrical stimulation of the presynaptic neuron (A), the candidate neurotransmitter is said to have identity of action at that synapse.

Jan and Jan, 1976). For example, if acetylcholine is added to the solution bathing a neuromuscular junction, the muscle contracts, just as it contracts after the neuromuscular junction is electrically stimulated to induce chemical transmitter release. Thus, we say acetylcholine has identity of action at the neuromuscular junction.

3. *Neurotransmitter release after nerve stimulation:* A candidate neurotransmitter is typically expected to be released from the presynaptic neuron following an action potential. However, when only a small number of neurotransmitter molecules are released in response to an action potential, it can be difficult to detect them experimentally. One way to determine that a candidate neurotransmitter is released by a neuron is by labeling the neurotransmitter with a radioactive tag, since radioactivity is easier to detect in small quantities (see, e.g., Yau and Youther, 1982; Rowell and Winkler, 1984). This can be accomplished by incubating the synapse in a radioactively labeled precursor to the candidate neurotransmitter (e.g., tritiated choline, a precursor for acetylcholine). The presynaptic neuron uses this precursor to synthesize a radioactive form of the candidate neurotransmitter. When the nerve is stimulated, it releases the radioactive neurotransmitter into the solution bathing the synapse. An experimenter can then collect this solution, separate precursor molecules from neurotransmitter molecules using **electrophoresis**, and determine whether any radioactively labeled neurotransmitter is present. The presence of radioactively labeled neurotransmitter confirms that the candidate neurotransmitter (synthesized from the radiolabeled precursor) was released from the nerve terminal following stimulation.

4. *Transmitter synthesis:* It has been proposed that a candidate neurotransmitter should be synthesized by the presynaptic neuron, which can occur either in the presynaptic terminal or in the cell body. One way to determine where a candidate neurotransmitter is synthesized is by using antibodies to label enzymes used in the neurotransmitter synthesis reaction (see, e.g., Armstrong et al., 1983). Because some neurotransmitters are synthesized by enzymes that are only expressed in neurons that use that neurotransmitter for synaptic communication (so-called "unique synthetic enzymes"), if those unique synthetic enzymes are found within a neuron, this can be used as evidence to identify the specific neurotransmitter that is released following nerve stimulation. Other methods include looking for the presence of neurotransmitter precursors within nerve terminals, looking for unique proteins expressed only in neurons that use these precursors to synthesize neurotransmitters, or looking for proteins that selectively transport the precursors into the nerve terminal. An example of such a transporter protein is the high-affinity choline uptake transporter, which brings choline into the axon terminals of neurons that release acetylcholine (see Chapter 16 and Misawa et al., 2001).

5. *Termination of action:* Neurotransmitter release is usually fast, and many synapses have evolved mechanisms that keep the effects of neurotransmitter release very brief. For some neurotransmitters, this is achieved when they are cleaved by specific enzymes within the synaptic cleft. In such a case, one might be able to identify enzymes that inactivate a candidate neurotransmitter at sites in the nervous system where that neurotransmitter is used to communicate between neurons. If these enzymes are located in the synaptic cleft of a given synapse, this is a good indication that the neurotransmitter on which they act is released at that synapse. For example, the action

of acetylcholine is terminated by the enzyme acetylcholinesterase, which is plentiful at peripheral cholinergic synapses and can be used to identify them (see, e.g., Silver, 1963).

6. *Predictable pharmacological action:* Candidate neurotransmitters are expected to have target receptors that mediate their action. How can one test for the presence of receptors? One method is to use pharmacological compounds that alter binding of a candidate neurotransmitter to its receptors. After applying a pharmacological blocker of a neurotransmitter receptor, the effect of stimulating the presynaptic neuron (which releases the candidate chemical transmitter to act on receptors) should also be blocked. For example, when curare (a naturally occurring nicotinic acetylcholine receptor antagonist; see Box 6.2) is applied to a neuromuscular junction, stimulating the presynaptic neuron does not lead to muscle contraction because the acetylcholine receptors have been blocked by the curare molecule (first shown by Langley, 1905; reviewed by Bowman, 2006). Receptors can also be identified by labeled antibodies, but the localization of receptors does not always identify the sites of release for chemicals that bind to those receptors. For example, in Chapter 14, we discussed the effects of neurotransmitters within the sympathetic chain ganglia and identified the presence of LHRH receptors on the large B cells. However, there were no LHRH-releasing nerve terminals onto B cells, because, in this case, the LHRH diffused some distance from release sites to act on B cells.

These six criteria provide a framework with which to organize our discussions of candidate neurotransmitter molecules. However, an important caveat to these criteria is that they were developed based on observations of the first chemical transmitters that were discovered (e.g., the classical transmitters like acetylcholine and the monoamines). As we detail how different chemicals are used as neurotransmitters in Chapters 16–20, it will become clear that the criteria listed above often do not hold true for all neurotransmitters. More and more chemicals have been identified in the nervous system that appear to serve as neurotransmitters, and their diversity makes it very difficult to identify a single, definitive list of properties that define a chemical as a neurotransmitter. A more inclusive definition of a neurotransmitter might be as broad as "A chemical substance that conveys information between cells in the nervous system."

It is also important to note that the same molecules that we might define as neurotransmitters based on their actions in the nervous system may also be used by other tissues in the body (e.g., the endocrine system) as nonsynaptic intercellular messengers. As a result, some of the chemical neurotransmitters discussed in the chapters that follow have important roles in other systems as well.

For some candidate neurotransmitters, it is not entirely straightforward to test all six of the listed criteria. This can be due to limitations in available experimental techniques and/or a lack of tractable experimental preparations. For example, the amount of neurotransmitter released at some synapses can be quite small, typically $\sim 10{,}000$ molecules of neurotransmitter per action potential. This sounds like a large number, but it can be difficult to detect thousands of a specific type of molecule in a vast extracellular space that contains many millions of other molecules. Additionally, the anatomy of the nervous system makes investigating neurotransmitter molecules difficult because it is rare for homogeneous groups of a specific type of neuron to be anatomically separated from other types of neurons. As a result, it can be challenging to specifically stimulate only one neuron and to selectively record the postsynaptic effect of such stimulation. This difficulty is especially prevalent in the central nervous system, particularly in regions that receive input from many types of neurons, such as in the hippocampus and the neocortex. In the hippocampus, a single neuron can receive thousands of synaptic inputs, and these may not each release the same transmitter molecules. In fact, many neurotransmitters were first identified at peripheral synapses, which are often larger in size, and can be more readily isolated from other types of neurons. This is one of the reasons specialized experimental preparations such as the neuromuscular junction have been, and still are, so critical for learning about synapses and neurotransmitters.

Conveniently, and perhaps surprisingly, many neurotransmitters are highly conserved across species. There is a high degree of similarity in neurotransmitters and in their actions, even between lower-order animals like squid, sea slugs, and flies and higher-order animals like mammals. This means that, in many cases, results from experiments that were first conducted on experimental preparations from lower-order animals also apply to higher-order animals, making experiments in lower animals indispensable in this line of research.

# NEUROTRANSMITTER CHARACTERISTICS

When discussing neurotransmitters in the subsequent chapters, we will focus our presentation around common themes in transmitter synthesis and regulation, including the neurotransmitter synthesis reactions and their rate-limiting steps, the physiological regulation of synthesis, and the mechanisms neurons use to terminate the action of transmitters. These will be introduced here generally and presented in more detail when discussing individual neurotransmitters.

## Synthetic Pathways

In order for chemicals to be used as neurotransmitters, neurons must either synthesize these molecules or collect and concentrate them for use in synaptic transmission. While this may sound obvious, it is critical to consider where these neurotransmitter chemicals come from, since their sites of origin impact how they are used and regulated. One advantage of local synthesis of neurotransmitters is that it allows for an efficient and rapid replenishment of neurotransmitter that has been used within the nerve terminal to reload synaptic vesicles (although not all neurotransmitters can be synthesized locally—see below). Thus, we will discuss not only where each neurotransmitter comes from, but also how it is processed before and after its use as a neurotransmitter.

## Regulation of Neurotransmitter Synthesis

Some, but not all, neurotransmitter chemicals have specific mechanisms that regulate the amounts in which they are produced within a neuron. It is critical for neurons to have ways to increase or decrease the amounts of neurotransmitter they produce so they can keep up with demand in a regulated manner. For some neurotransmitters (type 1; see below), there is a protein (e.g., a synthetic enzyme or transporter) that serves as the first unique step in the synthetic pathway for that neurotransmitter, called the **first unique synthetic protein (or enzyme)**. When such proteins exist, they are important because they are rate-limiting, and therefore a regulatory step in their respective synthesis pathways.

For other transmitters, synthesis can be regulated by the amount of precursor available. Precursor might be limited by availability in the body or by the ability of the precursor molecule to get to the location where transmitter synthesis occurs. Precursor availability can be affected by factors such as diet, the rate at which transporter molecules in presynaptic membranes can move precursor molecules into the presynaptic terminal, or the number of precursor transporters present in the presynaptic membrane.

## Methods for Termination of Neurotransmitter Action

At most synapses, it is advantageous for neurotransmitter molecules to be cleared quickly from the synaptic cleft (or at least for their effects to be terminated rapidly), because this keeps the effects of the released neurotransmitter brief, allowing for precise signaling that can be repeated at a high rate. There are a number of mechanisms by which neurotransmitter action can be terminated, and several of the common ones include:

1. *Enzymatic degradation:* Some neurotransmitters can be broken down, or degraded, by enzymes in the synaptic cleft. This mechanism can be an efficient way to clear neurotransmitter from the synapse. For example, acetylcholine is released at the neuromuscular junction and then promptly degraded by the enzyme acetylcholinesterase.
2. *Uptake into the presynaptic terminal or into glia:* At some synapses, transporters in the membrane of the axon terminal actively move (or transport) neurotransmitter molecules back into the presynaptic neuron. These transporters are efficient and numerous enough to quickly clear neurotransmitter molecules from the synaptic cleft, thus preventing them from continuing to activate receptors there. Sometimes, neurotransmitter molecules can also be taken up into glia by transporters found in the membranes of these cells.
3. *Diffusion away from the synaptic cleft:* Some neurotransmitter systems lack degradation enzymes or reuptake transporters (e.g., peptide transmitters, see below). In these systems, neurotransmitters may simply diffuse away from receptors to terminate their response.
4. *Rapid receptor desensitization:* In some cases, receptors can desensitize soon after they are activated by a neurotransmitter, becoming unresponsive despite the continued presence of the neurotransmitter. This effectively terminates the neurotransmitter action.

# TYPES OF NEUROTRANSMITTERS

There are about 100 molecules that are considered neurotransmitters. Neurons use only about 10 of these molecules for fast synaptic neurotransmission, with the rest acting in a neuromodulatory capacity. Most of these 100 neurotransmitters can be divided into four categories, which are defined by several characteristics, as described below (also see Table 15.1). Note that these categories are defined using characteristics of the neurotransmitter molecules themselves, their synthesis, and the process by which they are released from neurons, not by the effects they have on cells. It is also important to note that the transmitter groupings presented in Table 15.1 omit some molecules that may represent unexpected categories of neurotransmitter (e.g., purines), which will be discussed individually in subsequent chapters.

Neurotransmitters are not categorized by their functions because they typically have more than one function, depending on the receptors present near their release sites. For example, in Chapter 16, we will discuss the "classical" neurotransmitter acetylcholine. It is common for acetylcholine to be referred to as an "excitatory" neurotransmitter, since in many cases it depolarizes neurons; all ionotropic acetylcholine receptors and some metabotropic acetylcholine receptors are excitatory. When neurons release acetylcholine at the neuromuscular junction, the muscle cells contract because they contain nicotinic acetylcholine receptors, which cause depolarization. However, other metabotropic acetylcholine receptors are inhibitory. For example, when acetylcholine is released onto the heart, the heart rate slows because pacemaker neurons in the heart contain a type of muscarinic receptor that indirectly increases potassium currents. The acetylcholine-evoked potassium currents hyperpolarize the pacemaker neurons, which slows the rate at which they depolarize and therefore decreases the heart rate. Thus, the effect that acetylcholine has on cells differs according to the type of receptor(s) found on the target tissue. This effect is but one example that highlights the importance of not defining transmitter molecules by their actions, which can be different depending on the receptors in the environment.

TABLE 15.1  Four Categories of Neurotransmitters and Their Properties

|  | Type 1: "Classical" Neurotransmitters | Type 2: Amino Acids | Type 3: Gaseous Messengers | Type 4: Neuropeptides |
|---|---|---|---|---|
| Neurotransmitters | ACh, GABA[a], monoamines | Glutamate, glycine, | Nitric oxide, carbon monoxide, hydrogen sulfide | Short peptides |
| Molecule size | Small | Small | Small | Large |
| Uniquely synthesized at the synapse | Yes | No | Yes | No |
| Type of vesicle | Small clear vesicles | Small clear vesicles | None | Large dense-core vesicles |

[a]GABA is considered a classical neurotransmitter, though it has the chemical structure of an amino acid.
ACh, acetylcholine; GABA, gamma-aminobutyric acid.

## Type 1: "Classical" Neurotransmitters (ACh, GABA, Monoamines)

Type 1 neurotransmitters are all small molecules that are uniquely synthesized in the nerve terminals that release them by enzymes that are uniquely expressed in those terminals. These neurotransmitters are most often released from **small clear synaptic vesicles** and their synthesis involves substantial local recycling of proteins and molecules. Type 1 neurotransmitters are expected to meet the six neurotransmitter criteria listed above, though these criteria can be complicated to evaluate in some cases, especially when a neurotransmitter has multiple effects. For example, acetylcholine acting at ionotropic receptors at the neuromuscular junction meets all of the criteria to be a type 1 transmitter, but when it acts on metabotropic receptors in the brain, its effects are harder to classify.

## Type 2: Amino Acid Neurotransmitters (Glutamate, Glycine)

Amino acid neurotransmitters are small molecules that are NOT uniquely synthesized in the neurons that release them. In fact, they are present in every cell. However, when these common molecules are used as neurotransmitters, they are packaged into vesicles by vesicular transporters that are uniquely expressed by the neurons that use them as transmitters. For example, to identify a glutamatergic neuron, it is necessary to look for the glutamate vesicular transporter instead of the synthetic enzymes that make glutamate, since the synthetic enzymes are not specific to glutamatergic neurons.

## Type 3: Gaseous Messengers: Nitric Oxide, Carbon Monoxide, and Hydrogen Sulfide

Many people don't think of gases as neurotransmitters, in part because they are so different from other transmitters. For example, carbon monoxide is known for its effects as a toxin. Nonetheless, some gases that can be made in the body are used by neurons as a neurotransmitter. Because gases diffuse freely across cell membranes, they are not packaged into vesicles. In lieu of calcium-triggered vesicle fusion, their presence is typically regulated by enzymes that synthesize them on demand.

## Type 4: Neuropeptides

Neuropeptides are short strings of amino acids that are synthesized in and released by neurons or glia and can affect the function of the nervous system. They are usually between 3 and 50 amino acids long, with most falling in the 8–15 amino acid range, which makes them shorter than many proteins in the body. Nevertheless, these peptides are relatively large compared to other types of neurotransmitters. Neuropeptides are synthesized in the cell body rather than in the presynaptic terminal, and their synthesis is regulated by gene expression, alternative mRNA splicing, and/or by posttranslational processing. After synthesis they are packaged by the Golgi apparatus into **large dense-core vesicles**, which are then transported to the synaptic terminal. At the terminal, large dense-core vesicles release their contents by fusion with the plasma membrane, similar to small clear vesicles. However, because neuropeptides are not synthesized locally at the nerve terminal, there is no reuptake of the released peptides, and they must be synthesized de novo in the cell body.

In summary, the four general categories of neurotransmitter we will use to organize our discussion in the chapters that follow are classical neurotransmitters, amino acid neurotransmitters, gaseous messengers, and neuropeptides. Keep in mind that there are other chemical neurotransmitters used by the nervous system that do not fit neatly into this classification system, some of which we will cover in the following chapters. As such, these categories act as a guide for studying neurotransmitters, but are not exhaustive.

# References

Armstrong, D.M., Saper, C.B., Levey, A.I., Wainer, B.H., Terry, R.D., 1983. Distribution of cholinergic neurons in rat brain: demonstrated by the immunocytochemical localization of choline acetyltransferase. J. Comp. Neurol. 216, 53—68.

Bowman, W.C., 2006. Neuromuscular block. Br. J. Pharmacol. 147 (Suppl 1), S277—S286.

Burger, P.M., Hell, J., Mehl, E., Krasel, C., Lottspeich, F., Jahn, R., 1991. GABA and glycine in synaptic vesicles: storage and transport characteristics. Neuron 7, 287—293.

Jan, L.Y., Jan, Y.N., 1976. L-glutamate as an excitatory transmitter at the Drosophila larval neuromuscular junction. J. Physiol. 262, 215—236.

Langley, J.N., 1905. On the reaction of cells and of nerve-endings to certain poisons, chiefly as regards the reaction of striated muscle to nicotine and to curari. J. Physiol. 33, 374—413.

Misawa, H., Nakata, K., Matsuura, J., Nagao, M., Okuda, T., Haga, T., 2001. Distribution of the high-affinity choline transporter in the central nervous system of the rat. Neuroscience 105, 87—98.

Rowell, P.P., Winkler, D.L., 1984. Nicotinic stimulation of [3H]acetylcholine release from mouse cerebral cortical synaptosomes. J. Neurochem. 43, 1593—1598.

Silver, A., 1963. A histochemical investigation of cholinesterases at neuromuscular junctions in mammalian and avian muscle. J. Physiol. 169, 386—393.

Yau, W.M., Youther, M.L., 1982. Direct evidence for a release of acetylcholine from the myenteric plexus of guinea pig small intestine by substance P. Eur. J. Pharmacol. 81, 665—668.

# CHAPTER 16

# Acetylcholine

## HISTORY OF THE DISCOVERY OF ACETYLCHOLINE AND ITS IDENTITY AS A NEUROTRANSMITTER

The first molecule to be identified as a neurotransmitter was **acetylcholine**, which is in the group of "classical" neurotransmitters introduced in Chapter 15. Acetylcholine was identified by the neuropharmacologist Sir Henry Dale, whose studies were focused on mechanisms for activation of the peripheral nervous system (Fishman, 1972). In 1904, Dale was working on a project to clarify the pharmacological effects of the enigmatic drug ergot, a fungus that grows on rye. He showed that ergot extracts antagonized the effects of adrenaline on blood pressure, which is controlled by catecholamines acting on the autonomic nervous system. In one experiment, he observed that when he applied the ergot extract he was testing to his experimental preparation, this caused an unexpected effect that mimicked the effects of muscarine, an agonist of muscarinic acetylcholine receptors (though these receptors had not yet been discovered or named). Dale later determined that the ergot extract he was working with had been contaminated by a bacterium that excreted acetylcholine as a waste product (Dale, 1953).

The acetylcholine molecule was first synthesized in a laboratory in 1906, and it induced the same effects on the autonomic nervous system as Dale's contaminated ergot derivative (Hunt and Taveau, 1906). However, because acetylcholine had not yet been isolated from animal tissue, there was doubt among researchers that it was a natural (endogenous) ligand, made in the body. It was later determined that acetylcholine is indeed made and released by neurons, not only in the autonomic nervous system, but also by motor neurons that activate skeletal muscle, and in multiple regions with in the central nervous system.

In 1921, Otto Löwi demonstrated that the vagus nerve releases a chemical transmitter that decreases the heart rate (see Box 16.1), but he did not know the identity of the molecule. This transmitter was later shown to be the same molecule, acetylcholine, which had been identified nearly 20 years earlier by Dale. We now know that acetylcholine has a

---

### BOX 16.1

### OTTO LÖWI: THE DISCOVERY OF CHEMICAL NEUROTRANSMISSION AND THE 1936 NOBEL PRIZE IN PHYSIOLOGY OR MEDICINE

As we discussed in Chapter 2, in the first part of the 20th century, there was a great deal of debate about whether synaptic communication was chemical or electrical. But, in 1921, a critical experiment was performed by the scientist Otto Löwi (see Fig. 16.1) that convinced many people that the nervous system does use chemical transmission.

At the turn of the 20th century, multiple researchers were studying the function of the autonomic nervous system. At this time, it was well known that stimulation of the vagus nerve, the tenth cranial nerve, slowed the heart rate, but it was not clear how it did so. Henry Dale (1875–1968) observed that acetylcholine mimicked the effects of parasympathetic nerve stimulation, but he did not go so far as to suggest that parasympathetic nerves secreted acetylcholine (Valenstein, 2002). He later indicated that this caution was due in part to the fact that there was no evidence that acetylcholine was found in the body, and in part due to the fact that there was no way at the time to isolate acetylcholine before it was inactivated by acetylcholinesterase or was otherwise broken down. It should also be noted that Dale, like many of his colleagues, was a pharmacologist, which meant that he was more interested in investigating the actions of drugs, rather than understanding how neurons communicate, per se.

FIGURE 16.1 Otto Löwi, date unknown. *Source: Wellcome Collection (CC BY 4.0).*

## BOX 16.1 (cont'd)

Otto Löwi (1873–1961) was studying glucose metabolism and nutrition when he first met Dale and other scientists who were studying the autonomic nervous system. In 1903, after discussions with these colleagues, Löwi speculated that the vagus nerve might decrease heart rate by secreting something like muscarine, which was known to mimic some of the effects of acetylcholine. However, Löwi was not studying "neurohumoral secretions," and he indicated later that while he occasionally considered that release of substances from neurons could occur, he did not know how to go about proving this. He said he did not think about it until many years later, when, in 1920, he had a dream about how to conduct just such an experiment. Löwi's description of this dream is below:

> "The night before Easter Sunday of [1920] I awoke, turned on the light and jotted down a few notes on a tiny slip of thin paper. Then I fell asleep again. It occurred to me at 6.00 o'clock in the morning that during the night I had written down something important, but I was unable to decipher the scrawl. The next night, at 3.00 o'clock, the idea returned. It was the design of an experiment to determine whether or not the hypothesis of chemical transmission that I had uttered 17 years ago was correct. I got up immediately, went to the laboratory, and performed a simple experiment on a frog heart according to the nocturnal design." (Loewi, 1960)

In his experiment, Löwi put two frog hearts in separate chambers (see Fig. 16.2). These were bathed in Ringer's solution, the composition of which is similar to the extracellular solution surrounding the frog heart in vivo, which kept the heart alive and able to beat. In one of these hearts, the vagus nerve remained intact, and in the second, it was removed. Löwi stimulated the vagus nerve on the first heart for several minutes, which, as expected, caused the heart rate to slow. When he transferred some of the Ringer's solution surrounding the first heart to the chamber containing the second heart, it too started to beat more slowly, without having a functioning vagus nerve at all. These results suggested that when the vagus nerve on the first heart was stimulated, a chemical must have been released into the solution surrounding the first heart that was responsible for slowing the heart rate. Löwi did not know the identity of the chemical that had been released by stimulation of the vagus nerve, so he referred to it as *"Vagusstoff,"* German for "vagus substance." We now know that this chemical is the transmitter acetylcholine.

While we now accept Löwi's experiment as a definitive demonstration of the presence of chemical synaptic transmission, at the time not all physiologists agreed that his data and conclusions were correct. There were several reasons for this skepticism (Valenstein, 2002). The first was conceptual: many physiologists were not willing to accept that nerves could secrete chemicals. One common theory at the time was that the chemicals that affected heart rate were derived from the heart itself, and that stimulation of the vagus nerve triggered their release from the heart, rather than the nerve releasing a substance that would then act on the heart muscle.

BOX 16.1 (cont'd)

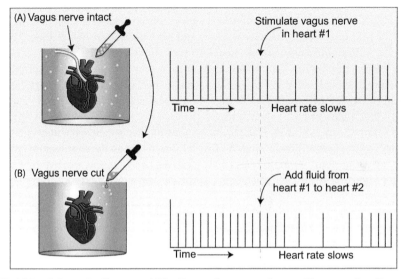

FIGURE 16.2  Schematic diagram of Löwi's two-heart experiment. Diagram of a dissected heart suspended in Ringer's solution, with the vagus nerve intact (A). The heart continues to beat, as indicated by the vertical lines on the graph to the right. When Löwi stimulated the vagus nerve, the heart rate slowed. Löwi then transferred some of the solution from the chamber containing the heart with the intact vagus nerve (A) to a chamber containing a second heart which had no vagus nerve (B). When he did so, the second heart also slowed, indicating that a diffusible substance from the solution around the first heart affected the rate of contraction of the second heart. This diffusible substance is acetylcholine.

Another reason for skepticism about Löwi's results was that researchers were unable to replicate his results in their studies of warm-blooded animals (Bain, 1932). However, there was a critical difference between Löwi's experiments on frogs and these later experiments. Namely, Löwi perfused (bathed) the frog hearts in Ringer's solution, whereas the other studies, attempting this experiment using dog hearts, perfused the hearts with blood. However, blood contains significant amounts of esterases, enzymes like acetylcholinesterase, which break down acetylcholine. As a result, bathing the dog hearts in blood meant that not enough acetylcholine remained intact for this experiment to work. In addition, acetylcholine breaks down faster at warmer temperatures, so the fact that dogs are warm-blooded and frogs are cold-blooded allowed the acetylcholine in Löwi's experiment to remain intact in the Ringer's solution for long enough for his experiment to work. In this sense, Löwi was somewhat lucky in his choice of experimental preparation!

There was initial resistance, including by Löwi himself, that this "neurohumoral" transmission could be applied to other peripheral nerves, let alone the central nervous system (Valenstein, 2002). It is critical to note that Löwi's experiment was

BOX 16.1 (cont'd)

FIGURE 16.3 Sir Henry Dale and Professor Otto Löwi outside the Grand Hotel, Stockholm, at the time of the presentation to them of the Nobel Prize for the Physiology and Medicine, 1936. *Source: Wellcome Collection (CC BY 4.0).*

performed in a model system that is distinct from cholinergic synapses in the brain. However, his observations were proven not only to be correct, but broadly applicable, and the concept of chemical neurotransmitter release fundamentally changed the fields of neuropharmacology and neuroscience. The general principle that resulted was that synapses transfer information by releasing chemicals, which is broadly true for the vast majority of synapses in the brain.

In 1936, Löwi and Dale were jointly awarded the Nobel Prize in Physiology or Medicine for "their discoveries relating to chemical transmission of nerve impulses" (see Fig. 16.3).

Even after Löwi's experiment, there was still a heated debate about whether synaptic transmission was mainly chemical or electrical, a controversy that has been referred to as the debate between the "Soups and the Sparks" (i.e., the pharmacologists vs the physiologists; see Chapter 2). As it turned out, both concepts were correct: both chemical and electrical transmission exist, though chemical transmission dominates signaling in most areas of the brain. Dale recounted that would not have done the experiment had he thought about it clearly in the morning, as it would have seemed to him an unlikely hypothesis that a chemical would survive long enough and not be too diluted by the extracellular solution to have an effect. Years later, Dale suggested that the account of Löwi's dream, and of him going immediately to the laboratory to perform the experiment in the middle of the night, was described by Löwi himself to sound somewhat more theatrical than it was in reality (Dale, 1962). Nonetheless, it was a significant experiment that shifted the way researchers approached neuronal communication.

# 16. ACETYLCHOLINE

wide range of seemingly disparate functions in the body, including motor and autonomic functions in the peripheral nervous system (as shown by Löwi), and modulatory functions in the central nervous system.

## SYNTHESIS, RELEASE, AND TERMINATION OF ACTION OF ACETYLCHOLINE

### Acetylcholine Synthesis

Acetylcholine is a small, positively charged molecule (see Fig. 16.4) that is synthesized within the presynaptic terminal of cholinergic neurons.

As is true for all type 1 transmitters, acetylcholine is synthesized within the presynaptic terminal. It is synthesized from the precursors **choline** and **acetyl CoA** (see Fig. 16.5). Acetyl CoA is found in all cells in the body and is generated in mitochondria via three pathways: (1) from glucose, via glycolysis and the pyruvate dehydrogenase system; (2) from citrate, either through reversal of the condensing enzyme citrate synthase or the citrate cleavage enzyme citrate lysate; and (3) from acetate, by acetatethiokinase (Cooper et al., 1996). Once synthesized, the acetyl CoA moves from mitochondria into the cytoplasm, which is where the acetylcholine synthesis reaction occurs. The acetylcholine synthesis reaction occurs in a single step, which is catalyzed by the enzyme **choline acetyltransferase** (ChAT; see Fig. 16.5). This is a reversible reaction that is driven by the concentrations of the molecules on either side of the reaction equation (in other words, it can work in either direction, depending on the concentrations of precursors and product). Choline and acetyl-CoA are not unique to cholinergic neurons, but the ChAT enzyme is. The fact that ChAT is a unique enzyme means it can be used to unambiguously identify neurons that release acetylcholine.

FIGURE 16.4 Molecular structure of the acetylcholine molecule.

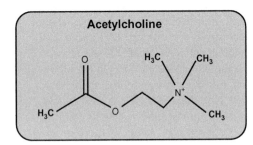

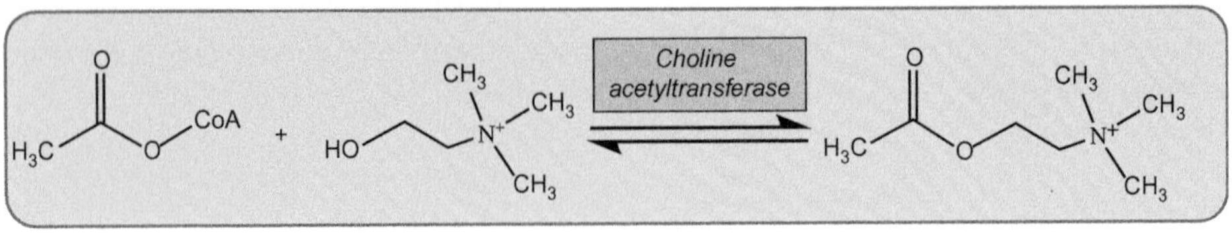

FIGURE 16.5 Acetylcholine synthesis reaction.

Choline is an essential nutrient, and most of it comes from the diet. It can be found free in the blood or stored in the body as phosphotidylcholine, a lipid-bound form essential for other important functions of the cell. In fact, choline is involved in many functions throughout the body that are not necessarily directly related to the nervous system (e.g., synthesis of choline-containing lipids), and it is present in the plasma at a concentration of about 10 μM. Choline enters neurons via transporters in the plasma membrane. All tissues in the body, not just cholinergic neurons, have a low-affinity, high-efficiency transporter for choline ($K_m = 10–100$ μM). However, acetylcholine releasing neurons also have a **high-affinity, low-efficiency choline uptake transporter** (HACU; see Fig. 16.6), which allows them to transport choline even when it is present at low concentrations ($K_m = 1–5$ μM). The HACU is driven by the sodium gradient across the neuronal plasma membrane.

The HACU is not a high-efficiency transporter like the choline transporter used by all cells. It might seem counterintuitive that the HACU does not have high transport efficiency. The reason it does not need to have high transport efficiency is that it is closely linked with the ChAT enzyme. It is thought that ChAT is not evenly distributed throughout the nerve terminal in cholinergic neurons but instead is found in close proximity to

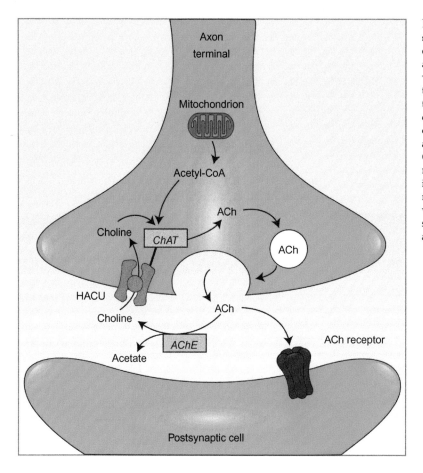

FIGURE 16.6 **The synthesis and breakdown of acetylcholine at a synapse.** ACh is synthesized within the presynaptic terminal and released into the synaptic cleft by vesicles. It is then broken down by the enzyme acetylcholinesterase (AChE) and the choline molecule is taken back into the presynaptic terminal by the HACU, where it is recycled for synthesis of more acetylcholine.

the HACUs. This close proximity means that choline molecules brought into the synaptic terminal by an HACU can rapidly be used by ChAT to synthesize acetylcholine. In fact, 50%–85% of choline transported into the presynaptic terminal by the HACU is converted into acetylcholine.

## Packaging of Acetylcholine Into Synaptic Vesicles

Once acetylcholine has been synthesized in the cytoplasm of the presynaptic terminal, it needs to be packaged into vesicles so it can be released in response to an action potential. Because acetylcholine is a charged molecule, it is not able to cross lipid membranes, including the vesicular membrane, on its own. Therefore, acetylcholine must be moved into vesicles actively, which is accomplished by the **vesicular acetylcholine transporter (VAChT)**. The acetylcholine transport process also involves a vacuolar-type proton ATPase (V-ATPase), also known as a proton pump (or $H^+$-ATPase). The proton pump moves protons into the vesicle in an ATP-dependent manner, creating a proton gradient. This proton gradient results in an intravesicular pH of approximately 5.5, which is substantially lower than cytoplasmic pH, which is near 7.4. The VAChT uses this proton gradient to pump acetylcholine molecules into the vesicle in exchange for protons that are moved out of the vesicle (see Fig. 16.7). The VAChT can be blocked by a drug called vesamicol, and doing so will eventually diminish vesicular release of acetylcholine as vesicles fuse with plasma membrane, recycle via endocytosis, but cannot be refilled with acetylcholine.

The general mechanism described above by which acetylcholine is pumped into synaptic vesicles (exchanging protons for transmitter molecules) is very similar to mechanisms that will be discussed in later chapters regarding the vesicular pumping of monoamines, glutamate, ATP, GABA, and glycine. However, in all of these cases we have simplified our discussion of this process for the purposes of clarity and to focus on the basic principles involved. For a more detailed discussion of other factors that control the pumping of neurotransmitter into synaptic vesicles, see Box 16.2.

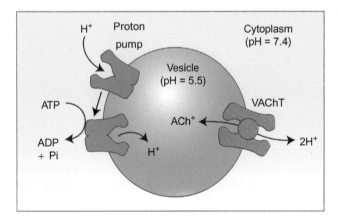

FIGURE 16.7 **The vesicular acetylcholine transporter (VAChT).** This transporter is found in the vesicle membrane and moves acetylcholine from the cytoplasm into the vesicle. It is dependent on a proton gradient, which is established by a vacuolar-type proton ATPase ("proton pump").

## BOX 16.2

## DETAILS OF VESICULAR NEUROTRANSMITTER TRANSPORT

When discussing "classical" and amino acid transmitters (Chapters 16–18), we discuss the transport of neurotransmitter molecules into small clear synaptic vesicles and simplify the presentation of this process to focus on the basic principles. Within these chapters, we present neurotransmitter transporters as pumps that simply exchange vesicular protons for cytoplasmic neurotransmitters across the vesicle membrane, with the proton ($H^+$) gradient across the synaptic vesicle membrane established by a vesicular $H^+$-ATPase. It is estimated that for each ATP hydrolyzed, about two protons are transported into the synaptic vesicle (Farsi et al., 2017). While this is true, it overlooks the complication that as protons are pumped into the synaptic vesicle by the vesicular $H^+$-ATPase, a positive electrical potential develops across the vesicular membrane, due to the accumulation of protons. In fact, this electrical potential can reduce the pH gradient across the vesicle membrane, which affects the ability of the neurotransmitter transporter to concentrate transmitter. Synaptic vesicles employ the $H^+$ electrochemical gradient across the vesicle membrane to accumulate and retain neurotransmitter cargo. The vacuolar-type $H^+$-ATPase generates this gradient, which is comprised of both a chemical $H^+$ gradient and an electrical potential (Blakely and Edwards, 2012). When considering how many free protons must be present within the lumen of a synaptic vesicle to generate the reported pH of 5.5, it should be noted that only one free proton within the very small volume of a synaptic vesicle would generate pH equal to about 4. Therefore, almost all of the protons that are pumped into a synaptic vesicle are thought to be bound up by buffers in the vesicle (including negatively charged vesicle lumen matrix proteins). Furthermore, the proton-buffering capacity of synaptic vesicles increases as the pH becomes more acidic. Therefore, there may be a large number of protons within synaptic vesicles, but they are thought to be protein-bound.

In order to compensate for the electrical potential that develops when protons are pumped across the vesicle membrane, the membrane potential of some synaptic vesicles is regulated by the expression of intracellular chloride-proton exchangers that pump two chloride ions into the synaptic vesicle in exchange for one proton out of the vesicle. In this case, there is one net negative charge pumped into the vesicle by the chloride-proton exchanger, and this can help dissipate the positive membrane potential generated by the $H^+$-ATPase. The action of the chloride-proton exchanger allows the vesicle pH to become more acidic by increasing pH, because it reduces the positive vesicular membrane potential that can inhibit proton concentration in the vesicle.

The electrical gradient and pH of synaptic vesicles can be regulated by the ratio of these two exchangers found on each type of synaptic vesicle (e.g., acetylcholine-containing or glutamate-containing; Edwards, 2007). Some transporters depend more on pH to pump neurotransmitter across the vesicle membrane, while others depend more on the electrical gradient, and this varies according to the charge on a given

## BOX 16.2 (cont'd)

neurotransmitter. For example, at neutral pH the predominant charge of synaptic vesicle cargo depends on the respective pKa values of the molecules. For example, in the cytoplasm (pH 7.4), acetylcholine and monoamines are typically positively charged, GABA and glycine are typically neutral, and glutamate and ATP are typically negatively charged. Therefore, neurons that use a particular neurotransmitter may express different levels of the chloride-proton exchanger, which can then neutralize charge to different extents in response to transport of these molecules from the neutral cytoplasm into the acidic vesicle lumen. Based on the charge on the neurotransmitters, one would expect that synaptic vesicles that load acetylcholine and monoamines would express the chloride-proton exchanger at the highest level, because this would keep the inside of the vesicle more negative (see Fig. 16.8). For acetylcholine and monoamine transmitters, vesicular transport depends more on pH than the membrane potential.

In contrast, synaptic vesicles in glutamate or ATP-releasing neurons might not express as many chloride-proton exchangers because they need to keep the inside of the vesicle more positive and thus be able to attract the negatively charged glutamate (see Fig. 16.8). As such, glutamate and ATP

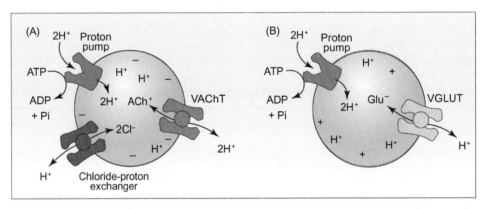

FIGURE 16.8 Synaptic vesicles may express a chloride-proton exchanger in addition to the $H^+$-ATPase to aid in the regulation of the membrane potential across the vesicle membrane. Depending on the charge on the neurotransmitter, transport into the synaptic vesicle may depend more on the vesicular pH gradient (acetylcholine and monoamines) than on the vesicle's membrane potential (glutamate and ATP). GABA and glycine have a neutral charge at cytoplasmic pH, and thus may not be as sensitive to the membrane potential across the vesicular membrane. (A) Acetylcholine (ACh) transport into synaptic vesicles is aided by a chloride-proton exchanger that helps to neutralize what would be a positive vesicular membrane potential. Under these conditions, loading of positively charged ACh into synaptic vesicles is primarily dependent on the vesicular pH gradient. (B) Glutamate transport into synaptic vesicles takes advantage of the positive membrane potential that builds up as protons are transported into the vesicle, as such the concentration of negatively charged glutamate into synaptic vesicles is primarily driven by the vesicle membrane potential.

## BOX 16.2 (cont'd)

vesicular transport depends primarily on membrane potential. Pumping of GABA or glycine into synaptic vesicles is not likely to be sensitive to voltage differences across the vesicle membrane, so the expression level of the chloride-proton exchanger in these vesicles may not be as important. The extent to which each type of synaptic vesicle expresses a chloride-proton exchanger is still a subject of study (Edwards, 2007).

The presence of other mechanisms hypothesized to move ions across the vesicle membrane may also contribute to transmitter loading into synaptic vesicles (e.g., a potassium-proton exchanger, and the movement of potassium ions by transmitter transporters), but these mechanisms will not be discussed here for the purpose of clarity (Farsi et al., 2017).

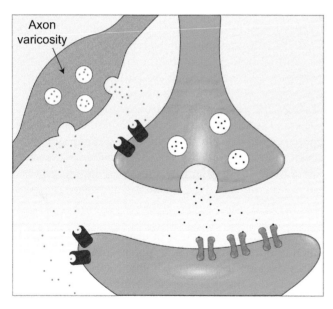

FIGURE 16.9 **Acetylcholine release from axon varicosities.** Vesicles containing acetylcholine (*green dots*) are released from a varicosity on the axon of a cholinergic neuron (*green*). The acetylcholine can diffuse and activate extrasynaptic acetylcholine receptors (*purple*) on nearby neurons (*blue*).

## Release of Acetylcholine

Acetylcholine can be released from vesicles into the synaptic cleft at synapses with a typical synaptic organization, as is the case at neuromuscular synapses described in Chapter 8. However, in the central nervous system, acetylcholine can be released both at typical synapses and also from varicosities (swellings) on axons (see Fig. 16.9). These varicosities are not located directly at synapses, and instead release acetylcholine into the extracellular space in the local area near the varicosities (Allen and Brown, 1996). From there, the acetylcholine molecules diffuse through the extracellular space to their receptors on neighboring cells. In this respect, acetylcholine released from varicosities signals in a **paracrine** manner.

## Regulation of Acetylcholine Synthesis

Work by Fatt and Katz demonstrated that cholinergic neurons continuously release single quanta of acetylcholine even at rest, and these are responsible for the mEPPs at cholinergic synapses, as discussed in Chapter 6. These spontaneous quantal events occur at about 3–4 per second, which means the amount of acetylcholine released by this spontaneous vesicle fusion within a minute constitutes about 0.1% of the total acetylcholine contained inside the axon terminal. Continuous spontaneous release at this rate suggests that in order to maintain required levels of acetylcholine in the synaptic terminal, cholinergic neurons might tune their acetylcholine synthesis rate to match the acetylcholine release rate. In an experimental setting, when neurons are stimulated at a high rate, the level of acetylcholine in the extracellular fluid increases by about 100-fold, which amounts to the release of about 10% of the internal acetylcholine stores per minute. This high rate of release can be sustained for more than an hour, suggesting cholinergic neurons can synthesize acetylcholine at a rapid rate under these conditions. Notably, despite releasing a large amount of acetylcholine, the overall concentration of acetylcholine in these neurons does not change appreciably.

Given the variable demand for acetylcholine synthesis during periods of low or high synaptic acetylcholine release, and because the concentration of acetylcholine in the nerve terminals does not change under these conditions, it is clear that acetylcholine synthesis must be regulated based on demand. One general theme, which we will revisit with other transmitter systems, is that the first unique synthetic enzyme in a neurotransmitter synthesis pathway is the rate-limiting step for the synthesis of that neurotransmitter. However, in the case of acetylcholine, there is one unique synthetic enzyme (ChAT), and this enzyme is not the rate-limiting step in acetylcholine synthesis. In fact, the first unique synthetic protein in this synthesis pathway is not an enzyme, but the HACU transporter. So, the general theme for regulating type 1 transmitters is that the first unique synthetic protein (enzyme or transporter) is the rate-limiting step, and is the protein that is physiologically regulated to control the synthesis rate during periods of demand (action potential activity).

The rate-limiting role of HACU transporters in the synthesis of acetylcholine has been demonstrated by using the drug hemicholinium-3, which is a potent inhibitor of the HACU. Applying hemicholinium-3 reduces the amount of acetylcholine released at cholinergic synapses because depleting choline entering the presynaptic terminal directly limits the amount of acetylcholine a cholinergic neuron can produce. Interestingly, inhibiting the low-affinity, high-capacity choline uptake transporter that all cells have has no effect on acetylcholine synthesis, proving that this transporter is not linked to acetylcholine synthesis.

How do nerve terminals that release acetylcholine know when to make more neurotransmitter? Specifically, during action potential activity, what is the physiological mechanism that increases the synthesis of acetylcholine so that the nerve terminal can have enough neurotransmitter to reload recycled synaptic vesicles? In the case of acetylcholine synthesis, the rate-limiting step is the HACU transporter moving choline into the nerve terminal, and experimenters have hypothesized that the mechanism by which HACU is regulated is by altering the number of HACU transporters in the nerve terminal plasma membrane (Ferguson et al., 2003; Ferguson and Blakely, 2004; Nakata et al., 2004). When studying the location of HACU transporters in the nerve terminal, they found that most were associated with the synaptic vesicle membrane, and some were in the plasma membrane. These results led to the hypothesis that insertion of these HACU transporters into the plasma membrane occurred as a result of synaptic vesicle fusion, representing a convenient mechanism to increase acetylcholine synthesis in times of need. That is, the activity that leads to the depletion of acetylcholine from the nerve terminal (fusion of synaptic vesicles) also inserts more HACU transporters into plasma membrane, to facilitate uptake of choline, and thus increase acetylcholine synthesis (Fig. 16.10; Ferguson et al., 2003).

Acetylcholine synthesis can also be downregulated, and this appears to occur in two ways. First, endocytosis of synaptic vesicle membrane retrieves HACU transporters from the plasma membrane, reducing the number of HACU transporters available to drive the synthesis of acetylcholine. These endocytosed HACU transporters are then in the synaptic vesicle membrane and are ready for reinsertion with the next round of action potential

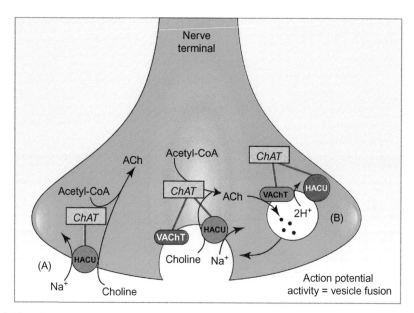

FIGURE 16.10  **Physiological regulation of acetylcholine synthesis.** The synthesis of acetylcholine is hypothesized to be regulated by the number of high-affinity choline uptake (HACU) transporters in plasma membrane. (A) At rest, there are some HACU transporters expressed in plasma membrane that are coupled to ChAT for ACh synthesis. (B) A large number of HACU transporters are thought to exist in the synaptic vesicle membrane and serve as a reserve pool of transporters that can be recruited when vesicle fusion occurs. With action potential activity, vesicle fusion inserts HACU transporters into plasma membrane where they now are exposed to a sodium gradient and will transport choline into the cytoplasm to increase ACh synthesis. It is important to note that synaptic vesicles have both vesicular ACh transporters, which depend on a proton gradient to transport ACh into the synaptic vesicle, and HACU transporters, which depend on a sodium gradient to transport choline into the nerve terminal. Therefore, each of these transporters is hypothesized to only be active (*green*) when the required co-transport molecule gradient exists, but inactive (*red*) when it does not.

activity. Second, the ChAT enzyme is driven by **mass action**, which means that the relative concentration of precursors (acetyl CoA and choline) versus end product (acetylcholine) determines the direction of the reaction (Potter et al., 1968). Therefore, when the concentration of acetylcholine builds up in the nerve terminal, this can effectively prevent the generation of more neurotransmitter because the reaction will run in reverse.

## Termination of Action for Acetylcholine

After acetylcholine molecules are released into the synaptic cleft, they are broken down by the enzyme **acetylcholinesterase** (AChE), which is one of the fastest and most efficient enzymes in the body (it can hydrolyze 5000 ACh molecules per second). Acetylcholinesterase is present in both the synaptic cleft and the cytoplasm of the axon terminal. In the synaptic cleft, it exists in a globular array and is anchored in the extracellular matrix between the pre- and postsynaptic membranes (Fig. 16.11). Acetylcholinesterase rapidly hydrolyzes these molecules into choline and acetic acid, neither of which is active at synapses. The choline molecules in the synaptic cleft can be taken directly back into the presynaptic

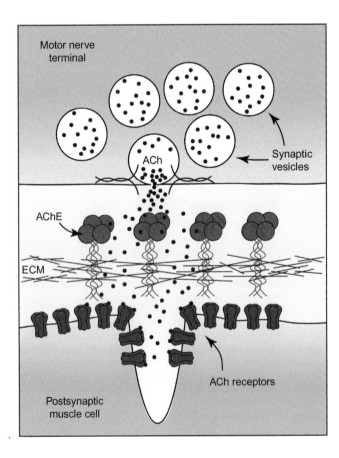

FIGURE 16.11 **Diagram of acetylcholinesterase in the synaptic cleft of the motor nerve terminal.** Acetylcholinesterase (AChE) is deposited into the synaptic cleft in globular clusters (*red spheres*) where it is tethered to the synaptic cleft by anchoring in the extracellular matrix (ECM). In this location, AChE is positioned to cleave acetylcholine (ACh) released by the nerve terminal and shortens the time that ACh has to interact with postsynaptic ACh receptors.

terminal by the HACU and reused to synthesize more acetylcholine. In fact, it is thought that uptake of choline from acetylcholine hydrolyzed in the synaptic cleft supplies about half of the choline required for synthesis of acetylcholine in the presynaptic terminal.

# ROLES OF ACETYLCHOLINE IN THE NERVOUS SYSTEM

Acetylcholine is utilized in both the peripheral and central nervous systems, where it can act as a neurotransmitter or neuromodulator, depending on the type of receptor (ionotropic or metabotropic) and where that receptor is located. This characteristic is a good example of the same signaling molecule playing a role in fast synaptic transmission at some synapses and acting as a neuromodulator at others. Acetylcholine has a wide range of effects, some of which are described below.

## Acetylcholine in the Peripheral Nervous System

All neurons that project out of the central nervous system are cholinergic. As discussed in previous chapters, all vertebrate neuromuscular junctions use acetylcholine. In addition, all preganglionic neurons in the autonomic nervous system are cholinergic, as are all postganglionic neurons of the parasympathetic division of the autonomic nervous system. Therefore, autonomic functions, such as pupillary constriction, salivation, heart rate, and blood pressure, are all directly or indirectly affected by acetylcholine. This fact is one of the main reasons many cholinergic drugs have wide-ranging and seemingly unrelated side effects.

## Acetylcholine in the Central Nervous System

While only about 1% of central nervous system neurons are cholinergic, they have a disproportionately large influence on the central nervous system, in large part because axons of many cholinergic neurons project widely throughout the forebrain (Woolf, 1991; Fig. 16.12). Additionally, the acetylcholine released at varicosities on these axons can diffuse away from the site of release to affect many neighboring synapses in a paracrine fashion (see Fig. 16.9). The presence and effects of acetylcholine in the peripheral nervous system were understood well before the effects of acetylcholine on the central nervous system were described, largely because of the difficulty in measuring acetylcholine content within the central nervous system and also because of the difficulty in measuring the sometimes subtle modulatory effects of metabotropic receptors on synaptic function. However, more modern techniques have allowed researchers to learn how and where acetylcholine has its central effects.

## Sources of Acetylcholine in the Central Nervous System

Cell bodies of cholinergic neurons originate in several regions within the brainstem and basal forebrain (see Fig. 16.12). Cholinergic neurons in the brainstem are found in the

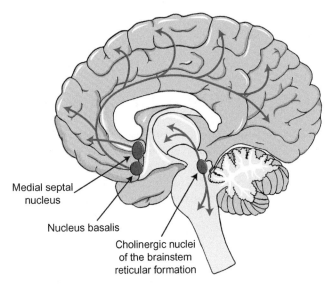

FIGURE 16.12 **Cholinergic neurons in the central nervous system originate in nuclei in both the brainstem and the forebrain.** Acetylcholine is made by groups of neurons in the brainstem reticular formation, and in two nuclei in the forebrain called the medial septal nucleus and the nucleus basalis. Acetylcholine-releasing neurons in these nuclei project widely to vast regions of the CNS where acetylcholine commonly modulates synaptic transmission carried primarily by other transmitters.

brainstem reticular formation, and they send axons to other structures in the brainstem as well as to the hypothalamus and thalamus. Cholinergic neurons are also found in two areas of the basal forebrain, the medial septal nucleus and the nucleus basalis (also called the nucleus basalis of Meynert). Cholinergic neurons in the medial septal nucleus send axons to the hippocampus and parahippocampal gyrus, whereas the nucleus basalis sends cholinergic projections to the neocortex and some parts of the limbic system, including the amygdala.

In the hippocampus and the neocortex, acetylcholine plays a modulatory role and is closely associated with memory function. It has been demonstrated that lesions of the nucleus basalis or administration of muscarinic antagonists within the central nervous system can disrupt memory abilities such as acquisition of learned behaviors. In contrast, treatment with acetylcholinesterase inhibitors (e.g., physostigmine), which reduce the rate of breakdown of acetylcholine increasing its concentration within the synaptic cleft (see Fig. 16.11), can enhance performance on learning and memory tasks. It is not surprising that a failure of cholinergic signaling within the hippocampus, an area of the brain that is critical for the formation of new memories, is one of the main pathologies involved in Alzheimer's disease and is hypothesized to play a prominent role in the memory-related symptoms of this disorder (see Box 16.3)

## Cholinergic Synaptic Transmission in the Autonomic Nervous System

The **sympathetic nervous system** is responsible for a number of activating effects on the body known as the "fight or flight response." In this system, preganglionic neurons release acetylcholine onto postganglionic neurons, which have nicotinic acetylcholine receptors (see Fig. 16.13). Most of these postganglionic neurons release norepinephrine onto their target effectors, such as blood vessels in skeletal muscle, ciliary muscles of the eye, and bronchioles in the lungs.

## BOX 16.3

## THE "ACETYLCHOLINE HYPOTHESIS" IN ALZHEIMER'S DISEASE

Many cholinergic neurons within the CNS are linked to learning, memory, and cognition, which is not surprising, given the robust cholinergic input to the forebrain. One area of the brain that is heavily involved in memory is the hippocampus, and this is one of the areas of the brain that is significantly compromised in Alzheimer's disease. While there are multiple hypotheses about what causes the death of neurons in Alzheimer's disease, there are multiple lines of evidence suggesting that cholinergic dysfunction occurs.

First, it has been shown in Alzheimer's disease patients that there is a decrease in the number of cholinergic neurons in nuclei that send cholinergic axons to the neocortex. Another related piece of evidence demonstrates that if a drug that blocks muscarinic acetylcholine receptors is given to healthy people, they lose the ability to create memories of recent events. Drachman and Leavitt (1974) administered the muscarinic receptor antagonist scopolamine (which can cross the blood–brain barrier to get into the central nervous system) to subjects and tested their memory ability on several tasks. They showed that subjects receiving scopolamine had impairments in memory storage, but subjects receiving either methscopolamine, a peripherally acting muscarinic receptor antagonist (which cannot cross the blood–brain barrier to get into the central nervous system), or physostigmine, a centrally acting acetylcholinesterase inhibitor (which increases the effects of acetylcholine), had no memory impairments. Additionally, in Alzheimer's disease patients, there is a dramatic decrease in HACU (Rylett et al., 1983; Bissette et al., 1996) and ChAT (Perry et al., 1977; Bowen et al., 1976), both of which are good markers of acetylcholine transmission. In addition, there is also evidence for a decrease in cholinergic receptors in patients with Alzheimer's disease (Burghaus et al., 2000; Schroder et al., 1991). These data led to the conclusion that cholinergic signaling is reduced in the neocortex and hippocampus of Alzheimer's patients.

These findings, among others, led to the cholinergic hypothesis of cognitive impairment in Alzheimer's disease [see Bartus et al. (1982) and Contestabile (2011) for reviews of this hypothesis]. Based on this general hypothesis, there were many attempts to develop treatments for Alzheimer's disease that enhanced the ability of acetylcholine to modulate synapses in the brain (also referred to as "cholinergic tone").

There are many exogenous compounds (i.e., natural or synthetic molecules not made by the human body) that affect cholinergic signaling, many of which have therapeutic value. Note that systemic administration of acetylcholine itself typically is not a useful therapeutic agent for at least two reasons. First, since acetylcholine is a charged compound, it cannot cross the blood–brain barrier well, and cannot get to the neurons in the central nervous system. Second, there is a nonsynaptic cholinesterase, butyrylcholinesterase, which is made by the liver and found largely in the blood plasma. This enzyme rapidly hydrolyzes acetylcholine in the bloodstream before it can have any significant effects on the nervous system or elsewhere. Many cholinergic

## BOX 16.3 (cont'd)

drugs, on the other hand, can be lipid-soluble and/or are not as readily degraded by cholinesterases, and can thus have longer-lasting effects and may affect the central nervous system even when administered peripherally (e.g., orally).

Some Alzheimer's disease symptoms can be partially addressed, at least early in the progression of this neurodegenerative disease, by enhancing cholinergic function. Such drugs include AChE inhibitors such as donepezil (sold as Aricept), Rivastigmine, galantamine; agonists of muscarinic acetylcholine receptors, which were largely discontinued because of side effects due to peripheral cholinergic stimulation; and nicotinic receptor agonists, for which there are ongoing clinical trials. Unfortunately, all of these drugs treat only some of the symptoms of Alzheimer's disease, and they are not especially effective or long-lasting. To date, there are no treatments of any kind that have been shown to significantly slow, halt, or reverse this devastating disease.

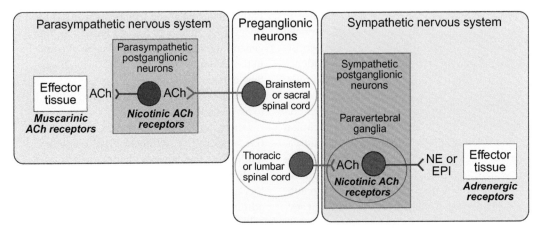

FIGURE 16.13 **Neurotransmitters used by the sympathetic and parasympathetic divisions of the autonomic nervous system.** Preganglionic neurons in both divisions of the autonomic nervous system release acetylcholine (ACh) onto postganglionic neurons, which contain nicotinic receptors. While postganglionic neurons in the parasympathetic nervous system release acetylcholine onto effector tissues, which have muscarinic acetylcholine receptors, those in the sympathetic nervous system release either norepinephrine or epinephrine (discussed in Chapter 17).

In contrast, the **parasympathetic nervous system** is responsible for functions that oppose the sympathetic nervous system and occur when the body is at rest (known as the "rest and digest" state). In this system, both preganglionic and postganglionic neurons release acetylcholine, but the postganglionic neurons have nicotinic acetylcholine receptors and the effector tissues (e.g., heart, salivary glands, ciliary muscles of the eye) have muscarinic receptors (see Fig. 16.13).

# DRUGS AND OTHER COMPOUNDS THAT AFFECT CHOLINERGIC SIGNALING

## Parasympathomimetic Drugs

Drugs that activate muscarinic receptors in the peripheral nervous system are called **parasympathomimetic drugs** because they mimic the effects of acetylcholine on the parasympathetic nervous system. An example of a parasympathomimetic drug is pilocarpine, which is a nonspecific muscarinic agonist. Pilocarpine can be used to treat some disorders of the eye, such as glaucoma, which is characterized by elevated intraocular pressure. Pilocarpine is an effective treatment for glaucoma because one effect is to contract the ciliary muscle, which allows for fluid drainage of the eye.

## Anticholinergic Drugs

Drugs that inhibit cholinergic action are referred to as **anticholinergics**. One such drug is atropine, a naturally occurring alkaloid that is found in the plant *Atropa belladonna* and is an antagonist for muscarinic receptors (thus, it is referred to as an antimuscarinic). One medical use of atropine is by ophthalmologists who apply it to the eye to dilate the pupil during an eye examination. Additionally, since activation of the parasympathetic nervous system normally slows the heart rate, blocking this system with atropine can be used clinically to treat bradycardia (slow heart rate). However, blocking muscarinic receptors can be fatal, which is why *Atropa belladonna* is also referred to as deadly nightshade (see Box 16.4).

---

### BOX 16.4

### DANGERS IN NATURE: THE DEADLY NIGHTSHADE PLANT AND MUSCARINIC ACH RECEPTORS

#### *Atropa belladonna* (Deadly Nightshade)

*Atropa belladonna* (Fig. 16.14) is a widely cultivated ornamental plant that contains neurotoxic alkaloids that act as competitive inhibitors of muscarinic acetylcholine receptors. Peripherally, *Atropa* poisoning reduces parasympathetic activation of smooth and cardiac muscle, resulting in sedation. The alkaloids in *Atropa* can cross the blood–brain barrier to act on central cholinergic synapses, causing ataxia, disorientation, short-term memory loss, coma, and death. Shakespeare's star-crossed lover, Juliet, most likely committed suicide by consuming a handful of the highly toxic *Atropa* berries. In antiquity, *Atropa* was used as an anesthetic but was suspended as a tool for surgery during the Middle Ages due to its association with witchcraft. This plant gets its species name, *belladonna*, which is Italian for "beautiful woman," because of its effect on parasympathetic innervation of the iris. Normally, parasympathetic neurons release acetylcholine onto the iris sphincter muscle, activating muscarinic receptors that cause iris muscle contraction, constricting the pupil. This reaction can be blocked by applying a muscarinic antagonist such as atropine, which is found in deadly nightshade.

> ### BOX 16.4 (cont'd)
>
> In 16th-century Italy, women applied eye drops prepared from deadly nightshade because it dilated the pupil, which was thought to make them look beautiful. Atropine eye drops are still used by ophthalmologists when they dilate a patient's pupils to get a good view of the retina during an eye exam.
>
>
>
> FIGURE 16.14 *Atropa belladonna* plant. *Source: H. Zell CC BY-SA 3.0 from Wikimedia Commons.*

## Compounds That Inhibit Acetylcholinesterase

As discussed earlier, acetylcholine is hydrolyzed by the enzyme acetylcholinesterase. There are multiple compounds that inhibit acetylcholinesterase. Some of these are reversible, and are used therapeutically (Box 16.5), whereas others are irreversible (or mostly irreversible) and are potent toxins. In order for cholinergic synapses to function properly, it is critical that the acetylcholine released there is quickly broken down by acetylcholinesterase. This is true both at peripheral cholinergic synapses (e.g., the neuromuscular junction) and at synapses in the central nervous system. When acetylcholinesterase is not able to play this role, the toxic effects can be very serious and sometimes deadly. Chemicals that interfere with the function of acetylcholinesterase are called **cholinesterase inhibitors**. Some nerve agents and pesticides are organophosphates (also known as phosphate esters), which are cholinesterase inhibitors. Organophosphate nerve agents (e.g., sarin, soman) are deadly toxins that have been used in warfare and terrorist attacks. Organophosphate insecticides are still used in many parts of the world, but there are efforts to ban use of them due to concerns about acute and prolonged exposure.

---

BOX 16.5

## TREATMENTS FROM NATURE: THE SNOWDROP PLANT, ACETYLCHOLINESTERASE, AND NICOTINIC ACETYLCHOLINE RECEPTORS

### *Galanthus nivalis* (Snowdrop)

*Galanthus nivalis* ("milk flower of the snow") (Fig. 16.15) is aptly named, as it is one of the first bulbs to penetrate the snowy ground in the springtime. The neuroactive ingredient in *Galanthus* bulbs is galantamine, which acts as an acetylcholinesterase inhibitor and an **allosteric** nicotinic acetylcholine receptor agonist. Galantamine has been used as a treatment for Alzheimer's, a disease characterized by cholinergic dysfunction (see Box 16.3). Galantamine improves cholinergic transmission by inhibiting the enzyme acetylcholinesterase, which breaks down acetylcholine in the synaptic cleft. It also acts as an allosteric modulator to acetylcholine receptors by improving the receptor's response to acetylcholine binding. Galantamine is readily absorbed into the body and has limited side effects at therapeutic doses. Because it is reversible and has a high affinity for acetylcholinesterase (it can outcompete other proteins that bind to acetylcholinesterase), it is considered a potential treatment for toxic nerve agent exposure.

FIGURE 16.15 *Galanthus nivalis* plant. *Source: Dominicus Johannes Bergsma, CC BY-SA 3.0.*

There are antidotes to cholinergic toxicity from organophosphate poisoning that can be used to treat these symptoms (Wiener and Hoffman, 2004). First, atropine, the competitive muscarinic antagonist discussed above, alleviates symptoms of organophosphate poisoning by preventing the excess acetylcholine from activating muscarinic receptors. Additionally, there is a type of drug called an oxime (e.g., pralidoxime) that can displace the organophosphate from the acetylcholinesterase molecule. Benzodiazepines, which enhance GABAergic transmission (see Chapter 18) can also be administered to prevent seizures that might otherwise occur.

# References

Allen, T.G., Brown, D.A., 1996. Detection and modulation of acetylcholine release from neurites of rat basal forebrain cells in culture. J. Physiol. 492 (Pt 2), 453–466.

Bain, W.A., 1932. A method of demonstrating the humoral transmission of the effects of cardiac vagus stimulation in the frog. Q. J. Exp. Physiol. 22, 269–274.

Bartus, R.T., Dean 3rd, R.L., Beer, B., Lippa, A.S., 1982. The cholinergic hypothesis of geriatric memory dysfunction. Science 217, 408–414.

Bissette, G., Seidler, F.J., Nemeroff, C.B., Slotkin, T.A., 1996. High affinity choline transporter status in Alzheimer's disease tissue from rapid autopsy. Ann. NY Acad. Sci. 777, 197–204.

Blakely, R.D., Edwards, R.H., 2012. Vesicular and plasma membrane transporters for neurotransmitters. Cold Spring Harb. Perspect. Biol. 4, a005595.

Bowen, D.M., Smith, C.B., White, P., Davison, A.N., 1976. Neurotransmitter-related enzymes and indices of hypoxia in senile dementia and other abiotrophies. Brain 99, 459–496.

Burghaus, L., Schutz, U., Krempel, U., de Vos, R.A., Jansen Steur, E.N., Wevers, A., et al., 2000. Quantitative assessment of nicotinic acetylcholine receptor proteins in the cerebral cortex of Alzheimer patients. Brain. Res. Mol. Brain. Res. 76, 385–388.

Contestabile, A., 2011. The history of the cholinergic hypothesis. Behav. Brain. Res. 221, 334–340.

Cooper, J.R., Bloom, F.E., Roth, R.H., 1996. The Biochemical Basis of Neuropharmacology, 7th ed. Oxford University Press, New York.

Dale, H.H., 1953. Adventures in Physiology. Pergamon Press, London.

Dale, H.H., 1962. Otto Loewi (1873–1961). Royal Society Obituaries and Memoirs 8, 67–89.

Drachman, D.A., Leavitt, J., 1974. Human memory and the cholinergic system. A relationship to aging? Arch. Neurol. 30, 113–121.

Edwards, R.H., 2007. The Neurotransmitter Cycle and Quantal Size. Neuron. 55 (6), 835–858.

Farsi, Z., Jahn, R., Woehler, A., 2017. Proton electrochemical gradient: Driving and regulating neurotransmitter uptake. Bioessays 39 (5).

Ferguson, S.M., Blakely, R.D., 2004. The choline transporter resurfaces: new roles for synaptic vesicles? Mol. Interv. 4, 22–37.

Ferguson, S.M., Savchenko, V., Apparsundaram, S., Zwick, M., Wright, J., Heilman, C.J., et al., 2003. Vesicular localization and activity-dependent trafficking of presynaptic choline transporters. J. Neurosci. 23, 9697–9709.

Fishman, M.C., 1972. Sir Henry Hallett Dale and acetylcholine story. Yale J. Biol. Med. 45, 104–118.

Hunt, R., Taveau, R.D.M., 1906. On the physiological action of certain cholin derivatives and new methods for detecting cholin. Brit. Med. J. 2, 1788–1791.

Loewi, O., 1960. An autobiographic sketch. Perspect. Biol. Med. 4, 3–25.

Nakata, K., Okuda, T., Misawa, H., 2004. Ultrastructural localization of high-affinity choline transporter in the rat neuromuscular junction: enrichment on synaptic vesicles. Synapse 53, 53–56.

Perry, E.K., Gibson, P.H., Blessed, G., Perry, R.H., Tomlinson, B.E., 1977. Neurotransmitter enzyme abnormalities in senile dementia. Choline acetyltransferase and glutamic acid decarboxylase activities in necropsy brain tissue. J. Neurol. Sci. 34, 247–265.

Potter, L.T., Glover, V.A., Saelens, J.K., 1968. Choline acetyltransferase from rat brain. J. Biol. Chem. 243, 3864–3870.

Rylett, R.J., Ball, M.J., Colhoun, E.H., 1983. Evidence for high affinity choline transport in synaptosomes prepared from hippocampus and neocortex of patients with Alzheimer's disease. Brain Res. 289, 169–175.

Schroder, H., Giacobini, E., Struble, R.G., Zilles, K., Maelicke, A., 1991. Nicotinic cholinoceptive neurons of the frontal cortex are reduced in Alzheimer's disease. Neurobiol. Aging. 12, 259–262.

Valenstein, E.S., 2002. The discovery of chemical neurotransmitters. Brain Cogn. 49, 73–95.

Wiener, S.W., Hoffman, R.S., 2004. Nerve agents: a comprehensive review. J. Intensive. Care. Med. 19, 22–37.

Woolf, N.J., 1991. Cholinergic systems in mammalian brain and spinal cord. Prog. Neurobiol. 37, 475–524.

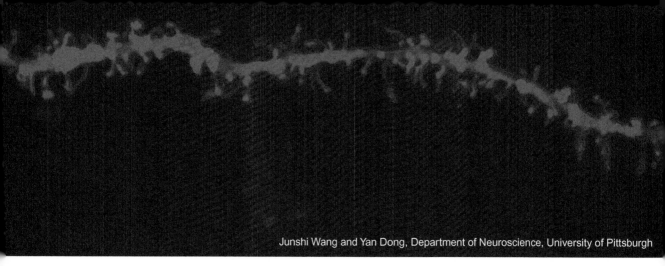

Junshi Wang and Yan Dong, Department of Neuroscience, University of Pittsburgh

# CHAPTER 17

# Monoamine Transmitters

Like acetylcholine, monoamine neurotransmitters are in the group of "classical" neurotransmitters. Though the molecular structures of amine neurotransmitters differ somewhat from one another, they all contain an amine chemical group. An **amine** is a functional chemical group containing a nitrogen atom bonded to a hydrocarbon group and two hydrogen atoms ($R\text{-}NH_2$; see Fig. 17.1). Amine groups are very common in biological systems, including being the common structure found in all amino acids.

There are five main neurotransmitters that contain amine groups: **dopamine**, **norepinephrine**, **epinephrine**, **serotonin**, and **histamine**. These are collectively referred to as the **monoamines** because of their chemical structure (see Fig. 17.1). Additionally, a subset of the monoamine neurotransmitters has an amine group plus a hydrocarbon moiety that contains a single **catechol** group. A catechol group is a benzene ring with two neighboring hydroxyl groups (1,2-dihydroxybenzene). This subset of monoaminergic neurotransmitters is referred to as the **catecholamines**, and includes dopamine, norepinephrine, and epinephrine.

The nomenclature used for monoamines, and the neurons that release them, can be confusing for a number of reasons. The first of these is that norepinephrine is also referred to as noradrenaline, and epinephrine is also referred to as adrenaline. The words "adrenaline" and "epinephrine" were chosen by separate research groups studying the same molecule who each named it based on the fact that it is secreted by the adrenal glands above the kidneys, but chose different linguistic derivations for these terms. Both sets of names have persisted in medical and scientific fields. A second point of confusion is related to the adjectives used to refer to monoaminergic neurons. The adjectives "aminergic" and

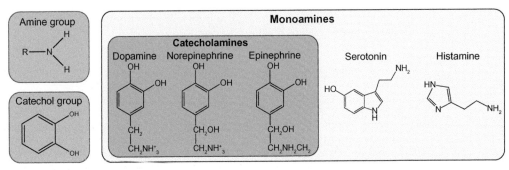

FIGURE 17.1 **Molecular structures of monoamines.** All monoamine neurotransmitters contain an amine group. A subset of these, the catecholamines, also includes a catechol chemical group.

"monoaminergic" refer to neurons that release any monoamine. Neurons that release any catecholamines are referred to as "catecholaminergic" neurons. The term "adrenergic" refers collectively to neurons that release either epinephrine or norepinephrine, though the term "noradrenergic" is also used to refer to neurons that release norepinephrine. In a more straightforward manner, serotonin-releasing neurons are referred to as serotonergic, and those that release histamine are called histaminergic. You will encounter all of these terms in this chapter and in the fields of neuroscience, pharmacology, and medicine.

## CATECHOLAMINE NEUROTRANSMITTERS

### Catecholamine Synthesis

Synthesis of the three catecholamine neurotransmitters, dopamine, norepinephrine, and epinephrine, starts with a common precursor molecule and progresses through a sequence of enzymatic steps. This synthetic process stops at a different point depending on the specific synthetic enzymes a particular neuron expresses. This neurotransmitter-specific stopping point is what dictates which neurotransmitter will be released by a particular type of catecholaminergic neuron (see Fig. 17.2).

There are four steps in the catecholamine biosynthetic pathway (see Figs. 17.2 and 17.3).

### Step 1: Tyrosine to L-Dihydroxyphenylalanine

The precursor molecule for all catecholamines is the amino acid **tyrosine**. Tyrosine is synthesized in the body from phenylalanine (an essential amino acid that must come from the diet). Tyrosine is a large, neutral amino acid that is transported across the blood–brain barrier by the large nonspecific neutral amino acid transporter. Tyrosine is then converted to L-**dihydroxyphenylalanine** (L-DOPA) by the cytoplasmic enzyme **tyrosine hydroxylase** (TH). This enzyme is only found in catecholaminergic neurons. It is the **first unique synthetic enzyme** in the catecholamine biosynthesis pathway, and requires three cofactors: $Fe^{2+}$, tetrahydrobiopterin (BH4), and $O_2$. Tyrosine hydroxylase is the rate-limiting step in

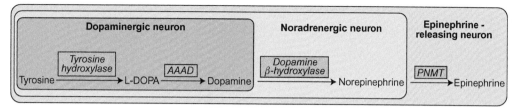

FIGURE 17.2 **Catecholamine synthesis reactions in dopaminergic, noradrenergic, and epinephrine-releasing neurons.** The neurotransmitter released by a given catecholaminergic neuron is dictated by the synthesis enzymes present in that neuron. Tyrosine hydroxylase is the first unique synthetic enzyme in this pathway. It is required for the synthesis of all catecholamine neurotransmitters, and is the only rate-limiting enzyme in this pathway. Aromatic amino acid decarboxylase (AAAD) is expressed in many areas of the nervous system. Dopamine β-hydroxylase is only expressed by norepinephrine- and epinephrine-releasing neurons. Phenylethanolamine-*N*-methyltransferase (PNMT) is only expressed by neurons that will use epinephrine as a neurotransmitter.

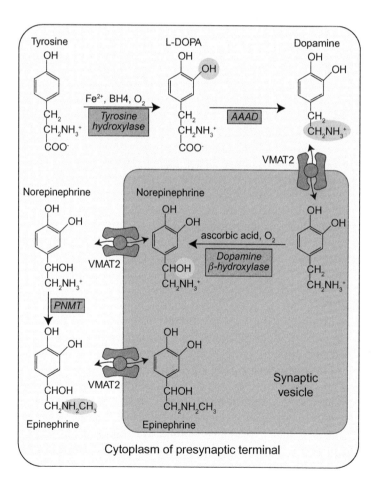

FIGURE 17.3 **Catecholamine synthesis pathways.** Catecholamine neurotransmitters are generated by a sequence of synthesis reactions that begins with the precursor tyrosine. The synthesis pathway in a given neuron ends with dopamine, norepinephrine, or epinephrine, according to the presence or absence of synthetic enzymes expressed by that neuron. This diagram also highlights the compartments within the nerve terminal within which each synthetic step in the pathway occurs. For example, dopamine and epinephrine are synthesized in the cytoplasm, but norepinephrine is synthesized in the synaptic vesicle. VMAT2 can transport any monoamine neurotransmitter. Green ovals highlight changes to the molecule structure that are catalyzed by each enzyme in the catecholamine reaction sequence.

catecholamine synthesis, which means it regulates the rate of synthesis for all three of these neurotransmitters (Kaufman and Kaufman, 1985).

### Step 2: L-Dihydroxyphenylalanine to Dopamine

L-DOPA is rapidly converted to dopamine via a cytoplasmic enzyme called **aromatic amino acid decarboxylase** (AAAD). This enzyme is not unique to catecholamine neurons (e.g., it is also used in the synthesis of serotonin). In a dopaminergic neuron, the biosynthesis process stops here because dopamine neurons do not express the enzymes DBH or PNMT (Fig. 17.2). After dopamine is synthesized in the cytosol, it is packaged into synaptic vesicles via the **vesicular monoamine transporter** (VMAT) and is ready to be released from the neuron.

VMAT is a transmembrane protein located in the vesicular membrane (see Fig. 17.3), and it utilizes a proton gradient to move monoamines across the membrane. There are two VMAT isoforms: VMAT1, which is preferentially expressed in neuroendocrine cells, and VMAT2, which is mainly expressed in neurons (Weihe et al., 1994). Among other things, VMAT2 plays a role in neuronal responses to a number of drugs of abuse, as well as in addiction (see below).

### Step 3: Dopamine to Norepinephrine

Once inside the vesicle, dopamine can be converted to norepinephrine, but only if the enzyme **dopamine β-hydroxylase** (DBH) is present. DBH is expressed within synaptic vesicles only in adrenergic neurons. This enzyme is unique to neurons that release norepinephrine or epinephrine. Importantly, DBH is *not* found in dopaminergic neurons. Without DBH, any dopamine that is transported into a vesicle will not be further altered, so the neuron will release dopamine. However, since other catecholaminergic neurons express DBH, the catecholamine synthesis pathway continues in these cells. DBH is not rate-limiting and is very efficient, so dopamine entering a synaptic vesicle in these neurons is rapidly and completely converted to norepinephrine. In a neuron that releases norepinephrine, the biosynthesis process stops here. Since norepinephrine is synthesized within the synaptic vesicle, it does not need to be transported anywhere in order for it to be released (see Fig. 17.3). DBH requires two cofactors, ascorbic acid and $O_2$ (which are both plentiful).

### Step 4: Norepinephrine to Epinephrine

Once norepinephrine has been synthesized, the transporter VMAT2 can move the norepinephrine molecules out of vesicles into the cytoplasm. Catecholaminergic neurons that release epinephrine express the third and final enzyme in the catecholaminergic synthesis pathway, **phenylethanolamine-N-methyltransferase** (PNMT). This enzyme is unique to neurons that release epinephrine, and converts norepinephrine to epinephrine. PNMT is located in the cytoplasm, so epinephrine is synthesized there, and the epinephrine molecules are then transported back into synaptic vesicles via VMAT2 so they can be released by the neuron. It is hypothesized that VMAT2 can move catecholamines in both directions, and that it must do so until the biosynthetic pathway reaches an endpoint at which the final neurotransmitter in a particular neuron's pathway is synthesized. At that point, the vesicle will contain the neurotransmitter that each neuron uses to communicate.

A critical concept here is that it is the presence or absence of the specific enzymes in the catecholaminergic synthesis pathway that determines which type of neurotransmitter will be released from a catecholaminergic neuron (see Fig. 17.2). All catecholaminergic neurons contain tyrosine hydroxylase, but dopaminergic neurons do not contain dopamine-β-hydroxylase or PNMT. In these neurons, the lack of dopamine-β-hydroxylase means any dopamine cannot be converted into norepinephrine (and thus not epinephrine either, since norepinephrine is the precursor to epinephrine and there is no PNMT present). Similarly, noradrenergic neurons do not contain the enzyme PNMT, so norepinephrine molecules in these neurons cannot be converted into epinephrine. Neurons that release epinephrine contain all three of the catecholaminergic synthetic enzymes.

## Regulation of Catecholamine Synthesis

As with each neurotransmitter system, we will focus on the physiological mechanisms that regulate the synthesis of neurotransmitters. In the case of catecholamines, we will discuss how the nerve terminal knows when to make more neurotransmitter, which step in the synthetic pathway is regulated, and how that step is altered. Since tyrosine hydroxylase is the rate-limiting step in the synthesis of catecholamines, the first question to ask is: What controls the rate of tyrosine hydroxylase activity? The answer is that tyrosine hydroxylase activity is controlled by one of its cofactors, BH4. The concentration of BH4 in catecholamine neurons is relatively low, and this limits how much is bound to tyrosine hydroxylase, thus limiting the activity of this enzyme. Phosphorylation of tyrosine hydroxylase regulates its activity, and thus, catecholamine synthesis (Iuvone et al., 1982; see Fig. 17.4). This was demonstrated experimentally by assaying the amount of radioactive phosphorus ($^{32}P$) incorporated into tyrosine hydroxylase produced in response to the addition of the protein kinase cAMP (Joh et al., 1978; Edelman et al., 1978). These researchers showed that when they incubated rat brain tissue with cAMP, relevant cofactors, and other required molecules, tyrosine hydroxylase was phosphorylated and its activity increased. It was later shown that this was due to a roughly 20-fold increase in the affinity of tyrosine hydroxylase for BH4.

The change in the phosphorylation state of tyrosine hydroxylase comes into play under conditions of high neuronal activity of catecholaminergic neurons. If the rate of action potentials generated by a catecholaminergic neuron increases, catecholamine synthesis also increases. In this case, action potential activity leads to elevated intracellular calcium in the presynaptic terminal. When action potentials occurs at high frequency, the calcium influx temporarily overcomes buffering and handling mechanisms (see Chapter 7), allowing calcium ions to diffuse deep into the nerve terminal (beyond the active zone region), which activates CaMKII. CaMKII then phosphorylates tyrosine hydroxylase, which increases its affinity for BH4. This regulatory mechanism allows a catecholaminergic neuron to adapt how much neurotransmitter it makes in response to changes in its firing rate [see Zigmond et al. (1989) for review]. This effect can be observed in response to direct stimulation of a catecholaminergic neuron (Alousi and Weiner, 1966), as well as in response to environmental signals that increase firing in these neurons, such as circadian

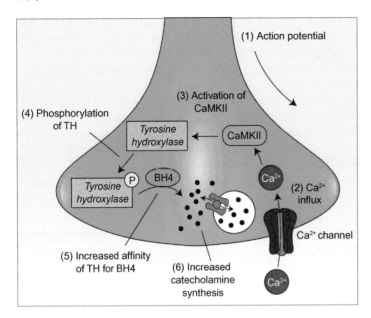

FIGURE 17.4 **Phosphorylation of tyrosine hydroxylase (TH) increases catecholamine synthesis.** Action potentials (1) open voltage-gated calcium channels, causing an influx of calcium ions (2). Calcium ions activate CaMKII (3). CaMKII phosphorylates TH (4), which increases its affinity for the cofactor BH4 (5), leading to increased synthesis of catecholamines (6).

rhythm (Craft et al., 1984), physiological stress such as cold or abnormal insulin levels (Fluharty et al., 1983), hypotensive agents, and immobilization (Kvetnansky et al., 1971).

As discussed above, the first unique synthetic enzyme (tyrosine hydroxylase) is the rate-limiting step in the catecholamine biosynthetic pathway. The precursor, tyrosine, is generally thought to be present in concentrations that are high enough not to limit catecholamine synthesis (Fernstrom, 1983). This hypothesis is consistent with the fact that tyrosine is not an essential amino acid in healthy adults (i.e., we usually do not need to acquire tyrosine from our diet) because our bodies can synthesize it from phenylalanine. However, under certain conditions (e.g., in premature infants and some pathological conditions) the body has a limited supply of tyrosine. As such, it is considered a **conditionally essential amino acid** (i.e., *most* people don't need it from their diet).

Tyrosine crosses the blood−brain barrier via an energy-dependent transporter for large neutral amino acids. However, since this transporter can also transport other large neutral amino acids such as tryptophan and phenylalanine, an abundance of these other large neutral amino acids could theoretically outcompete tyrosine for transport across the blood−brain barrier, reducing brain levels of tyrosine, and, consequently, catecholamine synthesis (but this is thought to be rare under normal dietary conditions).

Under conditions when catecholamines build up in the cytoplasm, a negative-feedback process called **end-product inhibition** can inhibit tyrosine hydroxylase activity. In catecholamine synthesis, dopamine, norepinephrine, or epinephrine that has accumulated within the cytoplasm of the presynaptic terminal decreases the activity of tyrosine hydroxylase (Alousi and Weiner, 1966; Spector et al., 1967). End-product inhibition of catecholamine synthesis was demonstrated by doing the following experiment (Nagatsu et al., 1964). In order to assay the activity of the catecholamine synthesis pathway, experimenters placed tyrosine hydroxylase in solution in a test tube and added tyrosine, the precursor molecule for catecholamine synthesis, as well as the cofactors required by tyrosine hydroxylase ($Fe^{2+}$, BH4, and $O_2$). In this control condition, tyrosine hydroxylase converted tyrosine to L-DOPA. The experimenters measured the amount of L-DOPA formed, which was used as a measure of the activity of tyrosine hydroxylase (see Fig. 17.5A). They then repeated this experiment, but in the presence of excess dopamine, norepinephrine, or epinephrine in the test tube, and measured the amount of L-DOPA formed. They showed that when dopamine, norepinephrine, or epinephrine was present in high concentrations, the amount of L-DOPA produced was reduced (see Fig. 17.5B). These results showed that all of the end products of the catecholamine biosynthesis pathway decrease tyrosine hydroxylase activity. This negative-feedback mechanism keeps catecholamine from being overproduced (see Fig. 17.5C).

## Termination of Action of Catecholamines

As with most neurotransmitters, the action of catecholamines must be terminated after vesicular release. There are two main mechanisms by which catecholamine action is terminated. First, they can be broken down by enzymes. The two main enzymes that break

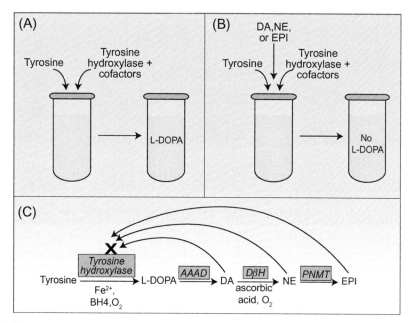

FIGURE 17.5 **Catecholamine synthesis is regulated by end-product inhibition.** (A) When tyrosine and relevant cofactors are combined with tyrosine hydroxylase, L-DOPA will be formed. (B) However, adding dopamine (DA), norepinephrine (NE), or epinephrine (EPI) to the reaction prevents or reduces L-DOPA synthesis. (C) All three catecholamine neurotransmitters can downregulate catecholamine synthesis by inhibiting the action of the enzyme tyrosine hydroxylase.

down catecholamines in the brain are **monoamine oxidase** (MAO) and **catechol-O-methyltransferase** (COMT). MAO is located both in mitochondrial membranes in the cytosol of presynaptic neurons, as well as extracellularly, and COMT is located mainly extracellularly. Whereas the enzyme acetylcholinesterase, which metabolizes acetylcholine (see Chapter 16), is extremely fast, metabolism of catecholamines by MAO and COMT is considerably slower. There are two isoforms of MAO, MAO-A and MAO-B, which both metabolize all monoamine neurotransmitters. Dopamine is broken down equally by MAO-A and MAO-B, but norepinephrine is preferentially metabolized by MAO-A (Finberg and Rabey, 2016). Drugs that inhibit these enzymes are often used to treat symptoms of mood disorders and Parkinson's disease. Some of these drugs are discussed later in the chapter.

The second way the action of catecholamines is terminated is when these molecules are taken up into the presynaptic terminal after release into the synaptic cleft (Hertting and Axelrod, 1961). This reuptake is accomplished by transporters located in the presynaptic membrane. There is a dopamine selective transporter (DAT) and a transporter that is equally selective for both norepinephrine and epinephrine across presynaptic membranes (NET). These transporters are expressed selectively by neurons that synthesize each type of catecholamine (see Box 17.1). Because of this specificity, these transporters can be used to uniquely label the types of catecholaminergic neurons in which they are found.

# SEROTONIN

Serotonin is a monoamine, but not a catecholamine, because while serotonin molecules have a single amine group, they lack the catechol group found on catecholamine molecules (see Fig. 17.1). The chemical name for the serotonin molecule is

---

### BOX 17.1

### JULIUS AXELROD AND THE DISCOVERY OF NEUROTRANSMITTER REUPTAKE

Julius Axelrod (1912–2004; see Fig. 17.6) was a molecular pharmacologist who played a significant role in understanding the functions of amines in the body. Some of his earlier research focused on the metabolism of exogenous amines (e.g., ephedrine, amphetamine) that affect the sympathetic nervous system, also referred to as sympathomimetic amines. During this work, he discovered the cofactor NADPH. Work by Axelrod and colleagues in the 1950s led to the discovery of the enzyme catechol-o-methyl transferase (COMT) and its role in metabolizing catecholamines. At the time, it was assumed that amine neurotransmitters were primarily inactivated was by enzymatic degradation, an assumption that was influenced by the earlier discovery that acetylcholinesterase terminates the action of acetylcholine at the synapse. However, neither COMT nor another catecholamine-metabolizing enzyme, monoamine oxidase, were sufficient to terminate the effects of injected norepinephrine or epinephrine.

## BOX 17.1 (cont'd)

FIGURE 17.6  **Julius Axelrod, March, 1973.** Dr. Axelrod checking a student's work on the chemistry of catecholamine reactions in nerve cells. Source: <https://profiles.nlm.nih.gov/ps/retrieve/ResourceMetadata/HHAABA>.

At the time, Axelrod was doing his work on COMT, and understanding the processes by which the action of catecholamine neurotransmitters was terminated was not his main area of research. However, a colleague studying epinephrine metabolism in people with schizophrenia happened to give him samples of tritiated ($^3$H) epinephrine and norepinephrine to work with in his lab. Both researchers found that when they administered the [$^3$H]norepinephrine and [$^3$H]epinephrine to animals, it accumulated in tissues in the body that were enriched in epinephrine and norepinephrine (e.g., the heart). This work led Axelrod to hypothesize that the action of norepinephrine was terminated by being taken up into sympathetic nerves. He went on to show that the accumulated norepinephrine and epinephrine were localized to nerve endings. Axelrod extended his research on neurotransmitter reuptake to understanding the action of sympathomimetic amine drugs, such as cocaine and amphetamine, and he demonstrated that these substances block amine reuptake.

From this pioneering work, it became clear not only that monoamine neurotransmitter action could be terminated by reuptake, but that this was, in fact, the main mechanism of termination for monoamines. Later work in Axelrod's lab showed that the therapeutic effects of tricyclic

> **BOX 17.1** (cont'd)
>
> antidepressants (TCAs) were correlated with a reduction in the reuptake of norepinephrine. It is now known that TCAs work by blocking monoamine reuptake (specifically, by preventing the reuptake of norepinephrine and serotonin). This observation bolstered the monoamine hypothesis of mood disorders, the theory that mood disorders are caused by an "imbalance" of monoamine neurotransmitters in the brain. This is now considered an overly simplistic view of the biochemical basis for mood disorders, but Axelrod's work in this area still influences the development of new antidepressant and other medications.
>
> In 1970, Axelrod shared the Nobel Prize in Medicine or Physiology with Sir Bernard Katz, of University College London, and Ulf von Euler, of the Karolinska Institute in Stockholm "for their discoveries concerning the humoral transmitters in the nerve terminals and the mechanism for their storage, release and inactivation."
>
> Julius Axelrod's students and fellow researchers noted that one of his characteristic skills was getting to the heart of a research question, and he disliked complex experiments and elaborate hypotheses. These qualities are reflected in a quote by Axelrod, reflecting on his own research process:
>
>> "One of the most important qualities in doing research, I found, was to ask the right questions at the right time. I learned that it takes the same effort to work on an important problem as on a pedestrian or trivial one. When opportunities came, I made the right choices."

**5-hydroxytryptamine**, which is commonly abbreviated as 5-HT. Serotonin is found in multiple cell types throughout the body. In fact, only ~ 1%–2% of all serotonin in the body is in the central nervous system, with the vast majority of it found in the intestine. However, serotonin cannot cross the blood–brain barrier, so in order for it to be used by neurons in the central nervous system, it must be synthesized by neurons there.

## Serotonin Synthesis

The precursor for serotonin is **tryptophan**, a large neutral amino acid. Tryptophan is an essential amino acid, meaning it cannot be made by the body and must come from the diet. For this reason, a lack of dietary tryptophan can dramatically reduce levels of serotonin in the brain. Tryptophan crosses the blood–brain barrier via the same transporter for large neutral amino acids described above for tyrosine, the precursor for catecholamine neurotransmitters. As is the case with tyrosine, large neutral amino acids, including tryptophan, must compete for transport by this carrier, which means that levels of tryptophan in the brain are determined not only by the blood concentration of tryptophan itself, but also by the blood concentration of competing large neutral amino acids.

The first step in the synthesis of serotonin is to convert tryptophan to **5-hydroxytryptophan** (5-HTP) via the enzyme **tryptophan hydroxylase** (see Fig. 17.7).

Tryptophan hydroxylase is the first and only unique synthetic enzyme for neurons that use serotonin. Therefore, in line with the general concept outlined above, the enzyme tryptophan hydroxylase is the rate-limiting step in serotonin synthesis. Tryptophan hydroxylase utilizes two cofactors, $O_2$ and BH4 (see Box 17.2). In the second step of serotonin synthesis, 5-HTP is rapidly converted to serotonin via AAAD, an enzyme that is also used in the catecholamine synthesis pathway, as well as other places in the body. AAAD is present in high concentrations and works efficiently to convert all 5-HTP into 5-HT (serotonin), and thus, is not rate-limiting. Serotonin is then packaged into vesicles using the vesicular monoamine transporter, VMAT (the same transporter that moves catecholamines into vesicles).

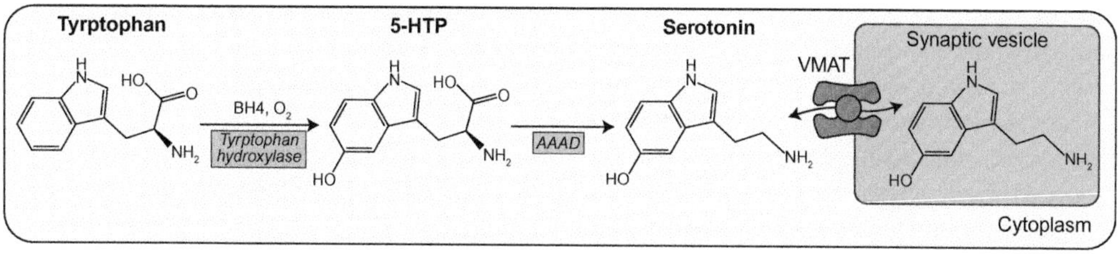

FIGURE 17.7 **Synthesis pathway for serotonin.** Serotonin is synthesized from tryptophan via two enzymes, tryptophan hydroxylase and AAAD. Once synthesized, it is transported into synaptic vesicles by VMAT.

---

### BOX 17.2

### HOW CRITICAL IS A COFACTOR? TREATING A GENETIC DISORDER WITH TWO MONOAMINE PRECURSORS

It can be easy to overlook the importance of cofactors in the neurotransmitter synthesis reactions we discuss, but they can be just as critical as the enzymes themselves. We have discussed the fact that catecholamine synthesis can be regulated by the firing rates of catecholaminergic neurons, which regulates phosphorylation of the rate-limiting enzyme (TH), increasing its affinity for the BH4 cofactor, and thus increasing the rate of neurotransmitter synthesis. But what if a person was not able to produce any BH4 at all? The case of the Beery twins illustrates just such a scenario and shows how basic research and medicine can come together to identify cures for genetic disorders.

At the age of 2, a set of fraternal twins named Alexis and Noah Beery suffered from a number of symptoms, including poor muscle tone, an inability to walk or sit up on their own, and seizures. These symptoms led doctors to diagnose them with cerebral palsy, but their symptoms did not abate with treatments that are usually effective for cerebral palsy. Later, at the age of 5, Alexis also developed significant problems breathing and was sometimes unable to swallow. The twins' mother noticed that Alexis' symptoms fluctuated throughout the day, being at their least severe just after waking and getting progressively worse until she slept (e.g., when she took a nap in the middle of the day), after which her symptoms were significantly reduced.

## BOX 17.2 (cont'd)

Finally, based on her collective symptoms, a neurologist at the University of Michigan correctly diagnosed Alexis with a rare genetic disorder called dopamine-responsive dystonia (DRD). This movement disorder is caused by an inability to produce sufficient dopamine. When Alexis was given a daily dose of L-DOPA (levodopa), the precursor to dopamine in the catecholamine synthesis pathway, her symptoms resolved within days. A few months later, Noah started showing signs of DRD as well, and, similarly, L-DOPA substantially reduced his symptoms.

However, at the age of 13 Alexis developed a severe cough and breathing problems that required daily doses of adrenaline in order for her to function. Doctors were not sure what was causing this, and her parents decided to have the twins' genomes sequenced to try to identify the specific problem that was causing their symptoms. Doctors at the Baylor College of Medicine Human Genome Sequencing Center determined that both of the Beery twins had mutations in the *SPR* gene, which encodes an enzyme called sepiapterin reductase. This enzyme had previously been linked to some cases of DRD, and catalyzes the production of BH4, a cofactor in the synthesis of catecholamines and serotonin. This discovery illuminated the fact that the Berry twins suffered from a deficiency in both dopamine and serotonin. Based on this genetic analysis, doctors added the serotonin precursor 5-hydroxytryptophan to the L-DOPA the twins had been receiving. With this additional treatment, all of their symptoms began to abate within 1–2 weeks.

By determining that the SPR gene was causing the Berry twins' symptoms, it was possible to identify a single synthetic cofactor that resulted in deficiencies in both catecholamine and serotonin. That said, it was not a single mutation that caused these symptoms in the Beery twins. It was determined that both twins had the same *two* separate mutations in the SPR gene. One of these mutations was a nonsense mutation that disrupted the secondary structure of sepiapterin reductase. The other mutation was a missense mutation that prevented the enzyme from binding to NADP. These mutations had been reported in the medical literature, but only when they had occurred as homozygous alleles. In the case of the Beery twins, both twins inherited one of these SPR mutations from their mother and the other SPR mutation from their father. This meant they both had compound heterozygous alleles, each generating a different mutant variant of sepiapterin reductase, and the combination was responsible for the symptoms they experienced.

Though not all genetic disorders have such a clear explanation and such a readily available treatment, in the case of the Beery twins, genetic sequencing was critical for being able to treat their disorder and allowing them to live largely symptom-free. This case also underscores the critical role that basic research plays in medical advances and treatments. If the synthesis pathways for catecholamines and serotonin were not as thoroughly understood as they are, it might not have been clear why the SPR mutations caused the symptoms they did, and critically, how to treat them. This convergence of the fields of genetics, chemistry, molecular neuroscience, and medicine led to a highly successful treatment.

## Regulation of Serotonin Synthesis

If most (80%) of the rate-limiting enzyme tryptophan hydroxylase is inhibited, there is a rapid decrease in serotonin levels. However, if most (80%) of the activity of AAAD is inhibited, serotonin levels do not change, underscoring the fact that tryptophan hydroxylase is the rate-limiting enzyme for serotonin synthesis. The cofactors used by tryptophan hydroxylase ($O_2$ and BH4) are the same as those required for tyrosine hydroxylase activity in the catecholamine pathway, and therefore one would predict that these should govern the rate of tryptophan hydroxylase activity and serotonin synthesis. Physiologically, $O_2$ is tightly regulated by the body, so it is not thought that this cofactor is normally responsible for regulation. However, in an experiment during which $O_2$ levels were artificially increased, there was a dramatic increase in 5-HT levels in the brain. Physiologically, BH4 binding to tryptophan hydroxylase is known to be regulated (as it is for tyrosine hydroxylase in catecholamine synthesis). As in the catecholamine synthesis pathway, increased action potential firing increases calcium influx, which leads to the phosphorylation of tryptophan hydroxylase, increasing its affinity for the cofactor BH4.

When we discussed catecholamine synthesis, we also explained that it can be regulated by end-product inhibition. However, this is not the case for serotonin synthesis. When the precursor tryptophan and tryptophan hydroxylase are added to a test tube (in the presence of cofactors), the resulting serotonin synthesis is unchanged by the addition of the end product (serotonin).

## Termination of Action of Serotonin

There are two primary mechanisms by which the action of serotonin is terminated. First, like acetylcholine, monoamines can be degraded by specific enzymes. One such enzyme is monoamine oxidase, which not only breaks down catecholamines, as discussed earlier, but also breaks down serotonin (see Fig. 17.8). Additionally, there is a serotonin-specific,

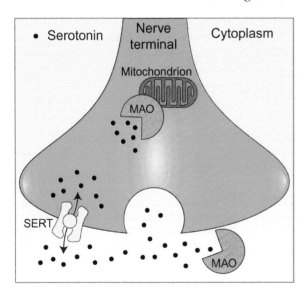

FIGURE 17.8 **Termination of action of serotonin.** Once serotonin (*black dots*) has been released into the synaptic cleft, it is either broken down by the enzyme monoamine oxidase (MAO) or taken up into the presynaptic membrane by the serotonin reuptake transporter (SERT).

high-affinity **serotonin reuptake transporter** (SERT; see Fig. 17.8) that removes serotonin from the synaptic cleft. The activity of SERT is driven by the serotonin concentration gradient and can work in either direction, pumping serotonin into or out of the presynaptic terminal, depending on the direction of the serotonin concentration gradient. When the concentration of serotonin in the synaptic cleft is high, SERT pumps serotonin back into the presynaptic terminal, and when cytoplasmic serotonin is high, SERT can work in reverse to pump serotonin out of the presynaptic terminal.

# HISTAMINE

Histamine is mostly known for its role in immune responses. But if you have ever taken an antihistamine drug to treat cold symptoms (which works by blocking histamine receptors), you might have experienced side effects such as drowsiness and hunger. That these side effects occur suggests that there are histamine receptors in the brain and that histamine is a neurotransmitter.

## Histamine Synthesis

The precursor to histamine is **histidine**, which is a semi-essential amino acid. It is considered semi-essential because while adults make enough of it, children need to get a sufficient amount from their diet from high-protein foods such as meat. In the brain, histidine is converted to histamine in a single-step reaction via the enzyme **histidine decarboxylase**, which uses pyridoxal phosphate as a cofactor (see Fig. 17.9). Histidine decarboxylase is the first unique synthetic enzyme for histamine synthesis, and it can be used to identify histamine-releasing neurons. AAAD can decarboxylate histidine to make histamine, but experimental evidence suggests that histamine synthesis is selectively coupled to the unique enzyme histidine decarboxylase. It was shown that blocking histidine decarboxylase reduces neuronal histamine levels, but blocking AAAD does not. Once synthesized, histamine is actively transported into synaptic vesicles via the VMAT, the transporter discussed earlier that also transports the other monoamines.

## Regulation of Histamine Synthesis

The mechanism for regulation of histamine synthesis is still somewhat controversial. Histidine decarboxylase does not appear to be saturated and will convert any histidine

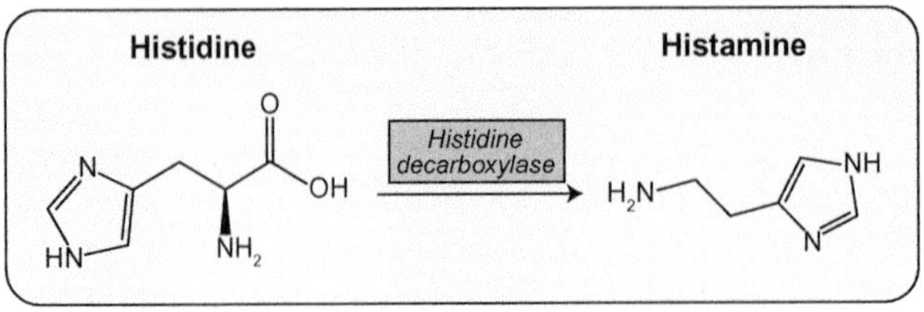

FIGURE 17.9 **Synthesis pathway for histamine.** The precursor histidine is converted to histamine by the enzyme histidine decarboxylase.

present in the presynaptic terminal. Histidine is considered a semi-essential amino acid, meaning healthy adults obtain plenty from their diet, but infants and those with an illness may not obtain enough. In some cases, the histidine precursor may be rate-limiting. Histidine decarboxylase is the first unique enzyme in the neurotransmitter synthesis pathway, and there is some uncertainty about whether it is rate-limiting. This is in contrast to catecholamine synthesis, in which it is clear that tyrosine hydroxylase is both the first unique enzyme and the rate-limiting factor in the synthesis pathway. For example, in the posterior hypothalamus, Hoffman and Koban (2016) provided evidence that histamine levels could be regulated by increasing the expression of histidine decarboxylase, thereby increasing synthesis of histamine. More studies will be required to confirm how histamine synthesis is up-regulated in other areas of the nervous system. Lastly, histamine does not inhibit histidine decarboxylase, so unlike with catecholamines, there is no end-product inhibition. Further studies of physiological regulation will be required to fully understand what controls histamine synthesis.

## Termination of Action of Histamine

Histamine can be degraded by the enzyme **histamine N-methyl transferase**, but it is unclear whether this degradation pathway is the main mechanism for terminating the action of histamine. One of the reasons it is challenging to study this is that all histamine receptors are metabotropic, which makes it difficult to accurately measure the time course of histamine action.

# PROJECTIONS OF MONOAMINERGIC NEURONS AND FUNCTIONS OF MONOAMINES IN THE NERVOUS SYSTEM

The cell bodies of neurons that release monoamines in the central nervous system are located in the brainstem and several areas of the forebrain. Despite the fact that the cell bodies are located only in these specific small nuclei/regions, the axons of these neurons project widely throughout many regions of the CNS. In addition to having widespread projections, monoaminergic neurons often release their neurotransmitter in a diffuse manner, rather than just into a single synaptic cleft. This diffuse release means these molecules are released into the extracellular space where they can diffuse to extrasynaptic receptors on multiple neurons near the area of release, rather than directly into a single synaptic cleft. This type of diffuse release is sometimes called **volume transmission, or paracrine transmission**.

At most synapses, monoamines play a neuromodulatory role. However, the effects that these neurotransmitters have on the nervous system are complex, in part because there are multiple subtypes of receptors for each monoamine neurotransmitter, and also because each of these receptor subtypes can be found in different combinations on different pre- and postsynaptic membranes. In addition, while there is a tendency to consider and/or study each monoamine neurotransmitter in isolation, the monoaminergic systems interact with one another and with other neurotransmitter systems such as the cholinergic system. Some of these functions are described below.

## Dopamine

Dopamine is synthesized mainly in the central nervous system, though it is also made by neurons in the enteric nervous system. While dopamine can be found outside of the central nervous system and affects multiple systems in the body, it plays a more dominant role as a neurotransmitter in the central nervous system. Cell bodies of dopaminergic neurons in the central nervous system are located in multiple areas of the midbrain, including the **substantia nigra pars compacta** (SNc) and the **ventral tegmental area** (VTA; see Fig. 17.10). The SNc projects to the striatum via the nigrostriatal pathway, and death of these dopaminergic neurons in the SNc causes the motor symptoms characteristic of Parkinson's disease. Dopaminergic neurons in the VTA project to the nucleus accumbens via the mesolimbic pathway, where they are involved in the brain's "reward system," which has been implicated in addiction. Additionally, dopaminergic neurons in the VTA project to widespread regions of the neocortex via the mesocortical pathway and are involved in functions such as reward, motivation, and addiction, among many others. Dopaminergic neurons are also found in the hypothalamus, from which they project to the pituitary gland via the tuberoinfundibular pathway (see Fig. 17.10). In humans, there are

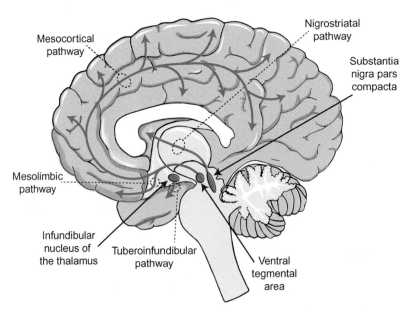

FIGURE 17.10 **Dopaminergic neurons project widely throughout the central nervous system.** Dopaminergic neurons originate in the substantia nigra pars compacta in the midbrain, the ventral tegmental area, also in the midbrain, and in the infundibular nucleus of the hypothalamus. These neurons project via four major pathways (nigrostriatal, mesolimbic, mesocortical, and tuberoinfundibular) to other areas of the brain, as described in the text.

roughly 400,000–600,000 dopaminergic neurons, with >70% of these located in the SNc (Björklund and Dunnett, 2007).

Dopamine is critically involved in the function of the basal ganglia, which regulate adaptive behaviors, including movement. Dopaminergic inputs to the limbic system and prefrontal corticies are involved in emotional responses and the reward system, and dopaminergic signaling has been strongly implicated in schizophrenia.

One role the basal ganglia play in the motor system is as a "gate" that either permits or restricts the initiation of movement. The reason patients with Parkinson's disease have difficulty initiating movements is because the neurodegenerative disease process kills dopaminergic neurons in the SNc. Without a source of dopamine, the basal ganglia are unable to aid in the initiation of movement. Not surprisingly, many drugs used to treat the motor symptoms of Parkinson's disease increase dopaminergic signaling, which can be accomplished in a number of ways. It seems intuitive that administering dopamine might increase dopaminergic signaling in the brain, but there is a complication with giving dopamine as an oral medication: dopamine itself cannot cross the blood–brain barrier and is also broken down rapidly by enzymes in the blood. However, it is possible to administer the direct precursor for dopamine synthesis, L-DOPA, as an oral medication (see Box 17.3). L-DOPA can cross the blood–brain barrier and is currently one of the most common pharmacological treatments for movement-related symptoms of Parkinson's disease. Other dopamine-related drugs for Parkinson's disease include inhibitors of COMT or MAO, dopamine receptor agonists, and dopamine reuptake inhibitors. However, these treatments only alleviate a subset of the motor symptoms of Parkinson's disease and they do not slow the neurodegeneration that causes this disorder. An interesting clinical case involving the death of dopamine neurons in the SNc after heroin use is presented in Box 17.4.

---

## BOX 17.3

### TREATMENT OF THE MOTOR SYMPTOMS OF PARKINSON'S DISEASE WITH THE CATECHOLAMINE PRECURSOR L-DOPA

The movement-related symptoms of Parkinson's disease include slowed movements (or a lack of movement), muscle rigidity, and tremor. These are caused by a lack of dopamine in the basal ganglia, due to the death of dopaminergic neurons in the substantia nigra that project to these regions. Since the 1960s, the primary, and most effective, pharmacological treatment for Parkinson's disease has been the drug L-DOPA (levodopa). L-DOPA is a naturally occurring isomer of the amino acid D,L-dihydroxyphenylalanine, which is made by the faba bean (*Vicia faba*) as well as other related plants. L-DOPA was first isolated from these beans in 1911, but it was many decades and multiple critical scientific discoveries later that led to its use for the motor symptoms of Parkinson's disease.

## BOX 17.3 (cont'd)

The first step was the discovery in 1938 of the enzyme aromatic-L-amino-acid decarboxylase (also referred to as L-DOPA decarboxylase), which converts the catecholamine precursor L-DOPA into dopamine in the catecholamine synthesis pathway (Holtz et al., 1938). This finding demonstrated that dopamine could be generated in the body, and it was later shown that administration of L-DOPA could increase catecholamine production. Subsequent research showed that L-DOPA administration has a wide range of effects on the body (e.g., reduced heart rate, decreased blood pressure), though early reports did not detect effects on movement, likely because it was administered to animals that did not have a deficit in dopamine, as occurs in Parkinson's patients.

In the late 1950s, a rapid series of discoveries showed that (1) dopamine is present in the brain (Montagu, 1957); (2) L-DOPA administration increased brain catecholamine levels (Pletscher, 1957); and (3) L-DOPA administration reversed the Parkinson's-like side effects of a drug called reserpine, which caused a loss of spontaneous movement in laboratory animals (Carlsson et al., 1957). During this time period, research on postmortem brains from Parkinson's patients demonstrated that the disease was characterized by a marked depletion of dopamine in the caudate and putamen (Ehringer and Hornykiewicz, 1960). Based on these discoveries, Hornykiewicz and others conducted trials in patients with Parkinson's disease, showing that administration of L-DOPA had a dramatic therapeutic effect on motor symptoms associated with this disorder (Birkmayer and Hornykiewicz, 1961).

L-DOPA is still the gold standard for alleviation of many of the motor symptoms of Parkinson's disease (though it is generally not effective for Parkinsonian tremor). However, this drug does not stop the progression of Parkinson's disease or cure it, and it can have side effects, including dyskinesias, which are involuntary, jerky movements that can only be reversed by stopping L-DOPA treatment. Additionally, after some time, usually several years, L-DOPA stops being effective for many patients. This is in part because the underlying pathology that causes the death of dopaminergic neurons in the substantia nigra continues.

It would be reasonable to wonder why L-DOPA is used as a drug, rather than simply administering dopamine itself. Why not give the final neurotransmitter (dopamine) as a drug, rather than its precursor? The answer lies in the fact that dopamine is unable to cross the blood−brain barrier, a requisite characteristic of drugs that reach the brain via the bloodstream, dopamine cannot do so. L-DOPA, on the other hand, is readily transported into the brain via the large neutral amino acid transporter. Because L-DOPA is rapidly decarboxylated to dopamine in the blood, giving it a short half-life in the body, it is usually administered along with a decarboxylase inhibitor that does not cross the blood−brain barrier and thus only reduces conversion of L-DOPA to dopamine in the periphery, not in the brain.

## BOX 17.4

## THE FROZEN ADDICTS

In 1982, six young heroin addicts were hospitalized with advanced Parkinsonian motor symptoms that had an incredibly rapid onset. Each patient had each injected synthetic heroin from a batch contaminated with a chemical byproduct called MPTP. This contaminant was responsible for their Parkinsonian symptoms (Langston et al., 1983). The group of patients was referred to as the "Frozen Addicts."

MPTP caused Parkinsonian symptoms because it is cleaved by MAO in the extracellular space in the brain, resulting in the formation of a chemical called $MPP^+$ (Bradbury et al., 1986). This compound is then taken up into presynaptic terminals via dopamine transporters, where it poisons the mitochondrial electron transport system (Ramsay et al., 1986; see Fig. 17.11). $MPP^+$ would be poisonous to any cell (if it could get inside), but because $MPP^+$ relies on DAT to gain entry into the cell cytoplasm, MPTP specifically leads to the death of dopaminergic neurons (Markey et al., 1984; Javitch et al., 1985). Note that this mechanism is not what normally causes Parkinson's disease; it simply replicates the motor symptoms of this disorder because it is a specific toxin for dopaminergic neurons. Nonetheless, understanding how and why MPTP caused Parkinsonian symptoms has been helpful to researchers studying the pathology involved in Parkinson's disease.

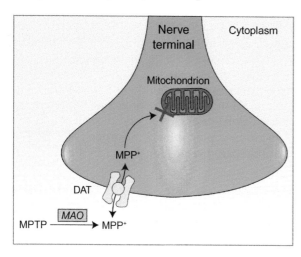

FIGURE 17.11 **Effect of MPTP on dopaminergic neurons.** MPTP is broken down by monoamine oxidase (MAO) into $MPP^+$, which is taken up into dopaminergic nerve terminals by the dopamine transporter (DAT). Once inside, $MPP^+$ impairs mitochondrial function, killing the neuron.

## Norepinephrine

Cell bodies of noradrenergic neurons are located in a number of brainstem nuclei. The main nucleus that contains noradrenergic neurons, the **locus coeruleus**, is at the level of the pons (see Fig. 17.12) and contains approximately 15,000 neurons in humans. Some

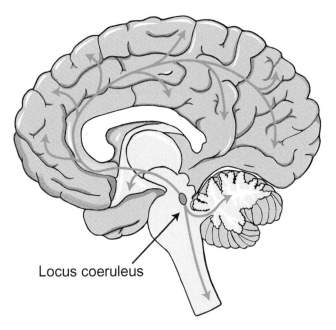

FIGURE 17.12 **Noradrenergic neurons project widely throughout the central nervous system.** Neurons that produce and release norepinephrine have cell bodies in brainstem regions such as the locus coeruleus. They send widespread projections throughout the central nervous system.

neurons in the locus coeruleus project to the hypothalamus, limbic system, and neocortex, where they affect functions such as attention, behavioral arousal, circadian rhythms, and memory. Other noradrenergic neurons from this region project to the cerebellum and spinal cord, where they regulate the respiratory, cardiovascular, and gastrointestinal systems.

Since norepinephrine is used by the sympathetic division of the autonomic nervous system (see discussion of the autonomic nervous system in Chapter 16), noradrenergic drugs can be used to treat a range of disorders involving autonomic function, including glaucoma, migraine, and low blood pressure. Along with dopamine and serotonin, norepinephrine has been implicated in psychiatric disorders such as mood disorders and anxiety. While it is not clear what roles these monoamines play in such disorders, affecting monoaminergic "tone" (the ambient amount of monoaminergic signaling) can alleviate some of their symptoms.

## Epinephrine

Epinephrine, also known as adrenaline, is a neurotransmitter that can also act as a hormone. Epinephrine is mainly known for its functions in the peripheral nervous system, and, in particular, for its effects on the sympathetic division of the autonomic nervous system (see discussion of the autonomic nervous system in Chapter 16). In fact, epinephrine is the active ingredient in auto-injectors used to rapidly reverse the anaphylaxis that can occur in severe allergic reactions.

Epinephrine does not cross the blood–brain barrier, so any epinephrine found in the central nervous system must be synthesized within neurons there. In the central nervous system, neurons that release epinephrine are located mainly in the lower brainstem and project primarily to other regions of the brainstem, the hypothalamus, and the spinal cord. Based on these projections, epinephrine is thought to regulate food and water intake, oxytocin secretion, blood pressure, respiration, thermoregulation, and neuroendocrine function, among other things.

## Serotonin

In the central nervous system, serotoninergic projections arise in the brainstem from a group of neurons called the **raphe nuclei**, which are distributed throughout multiple levels of the brainstem (see Fig. 17.13). As with the other monoaminergic brainstem nuclei we have discussed, neurons in the raphe nuclei send axons widely and diffusely throughout many areas of the central nervous system, including other brainstem nuclei, the spinal cord, hypothalamus, limbic system, basal ganglia, cerebellum, and neocortex. This wide-ranging, diffuse release makes it somewhat challenging to ascribe a specific function to serotonin. This task is also complicated by the fact that there are multiple subtypes of serotonin receptors that have been discovered to date. However, these receptors have subtly different effects on neurons, which is one reason it can be difficult to clearly define, let alone predict, how serotonin affects brain function. Serotonin is well known for its effects on mood and anxiety disorders, but it regulates a range of other functions in the body, including body temperature, blood pressure, sleep–wake states, and hormones. Serotonin is also produced by the gut and controls components of the enteric nervous system.

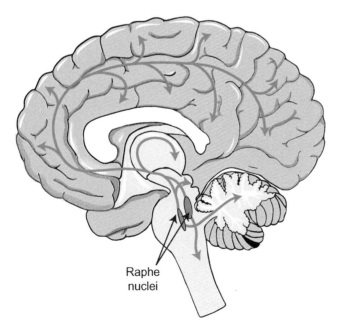

FIGURE 17.13 **Serotonergic neurons project widely throughout the central nervous system.** Cell bodies of neurons that produce and release serotonin are found in the raphe nuclei, which are located at multiple levels of the brainstem within numerous nuclei; two are represented here. Just like other monoaminergic neurons, neurons in the raphe nuclei send widespread projections throughout the central nervous system.

### Sources of Histamine and Its Roles in Brain Function

Cell bodies of neurons that release histamine are located in the hypothalamus. Axons of these neurons project broadly to many areas of the telencephalon, as well as to the brainstem and spinal cord. Histamine transmission in the brain modulates functions such as appetite, pituitary hormone secretion, vestibular function, and arousal. There are three types of histamine receptors, H1, H2, and H3, all of which are metabotropic receptors and have neuromodulatory roles. Older antihistamine drugs used to treat cold symptoms, which are H1 receptor antagonists, have effects on the central nervous system because they can cross the blood−brain barrier. This is why these drugs can cause drowsiness or hunger. Fortunately, newer "nondrowsy" antihistamines have been developed that are H1 antagonists that cannot cross the blood−brain barrier as easily. H1 receptor antagonists can also be used to treat motion sickness, but, like the older antihistamines, these drugs do cause drowsiness.

## THERAPEUTIC DRUGS RELATED TO MONOAMINE NEUROTRANSMITTERS

There are many drugs that affect monoaminergic systems in the brain (see Di Giovanni et al., 2016 for a review). Monoaminergic systems are closely related, not just in terms of the molecular structure of monoamine neurotransmitter molecules, but also in their many effects on the nervous system. In fact, since monoaminergic synthesis enzymes, transporters, receptors, and degradation enzymes can affect or be affected by multiple monoaminergic neurotransmitters, drugs that target these systems are not always specific to a single monoamine neurotransmitter. For this reason, we will present drugs that affect monoaminergic transmission according to their mechanisms of action, rather than by the specific drug or the disorder(s) they treat, which often overlap. As you read about these mechanisms of action below, note that systemic administration (e.g., by mouth) of the monoamine neurotransmitters themselves is not therapeutic because these molecules cannot cross the blood−brain barrier and are quickly degraded once they enter the bloodstream. Thus, the monoaminergic drugs discussed below utilize somewhat indirect methods to control levels of monoamines in the brain.

### Therapeutic Drugs That Stimulate Monoaminergic Receptors

Given that the motor symptoms of Parkinson's disease are caused by the death of dopaminergic neurons in the substantia nigra pars compacta, it is not surprising that dopamine receptor agonists can be used to help alleviate these symptoms (effectively mimicking the effects of the missing dopamine). Five types of dopamine receptors (D1−D5) have been discovered to date, all of which are G-protein-coupled receptors. Initially, it was shown that agonists of D2 and D3 receptors can treat some symptoms of

Parkinson's disease, and it is possible that targeting the D1 receptor may also be a viable strategy for drug development [see Butini et al. (2016) for a review].

## Therapeutic Drugs That Block Monoaminergic Receptors

While it is not completely clear exactly what role dopamine plays in schizophrenia, dopaminergic antagonists can reduce some of the symptoms of this disorder. Most antipsychotic drugs used to treat the positive symptoms of schizophrenia (e.g., hallucinations, delusions, and illogical thinking), are D2 receptor antagonists, and their clinical efficacy is correlated with the ability to block these receptors (Carlsson and Lindqvist, 1963; Seeman et al., 1975; Seeman and Lee, 1975). However, some antipsychotics can also affect other receptors. For example, clozapine is an antagonist of both dopamine and serotonergic receptors, and can alleviate the negative symptoms of schizophrenia (e.g., flattened affect, cognitive impairment). Note that some of the pharmacological strategies for treating Parkinson's disease (e.g., dopamine receptor agonists) and some of those for treating schizophrenia (e.g., D2 receptor antagonists) are the opposite of one another. As you might predict, this means the dopamine receptor agonists given to Parkinson's patients can induce psychotic symptoms in these patients, and the D2 receptor agonists given to people suffering from schizophrenia can lead to Parkinsonian symptoms such as rigidity and paucity of movement.

## Therapeutic Drugs That Affect VMAT2

Drugs that inhibit VMAT2 are no longer used clinically, due to the side effects they can cause. Nonetheless, we will discuss some of these older drugs here to illustrate their mechanisms of actions and effects. A drug called reserpine irreversibly blocks VMAT2, which prevents monoamines from being packaged into vesicles (see Box 17.5). This mechanism has at least two effects that decrease the release of monoamine neurotransmitters. First, once they are synthesized, monoamines cannot be transported into vesicles, which means they must remain in the cytoplasm, where they are subject to degradation by cytoplasmic MAO. Second, since vesicles cannot be filled with neurotransmitter, less neurotransmitter is released from the neuron. Reserpine has been used in the past to treat hypertension, though it has largely been replaced by safer drugs that have fewer side effects. It has also been used as an antipsychotic medication, primarily for schizophrenia. However, one of the side effects of reserpine is Parkinsonism, the development of symptoms of Parkinson's disease without having the actual disorder itself (i.e., the side effects can be similar to symptoms of Parkinson's disease, but they are not brought about by the death of dopamine neurons in the substantia nigra, as is the case in Parkinson's disease). These effects make some sense since the motor symptoms of Parkinson's disease are caused by a depletion of dopamine. Another side effect of reserpine is depression, a finding which supports the hypothesis that a shortage of monoamines causes depression (Frize, 1954). However, use of these drugs as therapeutic agents was discontinued because of their side effects and also because of the development of more selective ways to target monoamine systems.

## BOX 17.5

## TREATMENTS FROM NATURE: INDIAN SNAKEROOT AND THE VESICULAR MONOAMINE TRANSPORTER

### *Rauvolfia serpentine* (Indian Snakeroot)

*Rauvolfia serpentine* (see Fig. 17.14) has been used medicinally for centuries as a treatment for dysentery, snakebite wounds, scorpion stings, and more. *Rauvolfia* contains hundreds of alkaloids, though the most notable of these compounds for use in modern medicine is a chemical called reserpine. Reserpine has been widely used as an antihypertensive and antipsychotic drug due to its effects at monoaminergic synapses. Reserpine prevents the loading of monoamines into synaptic vesicles by irreversibly blocking the vesicular monoamine transporter (specifically, VMAT2). Because VMAT2 is a monoamine transporter, vesicular loading of norepinephrine, epinephrine, serotonin, and dopamine is reduced, and these neurotransmitters remain in the cytoplasm until they are degraded by MAO and COMT. In high doses, reserpine can cause Parkinson's-like effects, and several days or weeks are required to replenish VMAT function to subsymptomatic levels. At low doses, reserpine causes a reduction in catecholamine release from peripheral sympathetic nerve endings, resulting in reduced heart rate and cardiac contraction.

FIGURE 17.14 *Rauvolfia serpentine* plant. *Source: Vinayaraj CC BY-SA 3.0, from Wikimedia Commons.*

## Therapeutic Drugs That Inhibit Monoamine Oxidase

It was discovered serendipitously that compounds that block monoamine oxidase can ease symptoms of depression in some people. In the 1950s it was observed that the tuberculosis drug iproniazid could significantly elevate the moods of patients taking the drug, and that these effects were unrelated to its effects on tuberculosis. It was shown that iproniazid inhibits monoamine oxidase (Zeller and Barsky, 1952), which led to the hypothesis that mood disorders are caused by a decrease in monoamine neurotransmitters in the brain. It has since been accepted that this hypothesis is too simplistic to explain the causes and range of symptoms of mood disorders, but it was nonetheless influential in the development of later psychiatric drugs. However, more recently monoamine oxidase inhibitors have fallen out of favor as drugs, due to their side effects and the development of safer alternatives.

## Therapeutic Drugs That Inhibit the Reuptake of Monoamines

Another set of antidepressant medications acts by blocking plasma membrane transporters for monoamines, which inhibits the reuptake of monoamines back into the presynaptic terminal. As one would expect, the neurochemical effect of these reuptake blockers is to increase the time that monoamines can act on receptors in the synaptic cleft. Thus, decreasing the rate of reuptake increases the effect that neurotransmitter molecules have on their receptors.

The first of these drugs that were used clinically are called tricyclic antidepressants (TCAs), named because of their characteristic three-ring structure. These drugs inhibit NET and SERT, though different TCAs have different affinities for these transporters. However, like MAO inhibitors, TCAs can have dangerous side effects, and have largely been replaced by newer antidepressant medications. For a discussion of St. John's Wort, which also inhibits monoamine uptake and is sometimes taken in an effort to treat depression, see Box 17.6.

A second generation of antidepressant medications was developed to specifically reduce serotonin reuptake by blocking the action of SERT. These are referred to as "selective serotonin reuptake inhibitors," or SSRI's. The best known of these is fluoxetine, originally marketed under the trade name Prozac (Wong et al., 1975). These drugs are mainly used to treat symptoms of major depressive disorder and anxiety disorders. In addition to SSRIs, there are newer drugs that, like TCAs, block both SERT and NET. These drugs are referred to as serotonin and norepinephrine reuptake inhibitors, or SNRIs, and have similar, though not identical, therapeutic effects as SSRIs. Fortunately, the newer selective reuptake inhibitors are significantly safer than MAOIs and TCAs and have largely replaced them in the treatment of major depressive disorder and anxiety.

## BOX 17.6

## DRUGS FROM NATURE: ST. JOHN'S WORT AND MONOAMINERGIC REUPTAKE

### *Hypericum perforatum*

Named because its traditional flowering day coincides with St. John's Day, *Hypericum perforatum* (see Fig. 17.15) has been used as a traditional medicine for nearly 2000 years. In antiquity, this plant was mixed into a "cure-all" concoction that included tens of other ingredients, including viper flesh, opium, and honey. In modern times, St. John's Wort can be purchased in nearly any store that carries herbal supplements, and has been extensively used as a homeopathic treatment for depression.

*Hypericum perforatum* contains the constituent hyperforin, which can prevent monoamine reuptake. People who benefit from taking *Hypericum perforatum* supplements report symptom improvements similar to those caused by SSRIs, suggesting hyperforin affects serotonergic signaling. There is a danger of overdosing on hyperforin, especially when it is taken in combination with serotinergic drugs, as this can rapidly induce "serotonin syndrome." Serotonin syndrome symptoms include elevated heart rate, tremor, hyperactive bowels (due to the large serotonergic population in the enteric nervous system), seizures, and, in severe cases, coma. For this reason, people seeking relief from depression (or any illness) need to be careful about understanding potential interactions between their prescribed and homeopathic medications to prevent dangerous side effects.

FIGURE 17.15 *Hypericum perforatum* plant. *Source: Credit: C T Johansson, CC BY-SA 3.0, from Wikimedia Commons.*

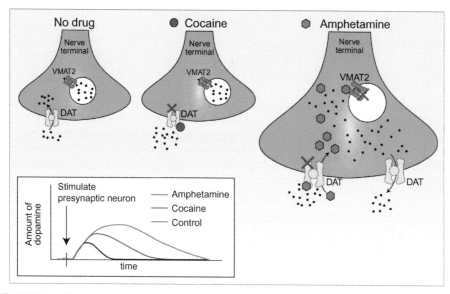

FIGURE 17.16  **Effects of cocaine and amphetamine on dopaminergic neurons.** Cocaine blocks dopamine transporters (DAT), preventing reuptake of dopamine, which then accumulates in the synaptic cleft and increases the magnitude and duration of dopaminergic signal (inset). Amphetamine affects the amount of dopamine in the synaptic cleft in two ways. First, it prevents the reuptake of dopamine by DAT. Second, amphetamine molecules are transported into the presynaptic terminal by DAT, but once inside the presynaptic terminal they block VMAT2, which prevents dopamine from being loaded into synaptic vesicles. Because of the high concentration of dopamine in the cytoplasm, the DAT works in reverse, moving dopamine molecules into the synaptic cleft. Thus, amphetamine increases the amount and duration of dopamine signaling, but more so than cocaine does (inset).

## MONOAMINERGIC DRUGS OF ABUSE

Many addictive drugs, including multiple drugs of abuse, directly or indirectly enhance dopaminergic activity. This is in large part because these drugs strongly activate the brain's reward pathway, and are addictive, which reinforces their use. Several of these drugs are discussed here, as well as some drugs whose effects seem to be primarily mediated by their action on serotonin receptors.

Cocaine is a natural compound found in leaves of the coca plant, which is native to South America. Cocaine blocks monoaminergic transporters, mainly DAT (see Fig. 17.16), though it also blocks NET and SERT to a lesser degree. However, it is thought that the main behavioral effects of cocaine (euphoria and increased alertness, as well as its addictive nature) are due to its inhibition of DAT. DAT knockout mice show hyperlocomotion, which is a characteristic behavioral effect of many stimulant drugs, and these mice stimulants such as cocaine and amphetamine (Giros et al., 1996).

Amphetamine is another addictive stimulant that blocks DAT, but this drug also affects dopaminergic signaling in several other ways (see Fig. 17.15; see Fleckenstein et al., 2007 for a review). The amphetamine molecule is similar in configuration to dopamine (see Fig. 17.17), which allows amphetamine to be transported into neurons

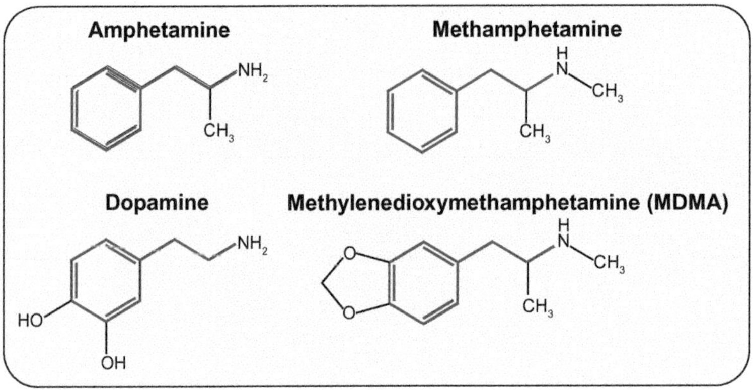

FIGURE 17.17 **Drugs of abuse that affect monoaminergic signaling and have a common chemical structural core.** The drugs shown here have a common set of structural features (benzene ring and a chain with two carbons and one nitrogen), but have slight variations outside of this core structure.

via DAT. Once in the cytosol, amphetamine molecules reduce the action of VMAT2, so dopamine can no longer enter synaptic vesicles causing it to build up inside the presynaptic terminal (Sulzer and Rayport, 1990). Finally, amphetamine also inhibits monoamine oxidase, preventing the breakdown of dopamine in the cytosol. The resulting high cytosolic concentration of dopamine causes dopamine transporters to work in reverse, pumping dopamine into the synaptic cleft (see Fig. 17.13). The net result of these actions is to substantially increase synaptic dopamine, which leads to the behavioral effects of amphetamine, including excessive alertness and energy, euphoria, and suppression of appetite. It should be noted that while the main behavioral effects of amphetamine appear to involve blocking DAT and VMAT2, amphetamine also affects SERT and NET. Amphetamine and related compounds also have therapeutic uses, such as for narcolepsy and ADHD (e.g., Adderall contains four structurally similar versions of the amphetamine molecule).

The synthetic drug methylenedioxymethamphetamine (MDMA), which is typically the main component in the street drug ecstasy, blocks the reuptake of serotonin by SERT (among other actions). MDMA can force SERT to work in reverse, pumping serotonin out of the presynaptic terminal, regardless of the cytoplasmic serotonin concentration (Rudnick and Wall, 1992). As a result, there is an increased amount of serotonin in the synaptic cleft. Interestingly, MDMA has both stimulant- and hallucinogen-like behavioral effects. These include some effects that are generally perceived as desirable, including increased energy, an enhanced sense of well-being, and emotional empathy toward others, as well as undesirable effects, including restlessness, panic attacks, and dangerously elevated body temperature, blood pressure, and heart rate. It is not clear why blocking serotonin reuptake leads to these diverse effects.

Interestingly, the effects of some hallucinogenic drugs appear to be fairly specifically related to activation of the 5-$HT_{2A}$ receptor. In fact, most 5-$HT_{2A}$ receptor agonists are hallucinogenic, and blocking these receptors will prevent the hallucinogenic effects these drugs have. An example of such a compound is lysergic acid diethylamide (LSD), an entirely synthetic compound developed by Albert Hoffman in 1938. LSD binds to multiple types of serotonin receptors, but its binding to the 5-$HT_{2A}$ receptors is thought to underlie its hallucinogenic effects. Similarly, mushrooms such as *Psilocybe semilanceata* contain the molecule psilocybin, which is phosphorylated by the body to make psilocin, the psychoactive compound responsible for the psychedelic effects of these so-called "magic mushrooms." Like LSD, the hallucinogenic effects of psilocin are attributed to its action as a 5-HT$_{2A}$ receptor agonist, but how 5-HT$_{2A}$ receptor agonists cause hallucinations is not well understood.

# References

Alousi, A., Weiner, N., 1966. The regulation of norepinephrine synthesis in sympathetic nerves: effect of nerve stimulation, cocaine, and catecholamine-releasing agents. Proc. Natl. Acad. Sci. USA 56, 1491–1496.

Birkmayer, W., Hornykiewicz, O., 1961. Der l-Dioxyphenylalanin (DOPA)–Effekt bei der Parkinson-Akinese. Wien Klin Wschr 73, 787–788.

Bjorklund, A., Dunnett, S.B., 2007. Dopamine neuron systems in the brain: an update. Trends Neurosci. 30, 194–202.

Bradbury, A.J., Costall, B., Domeney, A.M., Jenner, P., Kelly, M.E., Marsden, C.D., et al., 1986. 1-Methyl-4-phenylpyridine is neurotoxic to the nigrostriatal dopamine pathway. Nature 319, 56–57.

Butini, S., Nikolic, K., Kassel, S., Bruckmann, H., Filipic, S., Agbaba, D., et al., 2016. Polypharmacology of dopamine receptor ligands. Prog. Neurobiol. 142, 68–103.

Carlsson, A., Lindqvist, M., Magnusson, T., 1957. 3, 4-Dihydroxyphenylalanine and 5-hydroxytryptamine as reserpine antagonists. Nature 127, 471.

Carlsson, A., Lindqvist, M., 1963. Effect of chlorpromazine or haloperidol on formation of 3methoxytyramine and normetanephrine in mouse brain. Acta Pharmacol. Toxicol. (Copenh) 20, 140–144.

Craft, C.M., Morgan, W.W., Reiter, R.J., 1984. 24-Hour changes in catecholamine synthesis in rat and hamster pineal glands. Neuroendocrinology 38, 193–198.

Di Giovanni, G., Svob Strac, D., Sole, M., Unzeta, M., Tipton, K.F., Muck-Seler, D., et al., 2016. Monoaminergic and histaminergic strategies and treatments in brain diseases. Front. Neurosci. 10, 541.

Edelman, A.M., Raese, J.D., Lazar, M.A., Barchas, J.D., 1978. In vitro phosphorylation of a purified preparation of bovine corpus striatal tyrosine hydroxylase. Commun. Psychopharmacol. 2, 461–465.

Ehringer, H., Hornykiewicz, O., 1960. Verteilung von Noradrenalin und Dopamin (3-Hydroxytyramin) im Gehirn des Menschen und ihr Verhalten bei Erkrankungen des extrapyramidalen Systems. Klin Wschr 38, 1236–1239.

Fernstrom, J.D., 1983. Role of precursor availability in control of monoamine biosynthesis in brain. Physiol. Rev. 63, 484–546.

Finberg, J.P., Rabey, J.M., 2016. Inhibitors of MAO-A and MAO-B in psychiatry and neurology. Front. Pharmacol. 7, 340.

Fleckenstein, A.E., Volz, T.J., Riddle, E.L., Gibb, J.W., Hanson, G.R., 2007. New insights into the mechanism of action of amphetamines. Annu. Rev. Pharmacol. Toxicol 47, 681–698.

Fluharty, S.J., Snyder, G.L., Stricker, E.M., Zigmond, M.J., 1983. Short- and long-term changes in adrenal tyrosine hydroxylase activity during insulin-induced hypoglycemia and cold stress. Brain Res. 267, 384–387.

Freis, E.D., 1954. Mental depression in hypertensive patients treated for long periods with large doses of reserpine. N. Engl. J. Med. 251, 1006–1008.

Giros, B., Jaber, M., Jones, S.R., Wightman, R.M., Caron, M.G., 1996. Hyperlocomotion and indifference to cocaine and amphetamine in mice lacking the dopamine transporter. Nature 379, 606–612.

Hertting, G., Axelrod, J., 1961. Fate of tritiated noradrenaline at the sympathetic nerve-endings. Nature 192, 172–173.

Hoffman, G.E., Koban, M., 2016. Hypothalamic L-histidine decarboxylase is up regulated during chronic REM sleep deprivation of rats. PLOS ONE . Available from: https://doi.org/10.1371/journal.pone.0152252.

Holtz, P., Heise, R., Lüdtke, K., 1938. Fermentative degradation of l-dioxyphenylalanine (dopa) by kidney. Naunyn-Schmiedeberg's Archives of Pharmacology 191, 87–118.

Iuvone, P.M., Rauch, A.L., Marshburn, P.B., Glass, D.B., Neff, N.H., 1982. Activation of retinal tyrosine hydroxylase in vitro by cyclic AMP-dependent protein kinase: characterization and comparison to activation in vivo by photic stimulation. J. Neurochem. 39, 1632–1640.

Javitch, J.A., D'Amato, R.J., Strittmatter, S.M., Snyder, S.H., 1985. Parkinsonism-inducing neurotoxin, N-methyl-4-phenyl-1,2,3,6-tetrahydropyridine: uptake of the metabolite N-methyl-4-phenylpyridine by dopamine neurons explains selective toxicity. Proc. Natl. Acad. Sci. USA 82, 2173–2177.

Joh, T.H., Park, D.H., Reis, D.J., 1978. Direct phosphorylation of brain tyrosine hydroxylase by cyclic AMP-dependent protein kinase: mechanism of enzyme activation. Proc. Natl. Acad. Sci. USA 75, 4744–4748.

Kaufman, S., Kaufman, E.E., 1985. Tyrosine hydroxylase. In: Blakley, R.L., Benkovic, S.J. (Eds.), Folates and pterins, vol 2, Chemistry and biochemistry of the pterins. Wiley, New York, pp. 251–352.

Kvetnansky, R., Weise, V.K., Gewirtz, G.P., Kopin, I.J., 1971. Synthesis of adrenal catecholamines in rats during and after immobilization stress. Endocrinology 89, 46–49.

Langston, J.W., Ballard, P., Tetrud, J.W., Irwin, I., 1983. Chronic parkinsonism in humans due to a product of meperidine-analog synthesis. Science 219, 979–980.

Markey, S.P., Johannessen, J.N., Chiueh, C.C., Burns, R.S., Herkenham, M.A., 1984. Intraneuronal generation of a pyridinium metabolite may cause drug-induced parkinsonism. Nature 311, 464–467.

Montagu, K.A., 1957. Catechol compounds in rat tissues and in brains of different animals. Nature 180 (4579), 244–245.

Nagatsu, T., Levitt, M., Udenfriend, S., 1964. Tyrosine hydroxylase. The initial step in norepinephrine biosynthesis. J. Biol. Chem. 239, 2910–2917.

Pletscher, A., 1957. Wirckung von Isopropyl-Isonikotinsäurehydrazid auf den Stoffwechsel von Catecholaminen und 5-Hydroxytryptamin im Gehirn. Schweiz Med Wochenschr 87, 1532–1534.

Ramsay, R.R., Salach, J.I., Singer, T.P., 1986. Uptake of the neurotoxin 1-methyl-4-phenylpyridine (MPP+) by mitochondria and its relation to the inhibition of the mitochondrial oxidation of NAD+-linked substrates by MPP+. Biochem. Biophys. Res. Commun. 134, 743–748.

Rudnick, G., Wall, S.C., 1992. The molecular mechanism of "ecstasy" [3&methylenedioxymethamphetamine (MDMA)]: serotonin transporters are targets for MDMA-induced serotonin release. Proc. Natl. Acad. Sci. USA 89, 1817–1821.

Seeman, P., Lee, T., 1975. Antipsychotic drugs: direct correlation between clinical potency and presynaptic action on dopamine neurons. Science 188, 1217–1219.

Seeman, P., Chau-Wong, M., Tedesco, J., Wong, K., 1975. Brain receptors for antipsychotic drugs and dopamine: direct binding assays. Proc. Natl. Acad. Sci. USA 72, 4376–4380.

Spector, S., Gordon, R., Sjoerdsma, A., Udenfriend, S., 1967. End-product inhibition of tyrosine hydroxylase as a possible mechanism for regulation of norepinephrine synthesis. Mol. Pharmacol. 3, 549–555.

Sulzer, D., Rayport, S., 1990. Amphetamine and other psychostimulants reduce pH gradients in midbrain dopaminergic neurons and chromaffin granules: a mechanism of action. Neuron 5, 797–808.

Weihe, E., Schafer, M.K., Erickson, J.D., Eiden, L.E., 1994. Localization of vesicular monoamine transporter isoforms (VMAT1 and VMAT2) to endocrine cells and neurons in rat. J. Mol. Neurosci. 5, 149–164.

Wong, D.T., Bymaster, F.P., Horng, J.S., Molloy, B.B., 1975. A new selective inhibitor for uptake of serotonin into synaptosomes of rat brain: 3-(p-trifluoromethylphenoxy). N-methyl-3-phenylpropylamine. J. Pharmacol. Exp. Ther. 193, 804–811.

Zeller, E.A., Barsky, J., 1952. In vivo inhibition of liver and brain monoamine oxidase by 1-Isonicotinyl-2-isopropyl hydrazine. Proc. Soc. Exp. Biol. Med. 81, 459–461.

Zigmond, R.E., Schwarzschild, M.A., Rittenhouse, A.R., 1989. Acute regulation of tyrosine hydroxylase by nerve activity and by neurotransmitters via phosphorylation. Annu. Rev. Neurosci. 12, 415–461.

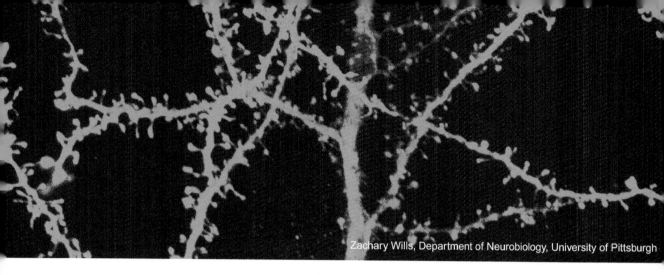

CHAPTER 18

# Amino Acid Neurotransmitters

This chapter will discuss three amino acid neurotransmitters, glutamate, glycine, and GABA. Because amino acids are found in all cells and are utilized in protein synthesis, early researchers did not consider them to be good candidate neurotransmitters. Specifically, it seemed unlikely that they could be used for the type of highly specific signaling messages the nervous system requires. However, it was subsequently shown that the most common excitatory (depolarizing) and inhibitory (hyperpolarizing) neurotransmitters in the mammalian nervous system are amino acids. The type 2 category of neurotransmitters we defined in Chapter 15, includes two amino acids: glutamate and glycine. We have included the discussion of GABA in this chapter, even though we classified it as a type 1 neurotransmitter (because it is uniquely synthesized by neurons that use it as a neurotransmitter), due to the fact that its chemical structure is that of an amino acid (Fig. 18.1).

## GLUTAMATE

More than 50% of all nerve terminals in the mammalian central nervous system use glutamate as a neurotransmitter, and it usually has an excitatory effect (when acting on its ionotropic receptors). However, remember that the effect of a neurotransmitter always depends on the receptors to which it binds. That is, the function of a neurotransmitter is not a characteristic of the specific neurotransmitter molecule itself. In Chapter 11, we discussed glutamate receptors, which have both ionotropic and metabotropic forms. While all ionotropic glutamate receptors are depolarizing (passing primarily sodium and potassium

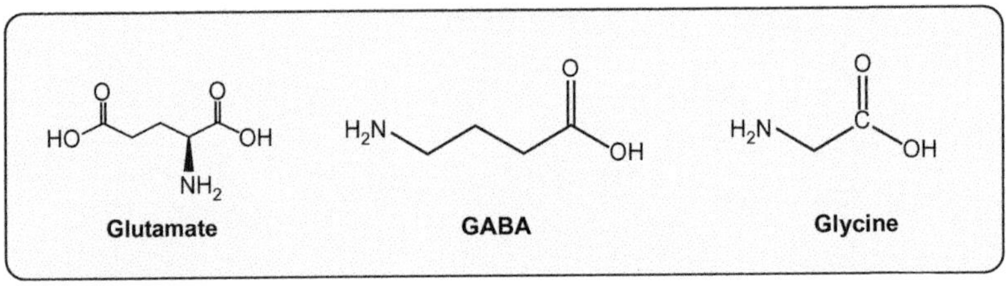

FIGURE 18.1  Molecular structures for the amino acids that are used as neurotransmitters.

ions), and thus excitatory, some metabotropic glutamate receptors are inhibitory (via their targeting of ion channels or neurotransmitter release machinery in the synapse). As such, it is not accurate to define glutamate as strictly an "excitatory" neurotransmitter.

## Glutamate Synthesis

When in a solution, such as extracellular fluid, glutamate is a negatively charged molecule (i.e., an anion), so it is not transported efficiently across the blood–brain barrier (Hawkins, 2009). In fact, the central nervous system has developed mechanisms to tightly control glutamate concentrations in the extracellular fluid (where it is $\sim 0.5-2\,\mu M$), relative to the amount in circulating blood (where it is $\sim 50-100\,\mu M$). As part of this control, it is necessary for glutamate to be synthesized in the brain (though it is made by all cells in the brain, not just neurons that use it as a neurotransmitter). There are several common pathways by which glutamate is synthesized (see Fig. 18.2). In one, **α-ketoglutarate**, a product of the Krebs (citric acid) cycle, is converted into glutamate by the enzyme **glutamate dehydrogenase**. In another, α-ketoglutarate and **aspartate** are converted into glutamate by the enzyme **aspartate aminotransferase** (AAT). In a third pathway, **glutamine** is converted into glutamate by the enzyme **glutaminase**. Glutamine can be obtained either from food or it can be synthesized in the body, and it is brought into neurons via glutamine transporters located in the neuronal plasma membrane. Additionally, glial cells metabolize glutamate into glutamine, which is then transported out of the glial cells and brought into neurons (via glutamine transporters) to be re-used in glutamate synthesis.

It is important to reiterate that none of the synthesis pathways described here is unique to glutamatergic cells. Unlike a number of other neurotransmitters (e.g., acetylcholine), glutamate is *not* uniquely synthesized by neurons that use it as a neurotransmitter. This means that neither glutamate nor the enzymes used to synthesize it can be used to uniquely identify glutamatergic neurons via histological labeling methods.

## Packaging of Glutamate Into Vesicles

Glutamate is loaded into synaptic vesicles via the **vesicular glutamate transporter** (VGLUT; see Fig. 18.3). There are three known types of VGLUTs, called VGLUT 1, 2, and 3. Like the vesicular acetylcholine transporter, vesicular glutamate transporters are driven by a proton gradient that is maintained by a vesicular proton-dependent ATPase (also called a "proton pump"; but see Box 16.2 for more details). Since VGLUTs are found only in glutamatergic neurons, they (unlike glutamate and its synthetic enzymes) can be used to uniquely identify these neurons using histological labeling techniques.

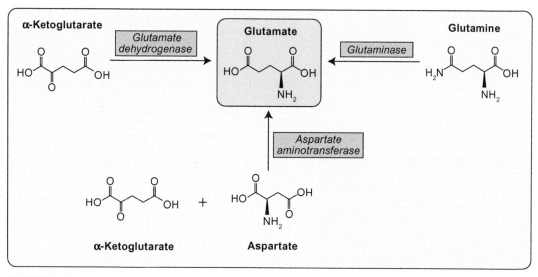

FIGURE 18.2 **Synthesis pathways for glutamate.** Glutamate is synthesized by three pathways. In the first, glutamate dehydrogenase converts α-ketoglutarate into glutamate (left). In the second, aspartate aminotransferase converts α-ketoglutarate and aspartate into glutamate (bottom). In the third, glutamine is converted into glutamate by the enzyme glutaminase (right).

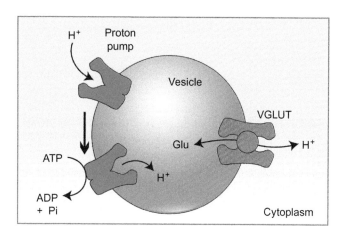

FIGURE 18.3 **The vesicular glutamate transporter (VGLUT) is uniquely expressed in the vesicular membranes of glutamatergic cells.** VGLUT utilizes a proton gradient, maintained by an ATP-dependent proton pump, to transport glutamate into the vesicle (see Box 16.2 for more details).

## Identity of Action for Glutamate

When glutamate is applied to a glutamatergic synapse in the central nervous system, it causes postsynaptic potentials (PSPs) that are similar to those resulting from stimulation of the presynaptic neuron. One way to show this effect is by applying **caged glutamate** to a synapse in a brain slice preparation. Caged glutamate is a glutamate molecule to which a chemical side group has been added, which makes the glutamate group inactive until the side group is broken off by UV light from a laser (this is similar to the caged calcium discussed in Chapter 7 see Fig. 18.4A). In the experiment shown in Fig. 18.4, a postsynaptic neuron was filled with a fluorescent dye so it would be visible under a microscope

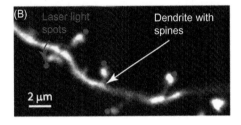
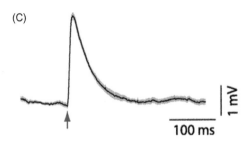
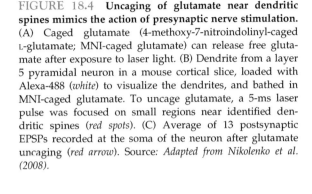

FIGURE 18.4 **Uncaging of glutamate near dendritic spines mimics the action of presynaptic nerve stimulation.** (A) Caged glutamate (4-methoxy-7-nitroindolinyl-caged L-glutamate; MNI-caged glutamate) can release free glutamate after exposure to laser light. (B) Dendrite from a layer 5 pyramidal neuron in a mouse cortical slice, loaded with Alexa-488 (*white*) to visualize the dendrites, and bathed in MNI-caged glutamate. To uncage glutamate, a 5-ms laser pulse was focused on small regions near identified dendritic spines (*red spots*). (C) Average of 13 postsynaptic EPSPs recorded at the soma of the neuron after glutamate uncaging (*red arrow*). Source: *Adapted from Nikolenko et al. (2008).*

(see Fig. 18.4B). When the experimenters applied caged glutamate to the solution surrounding the cell and flashed a narrowly focused laser light just outside a single spine, they recorded EPSPs at the soma (see Fig. 18.4C). These results suggest that application of glutamate at a single spine can cause EPSPs, triggered by the activation of ionotropic glutamate receptors. This is strong evidence of glutamate's identity of action, meaning that it mimics the postsynaptic effects of stimulating a neuron that uses it as a neurotransmitter.

## Regulation of Glutamate Synthesis

Glutamate is not synthesized uniquely by neurons that use it as a neurotransmitter, there are two mechanisms by which glutamate synthesis is hypothesized to be regulated, at the level of the whole cell and at the level of individual presynaptic terminals. In the first mechanism, an increase in neuronal activity leads to an increase in glucose utilization via the glycolysis cycle, producing pyruvate. Pyruvate then enters the Krebs cycle, increasing production of the glutamate precursor α-ketoglutarate, which leads to increased synthesis of glutamate. In the second mechanism, it is proposed that action potential activity may increase the activity of glutaminase. It is thought that this is caused by an accumulation of cytoplasmic calcium, which increases the activity of an inorganic phosphate

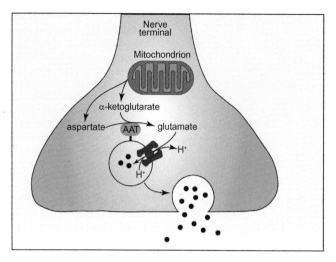

FIGURE 18.5 Aspartate aminotransferase (AAT) may be uniquely localized to synaptic vesicles in glutamatergic neurons to facilitate glutamate packaging. The glutamate synthetic enzyme AAT (*pink*) is located on synaptic vesicles and may use precursors α-ketoglutarate and aspartate derived from mitochondria (*green*) to synthesize glutamate and to preferentially "feed" this glutamate to the vesicular glutamate transporter (*purple*).

transporter. Since the activity of glutaminase is regulated by inorganic phosphate, an increase in the activity of this transporter would in turn increase the production of glutamate. Glutaminase is abundant on mitochondria, which are found in large numbers in synaptic terminals. According to this hypothesis, an action-potential-induced increase in cytoplasmic calcium concentration would lead to an increase in glutamate synthesis localized specifically within the synaptic terminal. The above two hypotheses describe activity-driven variations on cellular synthetic pathways that are not unique to neurons that use glutamate as a neurotransmitter.

It was recently proposed that there is a variation in the AAT enzyme pathway (which converts α-ketoglutarate and aspartate to glutamate, as described above) that only occurs in glutamatergic neurons. The enzyme AAT has been shown to exist in a bound form that is uniquely localized to synaptic vesicles in glutamatergic neurons. Does this localization of AAT mean that the AAT enzyme pathway is important for synthesizing glutamate that will be used as a neurotransmitter? To test this hypothesis, experimenters applied a specific inhibitor of AAT to glutamatergic neurons. They found that inhibiting AAT significantly reduced glutamate loading into synaptic vesicles, which supports the idea that the unique localization of the common cellular enzyme (AAT) to synaptic vesicles facilitates specialized synthesis and loading of glutamate for use as a neurotransmitter (Takeda et al., 2012). Therefore, while there is no unique expression of a synthetic enzyme for glutamate in neurons that will use glutamate as a neurotransmitter, there may be unique localization of an otherwise common synthetic enzyme to synaptic vesicles (Fig. 18.5).

## Termination of Glutamate Action

As with other neurotransmitters, it is important that glutamate is cleared efficiently from the synaptic cleft. To illustrate the importance of this clearance process, consider that different types of receptors for the same neurotransmitter can be activated at different neurotransmitter concentrations. For example, ionotropic glutamate receptors (AMPA and

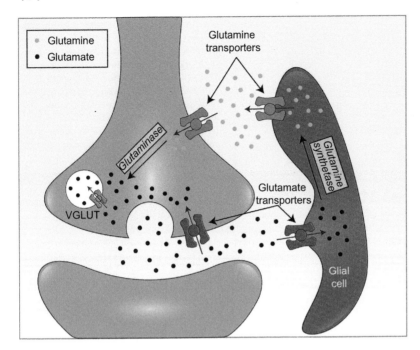

FIGURE 18.6 **The glutamine cycle and termination of glutamate action.** After glutamate is released into the synaptic cleft, most of it is taken up by glutamate transporters on glia, though some is also taken up into the presynaptic terminal by glutamate transporters there. Inside glia, glutamate is converted to glutamine by glutamine synthetase. Glutamine is then released into the cytoplasm via glutamine transporters on glial cells and taken up into the presynaptic neuron, also by glutamine transporters. This process is called the glutamine cycle.

NMDA receptors) are activated at glutamate levels of 200–300 μM, whereas metabotropic glutamate receptors only require about 1 μM glutamate to be activated. Given that circulating extracellular fluid typically has a glutamate concentration of about 1 μM, it is critical for the glutamate concentration in the synaptic cleft of glutamate synapses to be lower than in the surrounding circulating extracellular fluid, since otherwise metabotropic glutamate receptors would be activated even in the absence of presynaptic glutamate release. In addition, after glutamate is released into the synaptic cleft, it needs to be removed rapidly to return to the normal low levels that exist in the absence of neurotransmitter release.

The nervous system uses two mechanisms to keep the concentration of glutamate in the synaptic cleft very low and to terminate the action of released glutamate. First, in a process called **reuptake**, the presynaptic neuron and nearby glial cells use glutamate transporters in their membranes to take up glutamate molecules that have been released into the cleft (see Fig. 18.6). Uptake by glial cells is the predominant mechanism by which glutamate is removed from the synaptic cleft, and glial cells recycle the glutamate by converting it to glutamine (see below). It has been shown that the extracellular fluid contains about 500 μM glutamine, but only about 1 μM glutamate. Since glial cells are responsible for converting extracellular glutamate into extracellular glutamine, this large difference in concentration supports the idea that uptake by glial cells is efficient and contributes significantly to the clearance of glutamate from the synaptic cleft.

The process by which glutamate is recycled is called the **glutamine cycle** (Norenberg and Martinez-Hernandez, 1979). In the glutamine cycle, glutamate is brought into glial cells via glial glutamate transporters (see Fig. 18.6). Glial cells convert the glutamate to glutamine via the enzyme **glutamine synthetase**, which is unique to these cells. The newly created

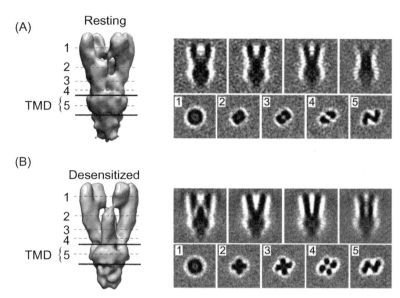

FIGURE 18.7 **AMPA receptor desensitization.** The structure of AMPA receptors differs between the resting state (A) and the desensitized state (B). Blue models on the left show the overall three-dimensional structure of the receptor in each state. Grayscale images on the right show side views (top row) and top-down cross-sections (bottom row, 1–5, at levels indicated by dotted red lines on the blue models to the left). *Source: Adapted from Schauder, D.M., Kuybeda, O., Zhang, J., Klymko, K., Bartesaghi, A., Borgnia, M.J., et al., 2013. Glutamate receptor desensitization is mediated by changes in quaternary structure of the ligand binding domain. Proc. Natl. Acad. Sci. USA 110, 5921–5926; and Fuenzalida, M., Fernandez de Sevilla, D., Buno, W., 2007. Changes of the EPSP waveform regulate the temporal window for spike-timing-dependent plasticity. J. Neurosci. 27, 11940–11948.*

glutamine molecules are then transported back across the glial cell membrane into the extracellular space by glutamine transporters. Finally, glutamine transporters in the presynaptic terminal pump glutamine into the axon terminals, where it is converted into glutamate by glutaminase and it is available to be packaged into vesicles by VGLUT for release by exocytosis.

The second mechanism for terminating glutamate action is called **receptor desensitization**. (Remember that a termination mechanism can be anything that shortens the time-course of neurotransmitter action, even if it does not actually remove the neurotransmitter from the cleft.) Some neurotransmitter receptors desensitize shortly after they are activated by ligand binding, meaning that they enter an inactive state/configuration (see Fig. 18.7A and B) in which they cannot be activated, even in the presence of an agonist. Receptors can desensitize at different rates and for different durations, depending on the type of receptor and receptor subtype (AMPA vs. NMDA). AMPA receptors desensitize rapidly, within milliseconds of the binding of glutamate, whereas NMDA receptors desensitize much more slowly. Additionally, applying a pharmacological agent that prevents desensitization of AMPA receptors increases the duration of AMPA-induced EPSCs by tens of milliseconds but does not increase the peak magnitude. These data indicate that although desensitization of AMPA receptors does not affect the initial postsynaptic response to glutamate (as indicated by the peak amplitude of the EPSCs), it is nonetheless a rapid mechanism for terminating the action of glutamate.

# GABA

GABA (gamma-aminobutyric acid) was first identified as a biological substance by researchers studying bacteria (Ackermann, 1910), and in the 1950s it was found to be a normal constituent of brain tissue (Awapara et al., 1950; Roberts and Frankel, 1950; Udenfriend, 1950). GABA is the major inhibitory neurotransmitter in the brain, where it is found at 25%–45% of all nerve terminals. It is found only in trace amounts in the spinal cord and is not used in the peripheral nervous system. GABA is classified as a type 1 "classical" neurotransmitter because it is synthesized using an enzyme uniquely expressed only in GABAergic neurons (see below), but its chemical structure is that of an amino acid. As with glutamate (discussed above), it has been shown that applying GABA to GABAergic synapses causes IPSPs that are similar to those resulting from stimulation of the GABAergic neuron, confirming GABA's identity of action. As discussed in earlier chapters, there are two types of GABA receptors. The $GABA_A$ receptor is ionotropic, and passes chloride when open. This means that, under most conditions, $GABA_A$ receptors are hyperpolarizing, though this is dependent on the chloride reversal potential and the membrane potential of the cell containing the GABA receptors. In contrast, $GABA_B$ receptors are metabotropic and are coupled to biochemical signaling systems in the neuron that might excite or inhibit cells, depending on the cascade of effects they initiate (see Chapter 12).

## GABA Synthesis

GABA is synthesized from glutamate (Roberts and Frankel, 1950) by the enzyme **glutamic acid decarboxylase** (GAD; see Fig. 18.8), which is also referred to as glutamate decarboxylase. GAD is the first unique synthetic enzyme in this pathway and is only expressed in GABAergic nerve terminals, which means GAD can be used to identify GABA nerve terminals using histological methods. The glutamate-to-GABA synthesis reaction requires a cofactor, **pyridoxal phosphate** (PLP), which is a derivative of vitamin $B_6$. PLP is not rate-limiting under normal conditions, since most adults get enough vitamin $B_6$ from their diet to produce more PLP than the GABA synthesis pathway needs. The only exception is in vitamin $B_6$-deficient infants, who develop seizures caused by insufficient GABAergic inhibition in the central nervous system. These seizures can be reversed rapidly by administration of the vitamin $B_6$ precursor pyridoxine, which is quickly converted to PLP in the brain, leading to a rapid increase in GAD activity and restoration of normal GABA levels. Therefore, under these particular circumstances, PLP is rate-limiting for GABA synthesis, but this is typically not the case. Under healthy conditions, consistent with the general scheme for type 1 neurotransmitter synthesis, the first unique synthetic enzyme in the pathway (GAD) is rate-limiting for the synthesis of GABA.

What normally regulates the activity of GAD? As it turns out, only about 35% of GAD is bound to the cofactor PLP at any given time, even though the concentration of PLP is normally high. Most GAD enzymes are not PLP-bound because ATP competes with PLP for binding to GAD, and since cells contain a lot of ATP, ATP binding predominates,

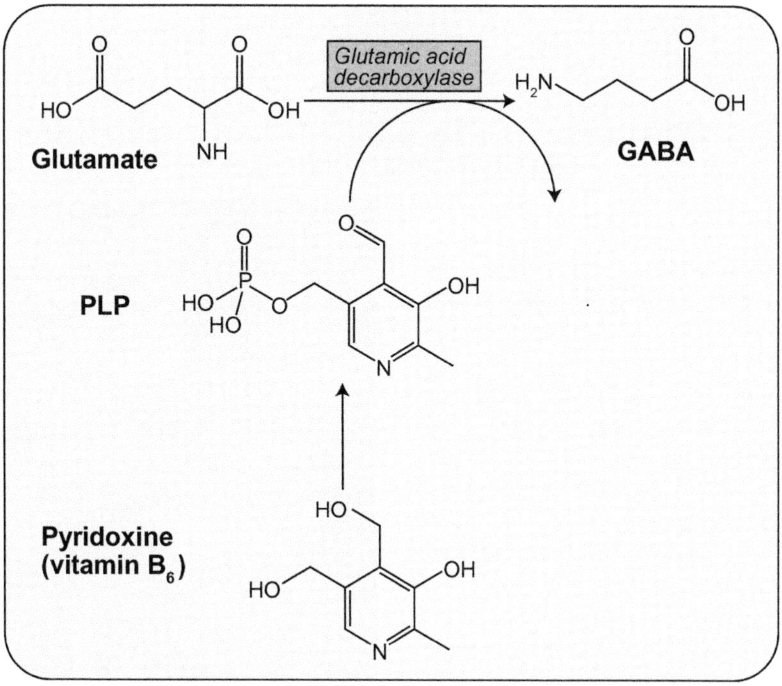

FIGURE 18.8 **Synthesis reaction for GABA.** GABA is synthesized from glutamate by glutamic acid decarboxylase. This enzyme uses PLP as a cofactor, which is a derivative of vitamin $B_6$ (pyridoxine).

reducing PLP binding. Interestingly, during the first two minutes after death, cellular ATP levels drop and GABA levels in brain tissue increase (Miller et al., 1977). However, ATP concentration is not regulated under normal physiological conditions in living cells. Therefore, nerve terminals must have another mechanism for influencing the competition between PLP and ATP for binding to GAD, and this mechanism is hypothesized to be phosphorylation of GAD by calcium calmodulin-dependent protein kinase (CaMKII). Action potential activity triggers calcium influx, which activates CaMKII. CaMKII then phosphorylates GAD, increasing GAD's affinity for PLP so that PLP can compete more effectively with ATP for binding (Jin et al., 2003). Increasing binding to PLP increases GAD activity, which in turn increases the synthesis of GABA (see Box 18.1 for a related natural product).

As previously discussed, CaMKII phosphorylation also regulates tyrosine hydroxylase activity in catecholamine synthesis and tryptophan hydroxylase activity in serotonin synthesis. CaMKII phosphorylation of synthetic enzymes appears to be a general mechanism for regulating transmitter synthesis, since it occurs in three different type 1 neurotransmitter systems.

After synthesis, GABA is transported into vesicles via the **vesicular inhibitory amino acid transporter** (VIAAT; see Fig. 18.10). Like VGLUT, VIAAT is dependent on a proton gradient, which is maintained by a proton-dependent ATPase.

## BOX 18.1

## TREATMENTS FROM NATURE: *GINKGO BILOBA* AND GABA AND GLYCINE TRANSMISSION

### Ginkgo biloba

*Ginkgo biloba* (see Fig. 18.9) is a unique tree in modern times, as it is the only member within its taxonomy that has survived past the Mesozoic era (all other species of *Ginkgo* became extinct over 66 million years ago). Despite the overwhelming "rotting" smell given off by the seeds (due to the malodorous butryric acid within the seeds, which is also the distinctive chemical odorant in human vomit), *Ginkgo* has been used in traditional Chinese medicine and is still commonly found on supermarket shelves today.

Lauded for its neuroprotective benefits, it can be taken by patients with Alzheimer's disease and has been reported to improve memory loss, dizziness, difficulty concentrating, and mood disturbances. *Ginkgo* leaf extracts are likely safe to consume, even though they contain unique constituents, ginkgolides and bilobalide, which act as glycine and $GABA_A$ receptor antagonists. However, another active compound found in *Gingko* seeds is called gingkotoxin, which decreases the activity of pyridoxal kinase [the kinase responsible for synthesizing pyridoxal phosphate (PLP; a cofactor for the GABA unique synthetic enzyme glutamate decarboxylase)]. Disruption of GABA synthesis due to gingkotoxin results in seizures due to an imbalance of excitatory–inhibitory signaling.

FIGURE 18.9 *Ginkgo biloba* plant. *By James Field (Jame), CC BY-SA 3.0, from Wikimedia Commons.*

## Termination of GABA Action

How is GABA action terminated? One mechanism that has been considered is enzymatic degradation. Biochemically, GABA can be broken down by the enzyme **GABA transaminase** (GABA-T), which converts GABA and alpha-ketoglutarate to succinic

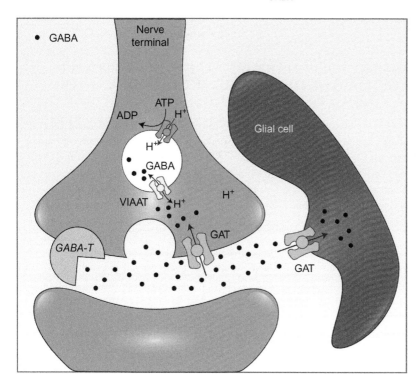

FIGURE 18.10 **GABA vesicular and reuptake transporters.** After synthesis, GABA is transported into synaptic vesicles by the VIAAT. After GABA is released into the synaptic cleft, it is either broken down by the enzyme GABA transaminase (GABA-T) or taken up into the presynaptic terminal or glial cell by the GABA transporter (GAT).

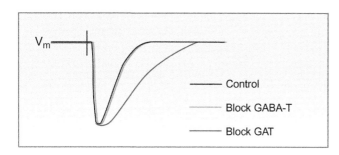

FIGURE 18.11 **Termination of GABA action.** When GABA-T is pharmacologically blocked (*blue line*), there is no change in the duration of an evoked IPSP (compare with control in *black*). In contrast, when GAT is blocked (*red line*), the duration of the IPSP is longer, indicating that GAT terminates the action of GABA.

semialdehyde and glutamate as part of the Krebs cycle. If GABA-T is responsible for clearing GABA from the synaptic cleft, then in the absence of GABA-T activity, GABA should remain in the cleft and continue to activate receptors for a longer period. Therefore, inhibition of GABA-T should result in prolonged IPSPs. However, in experiments on inhibitory neurons such as cerebellar Purkinje cells, pharmacologically blocking GABA-T does not change IPSP duration (see Fig. 18.11). These findings suggest that GABA-T-mediated breakdown is *not* the termination mechanism for GABA action in the cerebellum.

## BOX 18.2

## DRUGS THAT MISSED THEIR MARK BUT ARE STILL EFFECTIVE THERAPEUTIC AGENTS: THE DEVELOPMENT OF GABA ANALOGS AS ANTISEIZURE DRUGS

Seizures can be caused by many things, including brain injury and genetic alterations in ion channel function. Additionally, there are multiple types of seizures that originate in or involve different brain regions. Seizures can arise when neurons are too easily excited and/or when there are insufficient mechanisms to keep neuronal excitation in check (e.g., GABA signaling). One of the primary ways to treat seizures is to administer drugs that increase the amount of inhibition, or "inhibitory tone," in the brainstem and/or brain. This makes sense, since too many neurons firing action potentials simultaneously and repeatedly can lead to a seizure. However, administering GABA itself is not an effective way to increase inhibitory tone, since GABA does not readily cross the blood–brain barrier.

Is it possible to develop a GABA analog that could cross the blood–brain barrier and activate GABA receptors? This "rational drug design" approach, in which drugs are created to target specific molecular components (e.g., proteins) that are implicated in disease, is logical on paper. In practice, however, designing molecules to play a specific role in the nervous system can be surprisingly challenging. This is in part because it is very difficult to predict the effects and specificity of a synthetic molecule in the complex environment of the nervous system. In fact, synthetic drugs often exert effects not just on their target protein(s) but on other, unexpected sites, which is one of the main reasons that drugs often have side effects. Additionally, a drug that is developed in vitro to interact with a specific target may not interact with that target in vivo in the predicted manner, if at all.

Two examples of drugs that fit this pattern are gabapentin and pregabalin (see Sills, 2006; for a review). Both drugs were designed to activate GABA receptors and to cross the blood–brain barrier via the L-type amino acid transporter (Luer et al., 1999; Su et al., 2005). When tested on animal models of seizures, and later in people with epilepsy, gabapentin worked well as an antiseizure drug. Fortuitously, researchers and doctors noticed that gabapentin also reduced pain in patients taking it for epilepsy, and this drug is sometimes prescribed as an antinociceptive agent. Furthermore, gabapentin can also be used as an antianxiety agent and as an adjunct to antidepressant medications. Based on the success of gabapentin, a GABA analog called pregabalin was identified as another promising therapeutic compound and, like gabapentin, it is used as an antiseizure drug, an antinociceptive, and an antianxiety drug. However, the shared mechanism of action for gabapentin and pregabalin was not well-understood prior to their market releases, and the main physiological reason for their effects came as a surprise when it was discovered. It seemed intuitive to assume both drugs would activate GABA receptors, as they were intended to, but their mechanisms of action have turned out to be more complex and do *not* include significant action at $GABA_A$ receptors.

## BOX 18.2 (cont'd)

It has been shown that gabapentin does not bind to GABA receptors, does not directly affect GABA synthesis or transport, and does not increase brain levels of GABA in rodents. Despite extensive research, there is little evidence that gabapentin directly affects GABA signaling in any significant way. The effects of pregabalin on GABA signaling have not been tested as extensively. Other research on both drugs suggests that while they may affect glutamate transmission or voltage-gated ion channels to a certain degree, these effects do not play a significant role in the mechanism(s) of their therapeutic action.

As it turns out, the direct effects of both gabapentin and pregabalin on neurons are mediated by a lone binding site on the $\alpha_2\delta$-1 subunit of voltage-gated calcium channels (Belliotti et al., 2005; Gee et al., 1996; Taylor, 2004; Wang et al., 1999). Exactly how the binding of these drugs to $\alpha_2\delta$-1 alleviates seizure, pain, and anxiety is still up for debate. It is possible that gabapentin and pregabalin have their therapeutic effects mainly because they block calcium channels that contain the $\alpha_2\delta$-1 subunit. However, the $\alpha_2\delta$-1 protein is also expressed in the nervous system as a standalone protein (independent of calcium channels) and has other roles in nervous system function. So although the effects of gabapentin and pregabalin are mediated by binding to $\alpha_2\delta$-1, they do not necessarily have anything to do with calcium channels (Celli et al., 2017).

Although the specific therapeutic mechanism(s) of action for gabapentin and pregabalin are different from what they were anticipated to be, these drugs are both therapeutically useful for a range of neuropathological conditions. There are two messages to take from this story: (1) rational drug design for neurological disorders can be challenging, even when a significant amount is known about a neurotransmitter and/or neurotransmitter receptors, and (2) it is not strictly necessary to understand why a given drug has the effects that it does in order for it to be an effective treatment.

---

An experimentally supported mechanism of GABA termination is reuptake via a selective plasma membrane **GABA transporter** (GAT), which is found on both neurons and glial cells (see Fig. 18.10). GAT blockers, such as the antiseizure medication tiagabine (see Schachter, 1999; for a review), *do* prolong IPSPs, suggesting GAT functions as a termination mechanism for GABA (see Fig. 18.11). After action potential activity, it is difficult to measure GABA levels in the extracellular space using biochemical methods, suggesting that GAT clears GABA from the synaptic cleft very efficiently. However, when GAT is pharmacologically blocked, GABA can be measured in the extracellular space following nerve stimulation.

Despite understanding these aspects of GABAergic function, it has proven difficult to design drugs to control GABAergic tone to treat neurologic disorders (see Box 18.2).

## GABA AND THE NEUROLOGICAL DISEASE SCHIZOPHRENIA

$GABA_A$ receptors can be disrupted in a number of neurological diseases, including schizophrenia (Lewis and Sweet, 2009). Schizophrenia is characterized by impairments in social function, attention, perception, decision-making, and the expression and recognition of emotion. Cognitive impairments associated with schizophrenia, such as deficits in working memory and attention, have been a focus of research because these symptoms begin in childhood (before diagnosis), they often persist throughout patients' lives, independent of other symptoms such as psychosis (Davidson et al., 1999; Cosway et al., 2000; Keefe and Fenton, 2007), and they can be difficult to address with treatment options that are currently available.

According to the "disconnectivity hypothesis" of schizophrenia, cognitive deficits in this disorder are closely linked to impairment of neuronal function in the prefrontal cortex. In some cases, normal neuronal activity involves synchronizing the action potential firing of populations of neurons, and this synchronous firing can be measured in humans using scalp electrodes for electroencephalography (EEG) (Williams and Boksa, 2010). Using this approach, rhythmic activity in the brain can be observed to occur at various frequencies, including delta (1–4 Hz), theta (3–10 Hz), alpha (8–12 Hz), beta (12–25 Hz), and gamma (>25 Hz) (see Chapter 14). This rhythmic activity is thought to be critical for cognitive processes. In particular, basic research in animal models has implicated gamma oscillations in cognitive tasks such as working memory (Yamamoto et al., 2014). In humans, there is a strong correlation between gamma oscillations and working memory tasks that involve the prefrontal cortex (Howard et al., 2003). Accordingly, there has been a focus on impaired gamma oscillation as a possible cause of cognitive deficits in schizophrenia patients (Williams and Boksa, 2010). Reductions in gamma oscillations are known to be a common feature of schizophrenia (Cho et al., 2006; Minzenberg et al., 2010), and gamma oscillations have been shown to be greatly reduced in schizophrenia patients while they are trying to perform working memory tasks (Haenschel et al., 2009)

Gamma oscillations are thought to result from repetitive high-frequency synaptic transmission between inhibitory interneurons and excitatory pyramidal cells in the brain (Traub et al., 2004; Atallah and Scanziani, 2009). Lewis et al. (2008) showed that a benzodiazepine drug that increases $GABA_A$ receptor function increases gamma oscillations in the prefrontal cortex and can reverse working memory deficits in schizophrenia patients.

A major area of research on this topic is focused on determining how neural circuits in the prefrontal cortex of schizophrenia patients differ from unaffected tissue (Lewis, 2012, 2014). While a number of alterations have been detected within the excitatory pyramidal neurons in schizophrenia patients, studies have also reported alterations in markers of inhibitory GABA neurotransmission within the prefrontal cortex, leading to the hypothesis that dysfunction of inhibitory circuits is a critical feature of the disease (Lewis, 2014). For example, there are reports of lower gene expression and protein content for GAD, and altered amounts of $GABA_A$ receptors on pyramidal cells in layer 3 of prefrontal cortex (Gonzalez-Burgos et al., 2010; Curley et al., 2011). These changes would be expected to impair inhibition onto pyramidal cells and hinder the generation of gamma oscillations in prefrontal cortex (see Fig. 18.12).

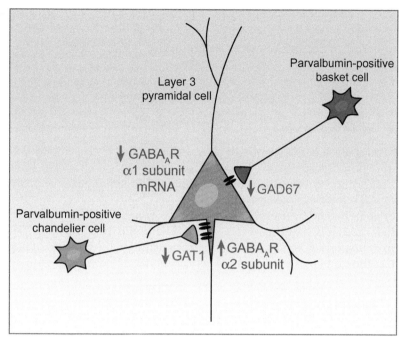

FIGURE 18.12 Diagram showing where alterations have been observed in $GABA_A$ receptor-mediated inhibitory control over layer 3 pyramidal cells in the human prefrontal cortex of schizophrenia patients. *Source: Adapted from Lewis (2014).*

# GLYCINE

Structurally, glycine is the simplest of the 20 amino acids, and it is found in high concentrations everywhere in the body. It is not an essential amino acid (i.e., it can be made by the body), and it is present in the amino acid sequences of many proteins. Aside from its normal functions in all cells, glycine plays two special roles in the brain and spinal cord. First, it acts as an inhibitory neurotransmitter in multiple regions of the central nervous system by activating ionotropic glycine receptors, which pass chloride (see Chapter 11). Applying glycine to a glycinergic synapse causes an IPSP similar to that resulting from stimulation of the presynaptic glycinergic neuron, confirming glycine's identity of action. Second, it plays an excitatory-like role as a co-agonist at NMDA receptors (see Chapter 11).

## Glycine Synthesis

There are two main hypotheses for how neurons acquire glycine, although this question is still debated. The first hypothesis is that neurons simply take up glycine that has already been synthesized and use it directly. The second is that glycine is synthesized in nerve terminals from the amino acid serine by the enzyme **serine trans hydroxymethyltransferase** (see Fig. 18.13). Unlike synthetic enzymes for some other neurotransmitters, serine trans

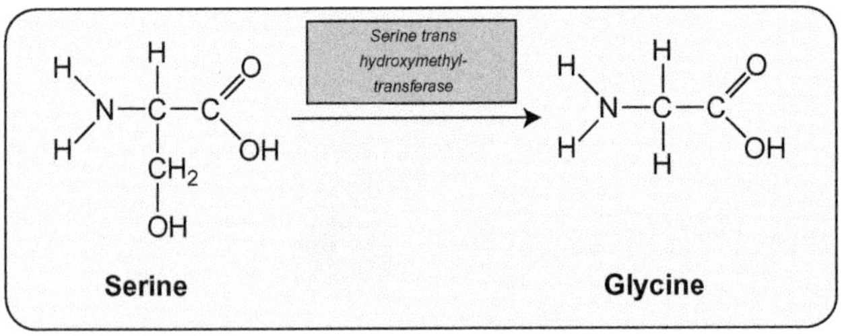

FIGURE 18.13 **Synthesis reaction for glycine.** The enzyme serine trans hydroxymethyltransferase converts serine to glycine. This reaction has been proposed as one mechanism by which glycine is acquired for release at glycinergic nerve terminals.

hydroxymethyltransferase is not unique to neurons and there is no evidence of biochemical regulation of this proposed pathway. Since glycine is a type 2 amino acid transmitter, it is not expected that glycine synthesis is uniquely regulated in nerve terminals that release glycine, but this is still a topic of investigation.

Once acquired (whether by uptake or synthesis), glycine is transported into synaptic vesicles via the VIAAT (Chaudhry et al., 1998; Dumoulin et al., 1999). VIAAT is also responsible for transporting the inhibitory neurotransmitter GABA into synaptic vesicles (see the above section on GABA synthesis).

## Termination of Glycine Action

The action of glycine can be terminated in two ways: diffusion and uptake/reuptake. Some glycine is removed from the synaptic cleft by diffusion into the surrounding extracellular fluid. Glycine is also removed from the synaptic cleft by specific, high-affinity transporters located in the cell membranes of both neurons and glial cells (see Fig. 18.14; Neal, 1971). There are two types of glycine transporters, **GLYT-1** and **GLYT-2**. These transporters are related to one another, sharing approximately 50% of their amino acid sequence, but they are located on different types of cells and appear to play very different roles in the nervous system.

GLYT-1 is closely associated with a type of glial cell known as an astrocyte (see Chapter 22), and is also found on glycinergic neurons in the retina. GLYT-1 can also be found in non-neuronal tissues such as the liver, pancreas, and intestine (Betz et al., 2006). GLYT-1 knockout mice are anatomically normal, but they have motor and respiratory deficits that are lethal on the first day after birth (Gomeza et al., 2003a). It is thought that GLYT-1 knockout is lethal because without this transporter to clear glycine from the synaptic cleft, constant overactivation of glycinergic receptors interferes with regulation of centrally generated rhythmic motor functions like breathing (Gomeza et al., 2003a). These findings underscore the critical role GLYT-1 plays in clearing glycine from the synaptic cleft.

In contrast, GLYT-2 plasma membrane transporters are found mainly on neuronal membranes in the spinal cord, brainstem, and cerebellum. GLYT-2 is a unique marker for glycinergic neurons (Poyatos et al., 1997), and is thought to be located outside of the active zone (Mahendrasingam et al., 2003). GLYT-2 knockout mice appear normal at birth, but soon develop a lethal motor neuron deficiency that causes spasticity, tremor, and an inability of the mice to right themselves when turned onto their backs (Gomeza et al., 2003b). The fact that GLYT-2 transporters are located extrasynaptically suggests that their primary role might be to replenish the cytoplasmic pool of glycine by transporting glycine

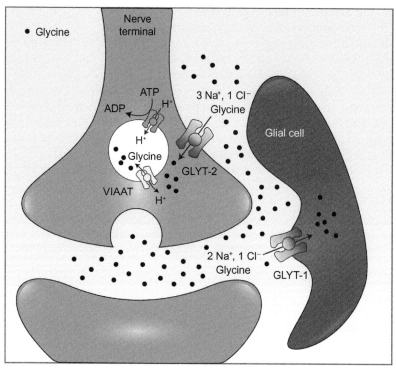

FIGURE 18.14 **Glycine vesicular and reuptake transporters.** After synthesis or uptake into the nerve terminal, glycine is transported into synaptic vesicles by the VIAAT. After glycine is released into the synaptic cleft, it is taken up into the presynaptic terminal by the glycine transporter GLY-2 or into neighboring glial cells via the glycine transporter GLY-1.

from outside the synaptic cleft (Gomeza et al., 2003b), whereas to GLYT-1 transporters clear glycine from the cleft immediately after transmitter release.

Glycine transporters belong to a family of transporters that are dependent on the co-transport of $Na^+$ and $Cl^-$ ions (Betz et al., 2006). This family also includes transporters for serotonin, norepinephrine, dopamine, and GABA. For every glycine molecule it moves into a cell, GLYT-1 also moves two sodium ions and one chloride ion into the cell, whereas GLYT-2 moves three sodium ions and one chloride ion into the cell (see Fig. 18.14; Roux and Supplisson, 2000). These transporters clear glycine from synapses so efficiently that it is difficult to even measure levels of glycine in the extracellular fluid after nerve stimulation. This efficiency is required in order to keep resting synaptic cleft glycine concentrations low enough to use glycine as a signaling molecule, because outside of the synaptic cleft regions that contain these glycine transporters, glycine concentration is rather high (5–10 μM; Fuchs et al., 2008).

## Functions of Glycinergic Neurons

Glycinergic neurons are found in multiple areas of the central nervous system, including the spinal cord, the retina, the brainstem, the superior olivary complex, and cerebellar

Golgi cells. In the retina, amacrine cells release glycine onto bipolar cells, which may contribute to the phenomenon of surround inhibition.

A well-understood function of glycine in the central nervous system is its role in a basic spinal cord reflex. In 1954, Eccles and colleagues demonstrated that a population of inhibitory interneurons in the spinal cord, known as Renshaw cells (Renshaw, 1946), are part of a negative feedback reflex that prevents overexcitation of skeletal muscles (Eccles et al., 1954). Motor neurons use acetylcholine to activate muscles, but they also send collateral axons that excite the Renshaw cells (see Fig. 18.15A). Renshaw cells in turn release glycine that inhibits the same or nearby motor neurons, reducing their output. This reflex decreases subsequent stimulation of relevant muscles for a short period of time.

Chapter 8, describes how tetanus toxin prevents vesicular release by cleaving vesicle-associated membrane protein (VAMP), a component of the CORE complex. Tetanus toxin can enter the body at the site of a wound (especially if it is a puncture wound) and travel in a retrograde direction up the axons of motor neurons. It then travels transynaptically to other neurons, including Renshaw cells, impairing the ability of Renshaw cells to release glycine. The loss of the critical Renshaw cell-mediated negative-feedback mechanism, which normally prevents overactivation of muscles, is what leads to the painful involuntary muscle contractions (spasms) that are the hallmark of tetanus toxin poisoning (see Fig. 18.15B).

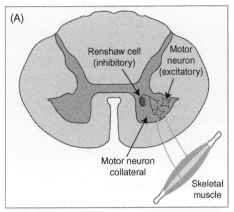

FIGURE 18.15 **The effects of tetanus toxin on neurons in the spinal cord and muscles.** (A) When tetanus toxin enters a wound, it can travel up the axons of motor neurons to the spinal cord, where it prevents Renshaw cells from releasing glycine. (B) This affects spinal cord circuits and results in the uncontrolled muscle spasms characteristic of tetanus toxin poisoning. *Source: Painting: "The Wounded following the Battle of Corunna: Tetanus Following Gunshot Wounds", by surgeon and neurologist Charles Bell, 1809.*

Glycinergic neurons in the dorsal horn of the spinal cord can also inhibit nociceptive (pain) signals, and dysfunction of glycinergic signaling in this region is thought to contribute to **allodynia** (the perception of pain in response to stimuli that are not normally painful) and **hyperalgesia** (the perception of disproportionately intense pain in response to a painful stimulus). Administration of glycine can reduce or prevent these types of pain (Simpson et al., 1997; Huang and Simpson, 2000), whereas application of the glycine receptor antagonist strychnine can induce both types of pain (Yaksh, 1989), raising the possibility that inhibitors of glycine transporters might be useful targets for drugs that reduce pain.

# References

Ackermann, D., 1910. Über ein neues, auf bakteriellem Wege gewinnbares, Aporrhegma. Hoppe-Seyler's Zeitschrift für physiologische Chemie 69, 273–281.

Atallah, B.V., Scanziani, M., 2009. Instantaneous modulation of gamma oscillation frequency by balancing excitation with inhibition. Neuron. 62, 566–577.

Awapara, J., Landua, A.J., Fuerst, R., Seale, B., 1950. Free gamma-aminobutyric acid in brain. J. Biol. Chem. 187, 35–39.

Belliotti, T.R., Capiris, T., Ekhato, I.V., Kinsora, J.J., Field, M.J., Heffner, T.G., et al., 2005. Structure-activity relationships of pregabalin and analogues that target the alpha(2)-delta protein. J. Med. Chem. 48, 2294–2307.

Betz, H., Gomeza, J., Armsen, W., Scholze, P., Eulenburg, V., 2006. Glycine transporters: essential regulators of synaptic transmission. Biochem. Soc. Trans. 34, 55–58.

Celli, R., Santolini, I., Guiducci, M., van Luijtelaar, G., Parisi, P., Striano, P., Gradini, R., Battaglia, G., Ngomba, R. T., Nicoletti, F., 2017. The $\alpha 2\delta$ Subunit and Absence Epilepsy: Beyond Calcium Channels? Curr Neuropharmacol. 15 (6), 918–925.

Chaudhry, F.A., Reimer, R.J., Bellocchio, E.E., Danbolt, N.C., Osen, K.K., Edwards, R.H., et al., 1998. The vesicular GABA transporter, VGAT, localizes to synaptic vesicles in sets of glycinergic as well as GABAergic neurons. J. Neurosci. 18, 9733–9750.

Cho, R.Y., Konecky, R.O., Carter, C.S., 2006. Impairments in frontal cortical gamma synchrony and cognitive control in schizophrenia. Proc Natl Acad Sci U S A 103, 19878–19883.

Cosway, R., Byrne, M., Clafferty, R., Hodges, A., Grant, E., Abukmeil, S.S., Lawrie, S.M., Miller, P., Johnstone, E. C., 2000. Neuropsychological change in young people at high risk for schizophrenia: Results from the first two neuropsychological assessments of the Edinburgh High Risk Study. Psychol Med. 30, 1111–1121.

Curley, A.A., Arion, D., Volk, D.W., Asafu-Adjei, J.K., Sampson, A.R., Fish, K.N., Lewis, D.A., 2011. Cortical deficits of glutamic acid decarboxylase 67 expression in schizophrenia: Clinical, protein, and cell type-specific features. Am J Psychiatry 168, 921–929.

Davidson, M., Reichenberg, A., Rabinowitz, J., Weiser, M., Kaplan, Z., Mark, M., 1999. Behavioral and intellectual markers for schizophrenia in apparently healthy male adolescents. Am J Psychiatry 156, 1328–1335.

Dumoulin, A., Rostaing, P., Bedet, C., Levi, S., Isambert, M.F., Henry, J.P., et al., 1999. Presence of the vesicular inhibitory amino acid transporter in GABAergic and glycinergic synaptic terminal boutons. J. Cell. Sci. 112 (Pt 6), 811–823.

Eccles, J.C., Fatt, P., Koketsu, K., 1954. Cholinergic and inhibitory synapses in a pathway from motor-axon collaterals to motoneurones. J. Physiol. 126, 524–562.

Fuchs, S.A., De Barse, M.M., Scheepers, F.E., Cahn, W., Dorland, L., de Sain-van der Velden, M.G., et al., 2008. Cerebrospinal fluid D-serine and glycine concentrations are unaltered and unaffected by olanzapine therapy in male schizophrenic patients. Eur. Neuropsychopharmacol. 18, 333–338.

Fuenzalida, M., Fernandez de Sevilla, D., Buno, W., 2007. Changes of the EPSP waveform regulate the temporal window for spike-timing-dependent plasticity. J. Neurosci. 27, 11940–11948.

Gee, N.S., Brown, J.P., Dissanayake, V.U., Offord, J., Thurlow, R., Woodruff, G.N., 1996. The novel anticonvulsant drug, gabapentin (Neurontin), binds to the alpha2delta subunit of a calcium channel. J. Biol. Chem. 271, 5768–5776.

Gomeza, J., Hulsmann, S., Ohno, K., Eulenburg, V., Szoke, K., Richter, D., et al., 2003a. Inactivation of the glycine transporter 1 gene discloses vital role of glial glycine uptake in glycinergic inhibition. Neuron 40, 785–796.

Gomeza, J., Ohno, K., Hulsmann, S., Armsen, W., Eulenburg, V., Richter, D.W., et al., 2003b. Deletion of the mouse glycine transporter 2 results in a hyperekplexia phenotype and postnatal lethality. Neuron 40, 797–806.

Gonzalez-Burgos, G., Hashimoto, T., Lewis, D.A., 2010. Alterations of cortical GABA neurons and network oscillations in schizophrenia. Curr Psychiatry Rep 12, 335–344.

Haenschel, C., Bittner, R.A., Waltz, J., et al., 2009. Cortical oscillatory activity is critical for working memory as revealed by deficits in early-onset schizophrenia. J Neurosci. 29, 9481–9489.

Hawkins, R.A., 2009. The blood-brain barrier and glutamate. Am J Clin Nutr 90 (3), 867S–874S.

Howard, M.W., Rizzuto, D.S., Caplan, J.B., Madsen, J.R., Lisman, J., Aschenbrenner-Scheibe, R., et al., 2003. Gamma oscillations correlate with working memory load in humans. Cereb. Cortex 13, 1369–1374.

Huang, W., Simpson, R.K., 2000. Long-term intrathecal administration of glycine prevents mechanical hyperalgesia in a rat model of neuropathic pain. Neurol. Res. 22, 160–164.

Jin, H., Wu, H., Osterhaus, G., Wei, J., Davis, K., Sha, D., Floor, E., Hsu, C., Kopke, R.D., Wu, J., 2003. Demonstration of functional coupling between $\gamma$-aminobutyric acid (GABA) synthesis and vesicular GABA transport into synaptic vesicles. PNAS 100 (7), 4293–4298.

Keefe, R.S., Fenton, W.S., 2007. How should DSM-V criteria for schizophrenia include cognitive impairment? Schizophr Bull. 33, 912–920.

Lewis, D.A., Cho, R.Y., Carter, C.S., et al., 2008. Subunit-selective modulation of GABA type A receptor neurotransmission and cognition in schizophrenia. Am J Psychiatry 165, 1585–1593.

Lewis, D.A., Sweet, R.A., 2009. Schizophrenia from a neural circuitry perspective: Advancing toward rational pharmacological therapies. J Clin Invest 119, 706–716.

Lewis, D.A., 2012. Cortical circuit dysfunction and cognitive deficits in schizophrenia: Implications for preemptive interventions. Eur J Neurosci 35, 1871–1878.

Lewis, D.A., 2014. Inhibitory neurons in human cortical circuits: substrate for cognitive dysfunction in schizophrenia. Current Opinion in Neurobiology 26, 22–26.

Luer, M.S., Hamani, C., Dujovny, M., Gidal, B., Cwik, M., Deyo, K., et al., 1999. Saturable transport of gabapentin at the blood-brain barrier. Neurol. Res. 21, 559–562.

Mahendrasingam, S., Wallam, C.A., Hackney, C.M., 2003. Two approaches to double post-embedding immunogold labeling of freeze-substituted tissue embedded in low temperature Lowicryl HM20 resin. Brain. Res. Brain. Res. Protoc. 11, 134–141.

Miller, L.P., Walters, J.R., Martin, D.L., 1977. Post-mortem changes implicate adenine nucleotides and pyridoxal-5′-phosphate in regulation of brain glutamate decarboxylase. Nature 266, 847–848.

Minzenberg, M.J., Firl, A.J., Yoon, J.H., Gomes, G.C., Reinking, C., Carter, C.S., 2010. Gamma oscillatory power is impaired during cognitive control independent of medication status in first-episode schizophrenia. Neuropsychopharm 35 (2010), 2590–2599.

Neal, M.J., 1971. The uptake of [14C]glycine by slices of mammalian spinal cord. J. Physiol. 215, 103–117.

Nikolenko, V., Watson, B.O., Araya, R., Woodruff, A., Peterka, D.S., Yuste, R., 2008. SLM microscopy: scanless two-photon imaging and photostimulation with spatial light modulators. Front. Neural Circuits . Available from: https://doi.org/10.3389/neuro.04.005.2008.

Norenberg, M.D., Martinez-Hernandez, A., 1979. Fine structural localization of glutamine synthetase in astrocytes of rat brain. Brain Res. 161, 303–310.

Poyatos, I., Ponce, J., Aragon, C., Gimenez, C., Zafra, F., 1997. The glycine transporter GLYT2 is a reliable marker for glycine-immunoreactive neurons. Brain. Res. Mol. Brain. Res. 49, 63–70.

Renshaw, B., 1946. Central effects of centripetal impulses in axons of spinal ventral roots. J. Neurophysiol. 9, 191–204.

Roberts, E., Frankel, S., 1950. gamma-Aminobutyric acid in brain: its formation from glutamic acid. J. Biol. Chem. 187, 55–63.

Roux, M.J., Supplisson, S., 2000. Neuronal and glial glycine transporters have different stoichiometries. Neuron 25, 373–383.

Schachter, S.C., 1999. A review of the antiepileptic drug tiagabine. Clin. Neuropharmacol. 22, 312–317.

Schauder, D.M., Kuybeda, O., Zhang, J., Klymko, K., Bartesaghi, A., Borgnia, M.J., et al., 2013. Glutamate receptor desensitization is mediated by changes in quaternary structure of the ligand binding domain. Proc. Natl. Acad. Sci. USA 110, 5921–5926.

Sills, G.J., 2006. The mechanisms of action of gabapentin and pregabalin. Curr. Opin. Pharmacol. 6, 108–113.

Simpson Jr., R.K., Gondo, M., Robertson, C.S., Goodman, J.C., 1997. Reduction in thermal hyperalgesia by intrathecal administration of glycine and related compounds. Neurochem. Res. 22, 75–79.

Su, T.Z., Feng, M.R., Weber, M.L., 2005. Mediation of highly concentrative uptake of pregabalin by L-type amino acid transport in Chinese hamster ovary and Caco-2 cells. J. Pharmacol. Exp. Ther. 313, 1406–1415.

Takeda, K., Ishida, A., kahashi, K., Ueda, T., 2012. Synaptic vesicles are capable of synthesizing the VGLUT substrate glutamate from α-ketoglutarate for vesicular loading. J Neurochem. 121 (2), 184–196.

Taylor, C.P., 2004. The biology and pharmacology of calcium channel alpha2-delta proteins Pfizer Satellite Symposium to the 2003 Society for Neuroscience Meeting. Sheraton New Orleans Hotel, New Orleans, LA November 10, 2003. CNS. Drug. Rev. 10, 183–188.

Traub, R.D., Bibbig, A., LeBeau, F.E., et al., 2004. Cellular mechanisms of neuronal population oscillations in the hippocampus in vitro. Annu Rev Neurosci 27, 247–278.

Udenfriend, S., 1950. Identification of gamma-aminobutyric acid in brain by the isotope derivative method. J. Biol. Chem. 187, 65–69.

Wang, M., Offord, J., Oxender, D.L., Su, T.Z., 1999. Structural requirement of the calcium-channel subunit alpha2-delta for gabapentin binding. Biochem. J. 342 (Pt 2), 313–320.

Williams, S., Boksa, P., 2010. Gamma oscillations and schizophrenia. J. Psychiatry Neurosci. 35, 75–77.

Yaksh, T.L., 1989. Behavioral and autonomic correlates of the tactile evoked allodynia produced by spinal glycine inhibition: effects of modulatory receptor systems and excitatory amino acid antagonists. Pain 37, 111–123.

Yamamoto, J., Suh, J., Takeuchi, D., Tonegawa, S., 2014. Successful execution of working memory linked to synchronized high-frequency gamma oscillations. Cell. 157, 845–857.

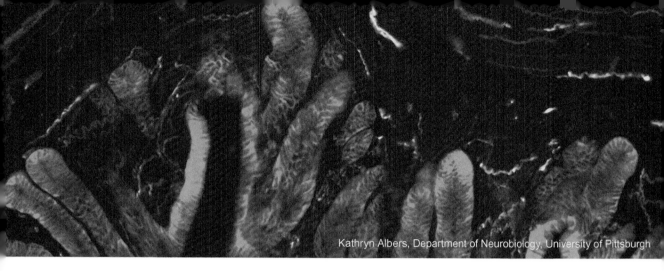

Kathryn Albers, Department of Neurobiology, University of Pittsburgh

# CHAPTER 19

# Neuropeptide Transmitters

**Neuropeptides** are small proteins that are distinguished from other peptides in the body by the fact that they are made by and released from neurons. Neuropeptides can affect both neurons and non-neuronal cells and can act on multiple time scales, affecting neuronal physiology on the order of seconds to minutes and modulating gene expression on the scale of hours to days. There are well over a hundred different neuropeptides that can be released from neurons, and it is important to emphasize that most, if not all of these neurons also release either a type 1 or 2 transmitter (i.e., classical transmitters or amino acid transmitters) as well as neuropeptides (Hökfelt et al., 2000; see Chapter 21). A small minority of neurons release only neuropeptides, including magnocellular neurons in the hypothalamus, which release peptides directly into the bloodstream.

Neuropeptides can affect the nervous system in a number of ways, including modulating the release and action of neurotransmitters, functioning as trophic factors, or upregulating the synthesis of neurotransmitter receptors (Fontaine et al., 1986; New and Mudge, 1986). They also play a role in homeostatic functions such as food and water intake, thermoregulation, reproductive behavior, and circadian rhythms. Additionally, they are involved in inflammatory responses and pain sensitivity/modulation, and some appear to have neuroprotective effects.

# HOW DO NEUROPEPTIDES DIFFER FROM CLASSICAL (TYPE 1) NEUROTRANSMITTERS?

There are several significant differences between neuropeptides and classical neurotransmitters, which are described below (also see Fig. 19.1).

1. *Precursor molecules:* Whereas classical neurotransmitters are synthesized from precursors available within the axon terminal, peptides are formed from amino acids on ribosomes on the rough endoplasmic reticulum in the cell body.
2. *Synthesis:* Classical neurotransmitters are synthesized in one or two specific enzymatic steps within the axon terminal, whereas the synthesis of neuropeptides requires gene expression and mRNA translation at the cell body.
3. *Vesicles:* After synthesis, classical neurotransmitters are actively moved into small clear vesicles by energy-driven vesicular transporters within the presynaptic terminal, and once these vesicles are released, their membranes are endocytosed and re-formed into vesicles that are again filled with transmitter (see Chapter 9). In contrast, precursors of neuropeptides

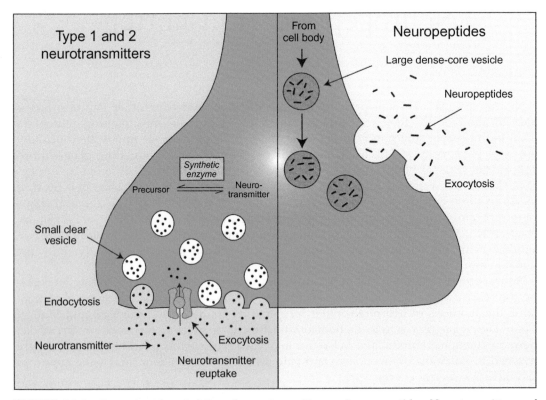

FIGURE 19.1 **Comparing characteristics of neurotransmitters and neuropeptides.** Neurotransmitters and neuropeptides differ in a number of ways, including the locations of synthesis and vesicular packaging, types of vesicles from which they are released, concentrations and receptor affinities, and the termination and recycling of the transmitter or its components. These characteristics are detailed in the text.

are packaged into large dense-core vesicles in the Golgi apparatus, and must be transported down the axon to the nerve terminal for secretion. After exocytosis, large dense-core vesicles are not recycled and re-filled with transmitter at the nerve terminal.

4. *Concentrations and affinities:* Inside small clear vesicles, classical neurotransmitters are present at a relatively high concentration (~100 mM), whereas neuropeptides within large dense-core vesicles are found at a concentration of about 3–10 mM (Mains and Eipper, 1999). However, there are also substantial differences in the affinity of these molecules for their receptors. Classical neurotransmitters can have a wide range of receptor affinities, but these are generally low [$K_D = 0.1–1$ millimolar (mM)], whereas receptor affinities for neuropeptides are typically very high [$K_D$ in the nanomolar (nm) to micromolar (μm) range].

5. *Sites of release:* While small clear vesicles are released within active zones at nerve terminals, large dense-core vesicles are released outside of active zone regions of the nerve terminal.

6. *Sites of action:* Whereas classical neurotransmitters typically act directly at or near the synapse at which they are released, neuropeptides can diffuse and act at much longer distances.

7. *Termination/recycling:* After release, classical neurotransmitters either undergo re-uptake into the presynaptic terminal or glial cells, or are degraded by enzymes. In many cases the neurotransmitter molecules themselves, or their breakdown product(s), can be reused. In contrast, neuropeptide transmitters are not recycled and need to be synthesized anew from raw materials in the cell body.

## NEUROPEPTIDE SYNTHESIS, RELEASE, AND REGULATION

### Neuropeptide Synthesis

Some neuropeptides are available for release under normal circumstances, whereas some other neuropeptides are normally expressed at low levels and their expression is upregulated as needed (Hökfelt et al., 2000). Sometimes neuropeptides are normally expressed early in development (usually prenatally) and are not synthesized in mature neurons, except under conditions of neuronal injury, highlighting the fact that the use of neuropeptides can be governed by the regulation of gene expression.

Neuropeptide synthesis begins with transcription of a gene into messenger RNA (mRNA). Translation of peptide transmitter mRNA occurs on ribosomes on the endoplasmic reticulum (see Fig. 19.2). At this stage, the peptides are called **prepropeptides** because they are precursor peptides with an extra tail (denoted by the "pre" in the name), and extra segments that will not be part of the final peptide (denoted by "pro" in the name). The "pre" tail is a signaling sequence that allows molecules to be trafficked appropriately within the endoplasmic reticulum and Golgi, so that they will end up in the correct location within a cell (e.g., nucleus, mitochondria, plasma membrane, large dense-core vesicles). The prepropeptides then move from the endoplasmic reticulum to the lumen of the trans-Golgi network. There, they attach to binding sites on the walls of the lumen of the Golgi to be concentrated and packaged into large dense-core vesicles.

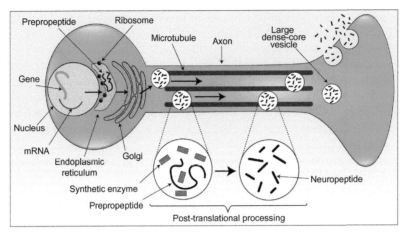

FIGURE 19.2  **Neuropeptide synthesis.** Proteins that will become neuropeptides are synthesized in the endoplasmic reticulum from mRNA following gene transcription, where they are referred to as prepropeptides. From there, they travel through the Golgi apparatus, and are packaged into large dense-core vesicles. These vesicles are transported down the axon via microtubules to the axon terminal, where they are released.

After prepropeptides are translated from mRNA and attached to the lumen in the trans-Golgi network, the lumen buds from the Golgi to form large dense-core vesicles containing the prepropeptides. Within the newly formed vesicles are synthetic enzymes that cleave the prepropeptides as the vesicles are transported down the axon along microtubules, toward the presynaptic terminal. This final step in the synthesis of neuropeptides is called **post-translational processing**, and it allows the prepropeptide encoded by a single gene to be converted into multiple transmitters, depending on the specific synthetic enzymes present in a vesicle. For example, the anterior and intermediate lobes of the pituitary gland process the same prepropeptide into different products, as can be seen in Fig. 19.3. Translation of mRNA from the pro-opiomelanocortin (POMC) gene generates pre-pro-opiomelanocortin in cells of the pituitary gland. Both lobes of the pituitary gland have enzymes that cleave the pre-pro-opiomelanocortin precursor peptide to generate the peptide transmitters adrenocorticotropic hormone (ACTH) and β-lipotropin (β-LPH). The anterior lobe does not process these proteins further, but the intermediate lobe has additional enzymes that cleave β-LPH into α-melanocyte-stimulating hormone (α-MSH), corticotropin-like intermediate lobe peptide (CLIP), γ-lipotropin (γ-LPH), and β-endorphin (β-ENDO). Thus, the single POMC gene codes for six distinct peptides, and the presence of these peptides in a given type of neuron is dictated by the enzymes that act on the precursor peptides.

Why would cells construct a large protein that is then cleaved into smaller parts to make multiple types of neuropeptides? One advantage to this process is that it allows a single gene to be used for multiple neurotransmitters, allowing for flexibility in how that gene is utilized by different cells. Second, some peptides need to be folded into a specific conformation in order to be functional. Often, these configurations require **disulfide bonds** to hold the peptide in a shape that is required for binding to a specific receptor. However, for some proteins this folding is only possible when starting as a larger protein.

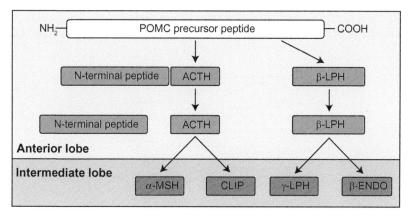

FIGURE 19.3 **Post-translational processing of ACTH.** The POMC precursor peptide undergoes post-translational processing in the pituitary gland that results in six different neuropeptides. Two of these (ACTH and β-LPH) are active in the anterior lobe of the pituitary and four (α-MSH, CLIP, γ-LPH, and β-ENDO) are active in the intermediate lobe of the pituitary. *ACTH*, adrenocorticotropic hormone; *α-MSH*, α-melanocyte-stimulating hormone; *β-LPH*, β-lipotropin; *β-ENDO*, β-endorphin; *CLIP*, corticotropin-like intermediate lobe peptide; *γ-LPH*, γ-lipotropin; *POMC*, pro-opiomelanocortin.

For example, insulin is a peptide hormone that has two chains bound together with disulfide bridges, and in order to get these bridges to form properly, it is necessary to start with a long protein that can fold on itself. Once bridges are formed between the folded parts, the components of the peptide that are only required for protein folding, but not the final product, can be cut off.

Another mechanism for generating peptide diversity from a single gene is **alternative mRNA splicing**. In this case, a precursor mRNA is transcribed from DNA and this precursor mRNA can be processed into one of several specific mature mRNAs, such that different prepropeptides can be translated from different final mRNAs.

For example, alternative splicing regulates which products are made from the calcitonin CALC1 gene, which codes for the neuropeptide **calcitonin gene-related peptide** (CGRP). In sensory neurons, CGRP is produced by splicing the CALC1 precursor mRNA (Amara et al., 1982); but in the thyroid, a different alternative splicing of the same CALC1 precursor mRNA produces the thyroid peptide calcitonin instead (see Fig. 19.4). This differential splicing of the calcitonin mRNA allows a single gene to produce two unique proteins that are utilized by different systems in the body.

CGRP is widely expressed in somatosensory and autonomic peripheral nerves, in the cardiovascular system, at the neuromuscular junction, and in the enteric and central nervous systems. Increased levels of CGRP are associated with inflammation, hyperalgesia, and arthritis, whereas decreased levels of CGRP have been associated with hypertension and poor wound healing. Additionally, this peptide seems to be protective against heart failure and ischemia (see Russell et al., 2014 for a review). CGRP has also attracted a lot of attention for its effects on peripheral sensory axons that transmit pain signals, where it has been heavily implicated in migraine pain (see Box 19.1) and neuropathic pain. At the neuromuscular junction, CGRP is hypothesized to contribute to the regulation of synapse formation, function, and maintenance (Changeux et al., 1992; Sala et al., 1995). Given the wide range of bodily systems influenced by CGRP, it is clear that this is a versatile peptide.

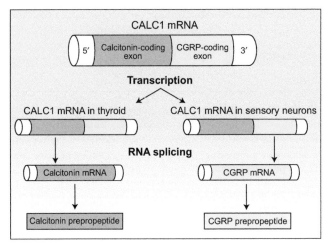

FIGURE 19.4 **Alternative splicing of the calcitonin CALC1 mRNA.** The CALC1 mRNA is spliced differently, according to whether it is being expressed in the thyroid, where it produces calcitonin, or in sensory neurons, where it produces CGRP.

---

BOX 19.1

## TREATMENT OF MIGRAINES BY BLOCKING THE ACTION OF CGRP

Migraines are often characterized by debilitating headache, increased sensitivity to light and sound, nausea/vomiting, and can often be preceded by a visual aura. It is not clear what triggers migraines, and most current drugs treat symptoms that occur only after the migraine has begun. A successful set of migraine drugs introduced in the 1990s, the triptans (e.g., sumatriptan), are agonists of specific subtypes of serotonin receptors. The physiological effects of these drugs include constricting blood flow in cerebral vasculature, inhibiting abnormal activation of peripheral nerves that transmit pain signals, and preventing leakage of vasoactive and neuroactive peptides through blood vessels. The precise mechanisms by which triptans stop migraine pain are debated, though their effects on cerebral arteries are heavily implicated in this effect (see Benemei et al., 2017 for a review).

While multiple neuropeptides (e.g., oxytocin, neuropeptide Y, corticotropin-releasing hormone, orexins) are thought to play a role in migraines, the neuropeptide CGRP (the synthesis of which is discussed in the main text) has several effects in the nervous and circulatory systems that have piqued the interest of multiple drug companies in the search for drugs to treat migraine headaches.

## BOX 19.1 (cont'd)

CGRP is particularly relevant to migraine headaches because it dilates cranial blood vessels and enhances the transmission of pain signals. In addition, CGRP and its receptors are expressed in areas of the nervous system that are related to for migraine pain, including the trigeminal ganglion, afferent neurons that transmit pain signals, and cerebral and meningeal blood vessels. It has been shown that CGRP levels can increase during migraines in some patients (Goadsby et al., 1990) and will decrease upon cessation of the migraine. Furthermore, intravenous administration of CGRP can induce migraine attacks in migraine patients (Hansen et al., 2010; Lassen et al., 2002). These findings led drug makers to investigate whether inhibiting CGRP might be a viable pharmacological method for treating migraines. The three main strategies developed to date for inhibiting the action of CGRP are CGRP receptor antagonists (referred to as "gepant" drugs), antibodies against CGRP, and antibodies against CGRP receptors.

A number of hurdles must be overcome in order for a drug to be a practical treatment option for migraine patients. First, a drug for acute migraine treatment needs to be administered soon after a migraine attack begins, which means drug molecules that must be administered by injection at a doctor's office are typically impractical. Some CGRP inhibitors require injection, but an oral or intranasal bioavailable drug is best so a patient can take it themselves as soon as a migraine begins. Second, any side effects must be tolerable. As indicated earlier in this chapter, CGRP affects many systems in the body, suggesting side effects could prove to be a significant problem. Fortunately, while triptan drugs are contraindicated (i.e., not recommended) for patients with cardiovascular disease, CGRP antagonists seem to have few cardiovascular side effects (Lynch et al., 2010), though some have been associated with liver toxicity.

There has been some success in clinical trials using monoclonal antibodies against CGRP and its receptors (see Bigal and Walter, 2014 for a review). These drugs do not cause liver toxicity, and are not associated with cardiovascular side effects. Additionally, since antibodies generally do not cross the blood—brain barrier, and thus do not have effects on the central nervous system, these drugs are not associated with side effects (such as drowsiness, fatigue, and memory problems), that are common with some other categories of drugs for migraine treatment (e.g., antiseizure drugs). The first antibody-based drug for migraines was approved by the FDA in May, 2018. This drug, Aimovig (erenumab-aooe), is a monoclonal antibody that binds to the CRGP receptor, blocking its function. It was approved for preventative treatment of migraine in adults and is administered by once-a-month subcutaneous injection. Unlike with CGRP inhibitors, anti-CGRP antibodies do not need to be administered "on demand" to treat migraines that have already begun, so it is practical to administer them by injection.

The long-term effects of anti-CGRP drugs are not yet known. However, their relatively specific effects mean they have fewer side effects than other migraine treatments, which makes them promising medications, both to stop migraine symptoms acutely and as prophylactic drugs to prevent migraines from beginning in the first place.

## Vesicular Release of Neuropeptides

One of the most commonly investigated mammalian model systems for neuropeptide release is neurons in the neurohypophysis division of the hypothalamus, which is also known as the posterior pituitary gland. Neurons from multiple hypothalamic areas release neuropeptides such as vasopressin and oxytocin into the vascular system, affecting functions such as water homeostasis and milk release during lactation, respectively. The prevalence of large dense-core vesicles in these neurons makes this an ideal model system for studying peptide release. However, it should be noted that these neurons are specialized for releasing only peptides and the large dense-core vesicles in these neurons are substantially larger than those in smaller axon terminals. Therefore, some of the principles discovered in the neurohypophysis do not necessarily apply to smaller nerve terminals in other areas of the nervous system.

A significant difference between the release of classical neurotransmitters (type 1) and the release of neuropeptides at smaller nerve terminals is that whereas vesicles containing classical neurotransmitters are usually released from the active zone, neuropeptides are released extrasynaptically (Thureson-Klein and Klein, 1990; see Fig. 19.5). While neurotransmitter release from small clear vesicles occurs in response to the relatively high concentrations of calcium ions entering the nerve terminal cytoplasm in close proximity to the vesicles (see Chapter 7), neuropeptide-containing large dense-core vesicles are triggered to release at levels of cytoplasmic calcium that are on the order of 10 times lower, but that represent an overall rise in cytoplasmic calcium rather than a transient calcium spike within a microdomain around calcium channels (Verhage et al., 1991; Huang and Neher, 1996). Increases in cytoplasmic calcium can come from influx through voltage-gated calcium channels or from release of intracellular stores.

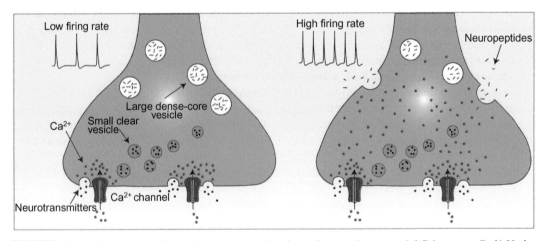

FIGURE 19.5  **The release of large dense-core vesicles depends on action potential firing rates.** (Left) Under conditions of low firing rates, calcium levels increase transiently near voltage-gated calcium channels, but not in the rest of the axon terminal. This means small clear vesicles are released, but not large dense-core vesicles. (Right) When a neuron is firing at a higher rate, calcium levels in the nerve terminal increase throughout the nerve terminal, which means large dense-core vesicles can also be released.

In general, neuropeptide release is triggered under conditions of high action potential firing. As discussed in Chapter 7, when an action potential reaches the axon terminal, the change in membrane voltage opens voltage-gated calcium channels. The calcium concentration increases significantly near the membrane and is locally high enough to trigger release of locally docked small clear vesicles. However, due to calcium buffering and handling, the calcium entering the cytoplasm after a single action potential does not usually diffuse far from the active zone, and thus does not trigger release of large dense-core vesicles. In contrast, when a neuron fires action potentials at a higher rate, cytoplasmic calcium levels can temporarily overcome local buffering and handling mechanisms and can diffuse deeper into the nerve terminal. Thus, cytoplasmic calcium concentration can increase to levels that are sufficient to trigger fusion of large, dense-core vesicles, leading to peptide release (Andersson et al., 1982; see Fig. 19.5). Additionally, it has been shown that high-freqency bursts of action potentials, with periods of silence between the bursts, are particularly effective for causing large dense-core vesicles to release neuropeptides (Dutton and Dyball, 1979; Bicknell and Leng, 1981). It should be noted that large dense-core vesicles require a high-affinity calcium sensor (in contrast to the low affinity synaptotagmin 1 or 2 sensor protein present on small clear vesicles) to respond to the lower concentrations of calcium that reach these vesicles positioned deeper into the nerve terminal and away from active zones (see Pinheiro et al., 2016 for a review).

## Regulation of Neuropeptide Synthesis

Because neuropeptide synthesis occurs in the cell soma, there is a significant delay of at least several hours between the release of neuropeptides and when they can be replenished in the axon terminal (van den Pol, 2012). Additionally, unlike the magnocellular neurons that synthesize and release peptide hormones from the posterior pituitary gland (neurohypophysis) on a more continuous basis, typical axon terminals have fewer and smaller large dense-core vesicles. As a consequence, under conditions of high peptide release from these neurons (e.g., after a long burst of action potential activity), the available peptide can be depleted and it requires some time to replenish.

How can neurons signal to the nucleus that more peptide transmitter is needed? As we discussed earlier, action potential activity increases cytoplasmic calcium concentration in the axon terminal. This increase in calcium concentration triggers a second-messenger cascade in the cell body that activates transcription regulators, which can lead to transcription of more prepropeptide genes. Rates of translation can also be modulated to change the amount of neuropeptides synthesized from mRNA.

## Neuropeptide Receptors and Their Effects on Neurotransmitter Release

Once released, some neuropeptides can diffuse long distances by volume transmission, which means they can act in a paracrine fashion at relatively large distances from their site of release. Such long-distance action is aided by the fact that some neuropeptides in the brain have a long half-life [e.g., ~20 minutes (Mens et al., 1983), compared with a half-life of ~5 ms for typical classical neurotransmitters], which means they remain intact long enough to diffuse through the extracellular space and act at distant sites.

In addition, all neuropeptide receptors are G-protein-coupled receptors, and neuropeptides typically bind to these receptors with high affinity (requiring only nanomolar concentrations for effective receptor activation). As a result, neuropeptides can bind these receptors even when their concentration has been diluted by long-distance diffusion. An advantage of neuropeptides' ability to diffuse long distances to act on high-affinity receptors is that neuropeptides released from one location can have widespread effects on multiple types of neurons, in contrast to the highly localized effects of neurotransmitters involved in fast synaptic transmission. However, volume transmission has the disadvantages of not being spatially or temporally restricted, making it more difficult to regulate.

Neuropeptide receptors are found throughout the brain and can be located on cell bodies, dendrites, and axon terminals. One way neuropeptides can affect neurons is by acting in a modulatory fashion to regulate the synaptic release of neurotransmitters such as glutamate and GABA. Some peptides tend to increase neurotransmitter release (e.g., hypocretin, glucagon-like peptide), whereas others have been shown to decrease neurotransmitter release (e.g., neuropeptide Y, somatostatin, and dynorphin). Adding to the complex effects of neuropeptides, some can activate a number of different neuropeptide receptors, allowing a single type of neuropeptide to play multiple roles within the same brain region.

## NEUROPEPTIDE Y AS A MODEL FOR NEUROPEPTIDE ACTION

The ubiquity and versatility of neuropeptides can be illustrated by considering an example neuropeptide such as **neuropeptide Y**. This peptide is expressed throughout the body, and is one of the most abundant neuropeptides in the central nervous system (Adrian et al., 1983). Neuropeptide Y has been associated with many physiological and behavioral functions, including feeding behaviors, homeostasis, circadian rhythm, stress, and anxiety (for a discussion of its effects in sympathetic neurons, see Chapter 21).

What allows this single peptide to play such diverse roles in the function of the nervous system? One factor is that neurons that release neuropeptide Y and express neuropeptide Y receptors are found in a wide range of diverse brain regions. They are particularly prevalent in the neocortex, hippocampus, amygdala, and striatum (Caberlotto et al., 2000), but are also found in the cerebellum, hypothalamus, olfactory bulb, and several brainstem areas. Note that despite this widespread distribution, it is not thought that neuropeptide Y diffuses as far or has as long a half-life as many other neuropeptides, which means neuropeptide Y may have effects that are localized to relatively small paracrine volumes within individual brain regions (van den Pol, 2012). Another factor that contributes to the range of effects of neuropeptide Y is that there are multiple types of neuropeptide Y receptors, five of which (Y1, Y2, Y4, Y5, and Y6) are found in mammals. Furthermore, these can be located either pre- or postsynaptically and can modulate various aspects of synaptic function through their varied G-protein-coupled signaling systems.

A study by Fu and colleagues (2004) delineated some of the physiological mechanisms by which neuropeptide Y can affect neurons. These authors tested the effects of neuropeptide Y on hypocretin/orexin neurons in the hypothalamus, which have been implicated in modulating sleep/arousal states and feeding behaviors. These experiments illustrate the range of physiological effects neuropeptide Y can have, which, in this case, includes at least three effects on a single type of neuron. In slices of hypothalamic tissue, administration of neuropeptide Y reduced the excitability of hypocretin/orexin neurons by at least three separate mechanisms, two of which were postsynaptic and the other presynaptic. First, neuropeptide Y hyperpolarized hypocretin/orexin neurons by activating Y1 receptors, which activated potassium currents (GIRK channels), hyperpolarizing the membrane potential. Second, neuropeptide Y depressed voltage-dependent calcium currents in these neurons, also due, at least in part, to Y1 receptors. Both of these were postsynaptic effects. Neuropeptide Y also reduced the excitability of these neurons by decreasing the frequency of glutamate release from the presynaptic terminal via activation of the Y2 and/or Y5 receptors on the presynaptic neuron.

## Effects of Neuropeptide Y on Behavior

Neuropeptide Y is plentiful in the nervous system, has widespread influence in the brain, and can affect multiple types of receptors with a range of physiological effects, what behaviors can it affect? There are too many effects of this neuropeptide to give an exhaustive list here, but several behavioral effects will be discussed as examples.

### Roles of Neuropeptide Y in Feeding Behavior

The involvement of neuropeptide Y in feeding behavior was established in 1984 by three groups who showed that when neuropeptide Y was administered in the brains of rats, their food intake increased acutely (Clark et al., 1984; Levine and Morley, 1984; Stanley and Leibowitz, 1985). Other studies also showed that chronic administration of neuropeptide Y caused obesity (Stanley et al., 1986; Zarjevski et al., 1993). Experiments involving neuropeptide Y receptor agonists, antagonists, and antisense oligonucleotides showed that the obesity induced by neuropeptide Y is caused mainly by the effects of this peptide in several subregions of the hypothalamus, and that these effects are mediated by activation of Y1 and Y5 receptors. It was also demonstrated that antagonists of these receptors could reduce feeding in rats (Kanatani et al., 1996; Criscione et al., 1998). Note that while neuropeptide Y is critical for the regulation of feeding behavior, it does so in conjunction with many other peptides (including α-MSH, orexins, and MCH, which were discussed earlier in this chapter), receptors, and involves brain regions, located mainly in the hypothalamus and brainstem.

Given that neuropeptide Y can regulate eating, researchers and clinicians have made an effort to develop anti-obesity drugs that target pathways involving this peptide. For example, multiple antagonists for Y1 and Y5 receptors have been tested. But despite promising results in rodent models, it has proven difficult to develop a drug that interferes with neuropeptide Y function without having problems either with the drug's inability to cross the blood–brain barrier or with significant side effects, due to the lack of specificity in peptidergic signaling. Given the prevalence of neuropeptide Y in the body, it is not surprising that changing neuropeptide Y signaling affects many systems, which can lead to a range of side effects. Another challenge is in achieving long-term pharmacological control of obesity, due to compensatory mechanisms the body undergoes as it adjusts to the effects of a given drug.

### Role of Neuropeptide Y in Seizures

Multiple lines of evidence have demonstrated that neuropeptide Y is an endogenous protective agent against seizures. For example, when the neuropeptide Y gene is deleted in transgenic animals, they are more prone to seizures (Erickson et al., 1996), whereas overexpression of neuropeptide Y can decrease both the number and duration of seizures (Vezzani et al., 2002). It is thought that neuropeptide Y exerts this effect by decreasing presynaptic release of glutamate, which reduces neuronal excitability and thus the likelihood of seizures (El Bahh et al., 2005; Guo et al., 2002). As with hypocretin/orexin neurons, neuropeptide Y can reduce calcium currents in glutamatergic neurons in the hippocampus via a presynaptic mechanism, which may underlie the reduction in glutamate release in this seizure-prone area of the brain. Woldbey et al. (1997) showed that administering neuropeptide Y can reduce chemically induced seizures, and blocking Y5 receptors prevented this effect.

Researchers have been investigating candidate anti-seizure drugs based on the protective effect of neuropeptide Y, including selectively blocking specific Y receptors. But the development of therapeutic tools has been elusive, and, akin to anti-obesity drugs, they can be limited by the ability of drug candidates to cross the blood−brain barrier.

## References

Adrian, T.E., Allen, J.M., Bloom, S.R., Ghatei, M.A., Rossor, M.N., Roberts, G.W., et al., 1983. Neuropeptide Y distribution in human brain. Nature 306, 584−586.

Amara, S.G., Jonas, V., Rosenfeld, M.G., Ong, E.S., Evans, R.M., 1982. Alternative RNA processing in calcitonin gene expression generates mRNAs encoding different polypeptide products. Nature 298, 240−244.

Andersson, P.O., Bloom, S.R., Edwards, A.V., Jarhult, J., 1982. Effects of stimulation of the chorda tympani in bursts on submaxillary responses in the cat. J. Physiol. 322, 469−483.

Benemei, S., Cortese, F., Labastida-Ramírez, A., Marchese, F., Pellesi, L., Romoli, M., et al., 2017. Triptans and CGRP blockade—impact on the cranial vasculature. J. Headache Pain 18, 103.

Bicknell, R.J., Leng, G., 1981. Relative efficiency of neural firing patterns for vasopressin release in vitro. Neuroendocrinology 33, 295−299.

Bigal, M.E., Walter, S., 2014. Monoclonal antibodies for migraine: preventing calcitonin gene-related peptide activity. CNS. Drugs 28, 389−399.

Caberlotto, L., Fuxe, K., Hurd, Y.L., 2000. Characterization of NPY mRNA-expressing cells in the human brain: co-localization with Y2 but not Y1 mRNA in the cerebral cortex, hippocampus, amygdala, and striatum. J. Chem. Neuroanat. 20, 327−337.

Changeux, J.P., Duclert, A., Sekine, S., 1992. Calcitonin gene-related peptides and neuromuscular interactions. Ann. NY Acad. Sci. 657, 361−378.

Clark, J.T., Kalra, P.S., Crowley, W.R., Kalra, S.P., 1984. Neuropeptide Y and human pancreatic polypeptide stimulate feeding behavior in rats. Endocrinology 115, 427−429.

Criscione, L., Rigollier, P., Batzl-Hartmann, C., Rueger, H., Stricker-Krongrad, A., Wyss, P., et al., 1998. Food intake in free-feeding and energy-deprived lean rats is mediated by the neuropeptide Y5 receptor. J. Clin. Invest. 102, 2136−2145.

Dutton, A., Dyball, R.E., 1979. Phasic firing enhances vasopressin release from the rat neurohypophysis. J. Physiol. 290, 433−440.

El Bahh, B., Balosso, S., Hamilton, T., Herzog, H., Beck-Sickinger, A.G., Sperk, G., et al., 2005. The anti-epileptic actions of neuropeptide Y in the hippocampus are mediated by Y and not Y receptors. Eur. J. Neurosci. 22, 1417–1430.

Erickson, J.C., Clegg, K.E., Palmiter, R.D., 1996. Sensitivity to leptin and susceptibility to seizures of mice lacking neuropeptide Y. Nature 381, 415–421.

Fontaine, B., Klarsfeld, A., Hokfelt, T., Changeux, J.P., 1986. Calcitonin gene-related peptide, a peptide present in spinal cord motoneurons, increases the number of acetylcholine receptors in primary cultures of chick embryo myotubes. Neurosci. Lett. 71, 59–65.

Fu, L.Y., Acuna-Goycolea, C., van den Pol, A.N., 2004. Neuropeptide Y inhibits hypocretin/orexin neurons by multiple presynaptic and postsynaptic mechanisms: tonic depression of the hypothalamic arousal system. J. Neurosci. 24, 8741–8751.

Goadsby, P.J., Edvinsson, L., Ekman, R., 1990. Vasoactive peptide release in the extracerebral circulation of humans during migraine headache. Ann. Neurol. 28, 183–187.

Guo, H., Castro, P.A., Palmiter, R.D., Baraban, S.C., 2002. Y5 receptors mediate neuropeptide Y actions at excitatory synapses in area CA3 of the mouse hippocampus. J. Neurophysiol. 87, 558–566.

Hansen, K.B., Furukawa, H., Traynelis, S.F., 2010. Control of assembly and function of glutamate receptors by the amino-terminal domain. Mol. Pharmacol. 78, 535–549.

Hökfelt, T., Broberger, C., Xu, Z.Q., Sergeyev, V., Ubink, R., Diez, M., 2000. Neuropeptides—an overview. Neuropharmacology 39, 1337–1356.

Huang, L.Y., Neher, E., 1996. Ca(2+)-dependent exocytosis in the somata of dorsal root ganglion neurons. Neuron 17, 135–145.

Kanatani, A., Ishihara, A., Asahi, S., Tanaka, T., Ozaki, S., Ihara, M., 1996. Potent neuropeptide Y Y1 receptor antagonist, 1229U91: blockade of neuropeptide Y-induced and physiological food intake. Endocrinology 137, 3177–3182.

Lassen, L.H., Haderslev, P.A., Jacobsen, V.B., Iversen, H.K., Sperling, B., Olesen, J., 2002. CGRP may play a causative role in migraine. Cephalalgia 22, 54–61.

Levine, A.S., Morley, J.E., 1984. Neuropeptide Y: a potent inducer of consummatory behavior in rats. Peptides 5, 1025–1029.

Lynch, J.J., Regan, C.P., Edvinsson, L., Hargreaves, R.J., Kane, S.A., 2010. Comparison of the vasoconstrictor effects of the calcitonin gene-related peptide receptor antagonist telcagepant (MK-0974) and zolmitriptan in human isolated coronary arteries. J. Cardiovasc. Pharmacol. 55, 518–521.

Mains, R., Eipper, B., 1999. Peptides. In: Siegel, G., Agranoff, B., Albers, R., Fisher, S., Uhler, M. (Eds.), Basic Neurochemistry, 6 Edition Lippincott-Raven, Philadelphia, PA.

Mens, W.B., Witter, A., van Wimersma Greidanus, T.B., 1983. Penetration of neurohypophyseal hormones from plasma into cerebrospinal fluid (CSF): half-times of disappearance of these neuropeptides from CSF. Brain Res. 262, 143–149.

New, H.V., Mudge, A.W., 1986. Calcitonin gene-related peptide regulates muscle acetylcholine receptor synthesis. Nature 323, 809–811.

Pinheiro, P.S., Houy, S., Sorensen, J.B., 2016. C2-domain containing calcium sensors in neuroendocrine secretion. J. Neurochem. 139, 943–958.

Russell, F.A., King, R., Smillie, S.J., Kodji, X., Brain, S.D., 2014. Calcitonin gene-related peptide: physiology and pathophysiology. Physiol. Rev. 94, 1099–1142.

Sala, C., Andreose, J.S., Fumagalli, G., Lomo, T., 1995. Calcitonin gene-related peptide: possible role in formation and maintenance of neuromuscular junctions. J. Neurosci. 15, 520–528.

Stanley, B.G., Leibowitz, S.F., 1985. Neuropeptide Y injected in the paraventricular hypothalamus: a powerful stimulant of feeding behavior. Proc. Natl. Acad. Sci. USA 82, 3940–3943.

Stanley, B.G., Kyrkouli, S.E., Lampert, S., Leibowitz, S.F., 1986. Neuropeptide Y chronically injected into the hypothalamus: a powerful neurochemical inducer of hyperphagia and obesity. Peptides 7, 1189–1192.

Sun, Q.Q., Huguenard, J.R., Prince, D.A., 2001. Neuropeptide Y receptors differentially modulate G-protein-activated inwardly rectifying K+ channels and high-voltage-activated Ca2+ channels in rat thalamic neurons. J. Physiol. 531, 67–79.

Thureson-Klein, A.K., Klein, R.L., 1990. Exocytosis from neuronal large dense-cored vesicles. Int. Rev. Cytol. 121, 67–126.

van den Pol, A.N., 2012. Neuropeptide transmission in brain circuits. Neuron 76, 98—115.

Verhage, M., McMahon, H.T., Ghijsen, W.E., Boomsma, F., Scholten, G., Wiegant, V.M., et al., 1991. Differential release of amino acids, neuropeptides, and catecholamines from isolated nerve terminals. Neuron 6, 517—524.

Vezzani, A., Michalkiewicz, M., Michalkiewicz, T., Moneta, D., Ravizza, T., Richichi, C., et al., 2002. Seizure susceptibility and epileptogenesis are decreased in transgenic rats overexpressing neuropeptide Y. Neuroscience 110, 237—243.

Woldbey, D.P., Larsen, P.J., Mikkelsen, J.D., Klemp, K., Madesn, T.M., Bolwig, T.G., 1997. Powerful inhibition of kainic acid seizures by neuropeptide Y via Y5-like receptors. Nat. Med. 3 (7), 761—764.

Zarjevski, N., Cusin, I., Vettor, R., Rohner-Jeanrenaud, F., Jeanrenaud, B., 1993. Chronic intracerebroventricular neuropeptide-Y administration to normal rats mimics hormonal and metabolic changes of obesity. Endocrinology 133, 1753—1758.

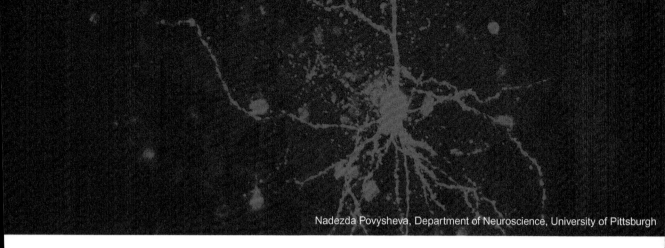

Nadezda Povysheva, Department of Neuroscience, University of Pittsburgh

# CHAPTER 20

# Gaseous Neurotransmitters

The final group of neurotransmitters we introduced in Chapter 15, is gas molecules that act as neurotransmitters, which are sometimes referred to as gasotransmitters. They include carbon monoxide, nitric oxide, and hydrogen sulfide. These are very small molecules that are not charged, though nitric oxide is a free radical with one unpaired electron (see Fig. 20.1). While it was known that the body produces gases as a product of the oxidation of organic matter, such gases initially seemed to be unlikely candidates to act as neurotransmitters. For example, it had been known for some time that carbon monoxide is a poisonous gas that can cause death when inhaled, due to its ability to bind to hemoglobin with a higher affinity than oxygen. Nonetheless, in the 1970s and 1980s researchers discovered that the body synthesizes gaseous messengers. They also discovered that these molecules are used by many systems in the body, including the immune system, the reproductive system, and the nervous system. Interestingly, these gases are bioactive (meaning they react with proteins in the cell to change their function) in many species from plants to mammals.

## NITRIC OXIDE

Nitric oxide (NO) was first identified as a gaseous molecule in 1772, though it was not thought to be bioactive until researchers in the 1980s began studying its effects on the body. At this time, Robert Furchgott and colleagues were investigating blood vessel dilation, and they determined that there was an unidentified factor released from endothelial

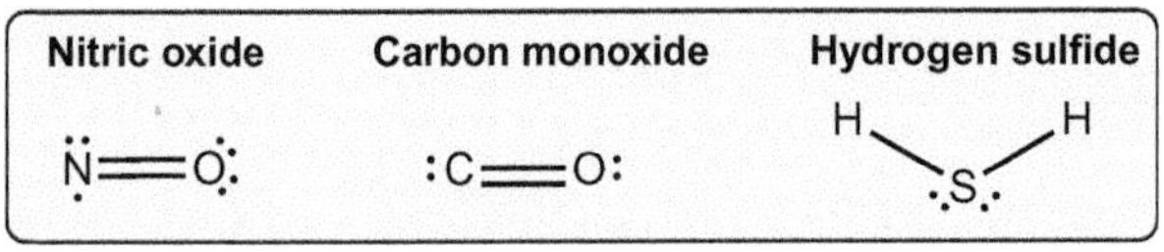

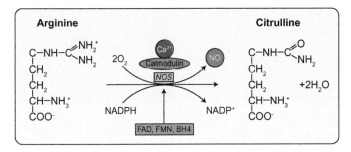

FIGURE 20.1 Molecular structures of three gaseous neurotransmitters.

FIGURE 20.2 Synthesis reaction for nitric oxide. Nitric oxide is a byproduct of the conversion of arginine to citrulline by the enzyme nitric oxide synthetase (NOS), which utilizes molecular oxygen and NADPH as cosubstrates. It also utilizes cofactors flavin adenine dinucleotide (FAD), flavin mononucleotide (FMN), and (6R-)5,5,7,8-tetrahydro-L-bioptein (BH4).

cells that was required for the relaxation of blood vessels (Furchgott and Bhadrakom, 1953). They showed that a diffusible, labile (short-lived) substance was involved in this reaction, and they called this substance "endothelial derived relaxing factor" (Cherry et al., 1982). However, it was not clear at the time what this factor actually was. Another researcher, Ferid Murad, showed that nitroglycerin causes vasodilation by releasing nitric oxide (Molina et al., 1987), and two other research groups, one led by Louis Ignarro and the other by Salvador Moncada, demonstrated that endothelial derived relaxing factor was actually nitric oxide (Palmer et al., 1987; Ignarro et al., 1987; see Box 20.1). It is now known that most types of mammalian cells produce nitric oxide, as do many other eukaryotic cells (Nathan, 1992).

## Synthesis of Nitric Oxide

Nitric oxide is synthesized from the amino acid arginine by the enzyme **nitric oxide synthase** (NOS; see Fig. 20.2). In this reaction, NOS oxidizes L-arginine to form L-citrulline, with nitric oxide as a byproduct. NOS is an oxidizing enzyme that requires an electron donor, which can be supplied by two cosubstrates: molecular oxygen and nicotinamide-adenine-dinucleotide phosphate (NADPH). Cofactors include flavin adenine dinucleotide (FAD), flavin mononucleotide (FMN), and (6R-)5,5,7,8-tetrahydro-L-biopterin (BH4).

There are three isoforms of NOS, which are named for the tissue in which they were first identified or for their biological activity, though all three isoforms are found in multiple types of cells and locations (Nathan and Xie, 1994). The first of these isoforms to be identified was **neuronal NOS** (nNOS). While NOS was first identified in neurons, it is also found in other tissues such as skeletal muscle, vascular smooth muscle, and the epithelial cells of multiple organs. Nitric oxide synthesized by nNOS affects physiological functions such as the central control of blood pressure, vasodilation in the periphery, gut peristalsis, penile erection, and synaptic plasticity. The second isoform is called

## BOX 20.1

## 1998 NOBEL PRIZE AWARDED FOR THE DISCOVERY OF NITRIC OXIDE

In 1998, three pharmacologists, Robert F. Furchgott, Louis J. Ignarro, and Ferid Murad (see Fig. 20.3), shared the Nobel Prize in Physiology or Medicine "for their discoveries concerning nitric oxide as a signaling molecule in the cardiovascular system." These researchers were instrumental in discovering that nitric oxide can act as a neurotransmitter in biological systems.

This discovery was built on a critical observation made by Furchgott and colleagues while they were studying the constriction and relaxation of tissue dissected from the aorta. They found that acetylcholine, which relaxes blood vessels in whole-animal preparations (i.e., with no dissection of the tissue), elicited constriction of aortic tissue when it was in the isolated preparation they were using (i.e., tissue dissected out of the animal; Furchgott and Bhadrakom, 1953).

Why would cholinergic agonists relax the aorta in the whole animal, but lead to vasoconstriction in the isolated preparation? Furchgott's group determined that when the aortic tissue was dissected out of the animal, if the inside surface of the blood vessel was rubbed during the dissection, the ability of acetylcholine to relax the blood vessel was eliminated. They also determined that such mechanical damage to the blood vessel caused the internal epithelial cells to be removed. They concluded that these epithelial cells were critical for cholinergic relaxation of the aortic tissue. Later work in this laboratory isolated a diffusible, labile factor, only present when the endothelial tissue was not disturbed during dissection, that was responsible for this effect. This molecule was termed "endothelium-derived relaxing factor" (EDRF; Cherry et al., 1982).

In the late 1980s, three separate research groups concluded that EDRF was in fact the diffusible molecule, nitric oxide (Ignarro et al., 1987; Kahn and Furchgott, 1987; Palmer et al., 1987). Work by Ferid Murad showed that various nitro compounds, including nitroglycerine, affected levels of cGMP in vascular tissue (Murad et al., 1978) and caused vasorelaxation (Molina et al., 1987), which is why nitroglycerine is effective in the treatment of angina. Murad showed that this effect was mediated by nitric oxide by simply bubbling nitric oxide

FIGURE 20.3 Robert F. Furchgott (left), Louis J. Ignarro (middle), and Ferid Murad (right). *Source: Left, Jzubrovich at English Wikipedia, CC BY 3.0; middle, Zé Carlos Barretta from São Paulo, Brasil, CC BY 2.0; right: Jeff Dahl CC BY-SA 4.0.*

### BOX 20.1 (cont'd)

through tissue containing guanylyl cyclase. This led to an increase in cGMP, which was subsequently shown to cause relaxation of myosin in vascular smooth muscle.

The discovery that nitric oxide is a signaling molecule in biological tissues has many physiological and therapeutic implications and rapidly opened up the study of multiple related clinical conditions. The research detailed above has yielded treatments for inflammation, hypertension, and erectile dysfunction, among other things.

---

**endothelial NOS** (eNOS), which was first described in endothelial tissue. Nitric oxide synthesized by eNOS is known for its multiple actions as a vasoprotector (e.g., vasodilation, control of the expression of genes involved in plaque deposition inside arteries, inhibition of platelet aggregation and their adhesion to the vascular wall; Förstermann and Sessa, 2012). The third isoform, **inducible NOS** (iNOS), is expressed in macrophages, glia, and tumor cells in response to proinflammatory cytokines and other compounds. Nitric oxide generated by iNOS plays a role in the symptoms of inflammation and septic shock.

### Regulation of Nitric Oxide Synthesis

The synthesis of nitric oxide is tightly controlled by the body, and this is important for several reasons. First, unlike chemical transmitters, nitric oxide cannot be stored in the body because it is a reactive gas that is rapidly degraded when it interacts with amino acids. Second, nitric oxide is not inactivated after release by typical mechanisms such as uptake by transporters. Third, at high intracellular concentrations, free cytoplasmic nitric oxide can damage cells. For example, abnormal nitric oxide signaling may contribute to neurodegenerative disorders such as Parkinson's disease and Alzheimer's disease. Nitric oxide can also be harmful to cells following other damage to brain tissue, such as after a stroke or in multiple sclerosis, when there can be a large influx of calcium into neurons via NMDA receptors (Bredt and Snyder, 1989; Garthwaite et al., 1989), leading to overactivation of nNOS. The resulting increase in nitric oxide production is thought to cause excitotoxicity, and this may account for much of the damage to neurons following stroke. In fact, there is evidence from animal studies that NOS inhibitors administered after stroke can prevent stroke-related damage, and stroke damage is substantially reduced in nNOS knockout mice. Finally, overproduction of nitric oxide is also thought to cause pathological changes in smooth muscle tone in the gastrointestinal tract. These potentially damaging effects make it critical for nitric oxide not to be present in high concentrations within cells for too long.

Since nNOS and eNOS are constitutively expressed, their activity must be closely regulated. One of the main mechanisms for this regulation is the level of intracellular calcium. Calcium that enters neurons through ion channels (such as voltage-gated calcium channels or NMDA receptors), or is released from intracellular stores, binds to calmodulin, and this complex then binds to nNOS or eNOS, increasing their enzymatic activity and, thus, the amount of nitric oxide produced. The increase in nitric oxide synthesis by nNOS and eNOS following intracellular calcium elevation is transient, lasting on the order of minutes (Malinski and Taha, 1992). The sensitivity of nNOS and eNOS to activation by calcium/calmodulin can itself be regulated by phosphorylation via protein kinases such as PKC, PKA, or PKG (Rameau et al., 2004; Mount et al., 2007). Nitric oxide synthesis can also be regulated by end-product inhibition (Assreuy et al., 1993). Finally, all three NOS isoforms can be regulated by gene expression in response to stimuli as diverse as hyperthyroidism, exercise, and estrogen.

As its name suggests, "inducible NOS" (iNOS) expression is induced by transcription factors, and unlike nNOS and eNOS, its expression is not directly regulated by intracellular calcium. Cellular activity that can lead to transcription factor expression includes the release of cytokines, immunologic or inflammatory stimuli, UV light, and ozone. Under these conditions, iNOS is capable of generating NO for up to five days after induction (Vodovotz et al., 1994).

## How Does Nitric Oxide Act as a Neurotransmitter, and What Roles Does It Play in the Nervous System?

Since nitric oxide is chemically unlike most neurotransmitters, it was not immediately clear how it could act as a cellular messenger. In particular, it does not satisfy some of the criteria we established for neurotransmitters in Chapter 15. First, unlike the effects of many other chemical neurotransmitters, because nitric oxide is not released from vesicles, its effects are not quantal. Second, there are no classic transmembrane receptors for nitric oxide (it freely diffuses across membranes). Do these characteristics disqualify nitric oxide from being considered a neurotransmitter? Recent evidence for the effects of nitric oxide at synapses suggest it can be a neurotransmitter, as long as we accept its unconventional mechanisms of action as acceptable for neurotransmitter action.

Researchers have now identified several physiological mechanisms by which nitric oxide affects neuronal signaling. First, nitric oxide can activate guanylyl cyclase, which generates cGMP. To do this, nitric oxide reacts with an iron molecule that is coordinated (bound) with a histidine amino acid in guanylyl cyclase (Lucas et al., 2000; Bellamy et al., 2002; Russwurm and Koesling, 2004; see Fig. 20.4), activating this enzyme to generate cGMP from GTP. cGMP activates kinases and phosphatases that change the phosphorylation states of proteins and can initiate many intracellular chemical cascades. Second, nitric oxide directly alters proteins by s-nitrosylation. In this reaction, nitric oxide reacts with free cysteines (those that are not already forming a disulfide bond) in the amino acid backbone of proteins, altering their structure and changing their conformation and function (see Fig. 20.4). For example, nitric oxide can affect signaling by altering the gating properties of receptors and ion channels. One of the ways the effects of nitric oxide were demonstrated was by creating NOS knockout mice and examining neuronal protein function in the absence of these s-nitrosylation reactions. In the knockout animals, there was no s-nitrosylation of NMDA receptors or the $Na^+/K^+$ ATPase, and, as a consequence, synaptic function was altered, suggesting that nitric oxide normally modulates these proteins.

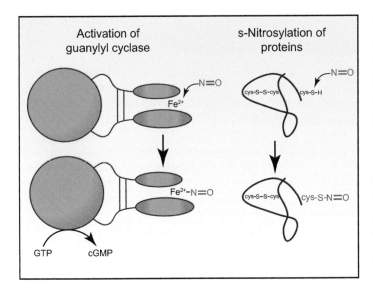

FIGURE 20.4 **Common nitric oxide reactions in the nervous system that mediate signaling.** (Left) Guanylyl cyclase can be activated by nitric oxide (N=O) when it reacts with an iron molecule that is coordinated between two domains of the guanylyl cyclase enzyme. This reaction activates the catalytic subunit of the enzyme, generating cGMP from GTP. (Right) Proteins can be covalently modified when nitric oxide reacts with free cysteine amino acids in a protein backbone. Cysteines within the protein that form a disulfide bond with another cysteine (cys-S-S-cys) do not react with nitric oxide. This s-nitrosylation of cysteines in the protein backbone changes the conformation of the protein, changing its function.

## Termination of Action of Nitric Oxide

Nitric oxide is a short-lived, reactive gas whose function terminates when it reacts with amino acids in surrounding proteins. As a consequence of its short life, nitric oxide molecules cannot be stored in storage pools (e.g., within vesicles), as can most neurotransmitters (Snyder et al., 1998). In addition to considering how nitric oxide itself is terminated, we also need to consider how its downstream actions are terminated. One such downstream action is the generation of cGMP (see above). Cyclic nucleotides like cGMP are degraded by phosphodiesterases, so the balance of activity of guanylyl cyclase and phosphodiesterases determines how long cGMP will remain in the cytoplasm. For example, the drug Viagra potentiates the effects of nitric oxide on blood flow by inhibiting the phosphodiesterase that breaks down cGMP, prolonging the downstream consequences of nitric oxide signaling. The other major downstream consequence of nitric oxide generation is s-nitrosylation of proteins. This process is reversed in an enzymatic manner through denitrosylation via a variety of enzymes (Gould et al., 2013).

### *Nitric Oxide Control of Autonomic Function*

While some experimental studies suggest a second-messenger-like function for nitric oxide, there is also experimental evidence indicating that nitric oxide is directly involved in the neural control of autonomic function, apparently functioning as a neurotransmitter. In the peripheral nervous system, synthesis of nitric oxide is not initiated via activation of NMDA receptors. Instead, it is mediated by calcium influx through voltage-gated ion channels, as well as calcium that is released from intracellular stores (Vincent, 2010).

An example of the role nitric oxide can play in the peripheral nervous system is the effects it has on motor neurons innervating the enteric nervous system (Bredt et al., 1990; Ward et al., 1992). These motor neurons help control parts of the intestine and are

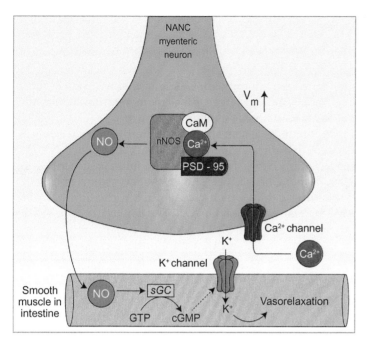

FIGURE 20.5 **Vasorelaxation triggered by nitric oxide in the enteric nervous system.** This schematic diagram shows the steps involved in the synthesis of nitric oxide within a presynaptic terminal, the movement of nitric oxide across the synaptic cleft to the postsynaptic smooth muscle, and the process by which nitric oxide causes vasorelaxation. The sequence of this signaling pathway is discussed in the text.

mostly found in the myenteric plexus. Catecholamine neurotransmitters and acetylcholine can cause vasorelaxation of blood vessels innervated by these neurons. However, vasorelaxation can occur even after cholinergic and catecholaminergic inputs are cut or otherwise blocked. Stimulation of the nonadrenergic, noncholinergic (NANC) nerves that remain causes vasorelaxation, as does application of nitric oxide (an experimental example of "identity of action"; Bult et al., 1990; Boeckxstaens et al., 1991). Additionally, NOS inhibitors block this induced vasorelaxation. These findings are supported by the fact that NOS is present in high concentrations in NANC nerve terminals. A proposed hypothesis for the vasorelaxation effect of nitric oxide involves the following sequence: (1) presynaptic action potential activity opens voltage-gated calcium channels, leading to calcium entry, (2) calcium binds to calmodulin and activated NOS, (3) NO is generated and diffuses to the postsynaptic cell, (4) soluble guanylyl cyclase (sGC) is activated and generates cGMP, (5) elevated cGMP opens postsynaptic K+ channels, and (6) smooth muscle relaxation occurs (see Fig. 20.5).

Another role nitric oxide plays in the autonomic nervous system is in the neural control of the pyloric sphincter, which is at the junction between the stomach and the small intestine. NOS knockout mice have enlarged stomachs, a symptom of reduced relaxation of the pyloric sphincter (Huang et al., 1993) that can be reversed by administering nitric oxide donors. It is now widely accepted that gas transmitters such as nitric oxide are common transmitters in the autonomic control of smooth muscle relaxation.

## *Effects of Nitric Oxide in the Central Nervous System*

In addition to its role in the peripheral neuronal control of autonomic function, 1%−2% of neurons in the central nervous system have detectable levels of NOS (Bredt and

Snyder, 1994). In a study of NOS immunoreactivity in Japanese macaques (also known as the snow monkey, *Macaca fuscata*), Satoh et al. (1995) found many strongly labeled NOS-containing neurons in multiple brainstem nuclei, the colliculi, striatum, nucleus accumbens, and the amygdaloid complex. Areas that had lightly labeled neurons included the globus pallidus, and lateral hypothalamic area, among others. Given that neurons in these regions have widespread projections, these findings suggest that many neurons in the central nervous system may be influenced by nitric oxide-containing neurons.

The first evidence that nitric oxide might be a neurotransmitter in the central nervous system came from experiments showing that nitric oxide synthesis increases when NMDA receptors are activated (Bredt and Snyder, 1989). This effect was initially described in experiments using slices of cerebellar tissue, where it was shown that activation of NMDA receptors stimulates the synthesis of nitric oxide, which leads to the formation of cyclic GMP (cGMP; Bredt and Snyder, 1989). The activation of soluble guanylyl cyclase (sGC) catalyzes the conversion of GTP to cGMP (see Fig. 20.6). Increased cGMP can modulate protein kinases, phosphodiesterases, ion channels, and transcription factors. Downstream effects of this cascade alter functions such as neurotransmitter release, synaptogenesis, apoptosis, and synaptic plasticity. Calling nitric oxide a neurotransmitter might be controversial in this case, as it is essentially acting as a second messenger downstream of glutamate, which is acting on NMDA receptors.

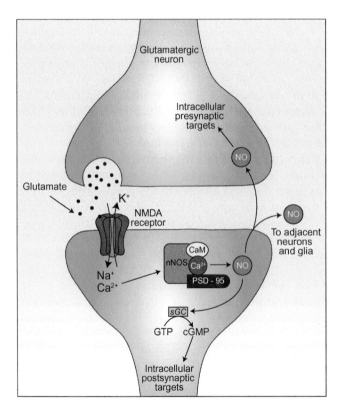

FIGURE 20.6 Nitric oxide can act as a retrograde messenger at some synapses after NOS is activated postsynaptically by calcium ions that enter through the NMDA receptor channel. Nitric oxide (NO) generated on postsynaptic spines can affect the postsynaptic spine and neighboring neurons and glial cells.

In addition to initiating a range of postsynaptic intracellular cascades, nitric oxide can also diffuse out of postsynaptic neurons and affect presynaptic nerve terminals (see Fig. 20.6). Because nitric oxide is a gas, it can diffuse directly through neuronal membranes and does not need to be released from vesicles. When nitric oxide diffuses out of a postsynaptic neuron and into a presynaptic one, this allows it to act as a **retrograde messenger**. As a retrograde messenger, nitric oxide can modulate the release of neurotransmitters, hormones, and peptides from the presynaptic neuron. In this context, nitric oxide acts more like a diffusible second messenger than a neurotransmitter.

## CARBON MONOXIDE

Carbon monoxide is best known as a poisonous gas, but, paradoxically, it is also generated in the brain under normal conditions. Carbon monoxide is synthesized by the enzyme **heme oxygenase** (see Fig. 20.7). There are three forms of heme oxygenase, HO1, HO2, and HO3. The major form found in the brain is HO2 (Sun et al., 1990). HO2 is often colocalized with NOS, so it is thought that nitric oxide and carbon monoxide act as cotransmitters in some systems.

As with nitric oxide, there may be second-messenger roles for carbon monoxide at some synapses that are distinct from their role as a neurotransmitter at other synapses. For example, when odorant molecules activate metabotropic receptors on olfactory receptor neurons, HO2 synthesizes carbon monoxide downstream of these receptors, which activates the enzyme guanylyl cyclase to generate the second messenger cGMP. Another example is in the nucleus tractus solitarius in the brainstem, which controls the heart rate in response to activation of chemo- and baroreceptors, among other things. Glutamate depresses blood pressure and heart rate by acting on a type of metabotropic glutamate receptor in the nucleus tractus solitarius, which appears to be linked to HO. Activation of the HO-linked receptor results in increased synthesis of carbon monoxide, which in turn increases cGMP production. In these examples, carbon monoxide seems to function more like a second messenger than a neurotransmitter.

However, carbon monoxide may act as a true neurotransmitter in the peripheral nervous system, especially in vasorelaxation (also called vasodilation) of smooth muscle in the digestive system. HO2 knockout mice have a similar phenotype in their digestive system as NOS knockout mice do (reduced relaxation of the pyloric sphincter, resulting in a distended stomach). A hypothetical signaling pathway by which carbon monoxide is proposed to promote smooth muscle vasorelaxation is as follows: (1) presynaptic action potential activity opens voltage-gated calcium channels, leading to calcium entry;

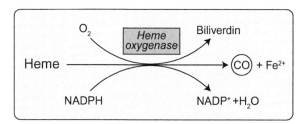

FIGURE 20.7 **Synthesis reaction for carbon monoxide.** Carbon monoxide is synthesized from heme by the enzyme heme oxygenase. NADPH and $O_2$ are cofactors in this reaction.

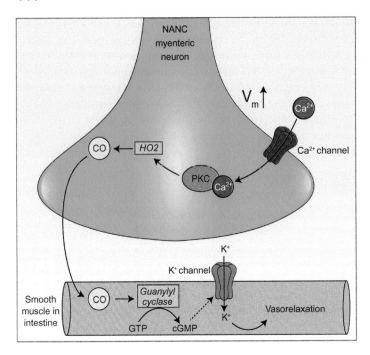

FIGURE 20.8 **Vasorelaxation triggered by carbon monoxide.** This schematic diagram shows the proposed steps involved in the synthesis of carbon monoxide (CO) within a presynaptic terminal, the movement of carbon monoxide to the smooth muscle at this synapse, and the process by which carbon monoxide causes vasorelaxation. In this scenario, carbon monoxide is hypothesized to act as a neurotransmitter. The details of this signaling pathway are discussed in the text.

(2) calcium activates PKC which phosphorylates the synthetic enzyme HO2, activating it; (3) activated HO2 generates carbon monoxide, which diffuses to the postsynaptic cell; (4) soluble guanylyl cyclase (sGC) is activated and generates cGMP; (5) elevated cGMP opens postsynaptic $K+$ channels; and (6) smooth muscle relaxation occurs (see Fig. 20.8).

## Regulation of Carbon Monoxide Synthesis

The synthesis of carbon monoxide by HO2 can be regulated by activation of PKC (Doré et al., 1999). PKC directly phosphorylates casein kinase 2 (CK2), which directly phosphorylates and activates HO2 (Boehning et al., 2003). Carbon monoxide synthesis can also be regulated by the end products of heme catabolism (Ryter and Tyrrell, 2000). It is likely that there are other mechanisms that regulate carbon monoxide synthesis, and this is currently an area of active research.

Carbon monoxide can have many physiological effects, largely because it leads to the production of cGMP. In addition to its role in vasorelaxation described above, carbon monoxide can prevent or reduce platelet aggregation, inflammation, and cell death (apoptosis), among other things.

It is important to emphasize that while carbon monoxide and nitric oxide can have beneficial effects in the body, they can also have deleterious effects, and the effects each has varies by tissue, physiological conditions, and concentration. For example, nitric oxide is damaging to the lung but can have protective effects in the liver. Similarly, while carbon monoxide can be lethal when inhaled, at low concentrations it can also have anti-inflammatory effects and can protect against cell death. Additionally, there are physiological interactions between the nitric oxide and carbon monoxide systems. For example, experiments have shown that nitric oxide can inhibit HO2 (Ding et al., 1999) and induce HO1 synthesis to produce carbon monoxide, which can then bind to iNOS and affect its production of nitric oxide. Therefore, the production and effects of these two gaseous neurotransmitters may be closely linked, at least in some tissues.

## HYDROGEN SULFIDE

Hydrogen sulfide ($H_2S$) is a toxic gas that smells like rotten eggs. It can lead to pathogenesis in many locations within the body, including the nervous system, the cardiovascular system, the retina, and the olfactory bulb. Nonetheless, while studying the effects of hydrogen sulfide poisoning on the brain, Warenycia et al. (1989) discovered that the brain contains endogenous hydrogen sulfide in concentrations ranging from 10 nM to 3 μM. However, the biological effects of hydrogen sulfide are somewhat dependent on its concentration. Lower levels promote beneficial processes such as mitochondrial function or antioxidant effects, whereas higher levels can lead to adverse effects, including cell death (Paul and Snyder, 2018).

### Synthesis and Regulation of Hydrogen Sulfide

In the central nervous system, hydrogen sulfide is synthesized by the enzyme cystathionine β-synthetase (CBS) or by 3-mercaptopruvate sulfurtransferase, in combination with cysteine aminotransferase (3MST and CAT, respectively; Shibuya et al., 2009). In other areas of the body such as blood vessels and the small intestine, hydrogen sulfide is synthesized by a different enzyme called cystathionine γ-lyase (CSE). The activity of CBS is regulated by S-adenosyl-L-methionine (SAM) and by calcium/calmodulin, and hydrogen sulfide production has been shown to be increased in response to neuronal excitation.

### Roles of Hydrogen Sulfide in the Brain

As more studies are done on the biological roles of hydrogen sulfide, it is becoming clear that this gaseous neurotransmitter may modulate multiple aspects of nervous system function, including long-term potentiation, cognition, and memory. For example, Abe and Kimura (1996) demonstrated that physiological concentrations of hydrogen sulfide can enhance NMDA receptor activity and facilitate the induction of LTP. Supporting these findings, there is also evidence that hydrogen sulfide signaling improves memory function in vivo (Li et al., 2017). There are also studies suggesting that hydrogen sulfide may be neuroprotective in neurological disorders such as Parkinson's disease and Alzheimer's disease, whereas elevated levels may be detrimental in amyotrophic lateral sclerosis (ALS) and Down's syndrome (see Paul and Snyder, 2018 for a review).

Despite the reported effects of hydrogen sulfide on the nervous system, its identity as a neurotransmitter is still debated and further studies will be required to better understand the role of this gas in the nervous system.

# References

Abe, K., Kimura, H., 1996. The possible role of hydrogen sulfide as an endogenous neuromodulator. J. Neurosci. 16 (3), 1066–1071.

Assreuy, J., Cunha, F.Q., Liew, F.Y., Moncada, S., 1993. Feedback inhibition of nitric oxide synthase activity by nitric oxide. Br. J. Pharmacol. 108, 833–837.

Bellamy, T.C., Wood, J., Garthwaite, J., 2002. On the activation of soluble guanylyl cyclase by nitric oxide. Proc. Natl. Acad. Sci. USA 99, 507–510.

Boehning, D., Moon, C., Sharma, S., Hurt, K.J., Hester, L.D., Ronnett, G.V., et al., 2003. Carbon monoxide neurotransmission activated by CK2 phosphorylation of heme oxygenase-2. Neuron 40, 129–137.

Bredt, D.S., Snyder, S.H., 1989. Nitric oxide mediates glutamate-linked enhancement of cGMP levels in the cerebellum. Proc. Natl. Acad. Sci. USA 86, 9030–9033.

Bredt, D.S., Hwang, P.M., Snyder, S.H., 1990. Localization of nitric oxide synthase indicating a neural role for nitric oxide. Nature 347, 768–770.

Boeckxstaens, G.E., Pelckmans, Pa, Ruytiens, I.F., Bult, H., De Man, J.G., Herman, A.G., Van Maercke, Y.M., 1991. Bioassay of nitric oxide released upon stimulation of non-adrenergic non-cholinergic nerves in the canine ileocolonic junction. Br. J. Pharmacol. 103 (1), 1085–1091.

Bredt, D.S., Snyder, S.H., 1994. Nitric oxide: a physiologic messenger molecule. Annu. Rev. Biochem. 63, 175–195.

Bult, H., Boeckxstaens, G.E., Pelckmans, P.A., Jordaens, F.H., Van Maercke, Y.M., Herman, A.G., 1990. Nitric oxide as an inhibitory non-adrenergic non-cholinergic neurotransmitter. Nature 345, 346–347.

Cherry, P.D., Furchgott, R.F., Zawadzki, J.V., Jothianandan, D., 1982. Role of endothelial cells in relaxation of isolated arteries by bradykinin. Proc. Natl. Acad. Sci. USA 79, 2106–2110.

Ding, Y., McCoubrey Jr., W.K., Maines, M.D., 1999. Interaction of heme oxygenase-2 with nitric oxide donors. Is the oxygenase an intracellular 'sink' for NO? Eur. J. Biochem. 264, 854–861.

Doré, S., Takahashi, M., Ferris, C.D., Zakhary, R., Hester, L.D., Guastella, D., et al., 1999. Bilirubin, formed by activation of heme oxygenase-2, protects neurons against oxidative stress injury. Proc. Natl. Acad. Sci. USA 96, 2445–2450.

Förstermann, U., Sessa, W.C., 2012. Nitric oxide synthases: regulation and function. Eur. Heart J. 33, 829–837. 837a–837d.

Furchgott, R.F., Bhadrakom, S., 1953. Reactions of strips of rabbit aorta to epinephrine, isopropylarterenol, sodium nitrite and other drugs. J. Pharmacol. Exp. Ther. 108, 129–143.

Garthwaite, J., Garthwaite, G., Palmer, R.M., Moncada, S., 1989. NMDA receptor activation induces nitric oxide synthesis from arginine in rat brain slices. Eur. J. Pharmacol. 172, 413–416.

Gould, N., Doulias, P.T., Tenopoulou, M., Raju, K., Ischiropoulos, H., 2013. Regulation of protein function and signaling by reversible cysteine S-nitrosylation. J. Biol. Chem. 288, 26473–26479.

Huang, P.L., Dawson, T.M., Bredt, D.S., Snyder, S.H., Fishman, M.C., 1993. Targeted disruption of the neuronal nitric oxide synthase gene. Cell 75, 1273–1286.

Ignarro, L.J., Buga, G.M., Wood, K.S., Byrns, R.E., Chaudhuri, G., 1987. Endothelium-derived relaxing factor produced and released from artery and vein is nitric oxide. Proc. Natl. Acad. Sci. USA 84, 9265–9269.

Khan, M.T., Furchgott, R.F., 1987. Additional evidence that endothelium-derived relaxing factor is nitric oxide. In: Rand, M.J., Raper, C. (Eds.), Pharmacology. Elsevier, Amsterdam, the Netherlands, pp. 341–344.

Li, Y.L., Wu, P.F., Chen, J.G., Wang, S., Han, Q.Q., Li, D., et al., 2017. Activity-dependent sulfhydration signal controls N-methyl-D-aspartate subtype glutamate receptor-dependent synaptic plasticity via increasing D-serine availability. Antioxid. Redox. Signal. 27, 398–414.

Lucas, K.A., Pitari, G.M., Kazerounian, S., Ruiz-Stewart, I., Park, J., Schulz, S., et al., 2000. Guanylyl cyclases and signaling by cyclic GMP. Pharmacol. Rev. 52, 375–414.

Malinski, T., Taha, Z., 1992. Nitric oxide release from a single cell measured in situ by a porphyrinic-based microsensor. Nature 358, 676–678.

Molina, C.R., Andresen, J.W., Rapoport, R.M., Waldman, S., Murad, F., 1987. Effect of in vivo nitroglycerin therapy on endothelium-dependent and independent vascular relaxation and cyclic GMP accumulation in rat aorta. J. Cardiovasc. Pharmacol. 10, 371–378.

Mount, P.F., Kemp, B.E., Power, D.A., 2007. Regulation of endothelial and myocardial NO synthesis by multi-site eNOS phosphorylation. J. Mol. Cell. Cardiol. 42, 271–279.

Murad, F., Mittal, C.K., Arnold, W.P., Katsuki, S., Kimura, H., 1978. Guanylate cyclase: activation by azide, nitro compounds, nitric oxide, and hydroxyl radical and inhibition by hemoglobin and myoglobin. Adv. Cyclic Nucleotide. Res. 9, 145–158.

Nathan, C., 1992. Nitric oxide as a secretory product of mammalian cells. FASEB J. 6, 3051–3064.

Nathan, C., Xie, Q.W., 1994. Nitric oxide synthases: roles, tolls, and controls. Cell 78, 915–918.

Palmer, R.M., Ferrige, A.G., Moncada, S., 1987. Nitric oxide release accounts for the biological activity of endothelium-derived relaxing factor. Nature 327, 524–526.

Paul, B.D., Snyder, S.H., 2018. Gasotransmitter hydrogen sulfide signaling in neuronal health and disease. Biochem. Pharmacol. 149, 101–109.

Rameau, G.A., Chiu, L.Y., Ziff, E.B., 2004. Bidirectional regulation of neuronal nitric-oxide synthase phosphorylation at serine 847 by the N-methyl-D-aspartate receptor. J. Biol. Chem. 279, 14307–14314.

Russwurm, M., Koesling, D., 2004. NO activation of guanylyl cyclase. EMBO J. 23, 4443–4450.

Ryter, S.W., Tyrrell, R.M., 2000. The heme synthesis and degradation pathways: role in oxidant sensitivity. Heme oxygenase has both pro- and antioxidant properties. Free Radic. Biol. Med. 28, 289–309.

Satoh, K., Arai, R., Ikemoto, K., Narita, M., Magai, T., Oshima, H., Kitahama, K., 1995. Distribution of nitric oxide synthase in the central nervous system of Macaca fuscata: subcortical regions. Neuroscience. 66 (3), 685–696.

Shibuya, N., Tanaka, M., Yoshida, M., Ogasawara, Y., Togawa, T., Ishii, K., et al., 2009. 3-Mercaptopyruvate sulfurtransferase produces hydrogen sulfide and bound sulfane sulfur in the brain. Antioxid. Redox. Signal. 11, 703–714.

Snyder, S.H., Jaffrey, S.R., Zakhary, R., 1998. Nitric oxide and carbon monoxide: parallel roles as neural messengers. Brain Res. Brain Res. Rev. 26, 167–175.

Sun, Y., Rotenberg, M.O., Maines, M.D., 1990. Developmental expression of heme oxygenase isozymes in rat brain. Two HO-2 mRNAs are detected. J. Biol. Chem. 265, 8212–8217.

Vincent, S.R., 2010. Nitric oxide neurons and neurotransmission. Prog. Neurobiol. 90 (2), 246–255.

Vodovotz, Y., Kwon, N.S., Pospischil, M., Manning, J., Paik, J., Nathan, C., 1994. Inactivation of nitric oxide synthase after prolonged incubation of mouse macrophages with IFN-gamma and bacterial lipopolysaccharide. J. Immunol. 152, 4110–4118.

Ward, S.M., Xue, C., Shuttleworth, C.W., Bredt, D.S., Snyder, S.H., Sanders, K.M., 1992. NADPH diaphorase and nitric oxide synthase colocalization in enteric neurons of canine proximal colon. Am. J. Physiol. 263, G277–G284.

Warenycia, M.W., Goodwin, L.R., Benishin, C.G., Reiffenstein, R.J., Francom, D.M., Taylor, J.D., et al., 1989. Acute hydrogen sulfide poisoning. Demonstration of selective uptake of sulfide by the brainstem by measurement of brain sulfide levels. Biochem. Pharmacol. 38, 973–981.

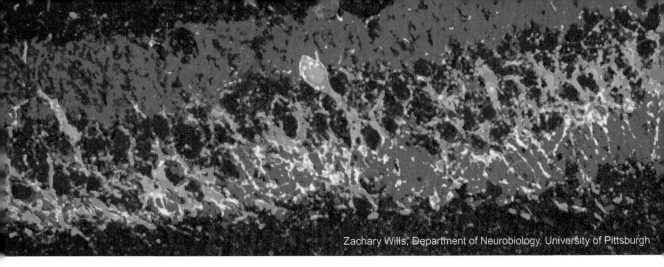

CHAPTER 21

# The Use of Multiple Neurotransmitters at Synapses

*Stephanie B. Aldrich*

Graduate Student PhD Program, Center for Neuroscience, University of Pittsburgh, PA, United States

## OVERVIEW AND HISTORICAL PERSPECTIVE

We often talk about neurons and synapses as though each one can only release a single type of neurotransmitter, and this "one neuron, one neurotransmitter" assumption was the dominant framework for conceptualizing neurotransmitter release for many decades. However, we now know that individual neurons can—and commonly do—release more than a single type of neurotransmitter (see Campbell, 1987; Gutierrez, 2009; Vaaga et al., 2014 for reviews). The idea that each neuron releases only one type of neurotransmitter is often referred to as "Dale's Principle" after Sir Henry Dale (see Fig. 21.1), a pharmacist, physiologist, and Nobel laureate, who was an early proponent of the idea that synaptic transmission is chemical rather than electrical. It was Dale who proposed classifying neurons by the type of neurotransmitter they release. Even today, we refer to neurons that release acetylcholine as **cholinergic**, neurons that release dopamine as dopaminergic, and so on.

However, "Dale's Principle" as originally written did not state that neurons release only a single neurotransmitter. The term was coined by Dale's close friend, neurophysiologist

FIGURE 21.1 (Left) Sir Henry Hallett Dale; (right) Sir John C. Eccles. *Source: Wellcome Collection. CC BY 4.0.*

John Eccles. In his 1957 book, *The Physiology of Nerve Cells*, Eccles referenced "the principle first enunciated by Dale (1935) that the same chemical transmitter is released from all the synaptic terminals of a neuron" (Eccles, 1957). The "one neurotransmitter" interpretation was a product of contemporary assumptions about neurotransmission. Because few neurotransmitters were known at the time, and no one had yet found evidence of multiple neurotransmitter expression within neurons, it was widely assumed that each neuron expresses only a single neurotransmitter. Therefore, it did not occur to Eccles to phrase his description of "Dale's Principle" in a way that would account for the possibility of multiple neurotransmitter release, and many authors interpreted it to mean that a neuron releases only one neurotransmitter from all of its synapses. [In a later publication, Eccles revised the principle to state instead that "at all the axonal branches of a neurone, there was liberation of the same transmitter substance or substances" (Eccles, 1982).]

As research into synaptic transmission continued and more advanced techniques were developed for investigating neurotransmitter expression and release, many laboratories began to report findings that implicitly challenged the "one neuron, one neurotransmitter" doctrine. Nonetheless, these findings were interpreted at the time to be consistent with that doctrine, as is frequently the case with scientific dogma that has been widely accepted. For example, in cases where the effects of more than one neurotransmitter were recorded in postsynaptic cells, researchers might propose that the neurotransmitters were each released by a separate population of presynaptic neurons, rather than that they were coreleased from the same neurons (Burnstock, 1972).

In response to mounting evidence of **cotransmission**, Geoffrey Burnstock published a seminal review in 1976 titled "Do some nerve cells release more than one neurotransmitter?" (Burnstock, 1976). In this review, Burnstock argued that Dale's Principle should be reevaluated due to new research findings that might contradict it, and that existing data (including his own) should be reexamined in light of the possibility of multiple neurotransmitter release. This proposal was controversial. In fact, a second review was soon released which directly challenged Burnstock's point of view, arguing that there was "little evidence against Dale's principle" (Osborne, 1979).

Although history has vindicated Burnstock and other early proponents of multiple neurotransmitter release, it is true that the evidence available at the time did not, on its own, prove that neurons could release more than one neurotransmitter. A number of experimental challenges had to be overcome in order to conclusively demonstrate multiple neurotransmitter release. For example, merely showing that a neuron *expresses* multiple neurotransmitters was not enough; it had to also be shown that it releases more than one of these substances in a stimulus-evoked manner at synaptic release sites. It was also not enough to merely demonstrate stimulus-evoked synaptic release of multiple substances; it had to be shown that these substances actually act as neurotransmitters, meaning that they bind to receptors and effect some change in the target cell. Over the past few decades, many elegant experiments have met these challenges, and it is now accepted that neurons can (and usually do) release more than one neurotransmitter. In addition, researchers have determined that neurons can change what type(s) of neurotransmitter(s) they release, and in some cases different sets of neurotransmitters have been found to be released at different synaptic terminals arising from the same neuron.

## Functional Implications of Multiple Neurotransmitter Release

The "one neuron, one neurotransmitter" doctrine that had persisted for so long in the study of synapses rested on three basic assumptions:

1. Each neuron expresses exactly one neurotransmitter.
2. A neuron's **neurotransmitter phenotype** (the set of neurotransmitters that it releases) is fixed and unchangeable.
3. A neuron releases the same neurotransmitter at each of its synapses.

But are these assumptions valid? How have they held up in light of discoveries that have been made in the field of neurotransmission since the days of Henry Dale? We now know that:

1. Most or all neurons express multiple neurotransmitters.
2. A neuron's neurotransmitter phenotype can change. Many neurons express a different set of neurotransmitters during development than they do in adulthood. Neurons can also switch between different neurotransmitter phenotypes in response to environmental or physiological changes.
3. It is usually true that a neuron releases the same neurotransmitter at all of its synapses. Nevertheless, some neurons have been found to release different neurotransmitters at different synapses.

These deviations from "one neuron, one neurotransmitter" are important for nervous system function, because they introduce additional complexity and plasticity potential into the nervous system circuitry. By differentially regulating the release of multiple neurotransmitters (varying the amount, timing, or sites of release of each neurotransmitter in relation to the others), or by changing their neurotransmitter phenotype, neurons can adjust the information carried by their chemical signals, which allows neural circuits to better adapt to a variety of situations in order to promote proper development and survival (Mayer and Baldi, 1991). This chapter will cover examples of neurons that release multiple neurotransmitters in both the developing and the mature nervous system, as well as examples of neurons that can express different neurotransmitters in different contexts.

## COTRANSMISSION AND CORELEASE OF NEUROTRANSMITTERS

Neurotransmitter cotransmission can take a variety of forms (Vaaga et al., 2014; see Fig. 21.2). In **corelease**, neurotransmitters are stored together in the same synaptic vesicles and released simultaneously when those vesicles fuse. In other cases, neurotransmitters are packaged into separate vesicles and **differentially released**, meaning that different neurotransmitters (or different ratios of neurotransmitters) are released under different conditions. Typically, this means that they are stored in separate vesicle pools at the same synapse. The vesicle pools for each neurotransmitter in a synapse may be sensitive to different patterns of electrical activity; for example, some neurons release only type 1 and 2 neurotransmitters (see Chapter 15) at low firing rates, while at higher firing rates they will also release neuropeptides (see Chapter 19). In other cases, a single neuron can make

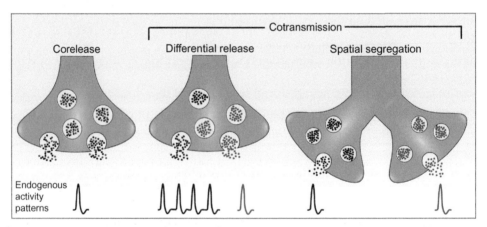

FIGURE 21.2 **Different mechanisms of cotransmission.** (Left) Corelease. A neuron indiscriminately packages two neurotransmitters (*red and black dots*, each representing one kind of neurotransmitter) into the same vesicles. When a stimulus triggers vesicle fusion, both neurotransmitters are released. (Center) Segregation of neurotransmitters into separate vesicle pools. A neuron packages its two neurotransmitters separately so that each vesicle contains only one of the neurotransmitters. Each vesicle pool is sensitive to different kinds of electrical stimulation. For example, a single stimulus might trigger the fusion of vesicles containing Neurotransmitter 1, while prolonged and/or high-frequency stimulation is required to trigger the fusion of vesicles containing Neurotransmitter 2. (Right) Segregation of neurotransmitters to different synapses. An axon branches into two distinct release sites, and two neurotransmitters are each isolated to one of the release sites and participate in transmission only from that site.

multiple synapses onto other neurons and release different neurotransmitters at each of these synapses. This section will cover examples of various types of cotransmission and their functional implications.

## Peptidergic Cotransmission

Some of the earliest coneurotransmitters discovered were neuropeptides (Lundberg and Hökfelt, 1983), short-chain polypeptide molecules that can act as neurotransmitters or neuromodulators. As discussed in Chapter 19, neuropeptides are synthesized by gene transcription in the neuron cell body and then transported to the axon terminal. It was originally thought that these neuropeptides had only endocrine functions, but many were found to also be released in a stimulus-dependent manner in the nervous system at synapses.

As discussed in Chapter 19, neuropeptides and small-molecule neurotransmitters are not usually packaged into the same type of synaptic vesicles. Instead, neuropeptides are packaged into large dense-core vesicles, and small-molecule transmitters are packaged separately into small clear vesicles. (Note that while large dense-core vesicles can also sometimes contain small-molecule neurotransmitters, such as catecholamines, small clear vesicles never contain peptides.) Because small-molecule and peptide neurotransmitters are stored in different vesicles, small-molecule and peptidergic transmission can be differentially regulated within a neuron. Small clear vesicles are released from active zones in response to high local concentrations of calcium, while large dense-core vesicles are released extrasynaptically in response to a milder overall increase in calcium concentration in the terminal—for example, when repeated stimulation of the presynaptic neuron causes calcium channels to open repeatedly, temporarily overcoming local buffering and handling mechanisms for calcium (see Chapter 7). Thus, peptide coneurotransmitters may or may not be released alongside small-molecule neurotransmitters, depending on the pattern of stimulation.

One important difference between these two types of transmitters is that neuropeptides tend to mediate slower, longer-lasting changes in cells than small-molecule neurotransmitters do. In addition, whereas small-molecule neurotransmitters are usually quickly removed from the synapse, neuropeptides can remain active for a long time, allowing them to diffuse away from the synapse where they were released to activate more distant receptors on postsynaptic cells in a paracrine manner. Through these and other mechanisms, peptide coneurotransmitters greatly enhance the complexity of a neuron's signaling (Nusbaum et al., 2017).

The first evidence of neuronal coexpression of a peptide with a small-molecule neurotransmitter was published in 1977 by Tomas Hökfelt and was based on studies of the postganglionic sympathetic neurons of guinea pigs (Hökfelt et al., 1977). These neurons were known to be almost universally noradrenergic, but Hökfelt's study showed that they also release a peptide called somatostatin, which acts as an endocrine hormone or as a synaptic neuropeptide depending on whether it is expressed in the endocrine system or the nervous system.

Hökfelt's discovery inspired further research aimed at identifying peptide coneurotransmitters. Another historically important example came only two years later with the discovery of the luteinizing-hormone-releasing hormone (LHRH)-like peptide, so named because of its similarity to LHRH. As discussed in Chapter 14, this peptide was identified in the sympathetic chain ganglia of bullfrogs, where "B cells" receive input from the superior ganglion in the chain and "C cells" receive input from the spinal cord. In Chapter 14, we focused on the mechanisms by which metabotropic receptors on B cells can modulate the ionotropic signaling on the B cell. Here, we are more concerned with the synapse onto C cells by the spinal input. At this synapse, both acetylcholine and LHRH are used as neurotransmitters, but they are released from different vesicles under different stimulus conditions. Acetylcholine is released from small clear vesicles during every presynaptic action potential, while LHRH is released from large dense-core vesicles only after a burst of presynaptic action potentials. Furthermore, these two transmitters have different effects on the postsynaptic C cells (acetylcholine causes fast postsynaptic potentials while LHRH causes slow postsynaptic potentials), and as discussed in Chapter 14, LHRH can also diffuse out of the synapse to bind to receptors on nearby B cells (see Jan and Jan, 1983 and Campbell, 1987 for reviews).

Many other peptides have since been found to play important roles throughout the nervous system. For example, vasoactive intestinal peptide (VIP) is characteristically coreleased with acetylcholine in parasympathetic neurons (Fahrenkrug and Hannibal, 2004). One of VIP's many functions is to mediate **vasodilation**, a function it performs in a wide variety of tissues. For instance, VIP is partially responsible for vasodilation in the skin that dissipates excess body heat (Kellogg et al., 2010). Conversely, neuropeptide Y is frequently coexpressed by noradrenergic sympathetic neurons, which counter the effects of the parasympathetic nervous system; for example, in contrast to VIP, neuropeptide Y is implicated in skin **vasoconstriction** that reduces heat loss in cold temperatures (Benarroch, 1994; Stephens et al., 2004; Hodges et al., 2009). Both VIP and neuropeptide Y are also widespread in the enteric nervous system (Furness et al., 1992; Furness, 2000), which contains many different classes of neuron, each with its own distinct neurotransmitter phenotype. These neuropeptides are coexpressed with a variety of neurotransmitters—most commonly with acetylcholine, but also with other neuropeptides, nitric oxide, and ATP—and are involved in the regulation of mucous secretion and the peristaltic reflex. Other well-known peptide coneurotransmitters include calcitonin gene-related peptide and substance P (Maggi, 2000; Salio et al., 2006).

## Segregation of Small-Molecule Neurotransmitters Into Separate Vesicle Pools

The previous section discussed neuropeptide/small-molecule neurotransmitter cotransmission and established that neuropeptides and small-molecule neurotransmitters are stored in different types of vesicles in neurons that cotransmit them. There are also many neurons that cotransmit two or more small-molecule neurotransmitters, and different small-molecule neurotransmitters can be located in separate pools of small clear vesicles within the same nerve terminal (Gutierrez, 2009). For example, catecholamines like norepinephrine can be stored in separate vesicles from small-molecule neurotransmitters like acetylcholine, and these separate noradrenergic and cholinergic vesicles can coexist at the same synapses in neurons that release both neurotransmitters (Potter et al., 1980).

## Spatial Segregation of Neurotransmitters

After it became apparent that most neurons in fact release more than one neurotransmitter, a modified version of Dale's Principle was proposed. The updated Dale's Principle, which was accepted for many years, states that neurons release the same *set* of neurotransmitters at each of their presynaptic terminals (Eccles, 1982). However, mounting evidence indicates that in at least some cases, different types of neurotransmitters can be released from separate presynaptic terminals of a single axon. In neurons that release different neurotransmitters from different terminals, the synthesis and packaging machinery for each neurotransmitter are apparently initially synthesized in the cell body, and then trafficked to distinct release sites (Sámano et al., 2012).

An early and well-known example of the segregation of signaling proteins to different release sites was identified in the nervous system of the sea slug *Aplysia californica*. This organism is known for laying millions of eggs at a time (Moroz, 2011), and its egg-laying behavior is controlled by the bag cell neurons of its abdominal ganglion, which have historically been a popular model system for studying neuropeptide synthesis, processing, and release (Conn and Kaczmarek, 1989). Bag cells are responsible for initiating the complex suite of stereotyped behavioral and physiological changes associated with egg-laying (Bernheim and Mayeri, 1995), and they accomplish this by releasing a variety of neuropeptides from their extensively branching axons. Neuropeptides released by bag cells vary in the set of target cells they act on, their duration of action, and whether their action is excitatory or inhibitory (Nagle et al., 1989).

The neuropeptides released by bag cells are derived from the cleavage and processing of a single precursor peptide, called the egg-laying hormone prohormone, or ELH prohormone. This prohormone is initially cleaved into two intermediate peptides representing its C-terminal and N-terminal regions, respectively. Each of these intermediates is further cleaved and processed into the active neuropeptides used for signaling. Among the C-terminal intermediate's products is egg-laying hormone (ELH), while the N-terminal intermediate's products are a group of structurally similar molecules called the bag cell peptides (Fisher et al., 1988).

Importantly, ELH and the bag cell peptides are packaged into separate vesicles, which are transported to different projections of the bag cell neuron for release (Sossin et al., 1990). At its release sites, ELH excites target abdominal ganglion neurons and is also released into the blood to act as a hormone in distant peripheral tissues. The bag cell peptides, at their own release sites, either excite or inhibit their respective target abdominal ganglion neurons and also excite the bag cells by autocrine action. ELH and bag cell peptides also differ in that ELH's actions last for several hours, while bag cell peptides are degraded after a few minutes (Rothman et al., 1983; Mayeri et al., 1985; Nagle et al., 1989). The fact that bag cells not only express a variety of signaling peptides with distinct actions, but can control where each of them is released in order to differentially regulate their actions on a large number of targets, is likely of functional importance to their role in coordinating the many behavioral and physiological changes that occur during egg-laying. Even if the modified Dale's Principle is true at most synapses, these findings in *Aplysia* bag cells show that it does not apply to all neurons, raising the possibility that segregation of signaling molecules could be found in other species and neuron types.

There is, in fact, evidence that neurotransmitter segregation also occurs in the mammalian nervous system. Preganglionic sympathetic neurons coexpress acetylcholine with a variety of neuropeptides, including VIP, neurotensin, somatostatin, calcitonin gene-related peptide, and the opioid peptide methionine enkephalin or met-enkephalin. Some synaptic terminals of preganglionic sympathetic neurons contain both cholinergic markers and these neuropeptides, but other terminals contain only the neuropeptides, indicating that neuropeptides are likely segregated from acetylcholine at these release sites (Sámano et al., 2006). Further investigation into the localization of met-enkephalin indicates that neurotrophic factors (growth factors that promote neuron survival) may be involved in regulating the segregation of met-enkephalin from acetylcholine, since reducing the availability of nerve growth factor increases the degree of segregation (Vega et al., 2016). The fact that neuropeptide segregation is apparently plastic, and subject (at least in this context) to regulation by cues from surrounding cells, suggests that it may increase the precision with which neurons can adapt to changes in the environment (Sámano et al., 2012).

Recently, investigators have identified putative examples of spatial segregation of two small-molecule neurotransmitters to separate release sites of the same neuron. Nishimaru et al. (2005) and Bhumbra and Beato (2018) have published electrophysiological and immunohistochemical data suggesting that spinal cord motor neurons cotransmit glutamate with acetylcholine from their projections to other spinal neurons, but not at the neuromuscular junction where only cholinergic transmission occurs. In another example, Zhang et al. (2015) have published electron microscope images that purport to show a segregation of dopamine and glutamate to separate release sites within a single axon. However, whether the observations made in these synapses genuinely reflect small-molecule neurotransmitter segregation is still under debate. For example, although it has been shown that some neurons communicate with different neurotransmitters at different synapses (e.g., the EPSPs motor neurons induce in spinal neurons are partially blocked by glutamate receptor antagonists, while the EPPs they induce in muscle cells may be completely blocked by acetylcholine receptor antagonists, indicating that glutamate cotransmission occurs only at motor neurons' central synapses and not at their peripheral synapses; Nishimaru et al., 2005), these observations could potentially be explained by segregation of postsynaptic receptors rather than of the neurotransmitters themselves (Lamotte d'Incamps et al., 2017). Methodical investigation will be necessary to definitively show whether small-molecule neurotransmitters are spatially segregated in some mammalian presynaptic neurons. Regardless of the underlying mechanism, however, the ability of neurons to transmit signals mediated by different neurotransmitters at different synapses is an important factor in the precise control of neural circuit activity.

## Neurotransmitter Corelease

This chapter has so far covered examples of multiple neurotransmitter expression in which coneurotransmitters are known (or suspected) to be segregated in different axon branches or vesicle pools at synapses within a neuron. However, in some neurons more than one neurotransmitter can be packaged into the same vesicles, meaning that both/all of these neurotransmitters are coreleased in response to the same stimuli and from the same neurotransmitter release sites (Vaaga et al., 2014).

For example, one of the earliest demonstrations of cotransmission in the central nervous system was of spinal cord interneurons coreleasing GABA and glycine (Jonas et al., 1998). Neurons that express the unique proteins needed to use these neurotransmitters typically corelease them, because the vesicular inhibitory amino acid transporter, VIAAT, transports both GABA and glycine into the same synaptic vesicles (Sagné et al., 1997). This shared transporter has the advantage of making GABA/glycine corelease relatively "inexpensive" in terms of protein expression. It is normally costly for a neuron to use multiple neurotransmitters, since it must express separate proteins for the transport, synthesis, and packaging of each. However, GABA/glycine corelease only requires the presynaptic neuron to express the glycine transporter GlyT2, the GABA synthetic enzyme glutamate decarboxylase, and VIAAT (see Fig. 21.3.) GABA/glycine corelease is most widespread during development, but persists at some synapses in the adult, especially in the brainstem and spinal cord (Todd et al., 1996; Lu et al., 2008).

Even though GABA and glycine both act on ionotropic receptors that pass chloride ions, they each have different effects on postsynaptic cells. Glycine receptors mediate fast, transient IPSPs, while GABA receptors mediate slower, longer-lasting IPSPs (Chéry and de Koninck, 1999). It has been possible to demonstrate the release of these neurotransmitters from shared vesicles by using strychnine (a glycine receptor blocker) and bicuculline (a GABA receptor blocker) to isolate fast glycine-receptor-mediated and slow GABA-receptor-mediated components of mIPSCs, as shown in Fig. 21.4. Since mIPSCs represent the spontaneous release of single vesicles, it follows that a compound GABA/glycine mIPSC represents the release of a single vesicle containing both neurotransmitters (Jonas et al., 1998; Russier et al., 2002). Synapses that coexpress these neurotransmitters have

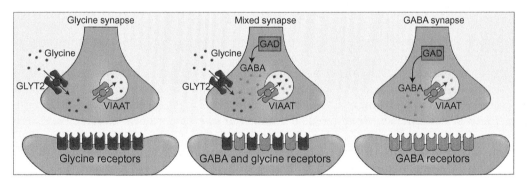

FIGURE 21.3 Glycine and GABA can be efficiently coexpressed in inhibitory terminals and packaged together into vesicles by their shared vesicular transporter, VIAAT. (Left) A glycine synapse. Glycine (*purple dots*) is transported into the nerve terminal from the extracellular space by the glycine transporter GLYT2. It is then packaged into a synaptic vesicle by the vesicular inhibitory amino acid transmitter VIAAT. The postsynaptic receptor site contains glycine receptors. (Center) A mixed GABA/glycine synapse. Glycine is again transported into the terminal by GLYT2. GABA (*orange dots*) is synthesized by the enzyme GAD. Both neurotransmitters are packaged into the same vesicle by VIAAT. The postsynaptic receptor site contains both GABA and glycine receptors, so it is sensitive to the corelease of GABA and glycine from the mixed inhibitory presynaptic terminal. (Right) A GABA synapse. GABA is synthesized by GAD and packaged into a vesicle by VIAAT. The postsynaptic receptor site contains GABA receptors.

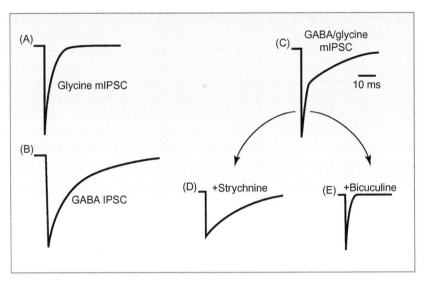

FIGURE 21.4 **Mixed GABA/glycine mIPSCs contains fast and slow components that can be isolated by receptor blockers.** The observation of mixed mIPSCs at a synapse, combined with the use of receptor blockers to identify their components, provides evidence of corelease of inhibitory neurotransmitters from shared vesicles. (A) The typical shape of a fast, transient mIPSC mediated by glycine receptors. (B) The typical shape of a slower, longer-lasting mIPSC mediated by GABA receptors. (C) The typical shape of a compound GABA/glycine mIPSC, which contains fast and slow components. This compound mIPSC is caused by the spontaneous release of a vesicle containing both GABA and glycine at a synapse containing both GABA and glycine receptors. (D) The slow component of the GABA/glycine mIPSC can be isolated by applying the glycine receptor blocker strychnine. (E) The fast component of the GABA/glycine mIPSC can be isolated applying the GABA receptor blocker bicuculine.

been shown to release a combination of vesicles containing GABA alone, glycine alone, or GABA and glycine together (Katsurabayashi et al., 2004).

While the release of two inhibitory neurotransmitters at the same synapse may seem redundant, there is substantial evidence of the functional importance of GABA/glycine corelease. For example, because glycine's fast inhibition complements GABA's slow inhibition, the two neurotransmitters together inhibit the target cell more strongly than either does alone. Importantly, this does not mean that a GABA/glycine-mediated IPSP is necessarily equivalent to a GABA-mediated and glycine-mediated IPSP added together. In brainstem abducens motor neurons that control movement of the eye, glycine and GABA have each been found to regulate each other's effects on postsynaptic cells by various mechanisms. For instance, glycine's fast hyperpolarization of the postsynaptic membrane reduces the driving force on $Cl^-$, which reduces the effect of ionotropic GABA receptor activation on the postsynaptic potential, thereby modifying the time course of the IPSP. The shape of the dual GABA/glycinergic IPSPs depends upon the relative contribution of each neurotransmitter to the IPSP, which is a function of the amount of each neurotransmitter released and the relative levels of GABA and glycine receptor expression (Russier et al., 2002). Therefore, in these neurons, GABA/glycine corelease both enhances inhibition and regulates the shape of the IPSP in these neurons, allowing more reliable and complex signaling.

GABA/glycine coregulation of IPSP characteristics has also been observed in the auditory circuits of the brainstem. One of the important tasks of the auditory system is to determine from which direction a sound originated. To do this, the auditory system takes advantage of the fact that the timing and intensity of a sound differ slightly between each ear, depending on which ear is closer to the sound's source. Brainstem nuclei specialized to integrate auditory signals receive excitatory input from the **ipsilateral** ear and integrate it with inhibitory input from the **contralateral** ear, and precise timing of these excitatory and excitatory inputs is required in order to compute differences in sounds arriving at both ears (Grothe et al., 2010). In adult animals, neurons of the medial nucleus of the trapezoid body (MNTB, an auditory brainstem nucleus), which carry information about sound intensity, trigger IPSCs that are mediated only by glycine receptors (Awatramani et al., 2005). Therefore, it was previously believed that these neurons only release glycine in adulthood.

However, there is evidence that adult MNTB neurons do release some GABA alongside glycine, and that IPSCs in their postsynaptic targets decay much faster when mediated by GABA and glycine together than when mediated by glycine alone. This faster decay greatly increases the temporal precision of the inhibitory signal (Lu et al., 2008). These effects of GABA are not mediated by GABA receptors, but by GABA binding to the glycine-binding subunit of glycine receptors, because GABA is a weak, quickly deactivating agonist at glycine receptors (Fucile et al., 1999; De Saint Jan et al., 2001). In other words, the corelease of GABA modulates the glycinergic IPSP. Although this phenomenon has not been explored at all GABA/glycine coreleasing synapses, Lu and colleagues suggest that it is possible—given GABA's affinity for a widely expressed glycine-binding subunit—that GABA corelease represents a widespread mechanism for regulating the timing of inhibitory transmission.

Regulation of IPSP shape is not the only function of GABA/glycine corelease. In the cerebellum, Golgi cells were found to corelease GABA and glycine onto two distinct populations of postsynaptic neurons, each of which is sensitive either only to GABA or only to glycine (Dugué et al., 2005). Therefore, even though GABA and glycine are not segregated in separate vesicle pools or at separate release sites, the effects of neurotransmitter release by Golgi cells are separated based on the type of target cell. Another functional advantage of corelease can be observed in a population of spinal inhibitory interneurons that corelease GABA and glycine onto postsynaptic GABA and glycine receptors, but also express GABA **autoreceptors** at their own axon terminals. The GABA released at these terminals can bind to the autoreceptors and induce short-term depression of the synapse, suggesting a role of GABA corelease in feedback inhibition of both glycine and GABA release from neurons (Chéry and De Koninck, 2000).

## Vesicular Synergy as a Function of Neurotransmitter Corelease

In addition to influencing the pre- or postsynaptic membrane potential, some coreleased neurotransmitters can affect the availability of their coneurotransmitter for synaptic release. This phenomenon has been observed in a population of neurons in the rodent striatum that were originally identified as cholinergic, but also release glutamate. The synaptic vesicles of these neurons contain both the vesicular acetylcholine transporter (VAChT) and the vesicular glutamate transporter (VGLUT). VGLUT's primary function is to mediate the packaging of glutamate into vesicles; however, it also enhances the packaging of acetylcholine via VAChT, in a phenomenon termed "vesicular synergy." When glutamate is present, uptake of acetylcholine by VAChT/VGLUT-coexpressing vesicles is increased. Conversely, knocking out the VGLUT subtype expressed in these neurons impaired the packaging and release of acetylcholine in the mouse striatum, and this impairment of striatal cholinergic transmission causes hyperactive behavior (Gras et al., 2008).

It has been proposed that when glutamate is pumped into vesicles, it indirectly causes an increase in the pH gradient across the vesicle membrane, allowing VAChT to more efficiently exchange protons for acetylcholine molecules and achieve a higher maximum intravesicular concentration of acetylcholine. This may add to the complexity of vesicular pumping of neurotransmitters discussed in Box 16.2 (Chapter 16). If this proposed mechanism is correct, then vesicular synergy is only possible under corelease conditions, when the vesicular transporters for each neurotransmitter are colocalized in synaptic vesicles. Vesicular synergy has also been hypothesized to occur in neurons expressing a dual serotonergic/glutamatergic phenotype, and there is evidence that glutamate coexpression may enhance dopamine packaging as well (Hnasko et al., 2010; Amilhon et al., 2010).

## Purinergic Cotransmission

ATP plays a variety of roles in synaptic transmission in its capacity as an energy source. As was discussed in earlier chapters, packaging of neurotransmitters into synaptic vesicles relies upon a proton gradient which is established by an ATP-fueled proton pump (see Chapters 16–18), and ATP is also required for the disassembly of SNARE complexes after vesicle fusion (see Chapter 8). However, ATP is also packaged into synaptic vesicles (using a vesicular nucleotide transporter) and released upon vesicle fusion. In fact, ATP is most likely present in *every* synaptic vesicle, though its concentration in vesicles and its contribution to signaling vary (Abbracchio et al., 2009).

The idea that ATP could act as a neurotransmitter was initially controversial. Since it is so ubiquitous and is involved in so many metabolic processes in cells, many researchers thought it unlikely that ATP could be useful as a signaling molecule (Burnstock, 2006). In reality, ATP may have been one of the earliest extracellular messengers, since it has been found to have extracellular signaling functions even in primitive organisms like bacteria (Burnstock, 1996).

It is now known that ATP can transmit signals by binding to purine receptors. There are two classes of purine receptors: P1, which bind the ATP breakdown product adenosine (discussed below), and P2, which bind ATP. The P2 class of receptors includes both ionotropic (P2X) and metabotropic (P2Y) receptors. These receptors are widely expressed in the body, with many types of cells likely expressing one or more P2 receptor subtypes

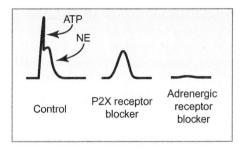

FIGURE 21.5 **The biphasic contractile response in vascular smooth muscle contraction is mediated by ATP and norepinephrine cotransmission.** (Left) Typical shape of the biphasic contraction in a vascular smooth muscle cell after sympathetic nerve firing. This contractile response includes a fast, transient "twitch" response and a slower, longer-lasting response. (Center) Applying a P2X receptor blocker eliminates the twitch component, suggesting that this component is mediated by ATP signaling. (Right) Applying an adrenergic receptor blocker in addition to the P2X receptor blocker eliminates the remaining slow component of the contractile response, suggesting that this component is mediated by norepinephrine signaling.

(Burnstock and Knight, 2004). Therefore, it is no surprise that ATP, once thought to be released only by a distinct population of "purinergic nerves" (Burnstock, 1972), has been identified as a coneurotransmitter at a variety of synapses. Depending on the synapse, ATP can be coexpressed with other small-molecule neurotransmitters, neuropeptides, or both (Burnstock, 2006).

ATP is widely used as a neurotransmitter in both the peripheral and central nervous systems. It is present in cholinergic, dopaminergic, GABAergic, dopaminergic, and glutamatergic synaptic vesicles in various regions of the CNS (Burnstock, 2006). It has also been extensively characterized as a coneurotransmitter in the sympathetic, parasympathetic, and enteric divisions of the autonomic nervous system (Lundberg, 1996; Hoyle, 1996). Examples of ATP cotransmission in the autonomic nervous system are discussed below:

In sympathetic nerves that innervate the smooth muscle of blood vessels and mediate vasoconstriction, ATP is frequently coexpressed with norepinephrine and neuropeptide Y in what is called the "sympathetic triad." While neuropeptide Y has a primarily neuromodulatory role at these synapses, potentiating the effects of ATP and norepinephrine (Ekblad et al., 1984; Saville et al., 1990), both ATP and norepinephrine induce vascular smooth muscle contraction by raising the intracellular calcium concentration. As shown in Fig. 21.5, ATP mediates a fast, transient contractile response (also called the "twitch" response) by binding to ionotropic P2X receptors, which are ligand-gated cation channels that allow calcium influx while simultaneously causing a rapid subthreshold depolarization called an **excitatory junction potential**. Repeated firing of the sympathetic neuron may cause temporal summation (see Chapter 14) of the excitatory junction potentials to depolarize the membrane enough to activate voltage-gated calcium channels, allowing additional calcium to enter the smooth muscle cell. Norepinephrine, meanwhile, mediates a slower and longer-lasting contractile response by binding to G-protein-coupled adrenoceptors on smooth muscle cells, initiating a downstream pathway that leads to the release of calcium from intracellular stores (Lundberg, 1996; Huidobro-Toro and Donoso, 2004; Ralevic and Dunn, 2015).

One advantage of this multiple-neurotransmitter phenotype is that norepinephrine and ATP can function synergistically, meaning that their combined effect on vascular smooth muscle contraction is stronger than the sum of their individual effects (Ralevic and Burnstock, 1990; Smith and Burnstock, 2004). There is also substantial variation in the ratio of the purinergic and noradrenergic contributions to the postsynaptic response. ATP plays a larger role (compared to norepinephrine) in smaller blood vessels, whose arterial pressure is higher (Gitterman and Evans, 2001; Rummery et al., 2007). In addition, whereas the ratio of ATP to norepinephrine release is higher following lower-frequency, lower-duration stimuli, norepinephrine release predominates in response to longer stimulus trains (Kennedy et al., 1986; Sjöblom-Widfeldt and Nilsson, 1990), suggesting that ATP and norepinephrine may be stored in separate vesicle populations that are sensitive to different stimulation patterns (Todorov et al., 1996; Todorov et al., 1999). The use of two vasoconstrictive neurotransmitters with distinct response **kinetics**, whose effects on vascular smooth muscle can be differentially regulated, may increase the precision with which sympathetic neurons can control the timing and degree of vasoconstriction.

ATP also acts as a cotransmitter in some parasympathetic nerves, particularly in those that innervate the urinary bladder. Both acetylcholine and ATP mediate the contraction of the detrusor muscle, which releases urine from the bladder. At this synapse, acetylcholine acts via muscarinic receptors, while ATP acts via P2X receptors (Burnstock, 2014). In healthy human bladders, acetylcholine is overwhelmingly responsible for mediating detrusor muscle contraction, and the contribution of purinergic transmission is close to negligible. However, in the aging bladder, ATP transmission increases while acetylcholine transmission decreases (Yoshida et al., 2004). High levels of purinergic transmission are also associated with overactive bladder and other urinary pathologies (Burnstock, 2014). Therefore, the purinergic component of multiple neurotransmitter expression in parasympathetic nerves appears to be strongly associated with bladder dysfunction in humans. This makes purinergic transmission a potential therapeutic target in urinary disease, meaning that drug treatments designed to modify purinergic transmission may help patients suffering from bladder overactivity (Burnstock, 2011).

Adenosine, another purinergic molecule, is an unusual neurotransmitter because it is not synthesized within the presynaptic neuron or loaded into synaptic vesicles. Instead, it is generated extracellularly in the synaptic cleft after the release of ATP, violating some of the general expectations about neurotransmitters that we outlined in Chapter 15. Normally, ATP is removed from the synaptic cleft by transmembrane enzymes called ecto-ATPases that hydrolyze released ATP to AMP. However, other **ecto-nucleotidases** present in the membranes of some postsynaptic cells can catalyze the conversion of AMP to adenosine. Since most cells that participate in purinergic transmission express both P1 and P2 receptors, the newly synthesized adenosine can then act on postsynaptic P1 receptors, while ATP that has not yet been hydrolyzed acts on postsynaptic P2 receptors (Matsuoka and Ohkubo, 2004). Adenosine therefore participates in an unusual type of cotransmission (shown in Fig. 21.6), wherein only one neurotransmitter—ATP—is actually released from the presynaptic cell, but both ATP and adenosine are involved in neurotransmission at the same synapse.

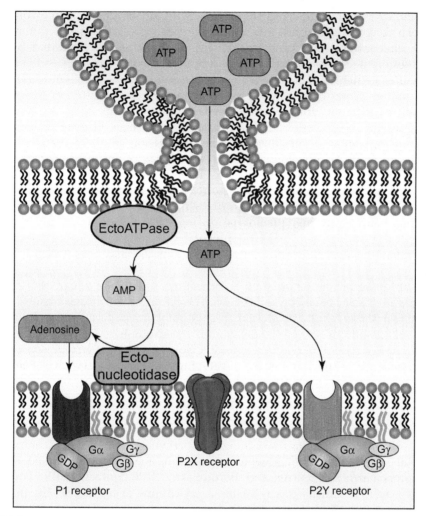

FIGURE 21.6  **ATP and adenosine cotransmission at a purinergic synapse.** ATP is released from the presynaptic terminal and into the synaptic cleft by vesicle fusion. Some ATP molecules diffuse across the cleft to act on ionotropic P2X or metabotropic P2Y receptors. Other ATP molecules are hydrolyzed first to AMP by a transmembrane enzyme called an ectoATPase, then to adenosine by a different ectonucleotidease in the postsynaptic membrane. Adenosine acts on metabotropic P1 receptors. In this way, the presynaptic neuron communicates to the postsynaptic neuron with both ATP and adenosine, even though it only releases ATP.

# NEUROTRANSMITTER SPECIFICATION AND SWITCHING

## The Neurotransmitter Phenotype of a Synapse can Change During Development

Dale wrote in 1935 that evidence available at the time "appear[ed] to indicate that the nature of the chemical function, whether cholinergic or adrenergic, is characteristic for each particular neurone, and is unchangeable." At the time, acetylcholine and epinephrine were the only known neurotransmitters, so this was not an unreasonable conclusion. It has since been discovered that just as a neuron can express two or more different neurotransmitters simultaneously, it can also express different neurotransmitters at different times or in different environmental or physiological contexts. In a phenomenon called **neurotransmitter switching**, neurons can switch from expressing one neurotransmitter to expressing another, gain expression of an additional neurotransmitter, or lose expression of one or more of their multiple neurotransmitters (Spitzer, 2015, 2017). Neurotransmitter switching is a complex process that entails long-term changes in gene expression, since functional transmission of a substance requires expression of all of the unique protein machinery necessary to acquire the substance (e.g., uptake transporters for the substance or its precursor, synthetic enzymes), prepare it for release (e.g., vesicular transporters), and terminate its action (e.g., reuptake transporters, inactivating enzymes).

Neurotransmitter switching events often occur in the developing animal, with neurons expressing one neurotransmitter phenotype during early development and switching to another phenotype as they mature. Developmental switching into the appropriate mature neurotransmitter phenotypes is critical for correct synapse and neural circuit formation. Neurotransmitter switching can also function as a mechanism of long-term plasticity in either the developing or the mature animal, occurring in response to experimental manipulation, external stimuli, behavior, or disease. These adaptive neurotransmitter switching events often switch the synapse from excitatory to inhibitory, or vice versa (Spitzer, 2017). In summary, the ability of neurons to change their neurotransmitter phenotypes is critical to nervous system development and plasticity. Below, we will cover representative examples of neurotransmitter switching in the developing and mature nervous system.

## Neurotransmitter–Receptor Matching

In most cases, neurotransmitter switching in a presynaptic neuron is accompanied by a matching change in postsynaptic receptor expression, a phenomenon called neurotransmitter–receptor matching (Spitzer and Borodinsky, 2008; Spitzer, 2012). For example, when a presynaptic neuron begins releasing acetylcholine at a synapse, the postsynaptic cell upregulates expression of acetylcholine receptors at that synapse (Borodinsky and Spitzer, 2007). Neurotransmitter–receptor matching is important because, if receptor expression did not change in response to neurotransmitter switching, the postsynaptic cell would be insensitive to the new neurotransmitter signal and the synapse would not be functional. Both the presynaptic switch to acetylcholine release and the postsynaptic switch to acetylcholine receptor expression are necessary to produce a functional, newly cholinergic synapse.

Mechanisms that underlie neurotransmitter–receptor matching have not been well characterized in most systems. However, there is evidence to support a model in which cells express receptors for multiple neurotransmitters and signaling molecules secreted by presynaptic neurons trigger the upregulation of receptors matching their neurotransmitter phenotype. Borodinsky and Spitzer (2007) found that in an embryonic frog neuromuscular model system (*Xenopus leavis*), muscle cells express receptors for acetylcholine, glutamate, GABA, and glycine prior to their innervation by cholinergic motor neurons. Following cholinergic innervation, acetylcholine receptor expression increases, while the other classes of receptor are eliminated. Borodinsky and Spitzer also found that experimentally inducing neurotransmitter switching in motor neurons caused the upregulation of receptors matching the new neurotransmitter phenotypes, resulting in the formation of aberrant, noncholinergic neuromuscular synapses.

The neurotransmitter−receptor matching process observed in *Xenopus* appeared to rely upon unknown diffusible signaling molecules released by presynaptic neurons in response to electrical activity. What might these signaling molecules be, and how do they coordinate postsynaptic receptor expression with presynaptic neurotransmitter phenotype? It is possible that the same paracrine signals regulate both presynaptic neurotransmitter switching and postsynaptic receptor matching. Alternatively, the neurotransmitters themselves may promote expression of their receptors at postsynaptic sites. This second hypothesis is supported by the finding that in *Xenopus* muscle cells, activating the ionotropic glutamate receptors expressed early in development is sufficient to upregulate their expression (Spitzer, 2017). Regardless of the mechanism, postsynaptic cells are clearly able to adapt to changes in presynaptic neurotransmitter phenotype in order to establish and maintain functional synapses.

## Neurotransmitter Switching in the Developing Nervous System

Neurons must complete three major developmental steps in order to be correctly integrated into mature neural circuits. They must proliferate via the division of neuronal precursor cells; they must migrate to the correct area of the body and innervate their target; and they must differentiate into their final neuronal subtype, adopting the **morphological** characteristics and neurotransmitter phenotype appropriate to their function in the nervous system (Francis and Landis, 1999). To ensure that neurons express the correct neurotransmitter phenotypes in maturity, neuronal differentiation is regulated by genetic "blueprints," extracellular signals (such as neurotrophic factors), electrical activity, and sensory stimuli (Spitzer, 2006; Dulcis and Spitzer, 2008; Spitzer, 2015).

For various reasons, a neuron may express a different phenotype early in development than it does in adulthood (Spitzer, 2015). Some neurons express an early neurotransmitter phenotype that serves an important functional role in development and is later replaced with a different, mature phenotype. Other neurons initially express a variety of neurotransmitters, and the type of tissue they innervate determines which of these neurotransmitters they will retain in maturity. In some cases, the mature phenotype of a neuron is determined by external stimuli the organism experiences during development, allowing early adaptation to the environment. Several examples of developmental neurotransmitter switching are discussed below.

## Developmental Neurotransmitter Switching: The Noradrenergic-to-Cholinergic Switch in Sympathetic Neurons

Studies of sympathetic neurons in vitro provided some of the earliest evidence that neurons are capable of changing the type of neurotransmitter they release. When neurons from the superior cervical ganglion (SCG) of newborn rats are grown in the absence of any other type of cell, they have a noradrenergic phenotype and are capable of functional norepinephrine transmission. However, when the same neurons are grown with cardiac muscle cells, they undergo a striking change. Although they initially form excitatory noradrenergic synapses on each other and on the myocytes, monitoring these neurons over time shows that they gradually switch to a cholinergic phenotype, a phenomenon called "cholinergic switch." As part of this switching process, the cultured neurons express the proteins necessary for the uptake, synthesis, storage, and release of acetylcholine, and they form functional inhibitory cholinergic synapses on the myocytes (Landis, 1976; Furshpan et al., 1976; Landis, 1990). During the transition to the cholinergic phenotype, norepinephrine and acetylcholine are coexpressed and packaged into separate vesicles, with acetylcholine increasingly predominating over norepinephrine until the noradrenergic phenotype disappears completely (Potter et al., 1980).

In addition, the shift to the cholinergic phenotype is accompanied by a change in neuropeptide expression. VIP, which is cotransmitted with acetylcholine in parasympathetic neurons, is upregulated. Meanwhile, neuropeptide Y, which is cotransmitted with norepinephrine in noradrenergic sympathetic neurons, is downregulated (Nawa and Sah, 1990; Nawa and Patterson, 1990; Fahrenkrug and Hannibal, 2004). The fact that neuropeptide expression changes to "match" the new phenotype suggests that multiple transmitter release is important enough to autonomic nervous system function that there are mechanisms in place to couple acetylcholine and norepinephrine with appropriate coneurotransmitters.

Even after the discovery that neurons can undergo neurotransmitter switching in cell culture, the question remained: Does neurotransmitter phenotype ever change in living organisms? Is this capacity for neurotransmitter switching important to our understanding of nervous system function, or does it only occur when we manipulate neurons in a dish? To answer this question, neurotransmitter switching had to be observed in a living animal (in vivo).

There is indeed evidence that cholinergic switch occurs in vivo, as a normal step in the development of certain parts of the sympathetic nervous system (Landis, 1990; Apostolova and Dechant, 2009). Interestingly, it can also occur in the mature nervous system as a disease symptom, a phenomenon that will be discussed later in the chapter. This switching process was first observed and has been most extensively characterized in rodent sudomotor neurons, the postganglionic sympathetic neurons innervating the sweat gland.

Most mature postganglionic sympathetic neurons are noradrenergic (see Chapter 16), and rodent sudomotor neurons are initially capable of releasing norepinephrine (Habecker and Landis, 1994), express markers of both noradrenergic and cholinergic function (Schütz et al., 2008), and lack neuropeptides (like VIP) that are associated with cholinergic transmission (Landis et al., 1988). However, unlike other sympathetic neurons, rodent sudomotor neurons transition to a parasympathetic-like cholinergic phenotype in the weeks after birth (Landis and Keefe, 1983; Landis, 1990; Schotzinger et al., 1994). The mature neurons still express much of the machinery involved in norepinephrine transmission, such as the catecholamine synthetic enzyme tyrosine hydroxylase, as shown in Fig. 21.7. However, the vesicular monoamine transporter (VMAT) is downregulated, so that norepinephrine can no longer be packaged

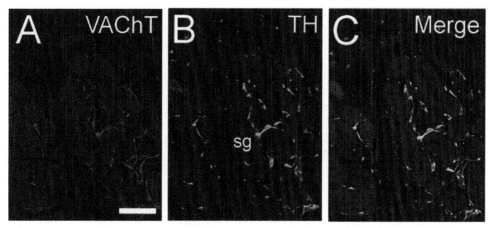

FIGURE 21.7 **Cholinergic and noradrenergic markers are coexpressed in mouse sudomotor nerve terminals even after cholinergic switch.** Although mature sudomotor neurons do not express VMAT and therefore do not release norepinephrine, they retain some protein markers associated with noradrenergic transmission. (A) Red fluorescent staining for the vesicular choline transporter VAChT in adult mouse sweat gland. VAChT is localized to sudomotor nerve terminals. (B) Green fluorescent staining for the catecholamine synthetic enzyme TH in the same mouse sweat gland preparation. TH is localized to sudomotor nerve terminals. (C) Fluorescence shown in (A) and (B) is merged to show colocalization of VAChT and TH at sudomotor nerve terminals. *Source: Used with permission from Weihe, E., Schütz, B., Hartschuh, W., Anlauf, M., Schäfer, M.K., Eiden, L.E., 2005. Coexpression of cholinergic and noradrenergic phenotypes in human and nonhuman autonomic nervous system. J. Comp. Neurol. 492(3), 370–379.*

into vesicles and noradrenergic transmission can no longer occur (Weihe et al., 2005). The loss of VMAT is reflected in the gradual disappearance of noradrenergic vesicles from rodent sudomotor axon terminals, as depicted in Fig. 21.8 (Landis and Keefe, 1983). Meanwhile, the machinery required for the synthesis, uptake, storage, and release of acetylcholine is upregulated, and the types of neuropeptides expressed change to complement the cholinergic phenotype; for example, VIP is upregulated (Schotzinger et al., 1994; Landis et al., 1988).

A growing body of evidence suggests that the cholinergic switch is target-dependent (meaning it is triggered by factors secreted from the postsynaptic target tissue) and that the potential to undergo this neurotransmitter switch is intrinsic to most or all sympathetic neurons (Apostolova and Dechant, 2009). Sudomotor neurons undergo a cholinergic switch not because of some special property that sets them apart from other sympathetic neurons, but because their target tissue signals them to develop a cholinergic phenotype. This interpretation is supported by experiments in which tissue transplants early in development caused sympathetic neurons to innervate different target tissues than they normally would, and these neurons then developed mature neurotransmitter phenotypes matching the atypical target tissue. For example, when transplanted sweat gland tissue was innervated by nonsudomotor sympathetic neurons that would normally remain noradrenergic as they matured, the neurons became cholinergic (Schotzinger and Landis, 1988); and when sudomotor neurons were forced to innervate a transplanted salivary

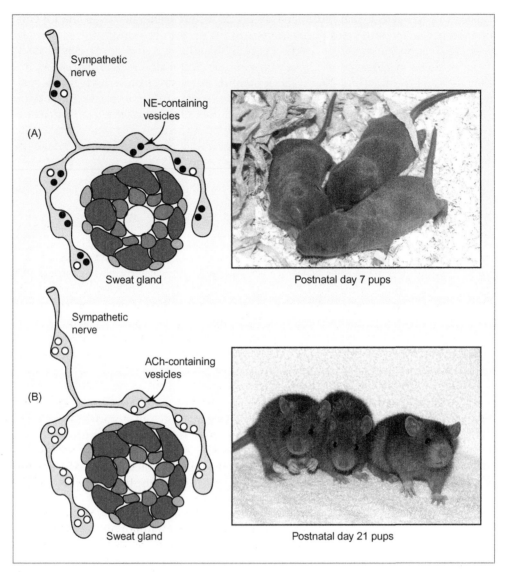

FIGURE 21.8  Rat sudomotor nerve terminals contain many catecholamine (norepinephrine)-containing vesicles for several days after birth, but these vesicles disappear by postnatal day 21. Vesicles that contain catecholamines, like norepinephrine, can be distinguished from cholinergic vesicles by applying a fixative that makes catecholamines visible under an electron microscope (Tomlinson, 1975). Under these conditions, noradrenergic vesicles appear "granular" while cholinergic vesicles appear clear. (A, left) Drawing of a rat sweat gland at postnatal day 7, surrounded by sudomotor nerve terminals that contain many noradrenergic vesicles (*black circles*). (A, right) Representative photograph of rat pups at postnatal day 7. (B, left) Drawing of a rat sweat gland on postnatal day 21, when the granular noradrenergic vesicles have disappeared completely from sudomotor nerve terminals, leaving only the clear cholinergic vesicles (*white circles*). (B, right) Representative photograph of rat pups at postnatal day 21. *Source: Both rat pup photos are of Russian Blue Standard rats, used with permission from http://www.afrma.org/babyratdevdaily.htm, 2011 Karen Robbins.*

gland instead of the sweat gland, they matured into a noradrenergic rather than a cholinergic phenotype (Schotzinger and Landis, 1990).

Interestingly, in primates, the neurons innervating adult eccrine sweat glands contain all of the proteins necessary for cotransmission of acetylcholine and norepinephrine, including VMAT (Weihe et al., 2005). Norepinephrine transmission is responsible for the phenomenon of nonthermogenic (not caused by temperature) "noradrenergic sweating" (Mack et al., 1985). This example of multiple transmitter release, in which the early noradrenergic/cholinergic dual neurotransmitter phenotype apparently persists into adulthood and is functionally important to sympathetic regulation of sweating, appears to be a unique adaptation in humans and other primates.

## Spontaneous Electrical Activity Drives Neurotransmitter Switching in Nervous System Development

In the developing nervous system, spontaneous activity is an important mechanism for mediating developmental neurotransmitter differentiation (Spitzer, 2006; Spitzer, 2012). The role of spontaneous activity in regulating early neurotransmitter switching has been characterized extensively in the embryonic *Xenopus* spinal cord (Rosenberg and Spitzer, 2011). At early stages of development, *Xenopus* spinal neurons widely coexpress GABA and glutamate, but different types of spinal neurons later begin to express different neurotransmitters. For example, later in development, motor neurons are cholinergic; mechanoreceptive neurons in the dorsal spinal cord are glutamatergic; dorsolateral interneurons are glycinergic; ventral interneurons are GABAergic; and dorsal interneurons coexpress GABA and glycine (Guemez-Gamboa et al., 2014). These findings raise the question: How does each type of neuron "know" the correct neurotransmitter phenotype to switch to? In this system, the answer centers on voltage-gated calcium channels.

In previous chapters, voltage-gated calcium channels were primarily discussed in terms of their role in triggering vesicle fusion at the presynaptic terminal. However, machinery that serves one function in the mature nervous system often has a completely different function during development, and voltage-gated calcium channels are no exception. Voltage-gated calcium channels are expressed in the developing nervous system even before synapses are formed, and calcium influx through these channels is an important regulator of neuron proliferation, migration, and neurotransmitter specification (Spitzer et al., 2000, 2004; Rosenberg and Spitzer, 2011). In mature neurons, action potentials are sodium-dependent, meaning that they occur when the neuronal membrane is rapidly depolarized by an influx of sodium ions through sodium channels. In contrast, immature neurons often generate slower, **calcium-dependent action potentials** (or "spikes") mediated by voltage-gated calcium channels (O'Dowd et al., 1988). During a sensitive period in the development of a *Xenopus* embryo (Fig. 21.9), calcium spiking activity causes neurons to release paracrine factors that induce surrounding cells to undergo changes in neurotransmitter phenotype (Borodinsky et al., 2004; Root et al., 2008; Guemez-Gamboa et al., 2014).

In neurons that are sensitive to activity-dependent neurotransmitter switching, calcium spikes are controlled by the paracrine action of GABA and glutamate on metabotropic receptors (Root et al., 2008). The timing and frequency of calcium spiking determine the

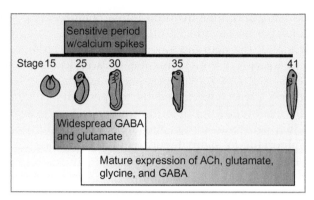

FIGURE 21.9 **Timing of activity-dependent neurotransmitter switching in the *Xenopus laevis* spinal cord.** In early stages of development, GABA and glutamate are widely expressed in the *Xenopus* spinal cord, where they regulate calcium spiking activity through paracrine signaling. During a transient sensitive period, calcium spikes induce different types of spinal neurons to transition into their mature cholinergic, glutamatergic, glycinergic, or GABAergic phenotypes. Representative drawings of *Xenopus* tadpoles show how their morphology changes during the developmental stages when neurotransmitter switching occurs.

neurotransmitter phenotype, via downstream pathways that culminate in increased expression of neurotransmitter synthetic enzymes and transport machinery (Gu and Spitzer, 1995; Rosenberg and Spitzer, 2011). Therefore, neurotransmitter switching in this context is termed "activity-dependent." Different types of neurons experience different patterns of calcium spike activity during the critical period for neurotransmitter specification. These distinct modes of calcium spiking are regulated by the different patterns of calcium channel localization and GABA/glutamate receptor expression in each of these neuron subtypes (Rosenberg and Spitzer, 2011). Inhibiting the expression of GABA or glutamate, or blocking their receptors, alters calcium spike patterns and consequently changes the number of neurons that differentiate into each neurotransmitter phenotype (Root et al., 2008). Therefore, the initial widespread GABA/glutamate expression in the embryonic spinal cord plays a critical role in regulating neurotransmitter specification in development.

Experimenters demonstrated the importance of calcium-dependent electrical activity for neurotransmitter specification during early development by manipulating calcium spiking activity, causing neurons to express the "wrong" neurotransmitter (Borodinsky et al., 2004). For example, non-mechanoreceptive neurons in the developing *Xenopus* spinal cord could be made to express a glutamatergic phenotype. These results indicate that *Xenopus* spinal neurons are not intrinsically "fated" to acquire a particular phenotype, but instead depend upon calcium spiking activity to differentiate correctly.

There is evidence that activity-dependent neurotransmitter switching is a mechanism of **homeostatic plasticity** in *Xenopus* spinal neurons. Homeostatic plasticity is a compensatory change in neuron excitability that helps to keep a network's overall electrical activity within a range appropriate for normal function (see Chapter 14). Notably, changing from excitatory to inhibitory transmission, or vice versa, in response to overactivity or suppression of activity have been observed in a variety of systems (Spitzer, 2012; Demarque and Spitzer, 2012), suggesting that activity-dependent neurotransmitter switching is a widespread and important mechanism of homeostatic plasticity in neurons. Accordingly, in developing *Xenopus* spinal neurons, early calcium spiking activity appears to regulate neurotransmitter switching in a homeostatic manner. Suppressing calcium spiking causes more neurons to switch to an excitatory glutamatergic or cholinergic phenotype, while fewer neurons switch to an inhibitory GABAergic or glycinergic phenotype. Meanwhile, enhancing calcium spiking has the opposite effect (Borodinsky et al., 2004; Guemez-Gamboa et al., 2014). Therefore, the dependence of spinal neuron neurotransmitter specification on calcium spiking activity may allow the developing *Xenopus* spinal cord to maintain a functional range of electrical activity through homeostatic control of neurotransmitter switching. If each class of neuron were "hard-coded" to adopt a particular neurotransmitter phenotype, regardless of the level of activity in early development, the *Xenopus* nervous system would have less flexibility to compensate for excessive or insufficient activity.

## Transient Glutamate Expression in Synaptic Refinement During Development

Neurons in the auditory brainstem receive excitatory input originating at the ipsilateral ear and inhibitory input originating at the contralateral ear, and they "compare" these inputs (representing the difference in sound intensity between the ears) to compute the direction from which the sound originated. Precise timing of excitatory and inhibitory signals is required for auditory processing, and the organization of auditory brainstem circuits needs to be just as precise. Most of the nuclei and pathways in these circuits are tonotopically organized, meaning that sound frequencies ranging from low to high are represented by the spatial organization of these neurons and their synaptic connections (Tollin, 2003). These tonotopic maps are established through a process of activity-dependent synaptic refinement, or synaptic pruning (Kim and Kandler, 2003; Kandler et al., 2009). Like other developing neurons, auditory brainstem neurons form many more synaptic contacts with their target cells than they will need when they mature. Prior to the onset of hearing, patterns of spontaneous neuronal activity generated by the cochlea strengthen a fraction of the synapses, while the rest are eliminated (Clause et al., 2014). The surviving synapses are those that are correctly arranged to encode tonotopic information.

Importantly, successful activity-dependent synaptic refinement in the auditory nuclei of the brainstem requires early expression of a neurotransmitter that is subsequently eliminated in a neurotransmitter switching event (Noh et al., 2010). Synapses in the auditory MNTB → LSO pathway (in which MNTB neurons project to neurons in the lateral superior olive) undergo a developmental switch from primarily GABAergic transmission in the newborn animal to primarily glycinergic transmission in the adult. Similar to the cholinergic switch, the GABA-to-glycine switch involves a transitional period during which neurons release both GABA and glycine, with the proportion of GABA in synaptic terminals and vesicles decreasing over time while the proportion of glycine increases (Nabekura et al., 2004). Early in development, some GABA/glycinergic neurons *also* release glutamate from the same terminals, which activates NMDA receptors, contributing to the early signaling that is critical for proper development. After the period of synaptic refinement ends, the glutamatergic phenotype disappears, since excitatory glutamate cotransmission would interfere with glycinergic inhibition at the adult auditory midbrain synapses. Disrupting early glutamate transmission by knocking out VGLUT in these nerve terminals impairs both the strengthening of "correct" synapses and the pruning of "incorrect" synapses, which in turn impairs the tonotopic organization of the synapses. Therefore, the early presence of glutamate transmission is critical for early synaptic refinement in the auditory brainstem.

How does this early glutamate transmission promote the refinement of auditory midbrain synapses? It is likely that synaptic refinement relies upon NMDA receptor-mediated plasticity (Noh et al., 2010). Recall from Chapter 11, that NMDA receptor function requires the coincident presence of agonists (glutamate and glycine) and membrane depolarization to remove the magnesium block in these receptor channels. During development, the chloride gradient in neurons is altered so that GABA and glycine ionotropic chloride channels are depolarizing (excitatory). Therefore, at these developing synapses, GABA/glycine neurotransmission is capable of removing the $Mg^{2+}$ block of NMDA receptors by depolarizing the membrane, allowing the spontaneous-activity-driven release of GABA, glycine, and glutamate from the same synapse to mediate NMDA receptor-dependent plasticity (which may also explain why GABA is coexpressed with glycine early in development). This plasticity is hypothesized to be critical for synaptic refinement in this circuit. The development of a tonotopic organization in the auditory brainstem showcases the importance of both multiple neurotransmitter release *and* developmental neurotransmitter switching in regulating the development of the extraordinary precision that is characteristic of auditory circuit activity.

## Sensation-Mediated Neurotransmitter Switching

In some cases, activity originating from a sensory stimulus can regulate neurotransmitter switching during development, which helps the developing organism adapt to its environment. When light hits the retina of a *Xenopus* tadpole, it excites retinal ganglion cells, which in turn excite dopaminergic neurons of the ventral suprachiasmatic nucleus (VSC). VSC neurons inhibit melanotrope cells, which regulate skin pigmentation by releasing a hormone (melanocyte-stimulating hormone) that stimulates the release of melanin by melanocytes in the skin (Dulcis and Spitzer, 2008). Inhibition of melanotrope cells therefore favors lighter skin, while excitation of these cells darkens skin. The skin of *Xenopus* tadpoles lightens in response to light (for example, in light-colored environments that reflect a lot of light) and darkens in the absence of light (e.g., in dark-colored environments; see Fig. 21.10), and these changes in skin pigmentation are primarily controlled by dopamine neuron regulation of the melanotrope cells (Ubink et al., 1998). Changing their skin color according to the surrounding light intensity helps *Xenopus* tadpoles to camouflage themselves in both bright and dark environments.

When *Xenopus* tadpoles are raised in bright light, VSC neurons expressing a GABA/neuropeptide Y phenotype undergo neurotransmitter switching and begin releasing dopamine, enhancing inhibition of melanocyte-stimulating hormone release from melanotrope cells. The dopaminergic phenotype of these VSC neurons disappears when the tadpoles are exposed to a dark environment, indicating that the dopaminergic switch is reversible. Suppressing electrical activity in the VSC blocks light-induced upregulation of the dopaminergic phenotype, which suggests that the dopaminergic switch is activity-dependent (Dulcis and Spitzer, 2008). The ability of *Xenopus* tadpoles to adapt their skin pigmentation

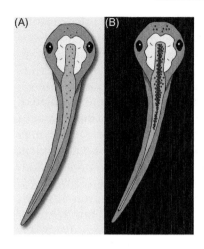

FIGURE 21.10 *Xenopus* **tadpole pigmentation changes in response to surrounding light levels.** The production of melanin by *Xenopus* skin melanocytes varies in an adaptive manner based on the background light intensity. *Xenopus* skin lightens for better camouflage in brightly lit or light-colored environments and darkens for better camouflage in dimly lit or dark-colored environments. (Left) Drawing of a *Xenopus* tadpole raised on a light-colored background. Few pigmented (melanin-producing) melanocytes are visible, represented by black dots. (Right) Drawing of a *Xenopus* tadpole raised on a dark background. Many of the tadpole's melanocytes are pigmented, making its body appear darker.

to match their surroundings is a useful example of how neurotransmitter switching can act as a mechanism of synaptic plasticity.

## Neurotransmitter Switching in the Adult Nervous System

As neurons mature, they often lose their sensitivity to stimuli that cause developmental neurotransmitter switching. However, under the right circumstances, neurotransmitter switching can occur in the adult nervous system, as an adaptation to changes in the environment or as a symptom of disease. Some examples are described below.

## "Stimulus-mediated" Neurotransmitter Switching in Mature Neurons

During winter, when the hours of daylight are shorter, many people suffer from a form of depression called seasonal affective disorder (SAD). Similar changes in daily **photoperiod** (the length of time that an organism is exposed to light) have been found to affect anxiety- and depression-like behaviors in rodents (Workman and Nelson, 2011).

The changes in behavior and physiology that occur in response to stressors (such as excessively long or short photoperiods) are collectively referred to as the **stress response**. The stress response is initiated by the release of corticotropin-releasing hormone (CRF) from neurons of the paraventricular nucleus of the hypothalamus (PVN), which activates other brain regions involved in stress (Smith and Vale, 2006). Dulcis et al. (2013) found that exposing rats to long-day photoperiods increases CRF levels in cerebrospinal fluid, while short-day exposure reduces them (remember that unlike humans, rats are nocturnal, so they are active at night and are stressed by long days). Consistent with this finding, rats display more anxiety- and depression-like behavior (such as reduced exploratory activity) after long-day exposure, while short-day exposure has the opposite effect. The photoperiod-dependent increase in CRF release, and the ensuing anxious and depressed behavior associated with the stress response, may be mediated by neurotransmitter switching.

Interneurons in the PVN, which synapse on CRF-releasing neurons, can release either dopamine or the neuropeptide somatostatin. Dulcis et al. (2013) found that exposing rats to long-day photoperiods caused dopaminergic interneurons to switch to a somatostatinergic phenotype, and this was accompanied by a decrease in dopamine receptor expression on CRF-releasing cells. Short-day exposure had the opposite effect: somatostatinergic neurons switched to a dopaminergic phenotype, and dopamine receptor expression increased. Furthermore, this neurotransmitter switch appeared to mediate the effects of long and short photoperiods on CRF release and behavior. Similar to developmental neurotransmitter switching in the *Xenopus* spinal cord, photoperiod-dependent switching appears to be activity-dependent, although it is regulated by photoperiod-induced rather than spontaneous changes in electrical activity (Meng et al., 2018).

There is some evidence to suggest that the photoperiod-dependent neurotransmitter switching observed in rodents may also occur in humans. Examining the brain tissue of humans who had died in the summer or winter showed that, in winter, fewer dopaminergic neurons and more nondopaminergic neurons were present in the midbrain, where the PVN is located (Aumann et al., 2016). There was no indication that dopaminergic neurons had died off, suggesting that—as in the rat—human midbrain dopaminergic neurons may switch to another neurotransmitter phenotype in response to stressful changes in photoperiod. Since humans are diurnal, the short days of winter are analogous to the long days that induce neurotransmitter switching in rats.

The discovery that changes in photoperiod can cause neurotransmitter switching, and that this switching phenomenon measurably affects behavior, highlights the importance of neurotransmitter switching as a mechanism of adult nervous system plasticity. In this case, the nervous system is able to use a neurotransmitter switching mechanism to adapt to a seasonal change in environment. By using a neurotransmitter switching mechanism to shift the ratio of dual glutamate/dopamine release from PVN neurons, the nervous system is able to adapt to a seasonal change in environment.

## Neurotransmitter Switching as a Compensatory Mechanism in Disease

Disease, by definition, disrupts the normal function of the body, but the body is exceptionally good at compensating for these disruptions. In the nervous system, the activity of synapses can be altered to compensate for disease effects—for example, by increasing inhibitory transmission to suppress dangerous levels of overactivity (as in seizure), or by upregulating the production of important synaptic proteins to replace those damaged by disease. Recent findings have shown that the nervous system can also exploit neurotransmitter phenotype plasticity as an apparent compensatory mechanism in disease. Below, we will describe an example of compensatory neurotransmitter switching in congestive heart failure, a chronic condition in which the pumping efficiency of the heart becomes too weak to supply the body with sufficient oxygen.

Both sympathetic and parasympathetic neurons innervate the heart, working together to regulate the heartbeat. Sympathetic nervous system activation ("fight or flight") increases heart rate, while parasympathetic nervous system activation slows it. In healthy hearts, sympathetic neurons release norepinephrine, which has an excitatory effect on cardiac muscle; in contrast, parasympathetic neurons release acetylcholine, which has an inhibitory effect on cardiac muscle (see Chapter 16, Box 16.1). Cardiac muscle contains receptors for both of these neurotransmitters. In congestive heart failure, as cardiac output (the amount of blood pumped per minute) decreases, sympathetic activity increases to help restore normal output. While this can improve cardiac output in the short term, chronic overstimulation of the cardiac muscle β-adrenergic norepinephrine receptors damages these cells and weakens the heart further (Floras, 1993; Triposkiadis et al., 2009; Lymperopoulos et al., 2013).

Paradoxically, the overall amount of norepinephrine present in heart tissue decreases even as sympathetic activity increases in congestive heart failure (Chidsey et al., 1964). Kanazawa et al. (2010) investigated this strange phenomenon in a rodent model of congestive heart failure and turned up a surprising explanation: Many of the sympathetic neurons innervating the heart had switched from a noradrenergic to a cholinergic phenotype, with some neurons expressing both neurotransmitters. Consistent with these observations in rats, cholinergic sympathetic neurons were discovered in the heart tissue of human patients who had died of congestive heart failure, suggesting that this neurotransmitter switch also occurs in humans with this disease. Interestingly, there is evidence that the cholinergic switch in congestive heart failure is controlled by the same mechanism (whose details are beyond the scope of this text) that induces the developmental cholinergic switch in sudomotor neurons (Geissen et al., 1998; Kanazawa et al., 2010). This suggests that the developmental cholinergic switch mechanism is conserved in adult sympathetic neurons, even though they are normally not exposed to stimuli that would trigger the switch.

Reduced β-adrenergic receptor activation and increased cholinergic signaling, both of which are effects of cholinergic switch in cardiac sympathetic neurons, have been found to improve patient outcomes in heart failure (Merit-HF Study Group, 1999; Rocha-Resende et al., 2012; Roy et al., 2015). This is why β-adrenergic receptor antagonists, or "beta blockers," are commonly prescribed for heart failure patients. Interestingly, Kanazawa et al. (2010) found that mice with an impaired ability to undergo cholinergic switch are less likely than wild-type control mice to survive heart failure induced by chronic **hypoxia** (oxygen deprivation). This result suggests that the cholinergic switch in cardiac sympathetic neurons may have a cardioprotective effect in heart failure, possibly as a compensatory response to mitigate noradrenergic signaling by overactive sympathetic neurons.

## SUMMARY

When the field of neurotransmission was in its infancy, the doctrine of "one neuron, one neurotransmitter" limited our understanding of the nuances of nervous system signaling. The field has since progressed from this early doctrine to widespread acknowledgment that most neurons release multiple neurotransmitters, and that neurons can change the types of neurotransmitter they release as a mechanism of synaptic plasticity. In this chapter, we covered representative examples of multiple neurotransmitter release and neurotransmitter switching, and described how these phenomena enhance the complexity and adaptability of nervous system circuitry. Further, we discussed how the transient expression of specific neurotransmitters in developing neurons can aid in refining circuits, how maturing neurons can adopt a neurotransmitter phenotype appropriate for their target tissue or environment, and how neurotransmitter switching in adulthood can act as an adaptive response to environmental or physiological changes, or a response to disease.

# References

Abbracchio, M.P., Burnstock, G., Verkhratsky, A., Zimmermann, H., 2009. Purinergic signalling in the nervous system: an overview. Trends Neurosci. 32 (1), 19–29.

Amilhon, B., Lepicard, E., Renoir, T., Mongeau, R., Popa, D., Poirel, O., et al., 2010. VGLUT3 (vesicular glutamate transporter type 3) contribution to the regulation of serotonergic transmission and anxiety. J. Neurosci. 30 (6), 2198–2210.

Apostolova, G., Dechant, G., 2009. Development of neurotransmitter phenotypes in sympathetic neurons. Auton. Neurosci. 151 (1), 30–38.

Aumann, T.D., Raabus, M., Tomas, D., Prijanto, A., Churilov, L., Spitzer, N.C., et al., 2016. Differences in number of midbrain dopamine neurons associated with summer and winter photoperiods in humans. PLoS ONE 11 (7), e0158847.

Awatramani, G.B., Turecek, R., Trussell, L.O., 2005. Staggered development of GABAergic and glycinergic transmission in the MNTB. J. Neurophysiol. 93 (2), 819–828.

Benarroch, E.E., 1994. Neuropeptides in the sympathetic system: presence, plasticity, modulation, and implications. Ann. Neurol. 36 (1), 6–13.

Bernheim, S.M., Mayeri, E., 1995. Complex behavior induced by egg-laying hormone in Aplysia. J. Comp. Physiol. A 176 (1), 131–136.

Bhumbra, G.S., Beato, M., 2018. Recurrent excitation between motoneurones propagates across segments and is purely glutamatergic. PLoS Biol. 16 (3), e2003586.

Borodinsky, L.N., Spitzer, N.C., 2007. Activity-dependent neurotransmitter-receptor matching at the neuromuscular junction. Proc. Natl. Acad. Sci. USA 104 (1), 335–340.

Borodinsky, L.N., Root, C.M., Cronin, J.A., Sann, S.B., Gu, X., Spitzer, N.C., 2004. Activity-dependent homeostatic specification of transmitter expression in embryonic neurons. Nature 429 (6991), 523–530.

Burnstock, G., 1972. Purinergic nerves. Pharmacol. Rev. 24 (3), 509–581.

Burnstock, G., 1976. Do some nerve cells release more than one transmitter? Neuroscience 1 (4), 239–248.

Burnstock, G., 1996. Purinoceptors: ontogeny and phylogeny. Drug. Dev. Res. 39 (3–4), 204–242.

Burnstock, G., 2006. Historical review: ATP as a neurotransmitter. Trends Pharmacol. Sci. 27 (3), 166–176.

Burnstock, G., 2011. Therapeutic potential of purinergic signalling for diseases of the urinary tract. BJU Int. 107 (2), 192–204.

Burnstock, G., 2014. Purinergic signalling in the urinary tract in health and disease. Purinergic. Signal. 10 (1), 103–155.

Burnstock, G., Knight, G.E., 2004. Cellular distribution and functions of P2 receptor subtypes in different systems. Int. Rev. Cytol. 240, 31–304.

Campbell, G., 1987. Cotransmission. Annu. Rev. Pharmacol. Toxicol. 27 (1), 51–70.

Chéry, N., de Koninck, Y., 1999. Junctional versus extrajunctional glycine and GABA(A) receptor-mediated IPSCs in identified lamina I neurons of the adult rat spinal cord. J. Neurosci. 19 (17), 7342–7355.

Chéry, N., De Koninck, Y., 2000. GABA(B) receptors are the first target of released GABA at lamina I inhibitory synapses in the adult rat spinal cord. J. Neurophysiol. 84 (2), 1006–1011.

Chidsey, C.A., Kaiser, G.A., Sonnenblick, E.H., Spann, J.F., Braunwald, E., 1964. Cardiac norepinephrine stores in experimental heart failure in the dog. J. Clin. Invest. 43 (12), 2386–2393.

Clause, A., Kim, G., Sonntag, M., Weisz, C.J., Vetter, D.E., Rübsamen, R., et al., 2014. The precise temporal pattern of prehearing spontaneous activity is necessary for tonotopic map refinement. Neuron 82 (4), 822–835.

Conn, P.J., Kaczmarek, L.K., 1989. The bag cell neurons of Aplysia. A model for the study of the molecular mechanisms involved in the control of prolonged animal behaviors. Mol. Neurobiol. 3 (4), 237–273.

Dale, H., 1935. Pharmacology and nerve endings. Proc. R. Soc. Med. 28 (3), 319–332.

De Saint Jan, D., David-Watine, B., Korn, H., Bregestovski, P., 2001. Activation of human alpha1 and alpha2 homomeric glycine receptors by taurine and GABA. J. Physiol. 535 (Pt 3), 741–755.

Demarque, M., Spitzer, N.C., 2012. Neurotransmitter phenotype plasticity: an unexpected mechanism in the toolbox of network activity homeostasis. Dev. Neurobiol. 72 (1), 22–32.

Dugué, G.P., Dumoulin, A., Triller, A., Dieudonné, S., 2005. Target-dependent use of co-released inhibitory transmitters at central synapses. J. Neurosci. 25 (28), 6490–6498.

Dulcis, D., Spitzer, N.C., 2008. Illumination controls differentiation of dopamine neurons regulating behaviour. Nature 456 (7219), 195–201.

Dulcis, D., Jamshidi, P., Leutgeb, S., Spitzer, N.C., 2013. Neurotransmitter switching in the adult brain regulates behavior. Science 340 (6131), 449–453.

Eccles, J.C., 1957. The physiology of nerve cells. Am. J. Phys. Med. Rehabil. 37 (6), 334.

Eccles, J.C., 1982. The synapse: from electrical to chemical transmission. Annu. Rev. Neurosci. 5 (1), 325–339.

Ekblad, E., Edvinsson, L., Wahlestedt, C., Uddman, R., Håkanson, R., Sundler, F., 1984. Neuropeptide Y co-exists and co-operates with noradrenaline in perivascular nerve fibers. Regul. Pept. 8 (3), 225–235.

Fahrenkrug, J., Hannibal, J., 2004. Neurotransmitters co-existing with VIP or PACAP. Peptides 25 (3), 393–401.

Fisher, J.M., Sossin, W., Newcomb, R., Scheller, R.H., 1988. Multiple neuropeptides derived from a common precursor are differentially packaged and transported. Cell 54 (6), 813–822.

Floras, J.S., 1993. Clinical aspects of sympathetic activation and parasympathetic withdrawal in heart failure. J. Am. Coll. Cardiol. 22 (4 Suppl A), 72A–84A.

Francis, N.J., Landis, S.C., 1999. Cellular and molecular determinants of sympathetic neuron development. Annu. Rev. Neurosci. 22, 541–566.

Fucile, S., de Saint Jan, D., David-Watine, B., Korn, H., Bregestovski, P., 1999. Comparison of glycine and GABA actions on the zebrafish homomeric glycine receptor. J. Physiol. 517 (Pt 2), 369–383.

Furness, J.B., 2000. Types of neurons in the enteric nervous system. J. Auton. Nerv. Syst. 81 (1–3), 87–96.

Furness, J.B., Bornstein, J.C., Murphy, R., Pompolo, S., 1992. Roles of peptides in transmission in the enteric nervous system. Trends Neurosci. 15 (2), 66–71.

Furshpan, E.J., MacLeish, P.R., O'Lague, P.H., Potter, D.D., 1976. Chemical transmission between rat sympathetic neurons and cardiac myocytes developing in microcultures: evidence for cholinergic, adrenergic, and dual-function neurons. Proc. Natl. Acad. Sci. USA 73 (11), 4225–4229.

Geissen, M., Heller, S., Pennica, D., Ernsberger, U., Rohrer, H., 1998. The specification of sympathetic neurotransmitter phenotype depends on gp130 cytokine receptor signaling. Development 125 (23), 4791–4801.

Gitterman, D.P., Evans, R.J., 2001. Nerve evoked P2X receptor contractions of rat mesenteric arteries; dependence on vessel size and lack of role of L-type calcium channels and calcium induced calcium release. Br. J. Pharmacol. 132 (6), 1201–1208.

Gras, C., Amilhon, B., Lepicard, E.M., Poirel, O., Vinatier, J., Herbin, M., et al., 2008. The vesicular glutamate transporter VGLUT3 synergizes striatal acetylcholine tone. Nat. Neurosci. 11 (3), 292–300.

Grothe, B., Pecka, M., McAlpine, D., 2010. Mechanisms of sound localization in mammals. Physiol. Rev. 90 (3), 983–1012.

Gu, X., Spitzer, N.C., 1995. Distinct aspects of neuronal differentiation encoded by frequency of spontaneous Ca2+ transients. Nature 375 (6534), 784–787.

Guemez-Gamboa, A., Xu, L., Meng, D., Spitzer, N.C., 2014. Non-cell-autonomous mechanism of activity-dependent neurotransmitter switching. Neuron 82 (5), 1004–1016.

Gutierrez, R., 2009. Co-Existence and Co-Release of Classical Neurotransmitters. Springer, New York.

Habecker, B.A., Landis, S.C., 1994. Noradrenergic regulation of cholinergic differentiation. Science 264 (5165), 1602–1604.

Hnasko, T.S., Chuhma, N., Zhang, H., Goh, G.Y., Sulzer, D., Palmiter, R.D., et al., 2010. Vesicular glutamate transport promotes dopamine storage and glutamate corelease in vivo. Neuron 65 (5), 643–656.

Hodges, G.J., Jackson, D.N., Mattar, L., Johnson, J.M., Shoemaker, J.K., 2009. Neuropeptide Y and neurovascular control in skeletal muscle and skin. Am. J. Physiol. Regul. Integr. Comp. Physiol. 297 (3), R546–R555.

Hökfelt, T., Elfvin, L.G., Elde, R., Schultzberg, M., Goldstein, M., Luft, R., 1977. Occurrence of somatostatin-like immunoreactivity in some peripheral sympathetic noradrenergic neurons. Proc. Natl. Acad. Sci. 74 (8), 3587–3591.

Hoyle, C.H., 1996. Purinergic cotransmission: parasympathetic and enteric nerves. Sem. Neurosci. 8 (4), 207–215.

Huidobro-Toro, J., Donoso, M., 2004. Sympathetic co-transmission: the coordinated action of ATP and noradrenaline and their modulation by neuropeptide Y in human vascular neuroeffector junctions. Eur. J. Pharmacol. 500 (1–3), 27–35.

Jan, Y.N., Jan, L.Y., Kuffler, S.W., 1979. A peptide as a possible transmitter in sympathetic ganglia of the frog. Proc. Natl. Acad. Sci. 76 (3), 1501–1505.

Jan, Y.N., Jan, L.Y., 1983. A LHRH-like peptidergic neurotransmitter capable of 'action at a distance' in autonomic ganglia. Trends Neurosci. 6, 320–325.

Jonas, P., Bischofberger, J., Sandkühler, J., 1998. Corelease of two fast neurotransmitters at a central synapse. Science 281 (5375), 419–424.

Kanazawa, H., Ieda, M., Kimura, K., Arai, T., Kawaguchi-Manabe, H., Matsuhashi, T., et al., 2010. Heart failure causes cholinergic transdifferentiation of cardiac sympathetic nerves via gp130-signaling cytokines in rodents. J. Clin. Invest. 120 (2), 408−421.

Kandler, K., Clause, A., Noh, J., 2009. Tonotopic reorganization of developing auditory brainstem circuits. Nat. Neurosci. 12 (6), 711−717.

Katsurabayashi, S., Kubota, H., Higashi, H., Akaike, N., Ito, Y., 2004. Distinct profiles of refilling of inhibitory neurotransmitters into presynaptic terminals projecting to spinal neurones in immature rats. J. Physiol. 560 (Pt 2), 469−478.

Kellogg Jr, D.L., Zhao, J.L., Wu, Y., Johnson, J.M., 2010. VIP/PACAP receptor mediation of cutaneous active vasodilation during heat stress in humans. J. Appl. Physiol. 109 (1), 95−100.

Kennedy, C., Saville, V.L., Burnstock, G., 1986. The contributions of noradrenaline and ATP to the responses of the rabbit central ear artery to sympathetic nerve stimulation depend on the parameters of stimulation. Eur. J. Pharmacol. 122 (3), 291−300.

Kim, G., Kandler, K., 2003. Elimination and strengthening of glycinergic/GABAergic connections during tonotopic map formation. Nat. Neurosci. 6 (3), 282−290.

Lamotte d'Incamps, B., Bhumbra, G.S., Foster, J.D., Beato, M., Ascher, P., 2017. Segregation of glutamatergic and cholinergic transmission at the mixed motoneuron Renshaw cell synapse. Sci. Rep. 7 (1), 4037.

Landis, S.C., 1976. Rat sympathetic neurons and cardiac myocytes developing in microcultures: correlation of the fine structure of endings with neurotransmitter function in single neurons. Proc. Natl. Acad. Sci. USA 73 (11), 4220−4224.

Landis, S.C., 1990. Target regulation of neurotransmitter phenotype. Trends Neurosci. 13 (8), 344−350.

Landis, S.C., Keefe, D., 1983. Evidence for neurotransmitter plasticity in vivo: developmental changes in properties of cholinergic sympathetic neurons. Dev. Biol. 98 (2), 349−372.

Landis, S.C., Siegel, R.E., Schwab, M., 1988. Evidence for neurotransmitter plasticity in vivo. II. Immunocytochemical studies of rat sweat gland innervation during development. Dev. Biol. 126 (1), 129−140.

Lu, T., Rubio, M.E., Trussell, L.O., 2008. Glycinergic transmission shaped by the corelease of GABA in a mammalian auditory synapse. Neuron 57 (4), 524−535.

Lundberg, J.M., 1996. Pharmacology of cotransmission in the autonomic nervous system: integrative aspects on amines, neuropeptides, adenosine triphosphate, amino acids and nitric oxide. Pharmacol. Rev. 48 (1), 113−178.

Lundberg, J.M., Hökfelt, T., 1983. Coexistence of peptides and classical neurotransmitters. Trends Neurosci. 6, 325−333.

Lymperopoulos, A., Rengo, G., Koch, W.J., 2013. Adrenergic nervous system in heart failure: pathophysiology and therapy. Circ. Res. 113 (6), 739−753.

Mack, G.W., Shannon, L.M., Nadel, E.R., 1985. Influence of beta-adrenergic blockade on the control of sweating in humans. J. Appl. Physiol. 61 (5), 1701−1705.

Maggi, C.A., 2000. Principles of tachykininergic co-transmission in the peripheral and enteric nervous system. Regul. Pept. 93 (1−3), 53−64.

Matsuoka, I., Ohkubo, S., 2004. ATP- and adenosine-mediated signaling in the central nervous system: adenosine receptor activation by ATP through rapid and localized generation of adenosine by ecto-nucleotidases. J. Pharmacol. Sci. 94 (2), 95−99.

Mayer, E.A., Baldi, J.P., 1991. Can regulatory peptides be regarded as words of a biological language? Am. J. Physiol. Gastrointest. Liver Physiol. 261 (2), G171−G184.

Mayeri, E., Rothman, B.S., Brownell, P.H., Branton, W.D., Padgett, L., 1985. Nonsynaptic characteristics of neurotransmission mediated by egg-laying hormone in the abdominal ganglion of Aplysia. J. Neurosci. 5 (8), 2060−2077.

Meng, D., Li, H.Q., Deisseroth, K., Leutgeb, S., Spitzer, N.C., 2018. Neuronal activity regulates neurotransmitter switching in the adult brain following light-induced stress. Proc. Natl. Acad. Sci. USA 115 (20), 5064−5071.

Merit-HF Study Group, 1999. Effect of metoprolol CR/XL in chronic heart failure: metoprolol CR/XL randomised intervention trial in-congestive heart failure (MERIT-HF). Lancet 353 (9169), 2001−2007.

Moroz, L.L., 2011. Aplysia. Curr. Biol. 21 (2), R60−R61.

Nabekura, J., Katsurabayashi, S., Kakazu, Y., Shibata, S., Matsubara, A., Jinno, S., et al., 2004. Developmental switch from GABA to glycine release in single central synaptic terminals. Nat. Neurosci. 7 (1), 17−23.

Nagle, G.T., Painter, S.D., Blankenship, J.E., 1989. The egg-laying hormone family: Precursors, products, and functions. Biol. Bull. 177 (2), 210−217.

Nawa, H., Patterson, P.H., 1990. Separation and partial characterization of neuropeptide-inducing factors in heart cell conditioned medium. Neuron 4 (2), 269−277.

Nawa, H., Sah, D.W., 1990. Different biological activities in conditioned media control the expression of a variety of neuropeptides in cultured sympathetic neurons. Neuron 4 (2), 279–287.

Nishi, S., Koketsu, K., 1968. Early and late after discharges of amphibian sympathetic ganglion cells. J. Neurophys. 31 (1), 109–121.

Nishimaru, H., Restrepo, C.E., Ryge, J., Yanagawa, Y., Kiehn, O., 2005. Mammalian motor neurons corelease glutamate and acetylcholine at central synapses. Proc. Natl. Acad. Sci. USA 102 (14), 5245–5249.

Noh, J., Seal, R.P., Garver, J.A., Edwards, R.H., Kandler, K., 2010. Glutamate co-release at GABA/glycinergic synapses is crucial for the refinement of an inhibitory map. Nat. Neurosci. 13 (2), 232–238.

Nusbaum, M.P., Blitz, D.M., Marder, E., 2017. Functional consequences of neuropeptide and small-molecule co-transmission. Nat. Rev. Neurosci. 18 (7), 389–403.

O'Dowd, D.K., Ribera, A.B., Spitzer, N.C., 1988. Development of voltage-dependent calcium, sodium, and potassium currents in Xenopus spinal neurons. J. Neurosci. 8 (3), 792–805.

Osborne, N.N., 1979. Is Dale's principle valid? Trends Neurosci. 2, 73–75.

Potter, D.D., Landis, S.C., Furshpan, E.J., 1980. Dual function during development of rat sympathetic neurones in culture. J. Exp. Biol. 89, 57–71.

Ralevic, V., Burnstock, G., 1990. Postjunctional synergism of noradrenaline and adenosine 5'-triphosphate in the mesenteric arterial bed of the rat. Eur. J. Pharmacol. 175 (3), 291–299.

Ralevic, V., Dunn, W.R., 2015. Purinergic transmission in blood vessels. Auton. Neurosci. 191, 48–66.

Rocha-Resende, C., Roy, A., Resende, R., Ladeira, M.S., Lara, A., de Morais Gomes, E.R., et al., 2012. Non-neuronal cholinergic machinery present in cardiomyocytes offsets hypertrophic signals. J. Mol. Cell. Cardiol. 53 (2), 206–216.

Root, C.M., Velázquez-Ulloa, N.A., Monsalve, G.C., Minakova, E., Spitzer, N.C., 2008. Embryonically expressed GABA and glutamate drive electrical activity regulating neurotransmitter specification. J. Neurosci. 28 (18), 4777–4784.

Rosenberg, S.S., Spitzer, N.C., 2011. Calcium signaling in neuronal development. Cold Spring Harb. Perspect. Biol. 3 (10), 004259.

Rothman, B.S., Mayeri, E., Brown, R.O., Yuan, P.M., Shively, J.E., 1983. Primary structure and neuronal effects of alpha-bag cell peptide, a second candidate neurotransmitter encoded by a single gene in bag cell neurons of Aplysia. Proc. Natl. Acad. Sci. USA 80 (18), 5753–5757.

Roy, A., Guatimosim, S., Prado, V.F., Gros, R., Prado, M.A., 2015. Cholinergic activity as a new target in diseases of the heart. Mol. Med. 20, 527–537.

Rummery, N.M., Brock, J.A., Pakdeechote, P., Ralevic, V., Dunn, W.R., 2007. ATP is the predominant sympathetic neurotransmitter in rat mesenteric arteries at high pressure. J. Physiol. 582 (Pt 2), 745–754.

Russier, M., Kopysova, I.L., Ankri, N., Ferrand, N., Debanne, D., 2002. GABA and glycine co-release optimizes functional inhibition in rat brainstem motoneurons in vitro. J. Physiol. 541 (Pt 1), 123–137.

Sagné, C., El Mestikawy, S., Isambert, M.F., Hamon, M., Henry, J.P., Giros, B., et al., 1997. Cloning of a functional vesicular GABA and glycine transporter by screening of genome databases. FEBS Lett. 417 (2), 177–183.

Salio, C., Lossi, L., Ferrini, F., Merighi, A., 2006. Neuropeptides as synaptic transmitters. Cell Tissue Res. 326 (2), 583–598.

Sámano, C., Zetina, M.E., Marín, M.A., Cifuentes, F., Morales, M.A., 2006. Choline acetyl transferase and neuropeptide immunoreactivities are colocalized in somata, but preferentially localized in distinct axon fibers and boutons of cat sympathetic preganglionic neurons. Synapse 60 (4), 295–306.

Sámano, C., Cifuentes, F., Morales, M.A., 2012. Neurotransmitter segregation: functional and plastic implications. Prog. Neurobiol. 97 (3), 277–287.

Saville, V.L., Maynard, K.I., Burnstock, G., 1990. Neuropeptide Y potentiates purinergic as well as adrenergic responses of the rabbit ear artery. Eur. J. Pharmacol. 176 (2), 117–125.

Schotzinger, R.J., Landis, S.C., 1988. Cholinergic phenotype developed by noradrenergic sympathetic neurons after innervation of a novel cholinergic target in vivo. Nature 335 (6191), 637–639.

Schotzinger, R.J., Landis, S.C., 1990. Acquisition of cholinergic and peptidergic properties by sympathetic innervation of rat sweat glands requires interaction with normal target. Neuron 5 (1), 91–100.

Schotzinger, R., Yin, X., Landis, S., 1994. Target determination of neurotransmitter phenotype in sympathetic neurons. J. Neurobiol. 25 (6), 620–639.

Schütz, B., von Engelhardt, J., Gördes, M., Schäfer, M.K., Eiden, L.E., Monyer, H., et al., 2008. Sweat gland innervation is pioneered by sympathetic neurons expressing a cholinergic/noradrenergic co-phenotype in the mouse. Neuroscience 126 (1), 129–140.

Sjöblom-Widfeldt, N., Nilsson, H., 1990. Sympathetic transmission in small mesenteric arteries from the rat: influence of impulse pattern. Acta Physiol. Scand. 138 (4), 523–528.

Smith, N.C., Burnstock, G., 2004. Mechanisms underlying postjunctional synergism between responses of the vas deferens to noradrenaline and ATP. Eur. J. Pharmacol. 498 (1-3), 241–248.

Smith, S.M., Vale, W.W., 2006. The role of the hypothalamic-pituitary-adrenal axis in neuroendocrine responses to stress. Dialogues. Clin. Neurosci. 8 (4), 383–395.

Sossin, W.S., Sweet-Cordero, A., Scheller, R.H., 1990. Dale's hypothesis revisited: different neuropeptides derived from a common prohormone are targeted to different processes. Proc. Natl. Acad. Sci. 87 (12), 4845–4848.

Spitzer, N.C., 2006. Electrical activity in early neuronal development. Nature 444 (7120), 707–712.

Spitzer, N.C., 2012. Activity-dependent neurotransmitter respecification. Nat. Rev. Neurosci. 13 (2), 94–106.

Spitzer, N.C., 2015. Neurotransmitter switching? No surprise. Neuron 86 (5), 1131–1144.

Spitzer, N.C., 2017. Neurotransmitter switching in the developing and adult brain. Annu. Rev. Neurosci. 40, 1–19.

Spitzer, N.C., Borodinsky, L.N., 2008. Implications of activity-dependent neurotransmitter-receptor matching. Philos. Trans. R. Soc. Lond. B. Biol. Sci. 363 (1495), 1393–1399.

Spitzer, N.C., Lautermilch, N.J., Smith, R.D., Gomez, T.M., 2000. Coding of neuronal differentiation by calcium transients. Bioessays 22 (9), 811–817.

Spitzer, N.C., Root, C.M., Borodinsky, L.N., 2004. Orchestrating neuronal differentiation: patterns of Ca2+ spikes specify transmitter choice. Trends Neurosci. 27 (7), 415–421.

Stephens, D.P., Saad, A.R., Bennett, L.A., Kosiba, W.A., Johnson, J.M., 2004. Neuropeptide Y antagonism reduces reflex cutaneous vasoconstriction in humans. Am. J. Physiol. Heart Circ. Physiol. 287 (3), H1404–H1409.

Todd, A.J., Watt, C., Spike, R.C., Sieghart, W., 1996. Colocalization of GABA, glycine, and their receptors at synapses in the rat spinal cord. J. Neurosci. 16 (3), 974–982.

Todorov, L.D., Mihaylova-Todorova, S., Craviso, G.L., Bjur, R.A., Westfall, D.P., 1996. Evidence for the differential release of the cotransmitters ATP and noradrenaline from sympathetic nerves of the guinea-pig vas deferens. J. Physiol. 496 (Pt 3), 731–748.

Todorov, L.D., Mihaylova-Todorova, S.T., Bjur, R.A., Westfall, D.P., 1999. Differential cotransmission in sympathetic nerves: role of frequency of stimulation and prejunctional autoreceptors. J. Pharmacol. Exp. Ther. 290 (1), 241–246.

Tollin, D.J., 2003. The lateral superior olive: a functional role in sound source localization. Neuroscientist 9 (2), 127–143.

Tomlinson, D.R., 1975. Two populations or granular vesicles in constricted post-ganglionic sympathetic nerves. J. Physiol. 245 (3), 727–735.

Triposkiadis, F., Karayannis, G., Giamouzis, G., Skoularigis, J., Louridas, G., Butler, J., 2009. The sympathetic nervous system in heart failure physiology, pathophysiology, and clinical implications. J. Am. Coll. Cardiol. 54 (19), 1747–1762.

Ubink, R., Tuinhof, R., Roubos, E.W., 1998. Identification of suprachiasmatic melanotrope-inhibiting neurons in Xenopus laevis: a confocal laser-scanning microscopy study. J. Comp. Neurol. 397 (1), 60–68.

Vaaga, C.E., Borisovska, M., Westbrook, G.L., 2014. Dual-transmitter neurons: functional implications of co-release and co-transmission. Curr. Opin. Neurobiol. 29, 25–32.

Vega, A., Cancino-Rodezno, A., Valle-Leija, P., Sánchez-Tafolla, B.M., Elinos, D., Cifuentes, F., et al., 2016. Neurotrophin-dependent plasticity of neurotransmitter segregation in the rat superior cervical ganglion in vivo. Dev. Neurobiol. 76 (8), 832–846.

Weihe, E., Schütz, B., Hartschuh, W., Anlauf, M., Schäfer, M.K., Eiden, L.E., 2005. Coexpression of cholinergic and noradrenergic phenotypes in human and nonhuman autonomic nervous system. J. Comp. Neurol. 492 (3), 370–379.

Workman, J.L., Nelson, R.J., 2011. Potential animal models of seasonal affective disorder. Neurosci. Biobehav. Rev. 35 (3), 669–679.

Yoshida, M., Miyamae, K., Iwashita, H., Otani, M., Inadome, A., 2004. Management of detrusor dysfunction in the elderly: changes in acetylcholine and adenosine triphosphate release during aging. Urology 63 (3 Suppl. 1), 17–23.

Zhang, S., Qi, J., Li, X., Wang, H.L., Britt, J.P., Hoffman, A.F., et al., 2015. Dopaminergic and glutamatergic microdomains in a subset of rodent mesoaccumbens axons. Nat. Neurosci. 18 (3), 386–392.

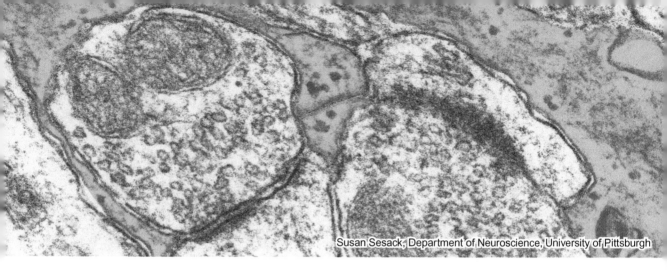

# CHAPTER 22

# Complex Signaling Within Tripartite Synapses

Throughout this book we have described the synapse as a communication site between a presynaptic neuron and its postsynaptic target cell. However, this is an incomplete picture of both synaptic structure and function because it does not include a critical third component: **glial cells** (also called glia). As the field of neuroscience developed, neurons received most of the attention as the main players in the nervous system, though scientists such as Virchow (1846) and Ramón y Cajal (1909) reported the presence of glial cells and speculated about their roles in nervous system function (see Fig. 22.1). The dogma for many decades was that glial cells were simply "support" cells that help maintain neurons by regulating their local environment. In fact, the word "glia" is derived from the Greek word for "putty or gum," intended to mean a "connective substance," reflecting the historical view that these cells simply helped to hold neurons together and support them. However, it is now appreciated that glial cell function is more complex than originally thought. Glial cells are critical not only for supporting the synaptic environment, but for their direct contributions to the communication between cells, including modulation of synaptic transmission.

Glial cells are 10–50 times more numerous than neurons, and although the ratio of neurons to non-neuronal cells varies substantially by brain region and species, glial cells make up ~20% of the total volume of the central nervous system (Hatton and Parpura, 2004). Glial cells in the peripheral nervous system are called **Schwann cells**. In addition to

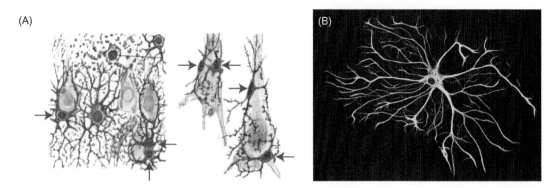

FIGURE 22.1 **Astrocytes.** (A) Drawing by Ramón y Cajal of astrocytes (*dark gray*, next to *blue arrows*) in the human hippocampus surrounding the somata of pyramidal neurons (*pale gray cell bodies*). (B) Astrocyte stained with antibodies to show the glial cell and its processes (*yellow*), as well as the nuclei of the glial cell and surrounding neurons (*blue*). Source: (A) Adapted from García-Marín et al. (2007); (B) By GerryShaw (CC BY-SA 3.0).

their well-recognized role of supplying myelin to peripheral axons, a subset of Schwann cells are specialized for surrounding the neuromuscular junction and the sensory endings of peripheral neurites, where they participate in synaptic communication. Glial cells in the central nervous system include **microglia**, **oligodendrocytes**, and **astrocytes**. Our discussion of glial cells in this chapter will focus largely on astrocytes.

The study of glial cells led researchers to define the concept of the **tripartite synapse**, which includes the presynaptic terminal, the postsynaptic cell, and the surrounding glial cells (see Fig. 22.2). This concept that synapses include glial cells as well as neurons underscores the fact that glial cells contribute significantly to synaptic function (Araque et al., 1999). In this chapter, we will discuss some of the roles glial cells play, both their historically acknowledged "supportive" roles in the nervous system and some of their more recently discovered active contributions to the function and operation of synapses.

## THE ROLE OF ASTROCYTES IN SYNAPTIC FUNCTION

Astrocytes are the most common type of glial cell in the central nervous system and are distributed throughout both gray and white matter. Astrocytes get their name because of their star-like appearance, which results from the many **astrocytic processes** emanating from their cell body (see Fig. 22.1). At synapses, these processes often wrap around and ensheath the pre- and/or postsynaptic neurons very tightly, surrounding the synaptic cleft and forming the tripartite synapse (see Fig. 22.2). Astrocytic processes have intricate branches that project between and around many components of brain tissue; as such, they have been described as having a spongiform shape (see Fig. 22.3).

Astrocytes and neurons do not communicate directly using synapses or gap junctions, so they can only interact via the narrow extracellular space between them. This space is about 20 nm across (compare with the average size of a synaptic cleft, which is about

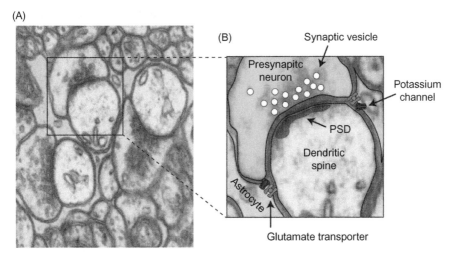

FIGURE 22.2  **Astrocytes surround synapses.** (A) Electron transmission micrograph showing three synapses with surrounding astrocyte processes (*blue*). The spines of the postsynaptic neurons are colored in yellow. (B) Enlargement of one of the synapses in (A) (*black box*), with components of the tripartite synapse indicated. Astrocytes (*blue*) can be seen on either side of the presynaptic neuron (*green*), the dendritic spine (*yellow*), and the synaptic cleft. Synaptic vesicles and the postsynaptic density (PSD) are apparent in (A) and have been shown schematically in (B). Astrocytic potassium channels and glutamate transporters are also indicated schematically, though they cannot be resolved in electron microscopic images like the one shown in (A). *Source: Atlas of Ultrastructural Neurocytology, http://synapseweb.clm.utexas.edu/atlas, Josef Spacek contributor. "SynapseWeb, Kristen M. Harris, PI, http://synapseweb.clm.utexas.edu/"*.

FIGURE 22.3  **Astrocyte processes are highly branched, and they surround synapses and other components of brain tissue.** 3D reconstruction showing astrocyte processes (*light green*) and three dendrites with dendritic spines (*gold, red, and dark blue*). *Source: Adapted from Witcher, M.R., Kirov, S.A., Harris, K.M., 2007. Plasticity of perisynaptic astroglia during synaptogenesis in the mature rat hippocampus. Glia 55, 13–23.*

30 nm across), though the exact width varies by location in the brain and type of tripartite synapse. At many synapses, the synaptic cleft is partially "sealed off" by perisynaptic astrocyte processes (see Fig. 22.2), which, among other things, control how many of the neurotransmitter molecules released into the synaptic cleft can spill over to other synapses. Astrocytes play a unique role in synaptic transmission by influencing the degree of crosstalk between neighboring synapses, which means they can also help to coordinate and/or regulate neuronal activity at multiple nearby synapses.

It has been shown that the shapes and locations of astrocytic processes surrounding synapses are not stable and in fact can change their morphology on a time course as short as minutes. Changes in how astrocytic processes ensheath the synaptic components can dynamically influence synaptic function. Interestingly, in 1895 Ramón y Cajal proposed that morphological changes in glial cells might occur across the sleep—wake cycle and even allow glial cells to help control sleep (reviewed in Garcia-Marin et al., 2007). This prediction was eventually borne out by research showing that during wakefulness, astrocytic processes move closer to the synaptic cleft, whereas during sleep they retract further from the synaptic cleft (Bellesi et al., 2015), which is predicted to influence the diffusion and uptake of neurotransmitter in and around the synapse.

## Uptake of Neurotransmitters by Astrocytic Transporters

One important function astrocyte processes perform at synapses is the uptake, and therefore removal, of neurotransmitter from the synaptic cleft. As in neurons, neurotransmitter is taken up into glial cells by transporters in the cell membranes (see Fig. 22.4).

Glutamate is removed from the synapse mainly by uptake into the presynaptic terminal, postsynaptic neuron, and astrocytes (see Fig. 22.4). Specifically, uptake by the astrocytes is the dominant mechanism for clearing glutamate from the synapse, whereas reuptake into neurons plays a lesser role. Glutamate uptake is critical to prevent overactivation of glutamate receptors, which can cause lethal excitotoxicity in the postsynaptic neuron.

In astrocytes, glutamate uptake is accomplished by excitatory amino acid transporters (EAATs) that are located in the cell membrane and utilize the transmembrane sodium gradient to move glutamate into the astrocyte cytoplasm (see Fig. 22.4). There are two such transporters found in astrocytes, which are called EAAT1 and EAAT2 in humans [the rodent forms of these transporters are called glutamate-aspartate transporter 1 (GLAST) and glutamate transporter 1 (GLT-1), respectively]. The expression of these two transporters varies by brain region, synapse type, and time during the development of the nervous system. EAATs bind glutamate rapidly enough to buffer released glutamate in perisynaptic areas almost instantly.

Glutamate that is taken up into astrocytes is converted to glutamine by the enzyme glutamine synthetase. Glutamine can then be transported out of astrocytes and subsequently into neurons by glutamine transporters on both cell types. This process is called the glutamine cycle and is discussed in more detail in Chapter 18.

To investigate the role of astrocytic glutamate transporters in glutamate reuptake, Bergles and Jahr (1997) used a hippocampal slice preparation in which they electrically

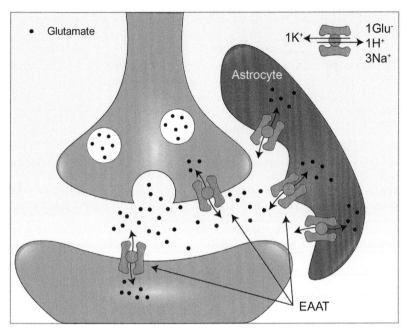

FIGURE 22.4  **Schematic diagram of astroglial and neuronal glutamate transporters.** Excitatory amino acid transporters (EAATs) are found in both astrocytes and presynaptic terminals. For every molecule of glutamate transported into the cytoplasm, three sodium ions and one proton are also moved in, and one potassium ion is moved into the extracellular space.

stimulated the axons of pyramidal cells and measured how this stimulation affected astrocytes near the axon terminals. They showed that release of glutamate from the stimulated axons caused an inward current in nearby astrocytes that was eliminated when both types of glial glutamate transporters were blocked (see Fig. 22.5). Pharmacologically blocking just the GLT-1 glutamate transporters (the rodent form of EAAT2; see above) decreased the size of the evoked glial current, showing that this current involves GLT-1. However, since blocking GLT-1 alone did not block all of the current, the GLAST transporter (the rodent form of EAAT1; see above) must also contribute to the glial glutamate current observed at this synapse.

In the extracellular solution used in this experiment to bathe cells (pH 7.4), glutamate is negatively charged, so why was an inward (positive) current recorded in the astrocyte when GLT-1 was active? This inward current occured because for every glutamate molecule transported into the cell, three sodium ions and one hydrogen ion also move in, and one potassium ion moves out (see Fig. 22.4). Therefore, the net electrogenic effect of GLT-1 is positive, with one net positive charge entering the cytoplasm per transporter cycle. When researchers clamped the astrocyte at different membrane voltages and stimulated the axon pathway, they were able to generate an $I-V$ plot showing how the glutamate transporter currents resulting from glutamatergic neuron stimulation changed as a function of membrane voltage (see Fig. 22.5C). They found that the glutamate

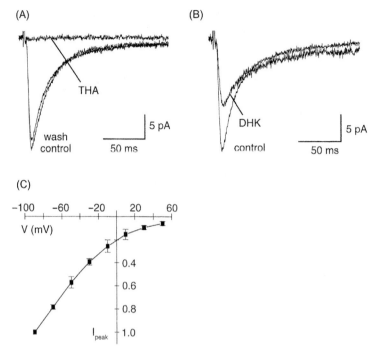

FIGURE 22.5 **Glutamate transporter currents in astrocytes.** (A) Under control conditions, stimulation of glutamatergic neurons caused a glutamate-mediated current in astrocytes. However, after application of a drug that blocks all glutamate transporters (D,L-threo-3-hydroxyaspartate; THA), this current was eliminated, but could be observed again once the drug was washed off. (B) Applying a more specific glutamate transporter antagonist that blocks only GLT-1 (dihydrokainate; DHK) only partially blocked the glutamate-mediated current, showing that both GLT-1 and GLAST contribute to this current in the absence of blockers. (C) The $I-V$ curve for the glutamate-mediated current shows that it is an inwardly-rectifying current that does not reverse at any membrane potential. *Source: Adapted from Bergles, D.E., Jahr, C.E., 1997. Synaptic activation of glutamate transporters in hippocampal astrocytes. Neuron 19, 1297–1308.*

transporter current decreased as the membrane potential increased, but it did not become a negative current at any of the holding potentials they tested, showing that this current is **inwardly rectifying**.

In addition to clearing released glutamate from glutamatergic synapses, glutamate uptake by astrocytes influences the ambient levels of glutamate in the circulating extracellular fluid, which is referred to as the "glutamatergic tone." Rats with reduced GLT-1 and GLAST expression have elevated levels of extracellular glutamate, and they suffer from progressive paralysis as well as neurodegeneration that is similar to damage resulting from excitotoxicity (Rothstein et al., 1996). Critically, reducing expression of *neuronal* glutamate transporters does not have such a severe effect, causing only mild neurotoxicity and seizures. These results suggest that under normal circumstances astrocytic glutamate transporters remove the bulk of synaptic glutamate (Rothstein et al., 1996; Arnth-Jensen et al., 2002).

Astrocytes also contain GABA transporters (GATs). Unlike glutamate transporters, astrocytic GATs play only a minor role in GABA uptake, removing about 20% of the GABA released into the synaptic cleft. The other 80% is removed by GABA transporters in the presynaptic terminal of the GABAergic neurons. Whereas GABA molecules that are taken back into the presynaptic terminal can be packaged into vesicles and released again, GABA that is transported into astrocytes is metabolized by GABA transaminase. Glycine transporters are also found in astrocytes, as was discussed in Chapter 18.

## Astrocytes Maintain Potassium in Extracellular Fluid

One of the support functions of astrocytes is to maintain homeostasis of neurons, and they can do this by regulating ion concentrations in the extracellular fluid. In Chapter 3, we discussed the types of ions found in neurons and the extracellular fluid and their relative concentrations. In order for neurons to generate action potentials and to release neurotransmitter, these ionic concentrations must be maintained at precise levels, despite changes in ion concentration caused by neuronal activity. For example, potassium accumulation in the extracellular space would cause neurons to fire action potentials more readily, which could lead to pathological conditions such as seizures. In order to help maintain the extracellular potassium concentration at a normal level, astrocytes sequester excess potassium from the extracellular fluid (Ransom and Sontheimer, 1992).

## Ion Channels in Astrocytes

Though astrocytes are not considered electrically excitable cells (that is, they can't generate action potentials), they nonetheless have a membrane potential, and it is possible to record ionic conductances in astrocytes using techniques such as patch clamp recordings. Astrocytes can also express a wide variety of ligand-gated ion channels to mediate such conductances, including ionotropic and metabotropic receptors for glutamate and GABA, as well as metabotropic receptors for monoamines, opioids, purines, and adenosine (see Verkhratsky and Nedergaard, 2018, for a review).

Potassium channels are the most common type of ion channel found in astrocytes. Astrocytes contain several types of potassium channels, including rapidly inactivating and inwardly rectifying potassium channels. The inwardly rectifying potassium channels are the main ion channels used by astrocytes to regulate the concentration of external potassium (see below), whereas resting two pore domain potassium channels are the main driver of the resting membrane potential in astrocytes (which is usually about −80 mV, or about the equilibrium potential for potassium).

Despite the presence of ligand-gated ion channels on astrocytes, it was initially assumed that because astrocytes are not excitable they would not have voltage-gated ion channels. Researchers working in the 1980s and 1990s were therefore surprised to discover a wide range of voltage-gated ion channels in these cells (see Verkhratsky and Steinhauser, 2000, for a review). For instance, astrocytes express voltage-gated sodium channels that are similar to those found in neurons. These sodium channels are thought to help maintain appropriate ionic concentrations in the extracellular fluid, in part by supplying sodium to the cytoplasm of astrocytes in order to maintain $Na^+/K^+$-ATPase activity. Astrocytes also contain voltage-gated chloride channels, which contribute to potassium buffering by astrocytes because cytoplasmic chloride concentration affects membrane transporters that couple the flux of chloride and potassium.

Additionally, astrocytes express multiple types of calcium channels. They have been shown to express N-type, L-type, and T-type voltage-gated calcium channels in the cell membrane, as well as several intracellular calcium channels that facilitate the release of calcium from intracellular stores. These channels can be activated by neurotransmitters, neuromodulators, neuropeptides, and other compounds. Astrocytes also have calcium-permeable mechanosensitive channels that allow calcium into the cell when the cell membrane is mechanically displaced or if it is stretched by osmotic swelling. As in neurons, the cytoplasmic calcium concentration in astrocytes modulates many cellular functions, and can be increased either by of an influx of calcium across the cell membrane or calcium release from intracellular stores.

# INTERACTIONS BETWEEN ASTROCYTES: GAP JUNCTIONS AND CALCIUM WAVES

Astrocytes are strongly connected by gap junctions, though they contain different connexin proteins from those found in neurons (see Chapter 5). Astrocytes can be coupled via gap junctions to other astrocytes or to oligodendrocytes, but not to neurons (Rash et al., 2001).

Charles et al. (1991) demonstrated that astrocytes communicate as part of a larger connected network by filling cultured astrocytes with the calcium-indicator dye fura-2 and monitoring their levels of intracellular calcium using fluorescent microscopy. These researchers took advantage of the fact that mechanical stimulation of astrocytes with the tip of a microelectrode increases their cytoplasmic calcium concentration (via mechanosensitive calcium channels; see above). Using this technique, they showed that when the concentration of calcium in one astrocyte increased, it also increased in adjacent astrocytes over the course of seconds (see Fig. 22.6). This spread of calcium through a network of astrocytes was termed a **calcium wave**, which is similar to the calcium waves described for networks of electrically coupled retinal and other neurons. Calcium waves can be propagated by more than one mechanism, including direct flow of calcium ions through gap junctions and by the movement of inositol triphosphate ($IP_3$) through gap junctions, which then initiates release of calcium from intracellular stores. The effect(s) and/or role(s) of astrocytic calcium waves are not yet completely understood.

The fact that ions and small molecules can travel between astrocytes via these gap junctions means that astrocytes do not function independently of one another. In fact, the membrane potentials and ionic concentrations of a group of 50–100 coupled astrocytes can be regulated in a coordinated fashion (Verkhratsky and Nedergaard, 2018). Coupling between astrocytes by gap junctions is not uniform. That is, neighboring astrocytes are not connected to neighboring astrocytes with the same probability. Instead, there appear to be

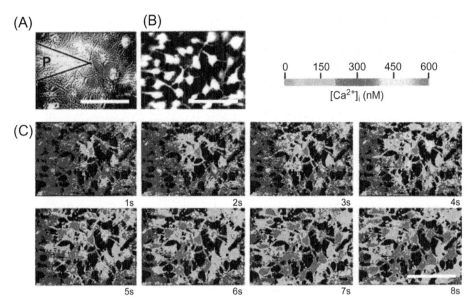

FIGURE 22.6  Calcium spread through networks of astrocytes coupled by gap junctions. (A) When an astrocyte (center of image) was mechanically stimulated with a micropipette (P), the intracellular calcium increased in that cell. (B) Astrocytes filled with the calcium-indicator dye fura-2. (C) Over the course of 8 s, the increased calcium in the stimulated astrocyte spread to neighboring astrocytes. The intracellular concentration of calcium is shown using the color scale indicated. Scale bar equals 100 μm. *Source: Adapted from Charles, A.C., Merrill, J.E., Dirksen, E.R., Sanderson, M.J., 1991. Intercellular signaling in glial cells: calcium waves and oscillations in response to mechanical stimulation and glutamate. Neuron 6, 983–992.*

local groups of about 50–100 astrocytes that are coupled to one another, but not to other nearby astrocytes. These groups of interconnected astrocytes tend to span about 200–300 μm and are thought to contribute to the existence of functional microdomains or territories within large groups of cells. The purpose of these microdomains is not known.

## RELEASE OF NEUROTRANSMITTERS FROM ASTROCYTES

Given early assumptions that glial cells only play passive/supportive roles in the nervous system, it was surprising to discover that astrocytes are capable of releasing neuroactive compounds (sometimes referred to as gliotransmitters), including glutamate, aspartate, taurine, and ATP. The release of these compounds is associated with both pathological and nonpathological conditions. We will focus our discussion of neurotransmitter release from glia on the release of glutamate from astrocytes, which has been well studied in a range of experimental preparations.

Glutamate can be released by astrocytes by multiple mechanisms. First, glutamate can be released from astrocytes in response to an increase in intracellular calcium levels, using the same conserved mechanisms and proteins discussed in Chapter 8, for calcium-triggered vesicular release of neurotransmitters from neurons. Astrocytes contain structures that are referred to as synaptic-like microvesicles (SLMVs), which contain vesicular transporters for glutamate and are released in a calcium- and SNARE-dependent manner (Bezzi et al., 2004). SLMVs are far less numerous in astrocytes (where they are observed in groups of 2–15) than synaptic vesicles are in neurons (where they usually number in the hundreds or thousands). As we discussed in Chapter 8, exocytosis of synaptic vesicles from nerve terminals occurs very rapidly (in less than a millisecond). In contrast, vesicle exocytosis in astrocytes occurs on the time scale of seconds, despite utilizing the same well-conserved SNARE protein mechanism for vesicle fusion.

Another mechanism by which astrocytes can release glutamate is through glutamate transporters operating in reverse. As described above, one of the main roles of astrocytes is to remove glutamate from the extracellular fluid using transporters in the cell membrane. When the extracellular concentration of glutamate is high (e.g., after a neuron releases neurotransmitter into the synaptic cleft), these transporters move glutamate into astrocytes. However, astrocyte glutamate transporters can work in reverse and transport glutamate out of the cell under pathological conditions such as ischemia.

Finally, astrocytes can also release glutamate through glutamate-conducting anion channels in their membrane, and this can occur following ischemia and other forms of trauma that cause astrocytes to swell. Astrocytes compensate for trauma-induced swelling by opening anion channels that are permeable to substances such as inorganic ions, glutamate, aspartate, and taurine. Unfortunately, excessive extracellular glutamate can lead to excitotoxicity, killing both neurons and astrocytes.

The effects of glutamate release from astrocytes can be observed by measuring glutamate currents in pre- and postsynaptic neurons. Neuronal responses to the released glutamate include increased frequency of spontaneous miniature synaptic currents, increased amplitude of evoked currents (see Fig. 22.7), membrane depolarization, presynaptic inhibition of synaptic release (mediated by mGluRs), and increased intracellular calcium in axon terminals due to membrane depolarization.

## DO ASTROCYTES PLAY A ROLE IN INFORMATION PROCESSING WITHIN THE BRAIN?

The characteristics of astrocytes discussed above allow them to play active roles in synaptic transmission. The fact that extracellular ligands such as glutamate, GABA, acetylcholine, norepinephrine, nitric oxide, histamine, and others can initiate calcium signaling in astrocytes means that release of neurotransmitters from neurons can directly affect the function of nearby astrocytes. In turn, increased calcium levels in astrocytes can cause the release of neurotransmitters such as glutamate and ATP from these astrocytes.

If astrocytes participate in synaptic communication, they may be involved in "information processing" within the brain (see Araque et al., 1999). This is a term usually reserved for neurons, but could astrocytes process information as well?

Evidence suggesting that astrocytes process information:

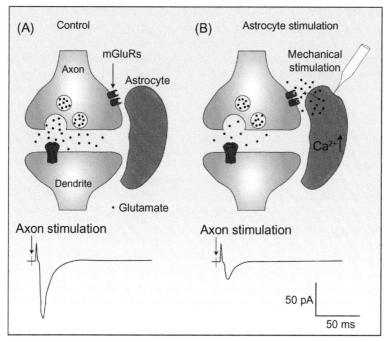

FIGURE 22.7 **Modulation of EPSC amplitude by glutamate release from astrocytes.** (A) A control EPSC was recorded in a dendrite in response to stimulation of a presynaptic axon. (B) Mechanical stimulation of the nearby astrocyte raised the intracellular calcium concentration and caused a release of glutamate from the astrocyte. The glutamate released from the glial cell activated presynaptic metabotropic glutamate receptors, which reduced the magnitude of the EPSC, caused by action potential-evoked release of glutamate from the nerve terminal an effect that lasted about one minute before the size of the EPSC recovered to control levels.

1. Astrocytes respond to external signals from other elements of the nervous system, including neurons.
2. Responses to external signals can change cytosolic calcium levels inside astrocytes rapidly and at high temporal precision.
3. Astrocytes release neurotransmitters in a calcium-dependent manner.
4. Release of neurotransmitter by astrocytes can elicit neuronal responses.

This cycle suggests that astrocytes and neurons are in close bidirectional communication at synapses, which could serve as a feedback mechanism. Since astrocytes are able to influence one another via gap junctions (e.g., via calcium waves), and can interact with multiple nearby neurons, they are in a position to integrate information from multiple sources so that their output can in turn affect many cells. These hypothesized functions reflect the importance of astrocytes in brain function and underscore the concept that synapses actually include three critical components (presynaptic neuron, postsynaptic neuron, and astrocyte): a "tripartite" structure. As scientists continue to study glial signaling in greater detail, the role of glial cells in the "tripartite synapse" will become clearer.

# References

Araque, A., Parpura, V., Sanzgiri, R.P., Haydon, P.G., 1999. Tripartite synapses: glia, the unacknowledged partner. Trends Neurosci. 22, 208−215.

Arnth-Jensen, N., Jabaudon, D., Scanziani, M., 2002. Cooperation between independent hippocampal synapses is controlled by glutamate uptake. Nat. Neurosci. 5, 325−331.

Bellesi, M., de Vivo, L., Tononi, G., Cirelli, C., 2015. Effects of sleep and wake on astrocytes: clues from molecular and ultrastructural studies. BMC. Biol. 13, 66.

Bergles, D.E., Jahr, C.E., 1997. Synaptic activation of glutamate transporters in hippocampal astrocytes. Neuron 19, 1297−1308.

Bezzi, P., Gundersen, V., Galbete, J.L., Seifert, G., Steinhauser, C., Pilati, E., et al., 2004. Astrocytes contain a vesicular compartment that is competent for regulated exocytosis of glutamate. Nat. Neurosci. 7, 613−620.

Charles, A.C., Merrill, J.E., Dirksen, E.R., Sanderson, M.J., 1991. Intercellular signaling in glial cells: calcium waves and oscillations in response to mechanical stimulation and glutamate. Neuron 6, 983−992.

Garcia-Marin, V., Garcia-Lopez, P., Freire, M., 2007. Cajal's contributions to glia research. Trends Neurosci. 30, 479−487.

Hatton, G.I., Parpura, V., 2004. Glial Neuronal Signaling. Kluwer Academic Publishers, Boston, MA.

Ramón y Cajal S., 1909. Histologie du système nerveux de l'homme & des vertébrés, Ed. française rev. & mise à jour par l'auteur, tr. de l'espagnol par L. Azoulay. Maloine, Paris.

Ransom, B.R., Sontheimer, H., 1992. The neurophysiology of glial cells. J. Clin. Neurophysiol. 9, 224−251.

Rash, J.E., Yasumura, T., Dudek, F.E., Nagy, J.I., 2001. Cell-specific expression of connexins and evidence of restricted gap junctional coupling between glial cells and between neurons. J. Neurosci. 21, 1983−2000.

Rothstein, J.D., Dykes-Hoberg, M., Pardo, C.A., Bristol, L.A., Jin, L., Kuncl, R.W., et al., 1996. Knockout of glutamate transporters reveals a major role for astroglial transport in excitotoxicity and clearance of glutamate. Neuron 16, 675−686.

Verkhratsky, A., Steinhauser, C., 2000. Ion channels in glial cells. Brain Res. Brain Res. Rev. 32, 380−412.

Verkhratsky, A., Nedergaard, M., 2018. Physiology of astroglia. Physiol. Rev. 98, 239−389.

Virchow, R., 1846. Uber das granulierte ansehen der Wandungen der Gerhirnventrikel. Allg. Z. Psychiatr 3, 242−250.

Witcher, M.R., Kirov, S.A., Harris, K.M., 2007. Plasticity of perisynaptic astroglia during synaptogenesis in the mature rat hippocampus. Glia 55, 13−23.

# Glossary

**Absolute refractory period**  the brief period of time after an action potential when a second action potential cannot be generated due to sodium channel inactivation.
**Acetylcholinesterase (AChE)**  an enzyme that degrades the neurotransmitter acetylcholine.
**Action potential**  a transient, all-or-none voltage spike that can travel very quickly along the length of an axon and is caused by the flow of ions through voltage-gated sodium and potassium channels.
**Action potential threshold**  the minimum membrane potential to which a cell must be depolarized in order to initiate an action potential.
**Activation**  the process of making something active. In neuroscience this is often used to describe the process by which an ion channel or receptor is made active (or in the case of an ion channel, opens) in response to the stimulus that normally activates it (e.g., voltage or a ligand).
**Active zone**  the site of neurotransmitter release within the presynaptic terminal.
**Adenylyl cyclase**  the cellular enzyme that synthesizes cAMP.
**Afterhyperpolarization**  the transient hyperpolarization of the membrane potential directly following an action potential to a voltage that is more negative than the resting membrane potential.
**Agonist**  a compound that elicits the same biological effect(s) as the natural ligand when it binds to a receptor.
**Allosteric**  a change in the shape and/or activity of a protein resulting from an interaction at a site on that protein other than the chemically binding site. For example, when a chemical binds to a particular site on a protein but causes a change in another location on that protein. In the context of ligand-gated neurotransmitter receptors, a neurotransmitter can bind to the receptor in one location, but its effects can be modulated when another molecule binds to the receptor at a different site on that receptor.
**Amine neurotransmitters**  neurotransmitters that have an amine group; includes dopamine, norepinephrine, epinephrine, serotonin, and histamine.
**Ampere**  one coulomb of electrical charge ($6.24 \times 10^{18}$ charged particles) moved in one second.
**Angstrom (Å)**  a measure of distance, equal to $10^{-10}$ m.
**Anion**  a negatively charged particle (such as an ion).
**Antagonist** (a.k.a. "blocker")  a compound that attenuates or prevents the effect of an agonist when bound to a receptor.
**Associative long-term potentiation**  a form of long-term potentiation (LTP) that is induced at weakly stimulated synapses positioned in close proximity to strongly stimulated synapses. The simultaneous activation of both weak and strong synapses relieves the magnesium block of NMDA receptors in the weakly stimulated synapses, leading to LTP.
**Autoreceptor**  a receptor that is activated by molecules released by the same cell that expresses the receptor.
**Binding affinity**  the strength of the binding interaction between two molecules (e.g., between a neurotransmitter and a neurotransmitter receptor).
**Calcium microdomain**  the local cytoplasmic volume within which the calcium ion concentration is elevated around open calcium channels (in the context of synapses, this concentration is often 10–100 μM).
**Calcium nanodomain**  a "cloud" or locally high concentration of calcium ions surrounding the pore of a single open calcium channel.
**Capacitor**  two conducting materials separated by an insulating material (e.g., a cell membrane).
**Catecholamines**  monoamine transmitters that contain a catechol chemical group (dopamine, norepinephrine, epinephrine).
**Cation**  a positively charged particle (such as an ion).
**Cerebrospinal fluid**  fluid produced in the ventricles that contains ions and nutrients needed by neurons and glia.

**Cofactor**  a substance [other than the substrate(s) for an enzymatic reaction] whose presence is essential for the activity of an enzyme.

**Concentration gradient**  exists when there is an unequal distribution of a given type of ion or molecule between two locations (e.g., on opposite sides of a plasma membrane).

**Conductance (g)**  the ability of a charged particle (such as an ion) to move from one point to another; measured in siemens (S), this is the inverse of resistance (R).

**Connexin**  the protein subunit found in connexons. Six connexins come together to form a hemichannel (connexon) at gap junctions.

**Connexon**  a hexamer composed of six connexin protein subunits; a connexon is a hemichannel that lines up with a connexon in another neuron to form a pore between two cells called a gap junction.

**Contralateral**  on the opposite side of the left–right division of the body.

**Corelease**  a type of cotransmission in which a neuron packages more than one neurotransmitter into the same synaptic vesicles so that these neurotransmitters are released simultaneously upon vesicle fusion.

**Cotransmission**  a neuron's use of two or more neurotransmitters in synaptic transmission.

**Coulomb**  a unit of electric charge containing $6.24 \times 10^{18}$ charged particles.

**Current (I)**  the movement of charged particles, measured in amperes (A), whose direction is defined as the direction of net positive charge movement.

**Cyclic AMP (3'5'-cyclic adenosine monophosphate; cAMP)**  an intracellular second messenger synthesized by adenylyl cyclase.

**Cyclic GMP**  an intracellular second messenger that is synthesized by guanylyl cyclase.

**Depolarization**  a change in membrane voltage toward more positive values.

**Differential release**  a type of co-transmission in which each of the neurotransmitters used by a neuron is released under different (though not necessarily mutually exclusive) conditions; e.g., in response to different stimulus patterns, or from different synaptic release sites.

**Dissociation constant ($K_D$)**  a measure of the strength of binding affinity, calculated as the ratio of the unbinding rate ($K_{off}$) divided by the binding rate ($K_{on}$). $K_D$ and affinity are inversely related.

**Downregulation**  the process by which a cell decreases the production and expression of a given cellular component or protein (e.g., neurotransmitter receptors).

**Driving force**  the difference between the equilibrium potential of an ion and a cell's membrane potential, measured in mV. This represents the force that moves an ion into a cell (negative driving force) or out of a cell (positive driving force). The equation used to calculate driving force is: $DF = V_m - E_{ion}$.

**Efflux**  the movement of charged particles (such as ions) out of a cell.

**Electrical potential (or voltage)**  the difference in net charge between two locations; this can exert a force on a charged particle, which is measured in volts (V).

**End-product inhibition**  a negative-feedback process in which the rate of a synthesis reaction is inhibited by its own product (e.g., a neurotransmitter).

**Endocytosis**  the process by which a patch of membrane (e.g., membrane from synaptic vesicles) is brought inside a cell.

**Endogenous ligand**  a natural molecule in the body that acts as an agonist for a specific type of neurotransmitter receptor.

**Endplate current (EPC)**  a current recorded in the postsynaptic cell at the neuromuscular junction (endplate) that occurs in response to neurotransmitter release by the presynaptic motor neuron.

**Endplate potential (EPP)**  a depolarization in the postsynaptic muscle cell at the neuromuscular junction (endplate) that occurs in response to neurotransmitter release by the presynaptic motor neuron.

**Equilibrium potential (also called the reversal potential or Nernst potential)**  the voltage across a cell membrane at which the diffusion force and the electrostatic force acting on a given type of ion are equal.

**-ergic (suffix)**  Relating to a specific type of neurotransmitter. For example, a neuron that releases acetylcholine, a vesicle that contains acetylcholine, and a synapse at which acetylcholine acts as a neurotransmitter can all be described as cholinergic.

**Excitatory postsynaptic potential (EPSP)**  a depolarization of the postsynaptic cell that occurs at a synapse in response to neurotransmitter release by the presynaptic neuron.

**Exocytosis**  the process by which the membrane of a synaptic vesicle or cell vacuole merges with the plasma membrane and releases its contents outside a cell.

**Extracellular matrix** a noncellular three-dimensional macromolecular network composed of collagens and other proteins forming a complex network that surround cells in tissues and organs. At synapses, adhesion and regulatory proteins are anchored in the part of the extracellular matrix that resides within the synaptic cleft.

**Gap junction** an intercellular pathway that connects the cytoplasm of two cells and is formed by the expression of a connexon in each cell that join together to form a pore. A gap junction is the site of electrical communication between neurons at electrical synapses.

**GTPase** an enzyme that converts GTP to GDP.

**Guanlylyl cyclase** the cellular enzyme that synthesizes cGMP.

**Heteromeric** containing more than one type of component (e.g., a neurotransmitter receptor made up of different types of protein subunits).

**Heteromeric connexon** a connexon that contains multiple different types of connexins.

**Heterosynaptic long-term potentiation** a form of long-term potentiation (LTP) induced in synapses that were not directly stimulated, but are adjacent to those that were stimulated to induce this form of LTP.

**Homosynaptic long-term potentiation** a form of long-term potentiation (LTP) induced at a synapse that was directly stimulated.

**Homomeric connexon** a connexon that contains six connexins of the same type.

**Homeostasis** a balance or equilibrium that is maintained in the face of external changes.

**Homomeric** containing only identical components (e.g., a neurotransmitter receptor made up of a single type of protein subunit).

**Hydrophilic** dissolvable in water ("water-loving").

**Hydrophobic** not dissolvable in water ("water-avoiding").

**I–V curve (I–V plot)** a plot of the relationship between ionic current (I) and voltage (V).

**in vitro** L. "in glass"; in neuroscience, this is used to refer to experiments conducted using biological components in a test tube or other artificial environment.

**in vivo** L. "in the living"; in neuroscience, this refers to experiments conducted in the intact nervous system of an animal.

**Ion channel** a protein pore that provides pathways for ions to flow across a membrane.

**Ion channel deactivation** the closing of an ion channel when the stimulus that opens it (e.g., a ligand or membrane voltage) has been removed.

**Ion channel gating** the opening and closing of ion channels.

**Ion channel inactivation** the closing of an ion channel despite the continued presence of a stimulus that can open it.

**Ionotropic receptor** a ligand-gated receptor that includes both a neurotransmitter binding site and an ion channel.

**Ipsilateral** on the same side of the left–right division of the body.

**Kinetics** the study of the rates of reactions or events. For example, in the context of synaptic transmission, this term can be applied to the speed of the onset and decay time-course of neurotransmitter-evoked currents in postsynaptic cells.

**Large dense-core synaptic vesicle** a synaptic vesicle larger than small clear vesicles that contains large peptide transmitters and has an electron-dense center.

**Length constant** a value that represents how far a change in membrane potential can spread passively down an axon until enough current leaks out of the axon that the depolarization is reduced to $\sim 37\%$ of its initial value.

**Ligand-gated ion channel** an ion channel that opens in response to ligand binding to the channel protein.

**Membrane potential ($V_m$)** the voltage difference across a cell membrane.

**Metabotropic receptor** a neurotransmitter receptor that triggers second messenger-mediated effects within a cell in response to neurotransmitter binding.

**Miniature endplate current (mEPC)** a current recorded in the muscle cell at a neuromuscular junction (the endplate) in response to the spontaneous fusion of a single synaptic vesicle with the plasma membrane, releasing the contents of one vesicle (a package or quantum of transmitter molecules).

**Miniature endplate potential (mEPP)** a depolarization in the muscle cell at a neuromuscular junction (the endplate) in response to the spontaneous fusion of a single synaptic vesicle with the plasma membrane, releasing the contents of one vesicle (a package, or quantum of transmitter molecules).

**Miniature excitatory postsynaptic potentials (mEPSPs)**  a depolarization in the postsynaptic cell at a chemical synapse that occurs in response to the spontaneous fusion of a single synaptic vesicle with the plasma membrane, releasing the contents of one vesicle (a package, or quantum of transmitter molecules).

**Monoamines**  neurotransmitters that contain one amine group (including dopamine, norepinephrine, epinephrine, serotonin, and histamine).

**Morphological**  relating to shape or structure (e.g., the shape of a neuron).

**Myelin**  an insulating material made from the membrane of glial cells that wraps around axons and increases action potential conduction velocity.

**Neuromodulator**  a molecule that is released from a cell in the nervous system and modifies the characteristics of fast synaptic transmission mediated by classical neurotransmitters.

**Neuromuscular junction**  a synapse between a motor neuron and a muscle cell.

**Neurotransmitter**  a molecule released from a neuron that transmits a signal to another cell.

**Neurotransmitter receptor**  a specialized protein on the surface of a cell to which neurotransmitters bind to exert their effects on the cell.

**Neurotransmitter switching**  a change in a neuron's neurotransmitter phenotype.

**Node of Ranvier**  the small gap between myelinated segments of an axon that contain high concentrations of voltage-gated sodium and potassium channels. At these sites, a propagating action potential can be regenerated.

**Ohm ($\Omega$)**  the unit of electrical resistance; inverse of conductance (g).

**Ohm's Law**  the relationship between voltage (V), current (I), and resistance (R); $V = IR$.

**Oligodendrocyte**  the type of glial cell that forms myelin around axons in the central nervous system.

**Oocyte**  an immature egg cell (often used in neuroscience experiments as a nonneuronal cell within which to experimentally express specific proteins for study in isolation).

**Paired-pulse plasticity**  a form of plasticity that results in a change in the magnitude of transmitter released after the second action potential in a pair as compared to the magnitude of transmitter released after the first action potential in the pair.

**Paracrine transmission**  a form of neurotransmission in which molecules diffuse away from their site of release and bind to receptors that are not immediately adjacent to the sites of release. See Volume transmission.

**Permeability**  the property of a porous material (e.g., a cell membrane) that permits substances to pass through it (or across it).

**Phosphodiesterase**  an enzyme that breaks a phosphodiester bond, converting cAMP to 5'-AMP or cGMP into 5'-GMP.

**Phosphorylation**  a process by which a phosphate group is added to an organic compound.

**Postsynaptic density**  the collection of transmitter receptors and associated anchoring proteins located in the postsynaptic cell at a synapse; essential for collecting postsynaptic proteins together and or mediating postsynaptic signaling.

**Potassium leak current**  a potassium current flowing through the types of neuronal membrane potassium channels that are open when the cell is at or near its resting membrane potential.

**Presynaptic density**  the collection of proteins at the site of neurotransmitter release in a presynaptic neuron that regulate synaptic vesicle fusion with the plasma membrane.

**Protein kinases**  enzymes that regulate the function of proteins by phosphorylating them.

**Quantal content (q)**  the number of synaptic vesicles that fuse with the plasma membrane during a presynaptic action potential.

**Quantum**  the amount of neurotransmitter contained in a single synaptic vesicle.

**Rate-limiting step**  the slowest step in a chemical reaction. The speed of this step dictates the speed of the reaction and can be regulated by a range of cellular processes.

**Receptor desensitization**  decreased responsiveness of a neurotransmitter receptor as a result of ongoing (long-term) exposure to an agonist.

**Rectifying ion channel**  an ion channel that passes current more readily in one direction than in the other.

**Relative refractory period**  a brief period of time following the absolute refractory period near the end of an action potential when a larger-than-normal depolarization is required in order to trigger a second action potential; this occurs because during this period only some voltage-gated sodium channels have recovered from inactivation.

**Resistance** the ability to prevent the flow of charged particles (such as ions), measured in Ohms; this is the inverse of conductance (g).

**Resting membrane potential (Vrest)** the membrane potential of a cell in when it is at rest, defined as not receiving substantial synaptic input or generating an action potential.

**Retrograde messenger** a signaling molecule that is released by the postsynaptic cell and diffuses "backwards" in order to affect the presynaptic cell.

**Reuptake** the process by which some neurotransmitters are transported from the synaptic cleft back into the presynaptic terminal after release.

**Reversal potential** see equilibrium potential.

**Saltatory conduction** the process by which an action potential is only regenerated at the nodes of Ranvier along myelinated axons. Using this process for conduction, action potentials are sometimes said to "jump" from one node of Ranvier to the next node along myelinated axons. This form of conduction allows an action potential to be conducted very rapidly over long distances.

**Schwann cell** the type of glial cell that forms myelin around axons in the peripheral nervous system.

**Second messenger** a molecule that is formed in the cytoplasm in response to an extracellular event (e.g., ligand binding to a receptor) and initiates various biochemical processes inside the cell.

**Small clear synaptic vesicle** a synaptic vesicle containing small-molecule chemical transmitters that is smaller in size as compared to large dense-core vesicles.

**Synapse** the specialized subcompartment of a neuron, often located at the end of an axon, that is the site of communication with neighboring cells.

**Synaptic cleft** the extracellular gap between pre- and postsynaptic neurons at the synapse.

**Synaptic plasticity** the ability of cells to change how they communicate in response to past experiences.

**Synaptic vesicle** a structure within the presynaptic cytoplasm that is enclosed in a lipid bilayer and contains neurotransmitters that can be released from the neuron via exocytosis.

**Transmembrane domain** the portion of a membrane protein that is embedded in the cell membrane.

**Upregulation** the process by which a cell increases the production and expression of a cellular component or protein (e.g., neurotransmitter receptors).

**Vesicle trafficking** the regulated movement of synaptic vesicles between different compartments within the nerve terminal.

**Volt (V)** a measure of the difference in electrical potential that can drive one ampere of current against one ohm of resistance; describes the force of an electrical current.

**Voltage (V)** see electrical potential.

**Voltage-gated ion channel** a channel that opens when the voltage across the membrane changes.

**Voltage-sensing domain** the region of a voltage-gated ion channel that is affected by membrane voltage and contributes to the gating of the channel.

**Volume transmission** a form of synaptic communication that is driven by diffusion gradients of transmitter molecules, affects widespread regions of the brain, and occurs within the extracellular space. See Paracrine transmission.

# Index

*Note:* Page numbers followed by "*f*," "*t*," and "*b*" refer to figures, tables, and boxes, respectively.

## A

AAAD. Aromatic-L-amino-acid decarboxylase
AAT. *See* Aspartate aminotransferase
Absolute refractory period, 57
Acetyl CoA, 350
Acetylcholine (ACh), 156, 300–302, 334–335, 341–342, 345, 440–441, 474–475, 484, 490
  hypothesis in Alzheimer's disease, 361*b*
  identity as neurotransmitter, 345–350
  molecular structure, 350*f*
  packaging of acetylcholine into synaptic vesicles, 352–354
  receptors, 105, 338
    electric fishes and eels as experimental source, 218*b*
  release, 355, 355*f*
    synaptic vesicles, 354*f*
  role in nervous system, 359–362
  synthesis, 350–352, 350*f*
    *Galanthus nivalis* plant, 365*f*
    regulation, 356–358, 357*f*
  termination of action for, 358–359
  vesicular neurotransmitter transport, 353*b*
Acetylcholinesterase (AChE), 103–104, 358–359, 358*f*
  inhibitors, 364–365
  snowdrop plant, and, 365*b*
ACh. *See* Acetylcholine
AChE. *See* Acetylcholinesterase
Acid-sensing ion channels (ASIC), 238
ACTH. *See* Adrenocorticotropic hormone
Action potential(s), 19, 53–61, 54*f*, 73
  calcium entry into presynaptic terminal during, 144–146, 145*f*, 146*f*
  inactivation and deactivation of voltage-gated ion channels, 55–57
  myelin and nodes of Ranvier, 60–61
  potassium channels role in shaping calcium entry during, 146–150
  propagation, 59–60
  refractory periods, 57–59
    distributions of sodium and potassium ions, 58–59
  threshold, 53–54
Active membrane properties, 279–280, 283–285
Active zone, 13–14
  coupling of voltage-gated calcium channels to, 176–177, 177*f*
  evidence for synaptotagmin as calcium sensor in, 179–183
  during synapse development, 16–17
Activity-dependent phosphorylation, 202
Adenosine, 462
  cotransmission, 463*f*
Adenylyl cyclase, 258–259
Adrenaline. *See* Epinephrine
Adrenergic neurons, 369–370
Adrenocorticotropic hormone (ACTH), 424
Aequorin, 102
2-AG. *See* 2-Arachidonoylglycerol
Agatoxin, 100
AHP, 56
AII amacrine cells, 84, 86–88
Alfred Gilman and Martin Rodbell Nobel prize in physiology or medicine (1994), 248*b*, 248*f*
Allodynia, 417
Allosteric effect, 40
Allosteric nicotinic acetylcholine receptor, 365
α subunits, 45–46, 220, 223, 225–226, 246–250, 255*f*
α-7 nicotinic receptors, 220–221
α-bungarotoxin, 100, 218–219
α-ketoglutarate, 400
α-latrotoxin, 102
α-melanocyte-stimulating hormone (α-MSH), 424
ALS. *See* Amyotrophic lateral sclerosis
Alternative mRNA splicing, 425
Alzheimer's disease, acetylcholine hypothesis in, 361*b*
Amacrine cells, 84, 85*f*, 87*f*
Amine, 369
Aminergic neurons, 369–370
Amino acid neurotransmitters, 342
  GABA, 406–412
  glutamate, 399–405
  glycine, 413–417
  molecular structures, 400*f*

Amino acid neurotransmitters (*Continued*)
  Schizophrenia, 412
Amino acids, 37, 413
AMPA receptors, 227, 234–235, 235f, 312, 318, 405f
Amperes (A), 20
Amperometry, 161b
Amphetamine, 377, 395–396
  effects on dopaminergic neurons, 395f
Amyotrophic lateral sclerosis (ALS), 445
Animal electric fluid, 66–67
Animal electricity, 66
Animal model systems for studying synapses, 2–4, 3f
Anion, 20
Anti-NMDA receptor encephalitis, 232, 232b
Anti-anxiety drugs, 225
Anticholinergic drugs, 363
Anti-seizure drugs, GABA analogs as, 410b
Antisense oligonucleotide technique, 252
AP2 protein, 195
*Aplysia*, 303, 304f, 305f, 308
Aquaporins, 42b
Arachidonic acid, 261
Arachidonic acid pathway, 261–265
  cytoplasmic phospholipase A2 activation, 263f
2-Arachidonoylglycerol (2-AG), 239, 318
Aromatic-L-amino-acid decarboxylase (AAAD), 372, 378–379, 382, 386
ASIC. *See* Acid-sensing ion channels
Aspartate, 400
Aspartate aminotransferase (AAT), 400, 403, 403f
Associative LTP, 312–313
Astrocytes, 414, 481–482, 482f
  EPSC amplitude modulation by glutamate release, 491f
  function, 482
  glutamate transporters in, 484, 485f, 486f
  interactions between, 488–489
  ion channels in, 487–488
  neurotransmitter
    release from astrocytes, 489–490
    uptake by astrocytic transporters, 484–487
  role in information processing within brain, 490–491
  roles in synaptic function, 482–488
  surround synapses, 483f
Asynchronous neurotransmitter release, 181
ATP, 30, 96, 124–125, 125f, 236, 460–462, 463f
Autocrine, 299, 300f
  inhibitory feedback, 299
Autonomic nervous system, cholinergic synaptic transmission in, 360–362
  anticholinergic drugs, 363
  compounds that inhibit AChE, 364–366
  deadly nightshade plant and muscarinic ACh receptors, 363b
  parasympathomimetic drugs, 363
  snowdrop plant, AChE, and nicotinic ACh receptors, 365b
Autonomic peripheral nerves, 425–426
Autophosphorylate, 266–268
Auxiliary proteins, 44–46
Axelrod, Julius, 376, 377f
  and discovery of neurotransmitter reuptake, 376b
Axial resistance, 277
Axon hillock, 59, 59f

# B

Back-propagating action potential, 313
Bacterial toxins, 253b
Bag cells, 455
"Ball and chain" mechanism, 55f, 56
BAPTA, 135–136, 135f
Barbiturates, 225
Basal ganglia, 385
Bassoon protein, 172
BDNF. *See* Brain-derived neurotrophic factor
Beery twins genetic sequencing, 380
Benzodiazepines, 225, 366
β blockers, 475
β subunits, 45–46
β-adrenergic receptors, 320
β-endorphin (β-ENDO), 424
β-γ subunits, 250
β-lipotropin (β-LPH), 424
BH4. *See* Tetrahydrobiopterin
Binding affinity, 124
Binomial theory, 108
Biochemical event, 210
Black mamba snakes (*Dendroaspis polylepsis*), 46b
Blood–brain barrier, 122–123, 122b, 427
*Bordetella pertussis*, 253
Botulinum toxin, 102, 168, 170b, 171f
Botulinum toxin A (BOTOX), 102
Brain-derived neurotrophic factor (BDNF), 78b, 211, 318–319
Bulk endocytosis, 195, 197, 198f
Bungarotoxin, 100

# C

"C-type" inactivation, 56–57, 57f
C2 domains. *See* Calcium-binding domains
CA1 region of hippocampus, 308
CA3 region of hippocampus, 308
Cable properties, 277–279
Caged calcium, 132–133, 132f

Caged glutamate, 401–402
Calbindin, 129, 296
CALC1 gene, 425, 425f
Calcitonin gene-related peptide (CGRP), 425
  migraines treatment by blocking action, 426b
Calcium
  calcium-activated potassium channels, 147–148, 148f
  calcium-binding domains, 179–180
  calcium-triggered synaptic vesicle fusion, 180f
  homeostasis, 128–129, 129f
  ionophore, 132
  ions, 97, 178
    binding, 288
    distribution across cell membrane, 121–124
  liposome technique, 132
  as trigger for neurotransmitter release, 121–129
    cellular mechanisms, 124–129
    neurotransmitter release control by calcium ions, 130–140
    VGCC in nerve terminals, 141–150
  waves in glial cells, 488–489
Calcium release-activated calcium (CRAC), 128
Calcium-binding domains (C2 domains), 168
Calcium-calmodulin-dependent protein kinase II (CaMKII), 86–88, 201–202, 236, 317, 373–374, 406–407
Calcium-dependent action potentials, 469
Calcium-dependent phosphatase calcineurin, 195
Calcium-release relationship, 133
Calcium-triggered synaptic vesicle fusion, biochemical mechanisms of, 163–184
  coupling of voltage-gated calcium channels, 176–177, 177f
  evidence for synaptotagmin as calcium sensor, 179–183
  experimental evidence, 168–171
    botulism and tetanus, 170b
  proteins in calcium-triggered vesicle release, 166–168
  recovery and disassembly of SNARE protein CORE complex, 183–184
  SNARE proteins, 174–176
    of CORE complex, 177–178
  study of vesicle fusion, 164–165
  synaptic vesicles moving to correct location, 172–174
  vesicle trafficking and fusion mechanisms in cells, 165b
Calmodulin, 124, 296–298
Calretinin, 129
CaMKII. *See* Calcium-calmodulin-dependent protein kinase II
cAMP. *See* Cyclic AMP
Capacitance, 20–22

Capacitor, 21–22
Capillaries, 122
Carbon monoxide (CO), 342, 435, 443–445
  synthesis
    reaction, 443f
    regulation, 444–445
  vasorelaxation triggered by, 444f
Carboxy tail, 296–298
Catechol group, 369
Catechol-O-methyltransferase (COMT), 375–376
Catecholamine (norepinephrine)-containing vesicles, 468f
Catecholamine neurotransmitters, 440–441
  catecholamine synthesis, 370–373, 371f, 407
    dopamine to norepinephrine, 372
    L-DOPA to dopamine, 372
    norepinephrine to epinephrine, 372–373
    regulation, 373–375, 375f
    tyrosine to L-DOPA, 370–372
  Julius Axelrod and discovery of neurotransmitter reuptake, 376b
  termination of action of catecholamines, 375–376
Catecholamine synthesis, 370–373, 371f, 407
Catecholaminergic neurons, 369–370, 372
Catecholamines, 162, 162f, 369, 375f, 453
Cation, 20
Cav3. *See* Low-voltage-activated calcium channels
CBS. *See* Cystathionine β-synthetase
Cell membrane, 21–22, 22f
  calcium ions distribution across, 121–124
  distributions of sodium and potassium ions, 58–59
  equilibrium distribution of potassium ions, 23f
  ion movement across, 23–33
Central nervous system, 220–221, 237, 275, 293, 413, 415–416
  acetylcholine in, 359–360, 360f
  effects of nitric oxide in, 441–443
Central synapses
  EPSCs, 115f
  EPSPs, 114f
  quantal analysis of action-potential-evoked EPSCs, 117f
  quantal theory of chemical transmitter release at, 113–116
Cerebral ischemia, 238
Cerebrospinal fluid, 122–123
CFTR. *See* Cystic fibrosis transmembrane conductance regulator
cGMP, 439–440
CGRP. *See* Calcitonin gene-related peptide
Channel deactivation, 56
Channel gating, 35–36
Channel inactivation, 55–57

Chaperone, 172
Charged amino acids, 37
ChAT. *See* Choline acetyltransferase
Chemical communication, costs and advantages of, 95–97, 96t
Chemical neurotransmitter, 287
Chemical synapse, 73, 88–89
   function, 97
      costs and advantages of chemical communication, 95–97, 96t
      electrical footprints of chemical transmitter release, 97–101
      spontaneous release of single neurotransmitter vesicles, 102–105
Chemical transmitter release
   costs and advantages of, 96t
   electrical footprints of, 97–101
      microelectrode technique and recording of synaptic events, 98b
      natural molecules as experimental tools, 100b
   quantal theory of, 105–109
Chemical transmitter systems
   neurotransmitter characteristics, 339–340
   neurotransmitter *vs.* neuromodulator, 333–335
   signaling molecule as neurotransmitter, 335–339
   types of neurotransmitters, 341–343
Cholera toxin, 253
Choline, 351
Choline acetyltransferase (ChAT), 350, 357–358
Cholinergic
   neurons, 359–360
   switch, 465–466
   synaptic transmission in autonomic nervous system, 360–362
   tone, 361–362
Cholinesterase inhibitors, 364–365
Chronic ziconotide treatment, 139–140
Classical (type 1) neurotransmitters, 422–423
Classical conditioning, 312–313
Classical neurotransmitters, 342
   acetylcholine, 341
Clathrin, 195
Clathrin-mediated endocytosis, 195–197, 196f
CLIP. *See* Corticotropin-like intermediate lobe peptide
*Clostridium botulinum*, 102
*Clostridium tetani*, 102
CMAP. *See* Compound muscle action potential
Coated pits, 193–195
Cocaine, 377, 395
   effects on dopaminergic neurons, 395f
Cognitive processes, 96
Coincidence detector, 229–231
Compound muscle action potential (CMAP), 291

COMT. *See* Catechol-O-methyltransferase
Concentration gradient, 20
Conditionally essential amino acid, 374
Conductance, 32
Cone snail toxins and chronic pain treatment, 139b
Coneurotransmitter, 460–461
*Conium aculatum*, 221b, 222f
Connexin (Cx), 70
   Cx36, 84, 86–88
Connexons, 70, 71f
Conotoxins, 100, 139–140
Contralateral ear, 459
*Conus magus* snails, 139–140, 139f
Conventional amperometry approach, 161
CORE complex of proteins, 168, 169f
   SNARE proteins of, 177–178
      priming of docked synaptic vesicles, 179f
      recovery and disassembly after synaptic vesicle fusion, 183–184
Corelease of neurotransmitters, 456–459
   vesicular synergy as function of, 459–460
Cortical synapses, electrical synapses in, 81–84
Corticotropin-like intermediate lobe peptide (CLIP), 424
Corticotropin-releasing hormone (CRF), 473
Cotransmission of neurotransmitters, 452–462, 452f
CRAC. *See* Calcium release-activated calcium
Crayfish giant motor synapse, 68
CRF. *See* Corticotropin-releasing hormone
Curare, 101, 103–104, 218–219
Current–voltage relationships ($I–V$ relationships), 49–53
Cx. *See* Connexin
Cyclic AMP (cAMP), 245–246, 442–444
   pathway, 258–259, 259f
Cyclic nucleotides, 440
Cys-loop receptors. *See* Pentameric ligand-gated ion channel family
Cystathionine β-synthetase (CBS), 445
Cysteine aminotransferase, 445
Cystic fibrosis transmembrane conductance regulator (CFTR), 253
Cytoplasmic calcium, 402–403, 428
   microdomains, 137–140
      imaging calcium entry during action potential activity, 140f
Cytoplasmic receptors, 210, 212–213, 212f

# D

D1 dopamine receptors, 86
DAG. *See* Diacylglycerol
Dale, Sir Henry Hallett, 346, 449, 450f
"Dale's Principle", 449–450, 455

Dalton, 71
DAP. *See* 3,4-Diaminopyridine
DAT. *See* Dopamine selective transporter
DBH. *See* Dopamine β-hydroxylase
Deadly nightshade (*Atropa belladonna*), 363
  plant and muscarinic ACh receptors, 363b
Deathstalker scorpion (*Leiurus quinquestriatus hebraeus*), 100
Declarative memories, 308
δ-Philanthotoxin (PhTX-433), 102
Delusions, 232
Dendrites, 280–283, 281f, 282f
Dendritic
  filtering process, 277
  ion channels, 283
  spikes, 284f, 285
  spines, 287
  tree, 275
Dense-core vesicle. *See* Large dense-core vesicles
Dentate gyrus, 308
Dephosphorylated synapsin, 202
Depolarization, 130–131
Depolarizing signal, 73
Desensitization, 221
DF. *See* Driving force
DGL. *See* Diacylglycerol lipase
Diacylglycerol (DAG), 259–260, 318
Diacylglycerol lipase (DGL), 318
3,4-Diaminopyridine (DAP), 150
Diffraction barrier, 41
L-Dihydroxyphenylalanine (L-DOPA), 370–372
  conversion to dopamine, 372
  Parkinson's disease treatment with catecholamine precursor, 385b
  tyrosine precursor of, 370–372
Direct ion channel pathway, 256–258, 258f
"Direct pathway", 252–253, 256f, 257f
"Direct" G-protein pathway, 256
"Disconnectivity hypothesis" for schizophrenia, 412
Disulfide bonds, 424–425
"Doctor Death", 27b
Domino effect, 254
Donepezil, 362
L-DOPA decarboxylase. *See* Aromatic-L-amino-acid decarboxylase
Dopamine, 84, 369, 384–386
  conversion to norepinephrine, 372
  dopamine-related drugs for Parkinson's disease, 385
  L-DOPA conversion to, 372
  receptors, 390–391
Dopamine selective transporter (DAT), 376, 395
Dopamine β-hydroxylase (DBH), 372
Dopamine-responsive dystonia (DRD), 380

Driving force (DF), 29–30
*Drosophila* neuromuscular junction, 321, 323f
Drugs from nature, 139b, 263b
Dynamin, 195–197

# E

EAATs. *See* Excitatory amino acid transporters
Earl Sutherland Nobel Prize in physiology or medicine (1971), 246b, 246f
Eastern Indian red scorpion (*Buthus tamulus*), 100
Eccles, Sir John C., 450f
Ecto-nucleotidases, 462
EDRF. *See* Endothelium-derived relaxing factor
EEG. *See* Electroencephalography
Effector proteins, 252–253
EGF. *See* Epidermal growth factor
Egg-laying hormone (ELH), 455
EGTA, 135–136
Electric current, 19–20
Electric fishes and eels as experimental source for acetylcholine receptors, 218b, 218f
Electric pile, 66–67, 67f
Electrical activity role in nervous system development, 469–471
Electrical potential. *See* Voltage
Electrical synapses, 76–85, 96
  in adult mammalian nervous system, 84–85
  conduction of action potential signals, 74f
  in cortical synapses, 81–84
  discovery, 67–69
  early evidence in electrical communication, 66–67
  fluorescent dyes, 76f
  history, 65–69
  laminar organization of retina, 87f
  in mammalian nervous system, 76–84
  in neuromuscular synapse, 77–81
  physiological characteristics, 71–76
  plasticity, 85–89
  rectifying and nonrectifying, 72f
  structure, 69–76
    gap junction, 70–71
  synaptic delays at chemical and, 75f
  transfer of electrical potential, 74f
Electrocytes, 218–219
Electroencephalography (EEG), 412
Electrogenic action, 126
Electrogenic neuron, 30
Electrophoresis, 337
Electrophysiologists, 65
ELH. *See* Egg-laying hormone
ENaC/DEG. *See* Epithelial sodium channel/degenerin
End-product inhibition, 374–375
Endfeet, 122–123

Endocannabinoids, 318
Endocrine, 299
Endocytosed synaptic vesicles, 190
Endocytosis, 183, 189, 191
　bulk, 197, 198f
　clathrin-mediated, 195–197, 196f
　kiss-and-run mechanism, 197–199, 199f
　mechanisms, 195
　occurring outside active zone, 193–194
　　endocytic zones, 194f
　synaptopHluorin used to visualize, 193f
Endogenous calcium buffer proteins, 129
Endogenous ligands, 333
Endophilin, 195
Endoplasmic reticulum, 127, 423
　calcium homeostasis mechanisms, 127f
Endosomes, 197
Endothelial cells, 122–123
Endothelial NOS (eNOS), 436–439
Endothelium-derived relaxing factor (EDRF), 435–438
Endplate current (EPC), 106f, 110
Endplate potentials (EPPs), 97, 99f, 130–132, 150f
eNOS. See Endothelial NOS
Entorhinal cortex, 308
Enzymatic degradation, 340
Enzyme-linked receptors, 211
EPC. See Endplate current
Ephrin receptors, 211
Epidermal growth factor (EGF), 239
Epinephrine, 369–370
　norepinephrine conversion to, 372–373
　sources and roles in brain function, 388–389
Epithelial sodium channel/degenerin (ENaC/DEG), 238
EPPs. See Endplate potentials
EPSCs. See Excitatory postsynaptic currents
EPSPs. See Excitatory postsynaptic potentials
Equilibrium dissociation constant ($K_D$), 124
Equilibrium potential for ion, 24–25
Erenumab-aooe, 427
Excitability, 19
Excitation-secretion coupling, 95
Excitatory amino acid transporters (EAATs), 484
Excitatory junction potential, 461
Excitatory neurotransmitter, 209–210
Excitatory postsynaptic currents (EPSCs), 115f
Excitatory postsynaptic potentials (EPSPs), 114, 114f, 277, 278f, 456
Excitatory neurotransmitter, 341, 399–400
Exocytosis
　biochemical mechanisms of calcium-triggered synaptic vesicle fusion, 163–184
　discovery of neurotransmitter release mechanisms, 155–162

Extracellular fluid, 400, 486–487
Extracellular ligand, 245–246
Extracellular matrix, 14–15
Extrasynaptic regions, 220–221
Exuberant synapses, 78b

## F
Faba bean (*Vicia faba*), 385
FAD. See Flavin adenine dinucleotide
Farad (F), 21–22
Fast-scan cyclic voltammetry, 161
Feeding behavior
　learning paradigm, 312–313
　neuropeptide Y effects, 431–432
Filtered synaptic events, 279
Flavin adenine dinucleotide (FAD), 436
Flavin mononucleotide (FMN), 436
Fluorescence-tagged toxin, 137
FM dyes, 192
FMN. See Flavin mononucleotide
Freeze-fracture electron microscopy, 158–159, 159f
Frozen addicts, 387b
Funnel web spider (*Agelenopsis aperta*), 100
Furchgott, Robert F., 436f, 437
Fusion protein, 192

## G
G-proteins, 248
　G-protein-coupled pathways
　　arachidonic acid pathway, 261–265, 262f
　　cyclic AMP (cAMP) pathway, 258–259
　　direct pathway, 256–258
　　metabotropic receptors, 251–252
　　phosphoinositol, 259–261
　transducers, 247–248
GABA, 209, 342, 399, 406–412, 407f, 457–458, 457f, 490
　effective therapeutic agents, 410b
　and glycine transmission, 408b
　synthesis, 406–407
　termination of GABA action, 408–411
　vesicular and reuptake transporters, 409f
GABA transaminase (GABA-T), 408–409
GABA transporter (GAT), 411, 487
GABA/glutamate receptor expression, 469–470
GABA/glycine
　coregulation of IPSP characteristics, 459
　corelease, 457, 459
　GABA/glycine-mediated IPSP, 458–459
$GABA_A$ receptors, 223–224, 223f
　antagonists, 408
　as therapeutic target, 224b
GABAergic neurons, 406

GAD. *See* Glutamic acid decarboxylase
Galantamine, 362, 365
*Galanthus nivalis* plant, 365f
Galvanism, 66
Gamma oscillations, 412
γ-lipotropin (γ-LPH), 424
Gap junction, 69, 70f, 488–489
   electron micrograph of neuronal, 69f
   pores, 72
   between spinal motor neurons, 77f
   structure, 70–71
GAPs. *See* GTPase-activating proteins
Gas molecules, 435
Gaseous messengers, 342. *See also* Gaseous neurotransmitters
Gaseous neurotransmitters
   carbon monoxide (CO), 443–445
   hydrogen sulfide ($H_2S$), 445
   molecular structures, 436f
   nitric oxide (NO), 435–443
Gasotransmitters, 435. *See also* Gaseous neurotransmitters
GAT. *See* GABA transporter
Gating of ion channels, 35–37, 36f
*Gazzetta Medica Italiana*, 8–9
GEF. *See* Guanine nucleotide exchange factor
Gene knockout technique, 181
Gepant drugs, 427
GFP. *See* Green fluorescent protein
*Ginkgo biloba* plant, 408b, 408f
GLAST. *See* Glutamate-aspartate transporter 1
Glaucoma, 363
Glia. *See* Glial cells
Glial cells, 61, 61f, 404, 481–482
Gliotransmitters, 489
GLT-1. *See* Glutamate transporter 1
Glutamate, 209–210, 215, 342, 399–405, 432, 484, 489–490
   dehydrogenase, 400
   identity of action for glutamate, 401–402, 402f
   ionotropic receptor family, 227–235, 228f
   packaging of glutamate into vesicles, 400
   receptor family, 217
   regulation of glutamate synthesis, 402–403
   synthesis, 400, 401f
   termination of glutamate action, 403–405
   transporters, 484, 485f, 486f
   uptake, 484, 486
Glutamate transporter 1 (GLT-1), 484–485
Glutamate-aspartate transporter 1 (GLAST), 484–485
Glutamatergic
   neurons, 403
   synapse, 401–402
   tone, 486

Glutamic acid decarboxylase (GAD), 406
Glutaminase, 400, 402–403
Glutamine, 400, 484
   cycle, 404–405, 484
   synthetase, 404–405
Glycine, 342, 399, 413–417, 457–458, 457f
   functions of glycinergic neurons, 415–417
   receptors, 225–226, 226f
   synthesis, 413–414, 414f
   termination of glycine action, 414–415
   transmission, 408b
   vesicular and reuptake transporters, 415f
Glycinergic neurons functions, 415–417
GLYT-1 knockout mice, 414
Goldman equation. *See* Goldman–Hodgkin–Katz voltage equation
Goldman–Hodgkin–Katz voltage equation, 25–28
Golgi, 423
   apparatus, 16–17
   golgi-stained neurons, 8–9
   method, 275
   stain, 10
Golgi, Camillo, 8–10, 9f
   drawings of brain section, 9f
Gram-positive anaerobic bacteria, 170
Green fluorescent protein (GFP), 102, 192
Growth cone, 15
Growth factor tyrosine kinase receptors, 266–268
GTP. *See* Guanosine 5′-triphosphate
GTP-binding protein-coupled receptors (G-protein-coupled receptors), 210–211, 211f, 430
GTPase, 254
GTPase-activating proteins (GAPs), 265
Guanine nucleotide exchange factor (GEF), 265
Guanosine 5′-triphosphate (GTP), 248, 249f, 439–440
   α-GTP hydrolysis, 254
Guanylate cyclase receptors, 211
Guanylyl cyclase, 318, 439
GV-58 compound, 151

# H

Habituation, 303–307
HACU. *See* High-affinity, low-efficiency choline uptake transporter
Hallucinations, 232
Hallucinogenic effects, 335
Hebbian plasticity, 78b, 81b, 82f, 308–311
Helical screw motion, 40
Henry Molaison (H.M.), 309b
Heteromeric neuronal nicotinic receptors, 220
Heterosynaptic plasticity, 315, 316f

Heterotrimeric G-proteins, 246–248
 families, 251–254
 receptor coupling to, 251–254
Heterotrimers, 251–252
High-affinity, low-efficiency choline uptake transporter (HACU), 351–352, 356
High-pass filtering, 294
Hippocampal slice preparation, 484–485
Hippocampus, 308–311, 338–339
Histamine, 369, 382–383, 490
 $N$-methyl transferase, 383
 synthesis, 382, 382f
  regulation, 382–383
 termination of action, 383
Histamine sources and roles in brain function, 390
Histaminergic neurons, 369–370
Histidine, 382
Histidine decarboxylase, 382
Homeostasis, 121–122, 320–321
Homeostatic functions, 421
Homeostatic plasticity, 323, 470–471
Homeostatic synaptic plasticity, 320–323
Homologs of yeast proteins, 164
Homomeric connexion, 70–71
Homomeric neuronal nicotinic receptors, 220
Homosynaptic depotentiation, 314–315
Homosynaptic plasticity, 308–311, 315, 316f
Horseradish peroxidase (HRP), 190, 191f
HRP. See Horseradish peroxidase
5-HT. See 5-Hydroxytryptamine
5-HT$_3$ serotonin receptor, 222
5-HTP. See 5-Hydroxytrytophan
Hydration of charged ions in solution, 41–43, 44f
Hydrogen sulfide (H$_2$S), 342, 445
 synthesis and regulation, 445
Hydrophilic protein structure (water loving), 37–38
Hydrophobic amino acids, 37
Hydrophobic protein structure (water avoiding), 37–38
Hydrophobicity plots, 37–38
Hydroxyapatite, 121–122
5-Hydroxytryptamine (5-HT), 222, 376–378
5-Hydroxytrytophan (5-HTP), 378–379
Hyperalgesia, 417
Hyperekplexia, 226
*Hypericum perforatum*, 394, 394f
Hyperpolarization-activated cation channels, 284–285
Hyperpolarizing signal, 73
Hypothalamus, 421, 428
Hypothesis development, 1–2
Hypoxia, 475

# I

Iberiotoxin (IbTX), 100
Ifenprodil, 231
Ignarro, Louis J., 436f, 437
Immediate early genes, 318–319
In vitro binding studies, 256
In vivo observations, 78b
Inactivation process, 53
Indian snakeroot (*Rauvolfia serpentina*), 392b
Inducible NOS (iNOS), 436–439
Information processing, astrocytes role in, 490–491
Inhibitory postsynaptic currents (IPSCs), 225–226
Inorganic phosphate transporter, 402–403
iNOS. See Inducible NOS
Inositol triphosphate (IP$_3$), 86, 239, 259–260, 488
Inositol trisphosphate receptor (IP$_3$R), 128
Insulin, 424–425
Interneurons in PVN, 474
Intracellular calcium, 86
Intracellular signaling cascades
 associated with NMDA receptors, 236
 associated with non-NMDA receptors, 235
Ion
 equilibrium potential for, 24–25
 hydration in ion permeability, 44
 ion-conducting pore, 215–217
 movement across cell membrane, 23–33
  fluxes at resting membrane potentials, 28–30
  maintaining ionic concentrations, 30–31
  membrane conductance and resistance, 32
  Ohm's law, 32–33
  multiple ions contribute to resting membrane potential, 25
  pore in potassium channel, 44, 45f
Ion channels, 3–4, 22. See also Voltage-gated ion channels
 in astrocytes, 487–488
 gating, 35–37, 36f
 subunits, 239
Ionic currents through voltage-gated ion channels, 47–53. See also Voltage-gated ion channels
 current–voltage relationships, 49–53
 potassium current, 50f
 sodium current, 51f
 voltage-depedent activation, 49
Ionotropic
 cannabinoid receptors, 239
 glutamate receptors, 403–404
 glycine receptors, 413
 metabotropic receptor-mediated plasticity of ionotropic signaling, 299–303, 301f

Ionotropic receptor, 209–210, 210f, 215, 460–461
  AMPA and kainate glutamate receptors, 234–235
  for ATP, 237–238
  comparison between metabotropic receptor and, 213–214, 213f, 213t
  families, 216f
  $GABA_A$ receptors, 223–224
  glutamate ionotropic receptor family, 227–235
  glycine receptors, 225–226
  $5-HT_3$ serotonin receptor, 222
  nAChRs, 218–221
  NMDA receptors, 228–233
  pentameric ligand-gated ion channel family, 217–227
  transient receptor potential channel family, 239
  trimeric receptor family, 236–238
  zinc-activated channel receptors, 226–227
$IP_3$. See Inositol triphosphate
$IP_3R$. See Inositol trisphosphate receptor
IPSCs. See Inhibitory postsynaptic currents
Ipsilateral ear, 459
IPSP, 406, 408–409, 413, 458–459

# J
Jellyfish (*Aequorea victoria*), 102

# K
K2P. See Two-pore-domain potassium channel
Kainate glutamate receptors, 234–235, 235f
Kainic acid, 227
Kandel, Eric, 307f
  Nobel Prize for studies elucidating cellular and signaling systems, 307b
KCC2. See Potassium–chloride extrusion cotransporter 2
Ketamine, 231
Kevorkian, Dr. Jack (assisted suicide device), 27b
"Kiss-and-run" mechanism, 195, 197–199, 199f
Kraits, 100

# L
Lambert–Eaton myasthenic syndrome (LEMS), 148b, 292
Large dense-core vesicles, 342–343, 422–423
Latrotoxin, 102
Leak current, 25
LEMS. See Lambert–Eaton myasthenic syndrome
Length constant, 60, 277–279
Levodopa (also called L-Dopa), 380
LHRH. See Luteinizing hormone-releasing hormone
Ligand-binding domain, 227

Ligand-gated ion channels, 36
  receptor, 254
Locus coeruleus, 387–388
Long-term depression (LTD), 308, 314–319
Long-term potentiation (LTP), 116, 236, 265, 308–311, 316–319
  associative, 312–313
  EEG activity patterns recorded from scalp, 311f
  hippocampal trisynaptic circuit, 310f
  induction, 313
Long-term synaptic plasticity, 308, 312. See also Short-term synaptic plasticity
  clinical case of Henry Molaison, 309b
  Nobel Prize to Eric Kandel for studies elucidating cellular and signaling systems, 307b
Low-pass filter, 294
Low-voltage-activated calcium channels (Cav3), 142–143
Löwi, Otto (discovery of chemical neurotransmission), 346b
LSD. See Lysergic acid diethylamide
LTD. See Long-term depression
LTP. See Long-term potentiation
Luteinizing hormone-releasing hormone (LHRH), 300, 453–454
Lysergic acid diethylamide (LSD), 396–397

# M
MacKinnon, Roderick (Nobel Prize winner in Chemistry), 42b
Magic mushrooms. See Mushrooms (*Psilocybe semilanceata*)
Magnetic resonance imaging (MRI), 309
Magnocellular neurons, 421
Malfunctioning synapses, 11–13
Mammalian central synapses, 321–323
Mammalian nervous system, 399
  electrical synapses in, 76–84
MAO. See Monoamine oxidase
MAP. See Mitogen-activated protein
MAP kinase. See Mitogen-activated protein kinase
MAPK. See Mitogen-activated protein kinase
Matrix metalloprotease (MMP), 78b
MCU. See Mitochondrial calcium uniporter
MDMA. See Methylenedioxymethamphetamine
Medial nucleus of trapezoid body (MNTB), 459
Memantine, 231
Membrane
  capacitance, 277
  conductance, 277
  fusion and fission, 164
  resistance, 32
  trafficking, 164

Membrane potential, 20–22, 25–28. *See also* Resting membrane potential
  impact of cell size and membrane resistance on changes, 33f
  recording using intracellular microelectrode, 26f
mEPCs. *See* Miniature endplate currents
mEPPs. *See* Miniature endplate potentials
mEPSPs. *See* Miniature excitatory postsynaptic potentials
3-Mercaptopruvate sulfurtransferase, 445
Messenger RNA (mRNA), 423
Metabotropic G-protein-coupled receptors, 239, 245–246
  Earl Sutherland Nobel prize in physiology or medicine (1971), 246b
  families of heterotrimeric G-proteins, 251–254
  metabotropic signaling termination, 254
  receptor coupling to heterotrimeric G-proteins, 251–254
  signaling pathways in nervous system, 255–268
  specificity of coupling between receptors and signaling cascades, 269–271
Metabotropic glutamate receptors, 318
Metabotropic receptors, 210, 213, 215–217, 250, 270, 299, 460–461
  activate G-proteins, 262
  comparison between ionotropic receptor and, 213–214, 213f, 213t
  metabotropic receptor-mediated plasticity of ionotropic signaling, 299–303
Metabotropic signaling termination, 254, 255f
Metaplasticity, 319–320
Methylenedioxymethamphetamine (MDMA), 396
Microdomain, 137–138
Microelectrode
  recordings, 102–103
  technique, 97, 98b, 98f
Microglia, 481–482
Migraines treatment by blocking action of CGRP, 426b
Miniature endplate currents (mEPCs), 110
Miniature endplate potentials (mEPPs), 102–103, 132
Miniature excitatory postsynaptic potentials (mEPSPs), 114
Mitochondria, 13–14, 128
Mitochondrial calcium uniporter (MCU), 128–129
Mitochondrial matrix, 128
Mitochondrial sodium–calcium exchanger (mNCX), 128–129
Mitochondrion, 128
Mitogen-activated protein (MAP), 265, 267f
Mitogen-activated protein kinase (MAPK), 262–266
Mixed GABA/glycine mIPSCs, 458f
MK-801, 231

MMP. *See* Matrix metalloprotease
mNCX. *See* Mitochondrial sodium–calcium exchanger
MNTB. *See* Medial nucleus of trapezoid body
MNTB → LSO pathway, 471–472
Molecule in synaptic vesicles of nerve terminals, 336
Monoamine oxidase (MAO), 375–376
Monoamine transmitters, 369
  catecholamine neurotransmitters, 370–376
  frozen addicts, 387b
  histamine, 382–383
  monoaminergic drugs of abuse, 395–397, 396f
  projections of monoaminergic neurons and functions, 383–390
    dopamine, 384–386
    epinephrine sources and roles in brain function, 388–389
    histamine sources and roles in brain function, 390
    Indian snakeroot and vesicular monoamine transporter, 392b
    norepinephrine, 387–388
    Parkinson's disease treatment with catecholamine precursor L-DOPA, 385b
    serotonin sources and roles in brain function, 389
    St. John's wort and monoaminergic reuptake, 394b
  serotonin, 376–382
  terminology, 369–370
  therapeutic drugs, 390–394
    affecting VMAT2, 391–392
    blocking monoaminergic receptors, 391
    inhibits monoamine oxidase, 393
    inhibits reuptake of monoamines, 393–394
    stimulating monoaminergic receptors, 390–391
Monoaminergic
  drugs of abuse, 395–397, 396f
  neurons, 369–370
    projections in nervous system, 383–390
Monoamines, 342, 369, 370f
  functions in nervous system, 383–390
  treating genetic disorder with monoamine precursors, 379b
Mood disorders, 232
MPTP, 387
MRI. *See* Magnetic resonance imaging
mRNA. *See* Messenger RNA
Multiple innervation, 78b
Multiple neurotransmitter release, 450–451. *See also* Neurotransmitter switching
  expression within neurons, 449–450
  functional implications, 451–452
  norepinephrine and ATP, 462
  proponents of, 451
Murad, Ferid, 436f, 437
Muscarine, 347

Muscarinic ACh receptors, deadly nightshade plant and, 363b
Mushrooms (*Psilocybe semilanceata*), 396–397
MUSK receptors, 211
Myelin, 60–61

# N

N-type inactivation. *See* "Ball and chain" mechanism
$Na^+/K^+$ ATPase. *See* Sodium–potassium ATPase
nAChRs. *See* Nicotinic acetylcholine receptors
NADPH. *See* Nicotinamide-adenine-dinucleotide phosphate
NALCN. *See* Sodium Leak Channel, Nonselective
NANC nerves. *See* Nonadrenergic, noncholinergic nerves
Nanodomain, 137–138
Natural molecules as experimental tools, 100b
NCX. *See* Sodium–calcium exchanger
Neocortex, 338–339
Nernst equation, 24–25
Nernst potential, 24–25
  calculator, 25
Nerve growth factor (NGF), 211
Nerve terminals, 13–14
  calcium channels located within, 134–137
    calcium buffer effects on neurotransmitter release, 135f
    calcium-sensitive dye fluorescence in squid giant synapse, 136f
    frog neuromuscular junction labeled with two fluorescent toxins, 138f
    imaging calcium at frog neuromuscular junction, 137f
  containing endogenous calcium buffer proteins, 129, 130f
  cytoplasm, 156
  voltage-gated calcium channels in, 141–150
NET. *See* Norepinephrine transporter
Neurohypophysis, 298–299
Neuromodulators, 333–335
Neuromuscular disorders, 291
Neuromuscular junctions, 14–15, 109–113
  synapse elimination at, 78b
Neuromuscular synapse, 109, 136
  electrical synapses in, 77–81
Neuron Doctrine, 7–11
Neuronal NOS (nNOS), 436–439
Neuropeptide Y
  effects on behavior, 431–432
  hyperpolarized hypocretin/orexin neurons, 430–431
  as model for neuropeptide action, 430–432

Neuropeptides, 342–343, 421, 453–455
  differences between classical (type 1) neurotransmitters and, 422–423, 422f
  receptors and effects on neurotransmitter release, 429–430
  synthesis, 423, 424f
    regulation, 429
  vesicular release, 428–429
Neuropharmacologists, 65
Neurotransmitter release
  calcium as trigger for, 121–129
  control by calcium ions, 130–140
    caged calcium, 132f
    calcium channels located within nerve terminal, 134–137, 135f
    cytoplasmic calcium microdomains, 137–140
    electrical response to nerve stimulation, 131f
    nonlinear relationship between calcium and neurotransmitter release, 133–134
    voltage recordings from presynaptic nerve terminal and postsynaptic cell, 133f
  mechanism discovery, 155–162
    amperometry, 161b
    through channel in presynaptic membrane, 156
    experimental evidence supporting synaptic vesicle fusion, 156–162, 160f
  neuropeptide receptors and effects, 429–430
Neurotransmitter switching, 464
  activity-dependent, 470–471, 470f
  in adult nervous system, 473
  as compensatory mechanism in disease, 474–475
  in developing nervous system, 465
  developmental, 465–469
  neurotransmitter phenotype change, 463–464
  neurotransmitter–receptor matching, 464–465
  sensation-mediated neurotransmitter switching, 472–473
  spontaneous electrical activity role in nervous system development, 469–471
  stimulus-mediated neurotransmitter switching in mature neurons, 473–474
  transient glutamate expression in synaptic refinement, 471–472
Neurotransmitters, 214, 236, 333–335, 369, 399, 428
  binding-induced channel, 215–217
  characteristics, 339–340
    methods for termination of neurotransmitter action, 340
    regulation of neurotransmitter synthesis, 339–340
    synthetic pathways, 339
  chemical, 209
  corelease, 456–459
    vesicular synergy as function of, 459–460

Neurotransmitters (Continued)
  cotransmission and corelease of, 452–462
  molecules, 215–217
  neurotransmitter-filled vesicles, 15
  nitric oxide act as, 439
  phenotype, 451
  receptors, 209–213, 333
    cytoplasmic receptors, 212–213, 212f
    enzyme-linked receptors, 211
    G-protein-coupled receptors, 210–211
  release from astrocytes, 489–490
  signaling molecule as, 335–339
  spatial segregation of, 455–456
  types, 341–343, 341t
    amino acid neurotransmitters, 342, 399–417
    classical neurotransmitters, 342, 345–366
    gaseous messengers, 342, 435–445
    neuropeptides, 342–343, 421–432
  uptake by astrocytic transporters, 484–487
Neurotrophin-3 (NT3), 211
Neurotrophin-4 (NT4), 211
NGF. See Nerve growth factor
Nicotiana tabacum, 335
Nicotinamide-adenine-dinucleotide phosphate (NADPH), 436
Nicotinic acetylcholine receptors (nAChRs), 218–221, 219f, 341
Nicotinic ACh receptors, 365b
Nitric oxide (NO), 318, 342, 435–443, 490
  acting as neurotransmitter, 439
  acting as retrograde messenger, 437f
  control of autonomic function, 440–441
  effects in central nervous system, 441–443
  Nobel Prize award for discovery of, 437b
  reactions in nervous system, 441f
  synthesis, 436–438
    regulation, 438–439
  termination of action, 440–443
  vasorelaxation triggered by, 442f
Nitric oxide synthase (NOS), 318, 436–438, 441–442
NKCC1. See Sodium–potassium–chloride cotransporter 1
NMDA receptors, 228–233, 229f, 230f, 231f, 232f, 233f, 285
  glutamate receptors, 86, 88f
  intracellular signaling cascades associated with, 236
  NMDA receptor-mediated plasticity, 472
nNOS. See Neuronal NOS
NO. See Nitric oxide
Nodes of Ranvier, 60–61, 62f
Non-neuronal cells, 421
Non-NMDA receptors, 235
Nonadrenergic, noncholinergic nerves (NANC nerves), 440–441
Nonmammalian animal model systems, 2
Nonrectifying gap junctions, 72–73
Noradrenaline. See Norepinephrine
Noradrenergic neurons, 369–370
Noradrenergic sweating, 469
Norepinephrine, 320, 369–370, 387–388, 388f, 461–462, 474–475, 490
  converted to epinephrine, 372–373
  dopamine converted to, 372
Norepinephrine transporter (NET), 376, 393, 395–396
NOS. See Nitric oxide synthase
NT3. See Neurotrophin-3
1-Oleoyl-2-acetyl-sn-glycerol (OAG), 261

## O

Oligodendrocytes, 60–61, 481–482
"One neuron, one neurotransmitter" doctrine, 449–452, 475
Optical quantal analysis, 116–117
Orai protein, 128
Organophosphates, 364–365
Oxime, 366
Oxygen ($O_2$), 381

## P

"p-loop", 39
P2X receptor, 237–238, 460–462
P2Y receptors, 460–461
p75 receptor, 78b
Paired-pulse
  depression, 290–291
  facilitation, 289–290
  plasticity, 288, 290–291
Palmitoyl side chains, 168
Palythoa corals, 31b
Palytoxin, 31b
Paracrine, 299
  transmission, 299, 383
Parasympathetic nervous system, 362
Parasympathomimetic drugs, 363
Paraventricular nucleus (PVN), 473–474
Parkinson's disease treatment with catecholamine precursor L-DOPA, 385b
Parvalbumin, 129
Passive membrane properties, 277–280
Pasteurella multocida toxin, 253
Patch amperometry, 161
Patch clamp techniques, 104–105
Pentameric (cys-loop) family, 217
Pentameric ligand-gated chloride channels, 223

Pentameric ligand-gated ion channel family, 217–227, 217f
Peptide neurotransmitter synthesis, 423–430
Peptide transmitter mRNA, 423
Peptidergic cotransmission, 453–454
Peptides, 421
Perforant path, 308
Pericytes, 122–123
Peripheral nervous system, acetylcholine in, 359
Peripheral neuromuscular diseases, 291b
Permeability
  porous membrane, 22
  values, 27
Permeate, 41–43
Pertussis toxin, 253
Phencyclidine, 231, 370–372
Phenylethanolamine-N-methyltransferase (PNMT), 372
Philanthotoxin, 102
Phosphate esters. See Organophosphates
Phosphatidylinositol (4,5)-bisphosphate (PIP2), 195, 259–260, 261f
Phosphodiesterases, 258–259
Phosphoinositide 3 Kinase pathway (PI3K pathway), 265, 266f
Phosphoinositol pathway (PI pathway), 259–261, 260f
Phospholipase A2 (PLA2), 261–262
Phospholipase C (PLC), 239, 259–260
Physiological effects, 426
Physiological mechanisms, 430–431
Physiological stimulus patterns, 311
Physiological stress, 373–374
Physostigmine, 360
PI pathway. See Phosphoinositol pathway
PI3K pathway. See Phosphoinositide 3 Kinase pathway
Piccolo protein, 172
Pilocarpine, 363
PIP2. See Phosphatidylinositol (4,5)-bisphosphate
PKA. See Protein kinase A
PLA2. See Phospholipase A2
Plasma membrane
  of cells, 21–22
  receptor, 246–247
  synaptic vesicle fusion with, 156–162, 160f, 165f
Plasma membrane calcium–magnesium ATPase (PMCA), 124, 125f
Plasticity, 308. See also Synaptic plasticity
  hebbian, 78b, 81b, 82f, 308–311
  heterosynaptic, 315, 316f
  homeostatic synaptic, 320–323
  homosynaptic, 308–311, 315, 316f
  modulation, 319–320
  paired-pulse, 288, 290–291
  spike timing-dependent, 313, 314f

of synaptic plasticity. See Metaplasticity
PLC. See Phospholipase C
PLP. See Pyridoxal phosphate
PMCA. See Plasma membrane calcium–magnesium ATPase
PNMT. See Phenylethanolamine-N-methyltransferase
Poison hemlock plant and ACh receptors, 221b
Poisson distribution, 108
Polar amino acids, 37
POMC. See Pro-opiomelanocortin
Pore-forming subunits, 45
Positive allosteric modulators, 224–225
Post-tetanic potentiation (PTP), 293
Post-translational processing, 424, 426f
Postsynaptic cells, 15, 73
Postsynaptic density (PSD), 13–14, 483f
Postsynaptic glutamate receptors, 321
Postsynaptic ionotropic receptors, 280
Postsynaptic potentials (PSPs), 401–402
Potassium channels, 55, 487
  crystal structure of, 42b
  role in shaping calcium, 146–150, 147f
Potassium ion equilibrium potential, 23–24
Potassium leak current, 22
Potassium–chloride extrusion cotransporter 2 (KCC2), 224
PP1. See Protein phosphatase 1
Preassembled active zone precursor vesicles, 17
Precursor of BDNF (proBDNF), 78b
Predictable pharmacological action, 338
Pregabalin, 410
Prepropeptides, 423
Presynaptic calcium ion plasma membrane transporters, 124–126
  NCX transporter, 126f
  PMCA, 125f
Presynaptic cell, 73
Presynaptic cellular organelles, 127–129
Presynaptic density. See Active zone
Presynaptic locus, 303–304
Presynaptic nicotinic acetylcholine receptors, 334–335
Primary effector (adenylyl cyclase), 252–253, 259
Primary protein structure, 37
Primary voltage sensor for ion channel, 40
"Primitive" types of electrical synapses, 67–68
Pro-opiomelanocortin (POMC), 424
proBDNF. See Precursor of BDNF
Prostaglandin hormones, 263b
Protein kinase A (PKA), 86–88, 250
Protein kinase B (PKB), 265
Protein kinase C (PKC), 261, 304–305, 318, 444
Protein phosphatase 1 (PP1), 236

Protein(s), 337
 in calcium-triggered vesicle release, 166–168
  average synaptic vesicle and proteins, 167f
  CORE complex proteins, 169f
  tangle of diverse proteins, 166f
 phosphorylation, 86, 307
 pores, 70
Proton pump. See Vacuolar-type proton ATPase
Prozac, 393
PSD. See Postsynaptic density
PSPs. See Postsynaptic potentials
Psychiatric disorders, 388
PTP. See Post-tetanic potentiation
Pulldown assay, 257–258
Purinergic cotransmission, 460–462
Purple foxglove (Digitalis purpurea), 31b
PVN. See Paraventricular nucleus
Pyramidal cells, 308
Pyridoxal phosphate (PLP), 406, 408

## Q

Quantal content (QC), 110
Quantal theory of chemical transmitter release, 105–109
 at central synapses, 113–116
 current–voltage relationship, 106f
 measurements of transmitter release, 107f
 at neuromuscular junction, 109–113, 112f
  of Drosophila, 118f
 optical quantal analysis, 116–117
Quantal transmitter release mechanism, 156–162
Quantum, 107–108
Quaternary protein structure, 44–45

## R

Rab proteins, 166–167, 176
Radioactive phosphorus ($^{32}$P), 373
Ramón y Cajal, Santiago, 10
 golgi-stained nervous system tissue, 11f
 portrait, 10f
Random process, 108
Raphe nuclei, 389
"Rational drug design" approach, 410
Readily releasable pool (RRP). See "Ready" pool
"Ready" pool, 200–201
Receptor
 comparison between ionotropic and metabotropic, 213–214, 213f, 213t
 coupling to heterotrimeric G-proteins, 251–254
  metabotropic G-protein-coupled receptors, 252f
 desensitization, 405
  AMPA, 405f
 neurotransmitter, 209–213
 receptor-mediated signaling systems, 269–271
 tyrosine kinases, 211
Reconstitution studies, 256
Rectifying gap junctions, 72–73
"Recycling" pool, 200–201
Reflex arc through spinal cord, 11, 12f
Regulation of neurotransmitter synthesis, 339–340
Regulators of G-protein signaling (RGS proteins), 254, 255f
Relative refractory period, 57
Renshaw cells, 416
Repeating domains, 39
Repetitive nerve stimulation test (RNS), 291b, 292f
Reserpine, 391–392
"Reserve" pool, 200–201
Residual bound calcium, 288, 294–295, 297f
Resistance, 32–33
Resting membrane potentials, 22
 ion fluxes at, 28–30
 multiple ions to, 25
Reticular Theory of connectivity in brain, 7–10, 8f
Retrograde messenger, 299
Reuptake process, 404
Reuptake transporters, 340
Reversal potential, 24
RGS proteins. See Regulators of G-protein signaling
Rho pathways, 265–266, 267f
Rivastigmine, 362
RNS. See Repetitive nerve stimulation test
Ryanodine receptor, 127–128

## S

S-adenosyl-L-methionine (SAM), 445
S4 segments acting as voltage sensors, 39–40
SAD. See Seasonal affective disorder
Safety factor, 291
Salix willow tree, 263–264, 264f
Saltatory conduction, 61
SAM. See S-adenosyl-L-methionine
Sarco-endoplasmic reticulum Ca$^{2+}$-ATPase (SERCA), 127
Sarcoma (Src), 266–268
 pathway, 266–268, 268f
SCG. See Superior cervical ganglion
Schekman, Randy, 165
Schizophrenia, 412
Schwann cells, 60–61, 481–482
SCLC. See Small-cell lung cancer
Scorpion toxins, 100
Sea slug (Aplysia californica), 68, 68f, 455
Seasonal affective disorder (SAD), 473
Sec/Munc proteins (SM proteins), 174
Second messengers, 245–246, 247f

Second-messenger cascade, 429
Secondary effector, 259
Secondary protein structure, 37
Seizures, neuropeptide Y role in, 432
Selective semipermeable membrane, 22
Selective serotonin reuptake inhibitors (SSRI), 393
Sensation-mediated neurotransmitter switching, 472–473
Sensitization, 303–307, 306f
Sensory axons, 61
Sensory neurons, 425
SERCA. See Sarco-endoplasmic reticulum Ca$^{2+}$-ATPase
Serine trans hydroxymethyltransferase, 413–414
Serine/threonine-specific protein kinases, 211
Serotonergic neurons, 369–370
Serotonin, 222, 306, 369, 376–382
  receptors, 86
  sources and roles in brain function, 389, 389f
  synthesis, 378–380, 379f
    regulation, 381
  termination of action, 381–382, 381f
  treating genetic disorder with two monoamine precursors, 379b
Serotonin and norepinephrine reuptake inhibitors (SNRIs), 393
Serotonin reuptake transporter (SERT), 381–382
sGC. See Soluble guanylyl cyclase
Short-term synaptic depression, 294
Short-term synaptic plasticity, 288–299, 295f. See also Long-term synaptic plasticity
  comparison of paired-pulse, tetanic, and post-tetanic plasticity, 290f
  Eric Kandel, 307f
Shunting inhibition, 223–224
Siemens (S), 32
Signaling
  molecule as neurotransmitter, 335–339, 336f
  pathways in nervous system, 255–268
    arachidonic acid pathway, 261–265
    cAMP pathway, 258–259
    direct ion channel pathway, 256–258
    MAP kinase and Rho pathways, 265–266
    phosphoinositol pathway, 259–261
    PI3K pathway, 265
    Src pathway, 266–268
Silent synapses, 234, 312
Silver nitrate, 8–9
Single ligand–receptor binding event, 250
Single neurotransmitter vesicles, spontaneous release of, 102–105
"Sliding filament" movement, 40
SLMVs. See Synaptic-like microvesicles

SM proteins. See Sec/Munc proteins
Small clear synaptic vesicles, 13, 172, 342
Small-cell lung cancer (SCLC), 149
Small-diameter axons, 60
Small-molecule neurotransmitters, 453
  segregation into separate vesicle pools, 454
SNARE proteins. See Soluble NSF-Attachment protein REceptor
SNc. See Substantia nigra pars compacta
S-nitrosylate proteins, 318
Snowdrop plant (*Galanthus nivalis*), 365
  AChE, and nicotinic ACh receptors, 365b
SNRIs. See Serotonin and norepinephrine reuptake inhibitors
Sodium channels, 53–54
Sodium Leak Channel, Nonselective (NALCN), 25
Sodium–calcium exchanger (NCX), 126, 126f
Sodium–potassium ATPase (Na$^+$/K$^+$ ATPase), 30, 30f, 31b
Sodium–potassium–chloride cotransporter 1 (NKCC1), 224
Soluble guanylyl cyclase (sGC), 440–444
Soluble NSF-Attachment protein REceptor (SNARE proteins), 168, 171f, 174–176, 198
  of CORE complex, 177–178
    recovery and disassembly after synaptic vesicle fusion, 183–184
Somatosensory, 425–426
South American vine plant (*Chondrodendron tomentosum*), 101
Spatial segregation of neurotransmitters, 455–456
Specificity of coupling between receptors and signaling cascades, 269–271, 269f
  unintended crosstalk between receptor-mediated signaling systems, 269–271
Specificity of coupling between receptors and signaling cascades, 269–271
Spike timing-dependent plasticity, 313, 314f
Spinal cord, 416, 416f
Spines, specialized postsynaptic compartments on dendrites, 280–283
*SPR* gene, 380
Src. See Sarcoma
SSRI. See Selective serotonin reuptake inhibitors
St. John's wort and monoaminergic reuptake, 394b
Staining technique, 8–9
STIM1, 128
Stimulus train, 293
"Stimulus-mediated" neurotransmitter switching in mature neurons, 473–474
Store-operated CRAC channels, 128
Stress response, 473
Stretch receptors, 36–37

Stretch-gated ion channels, 36–37
Strontium, 115
Substantia nigra pars compacta (SNc), 384–385
Super resolution methods, 201
Superior cervical ganglion (SCG), 465–466
Sympathetic nervous system, 360, 474–475
Sympathetic neurons, in vitro studies, 465–466
Sympathomimetic amines, 376
Synapses, 11
 active zones during synapse development, 16–17
 animal model systems to studying, 2–4, 3f
 in central nervous system, 14f
 early formation of chemical, 16f
 elimination at neuromuscular junction, 78b
 formation and structure
  artist's rendering of specialized compartments of neuron, 12f
  neurons sending signals, 7–13
 maturation after initial assembly, 17f
 neuron assemble cellular components required to creating, 15
 structure and organization, 13–15
Synapsin, 201–202
 phosphorylation, 202
Synaptic cleft, 13, 482–484
Synaptic events recording, 98b
Synaptic integration within postsynaptic neurons
 active membrane properties, 283–285
 Golgi stain of Purkinje neuron with extensive dendritic branches, 276f
 neuron emphasizing specialization, 276f
 passive membrane properties, 277–280
 spines specialized postsynaptic compartments on dendrites, 280–283
Synaptic plasticity, 97, 287
 associative long-term potentiation, 312–313
 habituation and sensitization, 303–307
 heterosynaptic plasticity, 315
 homeostatic synaptic plasticity, 320–323
 investigation of long-term synaptic plasticity, 308
 long-term, 308
 long-term depression (LTD), 314–315
 long-term potentiation (LTP), 308–311
 metabotropic receptor-mediated plasticity of ionotropic signaling, 299–303
 metaplasticity, 319–320
 physiological stimulus patterns, 311
 plasticity modulation, 320
 short-term, 288–299
 spike timing-dependent plasticity, 313
 synaptic signaling mechanisms, 316–319
Synaptic Protein Interaction site (SynPrInt site), 143, 176–177

Synaptic scaling, 321, 322f
Synaptic signaling mechanisms, 316–319, 317f
Synaptic transmission, 1–2
Synaptic vesicle membrane, 189
 retrieval and reuse of, 189–192
  changes in structure of frog motor nerve terminals, 190f
  electron micrograph of transmitter release site, 194f
  loading of recycled nerve terminal synaptic vesicles, 191f
  pH-sensitive synaptopHluorin proteins, 193f
  synaptopHluorin used to visualize exocytosis and endocytosis, 193f
 synaptic vesicle pools, 199–201
 synaptic vesicle trafficking in nerve terminal, 201–203
Synaptic vesicles, 13, 483f
 fusion
  biochemical mechanims of calcium-triggered, 163–184, 163f
  experimental evidence supporting, 156–162, 160f
 moving to correct location in nerve terminal, 172–174
  cytoskeletal cytomatrix proteins, 173f
 pools, 199–201, 200f
 trafficking in nerve terminal, 201–203
Synaptic-like microvesicles (SLMVs), 489–490
Synaptobrevin, 192
synaptopHluorin protein, 192
 pH-sensitive, 193f
 using to visualize exocytosis and endocytosis, 193f
Synaptotagmin, 133, 166–167, 177–178, 183, 192, 195, 298
 evidence for synaptotagmin as calcium sensor at active zones, 179–183
SynPrInt site. See Synaptic Protein Interaction site
Syntaxin, 168, 176–177
 proteins, 197
Synthetic pathways, 339
Synthetic receptor ligands, 333

# T

Tail current, 145–146
TCAs. See Tricyclic antidepressants
Temporal summation, 279, 279f
Termination of transmitter action, 337–338
Tertiary protein structure of voltage-gated ion channel, 39, 39f
Tetanic depression, 290–291
Tetanic potentiation, 290–291
Tetanus toxin, 102, 168, 170b, 171f

Tetrahydrobiopterin (BH4), 370–372, 381, 436
Tetrodotoxin (TTX), 78b, 100
TH. See Tyrosine hydroxylase
Theta burst stimulation, 311
Threshold voltage, 53–54
Traditional electron microscopy, 157
Transcription factors, 318–319
Transient glutamate expression, 471–472
Transient receptor potential (TRP) channel family, 239
   receptor family, 217
Transmembrane portion of protein structure, 37
Transmitter
   release process, 96
   synthesis, 337
   transmitter-containing synaptic vesicles, 13
Tricyclic antidepressants (TCAs), 377–378, 393
Trimeric receptor family, 217, 236–238, 237f
   ASIC, 238
   P2X receptors for ATP, 237–238
Tripartite synapse, 14–15, 482–484, 491
Trisynaptic loop, 308
TrkB receptor, 78–79
TRP. See Transient receptor potential
Tryptophan, 378
Tryptophan hydroxylase, 378–379
   activity, 407
TTX. See Tetrodotoxin
Tubocurarine, 101
"Twitch" response, 461
Two-electrode voltage-clamp technique, 47b
Two-pore-domain potassium channel (K2P), 22
Tyrosine hydroxylase (TH), 370–372, 374–375, 374f
Tyrosine to L-dihydroxyphenylalanine, 370–372

## U

Unconventional ionotropic glutamate receptor signaling, 235b
Unique synthetic enzymes, 337
Unmyelinated axons, 61

## V

V-ATPase. See Vacuolar-type proton ATPase
VAChT. See Vesicular acetylcholine transporter
VAChT/VGLUT-coexpressing vesicles, 459–460
Vacuolar-type proton ATPase (V-ATPase), 352, 400
VAMP. See Vesicle-associated membrane protein
Vasoactive intestinal peptide (VIP), 454, 466
Vasoconstriction, 454
Vasodilation, 454
VDAC. See Voltage-dependent anion channel
Ventral suprachiasmatic nucleus (VSC), 472
Ventral tegmental area (VTA), 384–385

Vesamicol, 352
Vesicle depletion, 288
Vesicle docking, 172, 175f
Vesicle mobility, 201
Vesicle trafficking and fusion mechanisms in cells, 165b
Vesicle-associated membrane protein (VAMP), 192, 416
Vesicular acetylcholine transporter (VAChT), 352, 352f, 400, 459–460, 467f
Vesicular glutamate transporter (VGLUT), 192, 400, 401f, 459–460
Vesicular inhibitory amino acid transporter (VIAAT), 407, 457
Vesicular monoamine transporter (VMAT), 372, 378–379, 392b, 466–467
   VMAT2, 391–392
Vesicular neurotransmitter transport, 353b
Vesicular proton-dependent ATPase, 400
Vesicular release of neuropeptides, 428–429, 428f
Vesicular synergy, 459–460
   as function of neurotransmitter corelease, 459–460
VGCC. See Voltage-gated calcium channels
VGLUT. See Vesicular glutamate transporter
VIAAT. See Vesicular inhibitory amino acid transporter
VIP. See Vasoactive intestinal peptide
Vitamin $B_6$ precursor pyridoxine, 406
VMAT. See Vesicular monoamine transporter
Voltage sensors, S4 segments acting as, 39–40
Voltage-clamp technique, 47–49, 47b, 110
Voltage-dependent activation, 49
Voltage-dependent anion channel (VDAC), 128–129
Voltage-gated calcium channels (VGCC), 131–133, 141–144, 258, 285, 321, 429
   coupling of, 176–177, 177f
   LEMS, 148b
   in nerve terminals, 141–150
      calcium entry into presynaptic terminal, 144–146, 145f, 146f
      potassium channels role in shaping calcium entry, 146–150, 147f
   structural organization and diversity of, 142f
Voltage-gated ion channels, 37–46, 288, 302–303
   auxiliary proteins, 44–46
   ion flux, 283–284
   mechanism, 40f
   permeation and selectivity, 41–44
   predicted membrane topology of classes, 38f
   S4 segments acting as voltage sensors, 39–40
   structure, 41
   tertiary structure, 39f

Voltage-gated potassium channels, 46b
Voltage-gated sodium channels, 285
Voltage-sensing domain, 40
Voltage-sensitive potassium channels, 22
Voltammetry, 161
Volume transmission, 299, 334, 334f, 383, 429–430
Voluntary euthanasia, 27b
VSC. *See* Ventral suprachiasmatic nucleus
VTA. *See* Ventral tegmental area

# W
Water homeostasis, 428
Willow trees and prostaglandin hormones, 263b

# X
X-ray crystallography, 41, 44
*Xenopus*
  muscle cells, 465
  spinal neurons, 469
  tadpoles, 472–473, 473f

# Y
Yeast cells, 164

# Z
Zinc-activated channels (ZACs), 226–227

CPI Antony Rowe
Eastbourne, UK
September 05, 2019